POWER PLANT
ENGINEERING

POWER PLANT
ENGINEERING

A.K. Raja
Director and Head, Department of Electrical Engineering
Saroj Institute of Technology & Management
Lucknow

Amit Prakash Srivastava
Sr. Lecturer and Head, Department of Mechanical Engineering
Saroj Institute of Technology & Management
Lucknow

Manish Dwivedi
Lecturer in Mechanical Engineering Department
Saroj Institute of Technology & Management
Lucknow

PUBLISHING FOR ONE WORLD

NEW AGE INTERNATIONAL (P) LIMITED, PUBLISHERS
(formerly Wiley Eastern Limited)
New Delhi • Bangalore • Chennai • Cochin • Guwahati • Hyderabad
Jalandhar • Kolkata • Lucknow • Mumbai • Ranchi
Visit us at www.newagepublishers.com

Copyright © 2006, New Age International (P) Ltd., Publishers
Published by New Age International (P) Ltd., Publishers
First Edition: 2006
Reprint: 2016

All rights reserved.
No part of this book may be reproduced in any form, by photostat, microfilm, xerography, or any other means, or incorporated into any information retrieval system, electronic or mechanical, without the written permission of the publisher.

BRANCHES

- **Bangalore** 37/10, 8th Cross (Near Hanuman Temple), Azad Nagar, Chamarajpet, Bangalore-560 018
  Tel.: (080) 26756823, Telefax: 26756820, **E-mail: bangalore@newagepublishers.com**

- **Chennai** 26, Damodaran Street, T. Nagar, Chennai-600 017, Tel.: (044) 24353401
  Telefax: 24351463, **E-mail: chennai@newagepublishers.com**

- **Cochin** CC-39/1016, Carrier Station Road, Ernakulam South, Cochin-682 016
  Tel.: (0484) 2377303, Telefax: 4051304, **E-mail: cochin@newagepublishers.com**

- **Guwahati** Hemsen Complex, Mohd. Shah Road, Paltan Bazar, Near Starline Hotel, Guwahati-781 008
  Tel.: (0361) 2513881, Telefax: 2543669, **E-mail: guwahati@newagepublishers.com**

- **Hyderabad** 105, 1st Floor, Madhiray Kaveri Tower, 3-2-19, Azam Jahi Road, Near Kumar Theater
  Nimboliadda, Kachiguda, Hyderabad-500 027, Tel.: (040) 24652456, Telefax: 24652457
  E-mail: hyderabad@newagepublishers.com

- **Kolkata** RDB Chambers (Formerly Lotus Cinema) 106A, 1st Floor, S.N. Banerjee Road, Kolkata-700 014
  Tel.: (033) 22273773, Telefax: 22275247, **E-mail: kolkata@newagepublishers.com**

- **Lucknow** 16-A, Jopling Road, Lucknow-226 001, Tel.: (0522) 2209578, 4045297, Telefax: 2204098
  E-mail:lucknow@newagepublishers.com

- **Mumbai** 142C, Victor House, Ground Floor, N.M. Joshi Marg, Lower Parel, Mumbai-400 013
  Tel.: (022) 24927869, Telefax: 24915415, **E-mail: mumbai@newagepublishers.com**

- **New Delhi** 22, Golden House, Daryaganj, New Delhi-110 002, Tel.: (011) 23262368, 23262370
  Telefax: 43551305, **E-mail: sales@newagepublishers.com**

ISBN: 978-81-224-1831-6
₹ 299.00
C-16-05-9419

Printed in India at Print 'O' Pack, Delhi.
Typeset at Monu Printers, Delhi.

PUBLISHING GLOBALLY
NEW AGE INTERNATIONAL (P) LIMITED, PUBLISHERS
7/30 A, Daryaganj, New Delhi-110002
Visit us at **www.newagepublishers.com**

Preface

There have been significant developments and advances in the field of power plant engineering, computer applications on energy audit and management, environmental audit and management, human development and environment. The authors have been encouraged to write this pioneer book for the benefit of students of engineering and researchers due to their contribution in power generation covering the syllabi of conventional power plants *i.e.,* Power Plant Engineering, at the international level in general as text cum reference book.

This book being pilot project of the authors specially in the area of conventional power plant will satisfy the engineering scholars as well as researchers in the field of direct energy conversion devices.

In the present book the syllabi enclosed has been covered in the most lucid manner from power plant point of view to avoid the unnecessary bulkiness and to reduce the cost of the price for the benefit of our beloved students of engineering in particular and others in general.

We have written this pioneering book on the basis of syllabi in the most lucid and compact manner for the benefit of the students and the readers.

The authors are greatly indebted to Ch. Sunil Singh, Chairman, SITM Lucknow, Mr. K.C. Mishra Suptt. Engineer, Saudia Electric Co. Saudi Arabia, for their great encouragement in writing this book. Without their support and help we would not have been able to accomplish this tough and challenging work.

In the end the authors will feel obliged for critical and useful suggestions since this pioneer book covers the syllabi in the most useful area of Mechanical Engineering in particular and is applicable for all branches of technology and engg. for all major Indian Universities, as well as at international level.

A.K. Raja
Amit P. Srivastava
Manish Dwivedi

Contents

Preface ... (v)

Chapter 1: Fundamental of Power Plant ... 1

1.1 Introduction ... 1
1.2 Concept of Power Plants ... 1
1.3 Classification of Power Plants ... 2
1.4 Energy ... 3
1.5 Types of Energy ... 4
1.6 Power ... 5
1.7 Power Development in India ... 5
1.8 Resources for Power Generation ... 7
1.9 Present Power Position in India ... 9
1.10 Future Planning for Power Generation ... 9
1.11 Power Corporations in India ... 11
    1.11.1 National Thermal Power Corporation ... 11
    1.11.2 National Hydro-Electric Power Corporation ... 11
    1.11.3 Rural Electrification Corporation ... 12
    1.11.4 Damodar Valley Corporation ... 12
    1.11.5 North-Eastern Electric Power Corporation Limited ... 12
    1.11.6 Bhakra Beas Management Board and Beas Construction Board ... 13
    1.11.7 Power Engineers Training Society (PETS) ... 13
    1.11.8 Central Power Research Institute (CPRI), Bangalore ... 13
    1.11.9 Naptha, Jhaicri Power Corporation Limited ... 13
1.12 Review of Thermodynamics Cycles Related to Power Plants ... 14
1.13 Classification of Power Plant Cycle ... 15
    1.13.1 Carnot Cycle ... 15
    1.13.2 Rankine Cycle ... 15
    1.13.3 Reheat Cycle ... 16
    1.13.4 Regenerative Cycle (Feed Water Heating) ... 16
    1.13.5 Binary Vapour Cycle ... 17
    1.13.6 Reheat-Regenerative Cycle ... 17
    1.13.7 Formula Summary ... 18
1.14 Fuels and Combustion ... 18

1.14	Fuels and Combustion	18
1.15	Steam Generators	18
1.16	Steam Prime Movers	19
1.17	Steam Condensers	19
	1.17.1 Surface Condensers	20
	1.17.2 Jet Condensers	22
	1.17.3 Types of Jet Condensers	22
1.18	Water (Hydraulic) Turbines	23
	1.18.1 Impulse and Reaction Turbines	23
1.19	Scienc Vs. Technology	25
	1.19.1 Scientific Research	25
	1.19.2 Science and Technology Infrastructure	26
1.20	Facts Vs. Values	26
1.21	Atomic Energy	26
1.22	Highlights of the Nuclear Power Programme	27
1.23	Nuclear Power Corporation of India Limited	27
1.24	Ocean Engineering Applications	28

Chapter 2: Non-Conventional Energy Resources and Utilisation — 33

2.1	Introduction	33
2.2	Energy Science	33
2.3	Various Energy Science	33
2.4	Energy Technology	35
2.5	Energy Technology and Energy Sciences	36
2.6	Law of Conservation of Energy	37
2.7	Facts and Figures about Energy	37
2.8	Indian and Global Energy Sources	38
	2.8.1 The Sun	38
	2.8.2 etroleum	39
	2.8.3 Natural Gas	43
	2.8.4 Coal	44
	2.8.5 Nuclear Energy	45
	2.8.6 LPG (Liquified Petroleum Gas)	46
	2.8.7 Alcohol	46
	2.8.8 Gasohol	46
	2.8.9 Hydro Power	46
2.9	Energy Exploited	46
2.10	Energy Demand	48

2.11	Energy Planning	49
2.12	Introduction to Various Sources of Energy	51
	2.12.1 Conventional Sources of Energy	51
	2.12.2 Non-Conventional Sources of Energy	51
2.13	Introduction to Various Non-Conventional Energy Resources	52
2.14	Bio-Gas	53
	2.14.1 Aerobic and Anaerobic Bio-Conversion Process	53
	2.14.2 Raw Materials	55
	2.14.3 Properties of Bio Gas	56
	2.14.4 Bio Gas Plant Technology	56
2.15	Wind Energy	58
	2.15.1 Wind Machine Fundamentals	59
	2.15.2 Aerofoil Design	60
	2.15.3 Wind Power Systems	62
	2.15.4 Economic Issues	63
	2.15.5 Selection of Wind Mill	64
	2.15.6 Recent Developments	65
2.16	Solar Energy	66
	2.16.1 Solar Radiations	67
	2.16.2 Solar Thermal Power Plant	68
	2.16.3 Solar Energy Storage	70
	2.16.4 Recent Developments in Solar Power Plants	72
2.17	Electrochemical Effects and Fuel Cells	73
	2.17.1 Reversible Cells	73
	2.17.2 Ideal Fuel Cells	74
	2.17.3 Other Types of Fuel Cells	75
	2.17.4 Efficiency of Cells	76
2.18	Thermionic Systems and Thennionic Emission	78
	2.18.1 Thermoionic Conversion	79
	2.18.2 Ideal and Actual Efficiency	80
2.19	Thermoelectric Systems	82
	2.19.1 Principle of Working	82
	2.19.2 Performance	83
2.20	Geo Thermal Energy	86
	2.20.1 Hot Springs	87
	2.20.2 Steam Ejection	87
	2.20.3 Site Selection	88

2.20.4 Geothermal Power Plants	...	88
2.20.5 Advanced Concepts	...	95
2.21 Ocean Energy		97
2.21.1 Power Plants Based on Ocean Energy	...	98
2.22 Other Energy Technology	...	101
2.22.1 Liquid Fuel	...	101
2.22.2 Fuel Cell Technology	...	103
2.22.3 Hydrogen Energy	...	104
2.22.4 Hydrogen Energy Technology	...	104
2.22.5 Battery Operated Vehicles	...	106
2.22.6 Bio Fuel Technology	...	107
2.22.7 Hydroelectric Power	...	107
2.22.8 Innovative Heat Exchanger to Save Energy	...	108

Chapter 3: Power Plant Economics and Variable Load Problem ... **120**

3.1 Terms and Factors	...	120
3.2 Factor Effecting Power Plant Design	...	122
3.3 Effect of Power Plant Type on Costs	...	122
3.3.1 Initial Cost		122
3.3.2 Rate of Interest	...	123
3.3.3 Depreciation		123
3.3.4 Operational Costs		124
3.3.5 Cost of Fuels		125
3.3.6 Labout Cost		125
3.3.7 Cost of Maintenance and Repairs		125
3.3.8 Cost of Stores	...	125
3.3.9 Supervision	...	126
3.3.10 Taxes	...	126
3.4 Effect of Plant Type on Rates (Tariffs or Energy Element)		126
3.4.1 Requirements of a Tariff		126
3.4.2 Types of Tariffs		126
3.5 Effect of Plant Type on Fixed Elements		129
3.6 Effect of Plant Type on Customer Elements		129
3.7 Investor's Profit		129
3.8 Economics in Plant Selection	...	130
3.9 Economic of Power Generation	...	131
3.10 Industrial Production and Power Generational Compared		132
3.11 Load Curves	...	132

3.12 Ideal and Realized Load Curves	...	133
3.13 Effect of Variable Load on Power Plan Design	...	133
3.14 Effect of Variable Load on Power Plant Operation	...	134

Chapter 4 : Steam Power Plant ... 142

4.1 Introduction	...	142
4.2 Essentials of Steam Power Plant Equipment	...	142
4.2.1 Power Station Design	...	144
4.2.2 Characteristics of Steam Power Plant	...	144
4.3 Coal Handling	...	145
4.3.1 Dewatering of Coal	...	149
4.4 Fuel Burning Furnaces	...	150
4.4.1 Types of Furnaces	...	150
4.5 Method of Fuel Firing	...	151
4.5.1 Hand Firing	...	151
4.5.2 Mechanical Firing (Stokers)	...	153
4.6 Automatic Boiler Control	...	156
4.7 Pulverized Coal	...	157
4.7.1 Ball Mill	...	158
4.7.2 Ball and Race Mill	...	159
4.7.3 Shaft Mill	...	161
4.8 Pulverized Coal Firing	...	161
4.9 Pulverized Coal Burners	...	163
4.9.1 Cyclone Fired Boilers	...	166
4.10 Water Walls	...	166
4.11 Ash Disposal	...	167
4.11.1 Ash Handling Equipment	...	168
4.12 Smoke and Dust Removal	...	172
4.13 Types of Dust Collectors	...	172
4.13.1 Fly Ash Scrubber	...	174
4.13.2 Fluidized Bed Combustion (FBC)	...	175
4.13.3 Types of FBC Systems	...	176

Chapter 5 : Steam Generator ... 179

5.1 Introduction	...	179
5.2 Types of Boilers	...	179
5.3 Cochran Boilers	...	181
5.4 Lancashire Boiler	...	183

5.5 Locomotive Boiler	...	185
5.6 Babcock Wilcox Boiler	...	187
5.7 Industrial Boilers	...	189
5.8 Merits and Demerits of Water Tube Boilers over Fire Tube Boilers Mertis	...	189
5.9 Requirements of a Good Boiler	...	190
5.10 High Pressure Boilers	...	190
5.10.1 La Mont Boiler	...	191
5.10.2 Benson Boiler	...	191
5.10.3 Loeffler Boiler	...	192
5.10.4 Schmidt-Hartmann Boiler	...	193
5.10.5 Velox Boiler	...	193

Chapter 6 : Steam Turbine ... **195**

6.1 Principle of Operation of Steam Turbine	...	195
6.2 Classification of Steam Turbine	...	196
6.3 The Simple Impulse Turbine	...	200
6.4 Compounding of Impulse Turbine	...	201
6.5 Pressure Compounded Impulse Turbine	...	202
6.6 Simple Velocity-Compounded Impulse Turbine	...	203
6.7 Pressure and Velocity Compounded Impulse Turbine	...	204
6.8 Impulse-Reaction Turbine	...	205
6.9 Advantages of Steam Turbine over Steam Engine	...	206
6.10 Steam Turbine Capacity	...	206
6.11 Capability	...	207
6.12 Steam Turbine Governing	...	207
6.13 Steam Turbine Performance	...	207
6.14 Steam Turbine Testing	...	208
6.15 Choice of Steam Turbine	...	208
6.16 Steam Turbine Generators	...	208
6.17 Steam Turbine Specifications	...	209

Chapter 7 : Fuels and Combustion ... **217**

7.1 Introduction	...	217
7.2 Coal	...	217
7.3 Coal Analysis	...	219
7.3.1 Proximate Analysis	...	219
7.3.2 Ultimate Analysis	...	219
7.3.3 Heating Value	...	219

7.4	Coal Firing	221
7.5	Mechanical Stokers	222
7.6	Pulverized-Coal Firing	224
7.7	Cyclone Furnaces	230

Chapter 8 : Diesel Power Plant ... 234

8.1	Introduction	234
8.2	Operating Principle	236
8.3	Basic Types of IC Engines	236
	8.3.1 Two-Stroke, Spark Ignition Gas Engines/Petrol Engines	237
	8.3.2 Diesel Engines/Heavy Oil Engines	237
	8.3.3 Duel Fuel Engines	237
	8.3.4 High Compression Gas Engines	238
8.4	Advantage of Diesel Power Plant	238
8.5	Disadvantage of Diesel Power Plant	238
8.6	Application of Diesel Power Plant	238
8.7	General Layout of Diesel Power Plant	239
8.8	Performance of Diesel Engine	239
	8.8.1 Indicated Mean Effective Pressure (IMEP)	239
	8.8.2 Indicated Hourse Power (IHP)	239
	8.8.3 Brake Horse Power (B.H.P.)	240
	8.8.4 Frictional Horse Power (F.H.P.)	240
	8.8.5 Indicated Thermal Efficiency	240
	8.8.6 Brake Thermal Efficiency (Overall Efficiency)	240
	8.8.7 Mechanical Efficiency	240
8.9	Fuel System of Diesel Power Plant	241
8.10	Lubrication System of Diesel Power Plant	242
	8.10.1 Liquid Lubicricants or Wet Sump Lubrication System	244
	8.10.2 Solid Lubricants or Dry Sump Lubrication System	246
	8.10.3 Mist Lubrication System	246
8.11	Air Intakes and Admission System of Diesel Power Plant	246
8.12	Supercharging System of Diesel Power Plant	247
	8.12.1 Types of Supercharger	248
	8.12.2 Advantages of Supercharging	249
8.13	Exhaust System of Diesel Power Plant	249
8.14	Cooling System of Diesel Power Plant	250
	8.14.1 Open Cooling System	251
	8.14.2 Natural Circulation System	251

8.14.3 Forced Circulation Cooling System	251
8.15 Diesel Plant Operation	252
8.16 Efficiency of Diesel Power Plant	253
8.17 Heat Balance Sheet	255

Chapter 9 : Gas Turbine Power Plant — 267

9.1 Introduction	267
9.2 Classification of Gas Turbine Power Plant	267
9.2.1 Open Cycle Gas Turbine Power Plant	268
9.2.2 Closed Cycle Gas Turbine Power Plant	269
9.3 Elements of Gas Turbine Power Plants	272
9.3.1 Compressors	272
9.3.2 Intercoolers and Heat Exchangers	273
9.3.3 Combustion Chambers	274
9.3.4 Gas Turbines	276
9.4 Regeneration and Reheating	276
9.4.1 Regeneration	276
9.4.2 Reheating	278
9.5 Cogeneration	279
9.5.1 Cogeneration—Why	279
9.5.2 Cogeneration Technologies	280
9.6 Auxiliary Systems	282
9.6.1 Starting Systems	282
9.6.2 Ignition Systems	283
9.6.3 Lubrication System	283
9.6.4 Fuel System and Controls	285
9.7 Control of Gas Turbines	286
9.7.1 Prime Control	286
9.7.2 Protective Controls	287
9.8 Gas Turbine Efficiency	289
9.8.1 Effect of Blade Friction	291
9.8.2 Improvement in Open Cycle	291
9.9 Operations and Maintenance Performance	292
9.9.1 Operation	292
9.9.2 Maintenance Performance	293
9.10. Troubleshooting and Remedies	294
9.11 Combined Cycle Power Plants	295
9.12 Applications of Gas Turbine	296

9.13	Advantages of Gas Turbine Power Plant	... 296
9.14	Disadvantages	... 297

Chapter 10 : Nuclear Power Plant ... 307

10.1	Introduction	... 307
10.2	General History and Trends	... 308
	10.2.1 Major Events	... 308
	10.2.2 What Might Change the Current Situation ?	... 309
	10.2.3 Technical History and Developments	... 311
	10.2.4 Developments After WW-2	... 311
10.3	The Atomic Structure	... 312
10.4	Summary of Nuclear Energy Concepts and Terms	... 314
	10.4.1 Summary of Features	... 314
	10.4.2 Fission	... 314
	10.4.3 Critical Mass	... 315
	10.4.4 Alpha Radiation	... 315
	10.4.5 Beta Particles	... 315
	10.4.6 Gamma Particles	... 315
	10.4.7 Uranium Fission	... 315
	10.4.8 Half Life, T	... 316
10.5	Ethical Problems in Nuclear Power Regulation	... 316
10.6	Chemical and Nuclear Equations	... 316
10.7	Nuclear Fusion and Fission	... 317
	10.7.1 Fusion	... 318
	10.7.2 Fission	... 319
10.8	Energy From Fission and Fuel Burn Up	... 320
10.9	Radioactivity	... 320
10.10	Nuclear Reactor	... 322
	10.10.1 Parts of a Nuclear Reactor	... 322
	10.10.2 Nuclear Fuel	... 322
	10.10.3 Moderator	... 324
	10.10.4 Moderating Ratio	... 324
	10.10.5 Reflector	... 325
	10.10.6 Reactor Vessel	... 325
	10.10.7 Biological Shieding	... 326
	10.10.8 Coolant	... 326
	10.10.9 Coolant Cycles	... 326
	10.10.10 Reactor Core	... 326

10.11	Conservation Ratio	327
10.12	Neutron Flux	327
10.13	Clasification of Reactors	327
10.14	Cost of Nuclear Power Plant	327
10.15	Nuclear Power Station in India	328
10.16	Light Water Reactor (LWR) and Heavy Water Reactor (HWR)	331
	10.16.1 Importance of Heavy Water	332
10.17	Site Selection	333
10.18	Comparison of Nulcear Power Plant and Steam Power Plant	334
10.19	Multiplication Factor	334
10.20	Uranium Enrichment	334
10.21	Reactor Power Control	336
10.22	Nuclear Power Plant Economics	336
10.23	Safety Measures for Nuclear Power Plants	337
10.24	Site Selection and Commissioning Procedure	338
10.25	Major Nuclear Power Disasters	338
10.26	Chernobyl Nuclear Power Plant	339
	10.26.1 Reactor Design: RBMK-1000	340
	10.26.2 Control of the Reactor	340
	10.26.3 Chernobyl Reactor Operations	340
	10.26.4 Accident/Safety Plans	340
	10.26.5 Evacuation	340
10.27	Safety Problems in Chernobyl Reactor Design	340
	10.27.1 System Dynamics	340
	10.27.2 Another Safety Problem with the Design	341
10.28	Other, Earlier, Soviet Nuclear Accidents	341

Chapter 11 : Hydro-Electric Power Plants — 343

11.1	Introduction	343
11.2	Run-Off	345
11.3	Hydrograph and Flow Duration Curve	347
11.4	The Mass Curve	347
11.5	Selection of Site for a Hydro-Electric Power Plant	348
11.6	Essential Features of a Water-Power Plant	349
11.7	Calculations of Water Power Plants	351
11.8	Classification of Hydro-Plant	352
	11.8.1 Storage Plants	352
	11.8.2 Run-of-River Power Plants	354

(xvii)

11.8.3 Pumped Storage Power Plants	...	354
11.9 Power House and Turbine Setting	...	355
11.9.1 Advantages and Disadvantages of Underground Power-House	...	359
11.10 Prime-Movers	...	362
11.11 Specific Speed of Turbine	...	367
11.12 Draft Tubes	...	370
11.12.1 Methods to Avoid Cavitation	...	375
11.12.2 Types of Draft Tubes	...	375
11.12.3 Different Types of Draft Tubes	...	377
11.13 Models and Model Testing	...	378
11.14 Selection of Turbine	...	381

Chapter 12 : Electrical System ... 386

12.1 Introduction	...	386
12.2 Generators and Motors	...	386
12.2.1 Rotors	...	387
12.2.2 Stators	...	389
12.2.3 Ventilation	...	390
12.2.4 Hight-Voltage Generators	...	392
12.3 Transformers	...	392
12.3.1 Constructional Parts	...	393
12.3.2 Core Constructions	...	393
12.3.3 Windings	...	397
12.4 Cooling of Transformers	...	402
12.4.1 Simple Cooling	...	403
12.4.2 Mixed Cooling	...	403
12.4.3 Natural Oil Cooling	...	403
12.4.4 Forced Oil Cooling	...	405
12.4.5 Internal Cooling	...	406
12.5 Bus-Bar	...	409
12.5.1 Single Bus-Bar System	...	409
12.5.2 Single Bus-Bar System with Sectionalisation	...	409
12.5.3 Duplicate Bus-Bar System	...	410
12.6 Busbar Protection	...	411
12.6.1 Differential Protection	...	411
12.6.2 Fault Bus Protection	...	412

Chapter 13 : Pollution and its Control — 414

- 13.1 Introduction — 414
- 13.2 Environment Pollution due to Energy Use — 414
- 13.3 Environment Pollution due to Industrial Trial Emissions — 415
- 13.4 Environment Pollution to Road Transport — 415
- 13.5 Harmful Effects of Emissions — 415
    - 13.5.1 Buildings and Materials — 416
    - 13.5.2 Soil, Vegetation and Animal Life — 416
    - 13.5.3 Human Beings — 416
- 13.6 Steps Taken so far and their Impact — 416
- 13.7 Noise Pollution and its Control — 417
- 13.8 Green House Gases and their Effects — 418
- 13.9 Fossil Fuel Pollution — 419
    - 13.9.1 Urban Air Pollution — 419
    - 13.9.2 Acid Rain — 420
    - 13.9.3 Global Climate Change — 420
    - 13.9.4 Stratospheric Ozone Depletion — 420
    - 13.9.4 Acid Fog — 420
- 13.10 Pollution due to Combustion of Fuel — 421
    - 13.10.1 Gas Fuel — 421
    - 13.10.2 Methane — 421
    - 13.10.3 Alkanes — 422
- 13.11 Pollution due to Gas Combustion — 422
    - 13.11.1 Unburned Hydrocarbons (UHCS) — 422
    - 13.11.2 Carbon Monoxide (CO) — 422
    - 13.11.3 Nitric Oxide (NO_x) — 422
    - 13.11.4 Soot — 424
- 13.12 Pollution due to Liquid Fuel — 424
    - 13.12.1 Atomization — 424
    - 13.12.2 Vaporization — 424
    - 13.12.3 Modes of Combustion — 424
- 13.13 Pollution due to Solid Fuel — 425
- 13.14 Air Pollution by Thermal Power Plants — 428
- 13.15 Water Pollution by Thermal Power Plants — 428
- 13.16 Environment Concerns and Diesel Power Plants — 429
- 13.17 Nuclear Power Plant and the Environment — 431

13.17.1 The Fuel Cycle	...	432
13.17.2 Wastes	...	433
13.18 Radiations from Nuclear-Power Plant Effluents	...	434
13.19 Impact on Pollution and Air Qality in Delhi	...	434
13.19.1 Environmental Concerns	...	435
13.19.2 Pollution Levels	...	435
13.19.3 Measurs to Combat Pollution	...	436
13.20 Method for Pollution Control	...	440
13.21 Control of Marine Pollution	...	441
Appendix	...	443
Glossary	...	445
Bibliography	...	455
Index	...	462

Chapter 1

Fundamental of Power Plant

1.1 INTRODUCTION

The whole world is in the grip of energy crisis and the pollution manifesting itself in the spiraling cost of energy and uncomforted due to increase in pollution as well as the depletion of conventional energy resources and increasing curve of pollution elements. To meet these challenges one way is to check growing energy demand but that would show down the economic growth as first step and to develop nonpolluting energy conversion system as second step. It is commonly accepted that the standard of living increases with increasing energy consumption per capita. Any consideration of energy requirement and supply has to take into account the increase conservation measures. On the industrial font, emphasis must be placed on the increased with constant effort to reduce energy consumption. Fundamental changes in the process, production and services can affect considerable energy saving without affecting the overall economy. It need not be over emphasized that in house hold commercial and industrial use of energy has considerable scope in energy saving. Attempt at understanding the integrated relationship between environment and energy have given shape due to development of R-134a, (an non pollutant refrirgent) to emerging descipling of environmental management. The government of India has laid down the policy "it is imperative that we carefully utilize our renewal (*i.e.*, non-decaying) resources of soil water, plant and animal live to sustain our economic development" our exploration or exploitation of these is reflected in soil erosion, salutation, floods and rapid destruction of our forest, floral and wild life resources. The depletion of these resources often tends to be irreversible since bulk of our population depends on these natural resources. Depletion of these natural resources such as fuel, fodder, and housing power plant;

1.2 CONCEPT OF POWER PLANT

A power plant is assembly of systems or subsystems to generate electricity, *i.e.*, power with economy and requirements. The power plant itself must be useful economically and environmental friendly to the society. The present book is oriented to conventional as well as non-conventional energy generation. While the stress is on energy efficient system regards conventional power systems *viz.*, to increase the system conversion efficiency the supreme goal is to develop, design, and manufacturer the non-conventional power generating systems in coming decades preferably after 2050 AD which are conducive to society as well as having feasible energy conversion efficiency and non-friendly to pollution, keeping in view the pollution act. The subject as a whole can be also stated as modern power plants for power viz electricity generation in 21st century. The word modern means pertaining to time. At present due to energy crisis the first goal is to conserve energy for future while the second step is to develop alternative energy systems including direct energy conversion devices, with the devotion, dedication and determination remembering the phrase, " Delve and Delve Again till wade into".

1.3 CLASSIFICATION OF POWER PLANTS

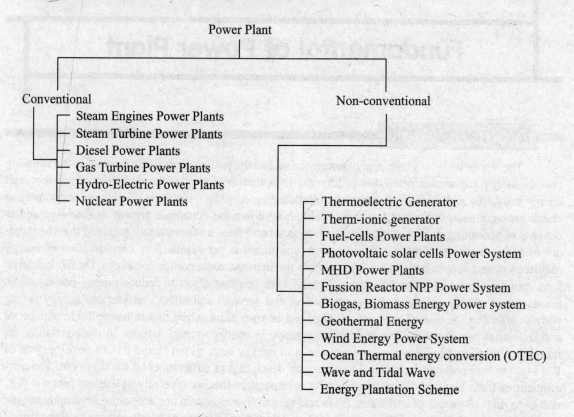

A power plant may be defined as a machine or assembly of equipment that generates and delivers a flow of mechanical or electrical energy. The main equipment for the generation of electric power is generator. When coupling it to a prime mover runs the generator, the electricity is generated. The type of prime move determines, the type of power plants. The major power plants, which are discussed in this book, are,

1. Steam power plant
2. Diesel power plant
3. Gas turbine power plant
4. Nuclear power plant
5. Hydro electric power plant

The Steam Power Plant, Diesel Power Plant, Gas Turbine Power Plant and Nuclear Power Plants are called **THERMAL POWER PLANT,** because these convert heat into electric energy.

1.4 ENERGY

Energy is the capacity for doing work, generating heat, and emitting light. The equation for work is the force, which is the mass time the gravity times the distance.

Heat is the ability to change the temperature of an object or phase of a substance. For example, heat changes a solid into a liquid or a liquid into a vapor. Heat is part of the definition of energy.

Another part of the definition of energy is radiation, which is the light and energy emitted in the form of waves traveling at the speed of light.

Energy is measured in units of calorie, quad, and joule. A kilocalorie is the amount of energy or heat required to raise the temperature of 1 kilogram of water from $14.5°C$ to $15.5°C$. The quad unit is used to measure energy needed for big countries. The final measurement of energy is joules.

Energy is an essential input for economic development and improving quality of life. India's per capita consumption of Commercial Energy (*viz.,* coal, petroleum and electricity) is only one-eighth of the Global Average and will increase with growth in Gross Domestic Production (GDP) and improvement in standard of living.

Commercial Energy accounts for a little over half of the total energy used in the Country, the rest coming from non-commercial resources like cow-dung, fuel wood and agricultural waste. Though the share of these non-commercial sources has been coming down, consumption has increased almost double since 1953.

These renewable, non-commercial sources have been used extensively for hundreds of years but in a primitive and ineffective way. Indiscriminate use of non-commercial energy sources is leading to an energy crisis in the rural areas. Seventh Plan laid emphasis on the development and accelerated utilisation of renewable energy sources in rural and urban areas. A major Policy of the Government is directed towards increasing the use of coal in household and of electricity in transport sector in order to reduce dependence on oil, which is becoming scarce gradually.

The Government has formulated an energy policy with objectives of ensuring adequate energy supply at minimum cost, achieving self-sufficiency in energy supplies and protecting environment from adverse impact of utilising energy resources in an injudicious manner. Main elements of the policy are:

1. Accelerated exploitation of domestic conventional energy resources-oil, coal, hydro and nuclear power;
2. Intensification of exploration to increase indigenous production of oil and gas;
3. Management of demand for oil and other forms of energy;
4. Energy conservation and management;
5. Optimisation of utilisation of existing capacity in the country;
6. Development and exploitation of renewable sources of energy to meet energy requirements of rural communities;
7. Intensification of research and development activities in new and renewable energy sources; and
8. Organisation of training far personnel engaged at various levels in the energy sector.

Development of conventional forms of energy for meeting the growing energy needs of the society at a reasonable cost is the responsibility of Government *viz.*, Department of Power, Coal and

Petroleum and Natural Gas. Development and promotion of non-conventional/alternate/new and renewable sources of energy such as Solar, Wind and Bio-energy, etc., are also getting sustained attention from the Department of Non-Conventional Energy Sources created in September, 1982. Nuclear Energy Development is being geared up by the Department of Atomic Energy to contribute significantly to overall energy availability in the Country.

Energy Conservation is being given the highest-priority and is being used as a tool to bridge the gaps between demand and supply of energy. An autonomous body, namely Energy Management Centre, has been set up on ten April, 1989, as a nodal agency for energy conservation projects.

1.5 TYPES OF ENERGY

There are various types of energy which, they include nuclear, electrical, thermal, chemical, and radiant energy. In addition, gravitational potential energy and kinetic energy that combines to produce mechanical energy.

Nuclear energy produces heat by fission on nuclei, which is generated by heat engines. Nuclear energy is the world's largest source of emission-free energy. There are two processes in Nuclear energy fission and fusion. In fission, the nuclei of uranium or plutonium atoms are split with the release of energy. In fusion, energy is released when small nuclei combine or fuse. The fission process is used in all present nuclear power plants, because fusion cannot be controlled. Nuclear energy is used to heat steam engines. A Nuclear power plant is a steam engine using uranium as its fuel, and it suffers from low efficiency.

Electricity powers most factories and homes in our world. Some things like flashlights and Game Boys use electricity that is stored in batteries as chemical energy. Other items use electricity that comes from an electrical plug in a wall socket. Electricity is the conduction or transfer of energy from one place to another. The electricity is the flow of energy. Atoms have electrons circling then, some being loosely attached. When electrons move among the atoms of matter, a current of electricity is created.

Thermal energy is kinetic and potential energy, but it is associated with the random motion of atoms in an object. The kinetic and potential energy associated with this random microscopic motion is called *thermal energy*. A great amount of thermal energy (heat) is stored in the world's oceans. Each day, the oceans absorb enough heat from the sun to equal the energy contained in 250 billion barrels of oil (Ocean Thermal Energy Conversion Systems).

Chemical energy is a form of energy that comes from chemical reactions, in which the chemical reaction is a process of oxidation. Potential energy is released when a chemical reaction occurs, which is called *chemical energy*. A car battery is a good example, because the chemical reaction produces voltage and current to start the car. When a plant goes through a process of photosynthesis, what the plant is left with more chemical energy than the water and carbon dioxide. Chemical energy is used in science labs to make medicine and to product power from gas.

Radiant energy exists in a range of wavelengths that extends from radio waves that many be thousands of meters long to gamma rays with wavelengths as short as a million-millionth (10^{-12}) of a meter. Radiant energy is converted to chemical energy by the process of photosynthesis.

The next two types of energy go hand and hand, *gravitational potential energy* and *kinetic energy*. The term energy is motivated by the fact that potential energy and kinetic energy are different aspects of the same thing, mechanical energy.

Potential energy exists whenever an object which has mass has a position within a force field. The potential energy of an object in this case is given by the relation $PE = mgh$, where PE is energy in joules, m is the mass of the object, g is the gravitational acceleration, and h is the height of the object goes.

Kinetic energy is the energy of motion. An object in motion, whether it be vertical or horizontal motion, has kinetic energy. There are different forms of kinetic energy vibrational, which is the energy due to vibrational motion, rotational, which is the energy due to rotational motion, and transnational, which is the energy due to motion from one location to the other. The equation for kinetic energy is $\frac{1}{2} mv^2$, where m is the mass and v is the velocity. This equation shows that the kinetic energy of an object is directly proportional to the square of its speed.

1.6 POWER

Power is the rate doing work, which equals energy per time. Energy is thus required to produce power. We need energy to run power plants to generate electricity. We need power to run our appliances, and heat our homes. Without energy we would not have electricity.

The units of power are watts, joules per second, and horsepower,

where ;   1 Watt = 1 joule per second

1 Kilowatt = 1,000 Watts

1 Megawatt = 1,000 kilowatts

= 1 horsepower

Electricity is the most convenient and versatile form of energy. Demand for it, therefore, has been growing at a rate faster than other forms of energy. Power industry too has recorded a phenomenal rate of growth both in terms of its volume and technological sophistication over the last few decades. Electricity plays a crucial role in both industrial and agricultural sectors and, therefore, consumption of electricity in the country is an indicator of productivity and growth. In view of this, power development has been given high-priority in development programme.

1.7 POWER DEVELOPMENT IN INDIA

The history of power development in India dates back to 1897 when a 200 kW hydro-station was first commissioned at Darjeeling. The first steam station was set up in Calcutta in 1899. By the end of 1920, the total capacity was 130 mW, comprising. Hydro 74 mW, thermal 50 mW and diesel 6 mW. In 1940, the total capacity goes to 1208 mW. There was very slow development during 1935-1945 due to Second World War. The total generation capacity was 1710 mW by the end of 1951. The development really started only after 1951 with the launching of the first five-year plan.

During the First Plan, construction of a number of Major River Valley Projects like Bhakra-Nangal, Damodar Valley, Hira Kund and Chambal Valley was taken up. These projects resulted in the stepping up of power generation. At the end of the First Plan, generation capacity stood at 34.2 lakh kW.

Emphasis in Second Plan (1956-61) was on development of basic and heavy industries and related need to step up power generation. Installed capacity at the end of Second Plan reached 57 lakh kw. comprising 3800 mW thermal and 1900 MW hydel.

During the Third Plan period (1961-66), emphasis was on extending power supply to rural areas. A significant development in this phase was emergence of Inter-state Grid System. The country was divided into Five Regions to promote power development on a Regional Basis. A Regional Electricity

Board was established in each region to promote integrated operation of constituent power system. Three Annual Plans that followed Third Plan aimed at consolidating programmes initiated during the Third Plan.

Fourth Plan envisaged need for central participation in expansion of power generation programmes at strategic locations to supplement activities in the State Sector. Progress during the period covering Third Plan, three Annual Plans and Fourth Plan was substantial with installed capacity rising to 313.07 lakh kW compression; 113.86 lakh kW from Hydro-electric Projects, 192.81 lakh kW from Thermal Power Projects and balance of 6.4 lakh kW from Nuclear Projects at the end of the Fifth Plan.

During the Sixth Plan, total capacity addition of 196.66 lakh kW comprising Hydro 47.68 lakh kW, Thermal 142.08 lakh kW and Nuclear 6.90 lakh kW was planned. Achievement, however, has been 142.26 lakh kW (28.73 lakh kW Hydro, 108.98 lakh kW Thermal and 4.55 lakh kW Nuclear) 72.3 per cent of the target.

The Seventh Plan power programme envisaged aggregate generating capacity of 22,245 mW in utilities. This comprised 15,999 mW Thermal, 5,541 mW Hydro and 705 mW Nuclear of the anticipated 22,245 mW additional capacity. Central Sector Programme envisaged capacity addition of 9,320 mW (7,950 mW Thermal, 665 mW Hydro and 705 mW Nuclear) during the Plan Period. During the Seventh Plan, 21401.48 mW has been added comprising 17104.1 mW Thermal 3,827.38 mW Hydro and 470 mW Nuclear. Year wise commissioning of Hydro, Thermal and Nuclear Capacity added during 1985-86 to 1989-90 is given in.

The Working Group on Power set up particularly the Planning Commission in the context of formulation of power programme for the Eighth Plan has recommended a capacity addition programme of 38,369 mW for the Eighth Plan period, out of which it is expected that the Central Sector Projects would add a capacity of 17,402 mW. The programme for the first year of the Eighth Plan (1990-91) envisages generation of additional capacity of 4,371.5 mW comprising 1,022 mW Hydro, 3,114.5 mW Thermal and 235 mW Nuclear.

The subject 'Power' appears in the Concurrent List of the Constitution and as such responsibility of its development lies both with Central and state governments. At the Centre, Department of Power under the Ministry of Energy is responsible for development of Electric Energy. The department is concerned with policy formulation, perspective planning, procuring of projects for investment decisions, monitoring of projects, training and manpower development, administration and enactment of Legislation in regard to power generation, transmission and distribution. The depart-ment is also responsible for administration of the Electricity (Supply) Act, 1948 and the Indian Electricity Act, 191() and undertakes all amendments thereto. The Electricity (Supply) Act, 1948, forms basis of administrative structure of electricity industry. The Act provides for setting up of a Central Electricity Authority (CEA) with responsibility, inter-alia, to develop a National Power Policy and coordinate activities of various agencies and State Electricity Boards. The act was amended in 1976 to enlarge scope and function of CEA and enable of creation of companies for generation of electricity.

The Central Electricity Authority advises Department of Power on technical, financial and economic matters. Construction and operation of generation and transmission projects in the Central Sector are entrusted to Central Power Corporations, namely, National Thermal Power Corpora-tion (NTPC), National Hydro-Electric Power Corporation (NHPC) and North-Eastern Electric Power Corporation (NEEPCU) under administrative control of the Department of Power. The Damodar Valley Corporation (DVC) constituted under the DVC Act, 1948 and the Bhitkra Beas, Management Board (BBMB) constituted under the Punjab Reorganization. Act, 1966, is also under administrative control of the Department of Power. In addition, the department administers Beas Construction Board (BCB) and National Projects Construction Corporation (NPCC), which are construction agencies and training and research organisations, Central Power Research Institute (CPRI) and Power Engineers Training Society (PETS). Programmes of rural electrification are within the purview of Rural Electrification Corporation (REC) which is a financing agency. "There are two joint venture Power Corporations under the administrative control of the Department of Power, namely, Nathpa jhakri Power Corporation and Tehri Hydro Development Corporation which are responsible for the execution of the Nathpa Jhakri Power Project and Projects of the Tehri Hydro Power Complex respectively. In addition to this, Energy Manage-ment Centre, an autonomous body, was established in collaboration with the European Economic Community, which is responsible for training, research, and information exchange between energy professionals. It is also responsible for conservation of energy programmes/activities in the Department of Power.

Significant progress has been made in the expansion of transmission and distribution facilities in the Country. Total length of transmission lines of 66 kV and above increased from 10,000 ckt (circuit) km in December 1950 to 2.02 lakh ckt Km in March, 1990. Highest transmission voltage in the Country at present is 400 kV and above 19800 ckt km of 400 kV lines had been constructed up to March, 1990 and about 18000 ckt km of these are in actual operation.

Prior to the Fourth Plan, Transmission Systems in the Country were developed more or less as state systems, as generating stations were built primarily in the State Sector. When State Transmission Systems had developed to a reasonable extent in the Third Plan, potentiality of inter-connected operation of individual state systems with other neighboring systems within the region (northern, western, southern, eastern and north-eastern) was thought of. Fairly well inter-connected systems at voltage of 220 kV with progressive overlay of 400 kV are presently available in all regions of the Country except North-eastern Region. With creation of Two Generation Corporations, namely National Thermal Power Corporation and National Hydro-Electric Power Corporation in 1975, the Centre had started playing an increasingly larger role in the development of grid systems.

The 400 kV transmission systems being constructed by these organisa-tions as part of their generation projects, along with 400 kV inter-state and inter-regional transmission lines would form part of the National Power Grid.

National Power Grid will promote integrated operation and transfer of power from one system to another with ultimate objective of ensuring optimum utilisation of resources in the Country. India now has well integrated Regional Power Systems and exchange of power is taking place regularly between a large numbers of state systems, which greatly facilitates better utilisation of existing capacity.

1.8 RESOURCES FOR POWER GENERATION

The hydel power source plays a vital role in the generation of power, as it is a non-conventional perennial source of energy. Therefore the French calls it "huile blanche"—white oil-the power of flowing water. Unlike black oil, it is a non-conventional energy source. A part of the endless cycle in which moisture is raised by the sun, formed into clouds and then dropped back to earth to feed the rivers whose flow can be harnessed to produce hydroelectric power. Water as a source of power is non-polluting which is a prime requirement of power industry today.

The world's total waterpower potential is estimated as 1500 million kW at mean flow. This means that the energy generated at a load factor of 50% would be 6.5 million kW-hr, a quantity equivalent to 3750 million tonnes of coal at 20% efficiency. The world hydel installed capacity (as per 1963 estimate is only 65 million kW or 4.3% of the mean flow.

India has colossal waterpower resources. India's total mean annual river flows are about 1675 thousand million cubic meters of which the usable resources are 555 thousand million cubic meters. Out of total river flows, 60% contribution comes from Himalayan rivers (Ganga, Indus and Brahmaputra). 16% from central Indian rivers (Narmada, Tapti and Mahanadi) and the remaining from the rivers drainning the Deccan plateau (Godavari, Krishna and Cauvery). India's power potential from hydel source as per the recent estimate is 41500 mW while its present hydel capacity is only 32000 mW. Still India has got enough hydel potential to develop to meet the increasing power needs of the nation. The abundant availability of water resources, its fairly even distribution and overall economy in developing this source of energy enhanced its development in India, The other factors responsible in its rapid development are indigenous technological skill, material and cheap labour. In the IX five-year plan; the Government considering the importance of this source has included a number of hydro-projects. The major difficulty in the development of hydroelectric projects is the relatively longer time required for it's hydrological, topographical and geological investigations. Lack of suitable. Site is an added problem for taking up hydro-projects.

Hydropower was once the dominant source of electrical energy in the world and still is in Canada, Norway and Switzerland. But its use has decreased in other countries since 1950s, as relatively less expensive fuel was easily available. In USA, only 10% of the total power production is water-generated. In the light of fuel scarcity and its up surging prices, the role of hydropower is again re-examined and more emphasis is being laid on waterpower development. As per Mr. Hays (Manager of Hydro Projects in USA), "It was less costly per mW to build a single 1000 mW thermal plant than 20 small hydro-plants. But, with the increased fuel cost and high cost of meeting environmental criteria for new thermal plants, interest in hydro is being revived". Small hydro-projects ranging from 10 to 1500 kW are becoming more feasible as standardization of major equipment reduces costs. India is yet to start in the field of micro-hydro projects, which is one major way for solving the present power problem.

Hydro-projects generate power at low cost, it is non conventional, easy to manage, pollution free and makes no crippling demands on the transportation system. But the major drawback is, it operates at the mercy of nature. Poor rainfall has on a number of occasions shown the dangers of over dependence on hydropower.

Let rivers flow and let rains shower the earth with prosperity is the ancient prayer chanted by Riches and continued to be chanted even now.

The development of hydropower systems as a back up for thermal systems has significant advantages. The flexible operation of hydraulic turbines makes them suitable for. Peak load operation. Therefore, the development of hydropower is not only economical but it also solves the major problem of peak load. The present Indian policy of power development gives sufficient importance for the hydel-power development. The next important source for power generation is fuel in the form of coal, oil or gas. Unfortunately, the oil and gas resources are very much limited in India. Only few power plants use oil or gas as a source of energy. India has to import most of the oil required and so it is not desirable to use it for power generation. The known resources of coal in India are estimated to be 121,000 million tonnes, which are localized in West Bengal, Bihar, Madhya Pradesh and Andhra Pradesh. The present rate of annual production of coal is nearly 140 million tonnes of which 40 million-tonnes are used for power generation. The coal used for power generation is mainly low-grade coal with high ash content (20-40%).

The high ash content of Indian coal (40–50%) is one of the causes for bad performance of the existing steam power plants and their frequency outages, as these plants have been designed for low ash coals. Due to the large resources of coal available in the country, enough emphasis has been given for thermal Power plants in the IX plan period.

The location of hydel-power plants is mostly determined by the natural topography available and location of thermal plants is dictated by the source of fuel or transportation facilities available if the, power plant is to be located far from coalmines. For nuclear power plant any site can be selected paying due consideration to safety and load. India has to consider nuclear generation in places remote from coal mines and water power sites. The states which are poor in natural resources and those which have little untapped conventional resources for future development have to consider the development of nuclear plants.

The nuclear fuel which is commonly used for nuclear power plants is uranium. Deposits of uranium have been located in Bihar and Rajasthan. It is estimated that the present reserves of uranium available in country may be sufficient to sustain 10,000 mW power plants for its thorium into nuclear Indian lifetime. Another possible nuclear power source is thorium, which is abundant in this country, estimated at 500,000 tonnes. But the commercial use of this nuclear fuel is tied up with development of fast breeder reactor which converts energy economy must wait for the development of economic methods for using thorium which is expected to be available before the end of twentieth century. The major hurdle in the development of nuclear power in this country is lack of technical facility and foreign exchange required to purchase the main component of nuclear power plant. Dr. Bhabha had envisaged 8000 mW of power from nuclear reactors by 1980–81 which was subsequently scaled down to a more realistic level of 2700 mW by Dr. Sarabhai out of this only 1040 MW has materialized which is less than 1.5% of the country's installed power capacity. Moreover the performance of nuclear plants has been satisfactory compared to thermal plants.

1.9 PRESENT POWER POSITION IN INDIA

The present power position in India is alarming as there are major power shortages in almost all states of the country leading to crippling of industries and hundreds of thousands of people losing jobs and a heavy loss of production.

The overall power scene in the country shows heavy shortages almost in all states. The situation is going to be aggravated in coming years as the demand is increasing and the power industry is not keeping pace with the increasing demand.

Many of the states in India depend to a large extent on hydro generation. The increase in demand has far outstripped the installation of new plants. Also there is no central grid to distribute excess energy from one region to another. The experience in the operation of thermal plants is inadequate. All these have led to heavy shortages and severe hardship to people.

Very careful analysis of the problem and proper planning and execution is necessary to solve the power crisis in our country.

Suitable hydrothermal mix, proper phasing of construction of new plants, training personnel in maintenance of thermal plants.

1.10 FUTURE PLANNING FOR POWER GENERATION

Considering the importance of power industry in the overall development of the country, power sector has been given high priority in the country's development plans. Energy sector alone accounts for about 29% of sixth plan investment. If investments in coal and oil transport and other infrastructures are taken into account, the total investment in the energy sector will account for about 40% of the plan investments. The fact alone is sufficient to exhibit the importance of power industry for the country's development. From a mere Rs. 149 crores in the First Plan, the outlay for power during sixth plan period has increased to Rs. 15750 crores. The installed generating capacity has grown ten-fold from 2300 mW in 1951 to 25900 mW in 1978. Of this, 11000 mW was in hydel, 14000 mW in thermal and less than 1000 mW in nuclear power stations. The total number of power stations of 20 mW capacities and above at the end of March 1978, was 127, of which 65 were hydel, 60 thermal and 2 nuclear. Power generation rose from 7514 million kWh in 1950–51 to 103754 million kWh in 1978–79, *i.e.*, nearly 15 times. The total users of electricity have risen from 15 lakhs in 1950 to 264 lakhs in 1978–79. The per capita consumption of electricity rose from 18 kWh in 1950–51 to 121 kWh in 1978–79.

In spite of these measures, this industry is unable to meet the demands. Power shortages have become a recurrent feature in the country. Against an estimated requirement of 108656 million kWh in 1978-79, the actual availability was only a 97588 million kWh a deficit of about 11070 million kWh or 10.2°C.

With the programme of large-scale industrialization and increased agricultural activity, the demand for power in the country is increasing at a rapid rate. If the present trend continues, the demand for power by the end of year 2000 would be about 125 to 150 million kW. Allowing for adequate reserve margins required for scheduled maintenance, a total generating capacity of about 175 to 200 million kW would be needed by the year 2000 to meet the anticipated demands. This would mean 8 to 10 fold increase of the existing capacity.

Only proper development of hydel, thermal and nuclear resources of the country can achieve the required growth. Out of total available hydel-potential (41,000 mW), only 16% has been developed, therefore there is sufficient scope to develop this source of power in future. The major hydel potential is available in the northern region. Even if all the hydel potential is developed, it will not be possible to meet the growing demand. Therefore, it is necessary to supplement the hydel potentials with thermal. The coal deposits are rich and ample, though in terms of per capita it is hardly 176 tonnes in India which is certainly poor compared with other countries as 1170 tonnes in China, 13500 tonnes in the U.S.A. and 22000 tonnes in the former U.S.S.R. The available coal is also unevenly distributed in the country (60°C only in Bihar and Bengal). This further requires the development of transportation facilities.

Therefore, it is also not possible to depend wholly on thermal power development. The consideration for the use of nuclear fuel for power production in future is equally essential particularly in those states, which are far away from coal resources and poor in hydel potential.

The future planning in the power development should aim at optimum exploitation of resources available so that power mix of hydel, thermal and nuclear is achieved.

Another step to be taken in the power development industry is setting up super-thermal power plants the central sector at different places in the country. The super-thermal power stations are at Farakka, Ramagundam, Korba and Singrauli and these are supplying power for the past 20 years. Presently all of them are supplying power through the national grid to deficit states.

In our country even 20 mW hydro potentials have not been developed, whereas it appears to be advantageous to develop even 20 kW units. Development of small hydro potentials as in China has, to a great extent, reduced the strain in existing plants.

The development of biogas can ease the strain on oil supply to domestic users, which can otherwise diverted to power generation.

Another suggestion to face the present alarming power situation in the country is Energy Plantation. India receives large amount of solar radiation and photosynthesis is the process by which solar energy is converted into food and fuel by green plants. Fast growing species of trees give a yield of about 15 to 35 tonnes/hectare/year. The land, which is presently not used either for agriculture or forest, can be used for energy plantation where average rainfall is 80 to 100 cm per annum. With present Forest Technology, planned production forestry offers an unusual opportunity. If the forest area is increased from present 22 to 30%, increase in forest area is 30 million hectares of land) it can yield sufficient energy after next 20 years. The Government does not seriously think this phase of energy production but it looks a fruitful proposition.

As per the present planning of the Government, the problem of increased power demand will be solved only by proper mixed development of hydel, thermal and nuclear atleast during one more decade.

The severity of the power problem can be partly solved by the conservation of power. The efficiency hest thermal power plant is 35%. In India, it is hardly 25%. If auxiliary consumption and line loss are taken into account, the efficiency still goes to hardly 16%. The problem can be partly solved by proper maintenance and good quality of fuel supply.

The efficiency of the power plant operation is also defined as kWh generated per kW installed. The maximum kWh per annum per kW is 8760. The average figure in India is hardly 4000, which shows that the utilisation is only 45%. If this utilisation is increased, need for new capacity for power generation will be reduced.

Increasing load factors can reduce the capacity of the power industry. The proper planning to develop hydel, thermal and nuclear resources in India in addition to measures taken to reduce outages and with proper load management will definitely go a long way in meeting the increasing power demand of the country.

1.11 POWER CORPORATIONS IN INDIA

1.11.1 NATIONAL THERMAL POWER CORPORATION

National Thermal Power Corporation (NTPC) was incorporated in November, 1975, as a public sector undertaking with main objective of planning, promoting and organising integrated development of Thermal Power in the Country. The Authorized Capital of the corporation is Rs. 6,000 crore.

NTPC is currently constructing and operating the Nine Super Thermal Power Projects at Singrauli (UP), Korba (MP), Ramagundam.(AP), Farakka (WB), Vindhyachal (MP), Rihand (UP), Kahalgaon (Bihar), Dadri (UP), Talcher (Orissa) and Four Gas-based Projects at Anta (Rajasthan), Auraiya (UP), Dadri (UP) and Kawas (Gujarat) with a total approved capacity of 15,687 mW. The corporation is also executing transmission lines of total length of about 20,200 ckt. km. NTPC has been entrusted with management of Badarpur Thermal Power Station (720 mW) which is a major source of power to Delhi.

Installed capacity of NTPC Projects stands at 9915 mW. The corporation has fully completed its projects at Singrauli (2,000 mW), Korba (2,100 mW) and Ramagundam (2,100 mW) and Rihand and Two Gas-based Projects at Anta (413 mW) and Auraiya (652 mW).

1.11.2 NATIONAL HYDRO-ELECTRIC POWER CORPORATION

The National Hydroelectric Power Corporation (NHPC) was incorporated in November 1975, with objectives to plan, promote and organize an integrated development of hydroelectric Power in the

Central Sector. NHPC is presently engaged in construction of Dulhasti, Uri and Salal (Stage-II) Hydro-electric Projects (all in Jammu and Kashmir), Chamera Stage-1 (Himachal Pradesh), Tanakpur Project (UP) and Rangit Project (Sikkim). NHPC is also responsible for operation and maintenance of Salal Project Stage-I (J & K), Baira Siul Project (Himachal Pradesh) and Loktak Project (Manipur).

NHPC has a shelf of projects ready with all statutory clearances awaiting Government Sanction for execution. These are Baglihar and Sawalkot (both in J & K), Chamera II (H.P.), Dhauliganga Stage-I (U.P.) and Koel Karo (Bihar). NHPC have completed investigation of Dhaleswari (Mizoram), Dhauliganga Intermediate Stage (U.P.) Goriganga Stage I and II (U.P.) and Kishenganga (J & K). These are under techno-economic appraisal by CEA. The Corporation is continuing investigations on Goriganga-III (U.P.). Two Mega Projects *viz.,* Teesta (Sikkim) and Katch Tidal Project (Gujarat) presently under techno-economic appraisal by CEA have also been entrusted to NHPC for execution.

The corporation has completed so far 3220 ckt kms of EHV transmission lines, along with the associated sub-stations. Besides, a giant transmission network encompassing 3170 ckt kms including 800 KV class is also under execution under World Bank Assistance for transfer for power in the Northern Region. In the snow-bound areas of J & K, a 400 kV Dulhasti Transmission Line is also under execution under Russian Assistance.

1.11.3 RURAL ELECTRIFICATION CORPORATION

The Rural Electrification Corporation (REC) was set up in July, 1969, with the primary objective of promoting rural electrification by financing rural Electrification Schemes and Rural Electric Cooperatives in the states.

REC have given loans aggregating to Rs. 4742.49 crore by 1989–90 to States and State Electricity Boards for the Rural Electrification Schemes. Loans during 1989–90 aggregated to Rs. 724.60 crore.

1.11.4 DAMODAR VALLEY CORPORATION

Damodar Valley Corporation (DVC) was established in 1948 under an Act of Parliament for unified development of Damodar Valley covering an area of 24,235 sq km in Bihar and West Bengal. Functions assigned to the corporation are: control of floods, irrigation, generation and transmission of power besides activities like navigation, soil conservation and afforestation, promotion of public health and agricultural, industrial and economic development of the valley. The corporation has Three Thermal Power Stations at Bokaro, Chandrapura and Durgapur with a total installed capacity (derated) of 1755 mw. It has four multi-purpose Dams at Tilaiya, Maithon, Panchet and Konar. There are three Hydel Stations appended to Tilaiya, Maithon and Panchet Dams with a capacity of 144 mW. DVC has also set up three Gas Turbine Units of 30 mW each at Maithon. The corporation is installing one more thermal units of 210 mW at Bokaro 'B', Three Thermal Units of 210 mW each at Mejia and the fourth units of 210 mW each at the right bank of Maithon.

1.11.5 NORTH-EASTERN ELECTRIC POWER CORPORATION LIMITED

The North-Eastern Electric Power Corporation Ltd., was constituted in 1976 under the Companies Act under the aim of developing the large electric power potential of the North-Eastern Region. The corporation is responsible for operation and maintenance of the 150 mW Kopili Hydro Electric Project which was commissioned in, June/July, 1988. The associated 220 kV and 132 kV transmission lines for supply power from this project to the constituent states of the region, namely; Assam, Manipur, Mizoram and Tripura, have also been completed.

The Corporation is presently executing the following projects: (*i*) Dovang Hydro-Electric Project (75 mW) Nagaland; (*ii*) Ranganadi Hydro Electric Project (405 mW) Arunachal Pradesh; (*iii*) Assam Gas Based Project (280 mW) Assam; (*iv*) Doyang Transmission Line Project; (*v*) Kanganaali Transmission Line Project; (*vi*) Gohpur-Itanagar Transmission Line Project and (*vii*) 400 kV Transmission Line System associated with the Assam Gas Based Project.

1.11.6 BHAKRA BEAS MANAGEMENT BOARD AND BEAS CONSTRUCTION BOARD

Under the Punjab Reorganization Act. 1966, Bhakra Management Board thereto managed management of Bhakra Darn and reservoirs and works appurtenant. The construction of Beas Project was undertaken by the Beas Construction Board. After completion of works of Beas Project, management of the project was taken over by Bhakra Management Board redesignated as Bhakra Beas Management Board (BBMB). BBMB now manages Hydro-electric Power Stations of Bhakra-Beas Systems, namely. Bhakra Right Bank (660 mW), Bhakra Left Bank (540 mW), Ganguwal (77 mW), Kotla (77 mW), Dehar Stage-I (660 mW), Debar Stage-II (330 mW). Pong Stage-I (240 mW) and Pong Stage-I1 (120 mW), all having a total installed capacity of 2,704 mW.

1.11.7 POWER ENGINEERS TRAINING SOCIETY (PETS)

The Power Engineers Training Society (PETS) was formed in 1980 as a autonomous body to function as an Apex National Body for meeting the training requirements of Power Sector in the Country. The society is responsible for coordinating training programmes of the various State Electricity Boards, Power stations, etc. and supplementing these with its own training activities. The society has Four Regional "Thermal Power Station Personnel Training Institutes at Neyveli, Durgapur, Badarpur (New Delhi) and Nagpur. These Training Institutes conduct regular induction courses, in-service refresher and short-term courses, on job and on plant training programmes for Power Engineers, operators and technicians of Thermal Power Stations/State Electricity Boards, etc. A Simulator installed at the Training Institute at Badarpur (New Delhi) provides training to engineers and operators of 210 mW Thermal Units.

1.11.8 CENTRAL POWER RESEARCH INSTITUTE (CPRI), BANGALORE

The Central Power Research Institute, which was set up in 1960 as a subordinate office under the erstwhile Central Water and Power Commission (Power Wing), was reorganised and registered as a Society under the Karnataka Societies Act, 1960, with effect from January, 1978. The CPRI functions as a National Laboratory for applied research in the field of Electric Power Engineering. While the Central Power Research Institute has a Switchgear Testing and Development Station at Bhopal, the main complex of its laboratories is at Bangalore. The institute is an Apex Body for Research and Development in the Power Sector and conducts tests of electrical apparatus in accordance with the National/International Standards so as to meet fully the research and testing needs of electrical, transmission and distribution equipment. The CPRI also serves as a National Testing and Certification Authority for transmission and distribution equipment. The institute possesses Highly Sophisticated Laboratories comparable to those in the Developed Countries.

1.11.9 NATHPA, JHAICRI POWER CORPORATION LIMITED

NJPC, a joint venture of the Centre and Government of Himachal Pradesh, was incorporated on May 24, 1988, for execution of Nathpa Jhakri Power Project (6 × 250 mW) with equity participation in the ratio of 3 : 1. The corporation has an authorized Share Capital of Rs 1,000 crore. It will also execute other Hydro-electric Power Projects in the region with consent of the state government.

The corporation has already taken up execution of Nathpa Jhakri Hydro-electric Project (6 × 250 mW) for which World Bank has agreed to extend financial assistance of 4370 lakh US dollar. The project is estimated to cost Rs. 1,678 crore (at September, 1988, price level). At present, infrastructure works on the project site are under execution. Drifts at Power House Site at Jhakri and Desalting Complex at Nathpa aggregating to a length of 2850 metres have been executed. Drill Holes at various locations totaling a length of 4300 metres as per recommendations of GSI have been made. About 76 hectares of land was acquired and acquisition proceedings for above 400 hectares are underway. About 26 kms of 22 kV double circuit HT Line and about 11 kms of 22 kV single circuit HT Line have also been completed for Construction Power. About 46 kms of roads have also been constructed. About 46250 sq. metres of buildings have been constructed. The project is expected to be completed within a period of about seven years and would yield benefits during the Eighth Plan.

1.12 REVIEW OF THERMODYNAMICS CYCLES RELATED TO POWER PLANTS

Thermodynamics is the science of many processes involved in one form of energy being changed into another. It is a set of book keeping principles that enable us to understand and follow energy as it transformed from one form or state to the other.

The zeroth law of thermodynamics was enunciated after the first law. It states that if two bodies are each in thermal equilibrium with a third, they must also be in thermal equilibrium with each other. Equilibrium implies the existence of a situation in which the system undergoes no net charge, and there is no net transfer of heat between the bodies.

The first law of thermodynamics says that energy can't be destroyed or created. When one energy form is converted into another, the total amount of energy remains constant. An example of this law is a gasoline engine. The chemical energy in the fuel is converted into various forms including kinetic energy of motion, potential energy, chemical energy in the carbon dioxide, and water of the exhaust gas.

The second law of thermodynamics is the entropy law, which says that all physical processes proceed in such a way that the availability of the energy involved decreases. This means that no transformation of energy resource can ever be 100% efficient. The second law declares that the material economy necessarily and unavoidably degrades the resources that sustain it. Entropy is a measure of disorder or chaos, when entropy increases disorder increases.

The third law of thermodynamics is the law of unattainability of absolute zero temperature, which says that entropy of an ideal crystal at zero degrees Kelvin is zero. It's unattainable because it is the lowest temperature that can possibly exist and can only be approached but not actually reached. This law is not needed for most thermodynamic work, but is a reminder that like the efficiency of an ideal engine, there are absolute limits in physics.

The steam power plants works on modified rankine cycle in the case of steam engines and isentropic cycle concerned in the case of impulse and reaction steam turbines. In the case of I.C. Engines (Diesel Power Plant) it works on Otto cycle, diesel cycle or dual cycle and in the case of gas turbine it works on Brayton cycle, in the case of nuclear power plants it works on Einstein equation, as well as on the basic principle of fission or fusion. However in the case of non-conventional energy generation it is complicated and depends upon the type of the system *viz.,* thermo electric or thermionic basic principles and

FUNDAMENTAL OF POWER PLANT

1.13 CLASSIFICATION OF POWER PLANT CYCLE

Power plants cycle generally divided in to the following groups,

(1) Vapour Power Cycle

(Carnot cycle, Rankine cycle, Regenerative cycle, Reheat cycle, Binary vapour cycle)

(2) Gas Power Cycles

(Otto cycle, Diesel cycle, Dual combustion cycle, Gas turbine cycle.)

1.13.1 CARNOT CYCLE

This cycle is of great value to heat power theory although it has not been possible to construct a practical plant on this cycle. It has high thermodynamics efficiency.

It is a standard of comparison for all other cycles. The thermal efficiency (η) of Carnot cycle is as follows:

$$\eta = (T_1 - T_2)/T_1$$

where, T_1 = Temperature of heat source

T_2 = Temperature of receiver

1.13.2 RANKINE CYCLE

Steam engine and steam turbines in which steam is used as working medium follow Rankine cycle. This cycle can be carried out in four pieces of equipment joint by pipes for conveying working medium as shown in Fig. 1.1. The cycle is represented on Pressure Volume P-V and S-T diagram as shown in Figs. 1.2 and 1.3 respectively.

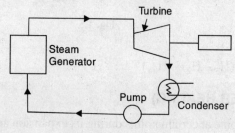

Fig. 1.1

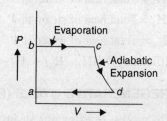

Fig. 1.2

Efficiency of Rankine cycle

$$= (H_1 - H_2)/(H_1 - H_{w2})$$

where,

H_1 = Total heat of steam at entry pressure

H_2 = Total heat of steam at condenser pressure (exhaust pressure)

H_{w2}= Total heat of water at exhaust pressure

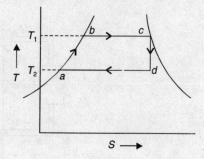

Fig. 1.3

1.13.3 REHEAT CYCLE

In this cycle steam is extracted from a suitable point in the turbine and reheated generally to the original temperature by flue gases. Reheating is generally used when the pressure is high say above 100 kg/cm². The various advantages of reheating are as follows:

(i) It increases dryness fraction of steam at exhaust so that blade erosion due to impact of water particles is reduced.

(ii) It increases thermal efficiency.

(iii) It increases the work done per kg of steam and this results in reduced size of boiler.

The disadvantages of reheating are as follows:

(i) Cost of plant is increased due to the reheater and its long connections.

(ii) It increases condenser capacity due to increased dryness fraction.

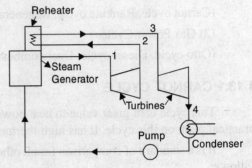

Fig. 1.4

Fig. 1.4 shows flow diagram of reheat cycle. First turbine is high-pressure turbine and second turbine is low pressure (L.P.) turbine. This cycle is shown on T-S (Temperature entropy) diagram (Fig. 1.5).

If,

H_1 = Total heat of steam at 1

H_2 = Total heat of steam at 2

H_3 = Total heat of steam at 3

H_4 = Total heat of steam at 4

H_{w4} = Total heat of water at 4

Efficiency = $\{(H_1 - H_2) + (H_3 - H_4)\}/\{H_1 + (H_3 - H_2) - H_{w4}\}$

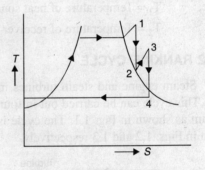

Fig. 1.5

1.13.4 REGENERATIVE CYCLE (FEED WATER HEATING)

The process of extracting steam from the turbine at certain points during its expansion and using this steam for heating for feed water is known as Regeneration or Bleeding of steam. The arrangement of bleeding the steam at two stages is shown in Fig. 1.6.

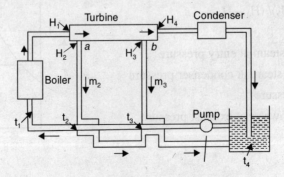

Fig. 1.6

Let,

m_2 = Weight of bled steam at a per kg of feed water heated

m_2 = Weight of bled steam at a per kg of feed water heated

H_1 = Enthalpies of steam and water in boiler

H_{w1} = Enthalpies of steam and water in boiler

H_2, H_3 = Enthalpies of steam at points a and b

t_2, t_3 = Temperatures of steam at points a and b

H_4, H_{w4} = Enthalpy of steam and water exhausted to hot well.

Work done in turbine per kg of feed water between entrance and a

$$= H_1 - H_2$$

Work done between a and $b = (1 - m_2)(H_2 - H_3)$

Work done between b and exhaust $= (1 - m_2 - m_3)(H_3 - H_4)$

Total heat supplied per kg of feed water $= H_1 - H_{w2}$

Efficiency (η) = Total work done/Total heat supplied

$$= \{(H_1 - H_2) + (1 - m_2)(H_2 - H_3) + (1 - m_2 - m_3)(H_3 - H_4)\}/(H_1 - H_{w2})$$

1.13.5 BINARY VAPOUR CYCLE

In this cycle two working fluids are used. Fig. 1.7 shows Elements of Binary vapour power plant. The mercury boiler heats the mercury into mercury vapours in a dry and saturated state.

These mercury vapours expand in the mercury turbine and then flow through heat exchanger where they transfer the heat to the feed water, convert it into steam. The steam is passed through the steam super heater where the steam is super-heated by the hot flue gases. The steam then expands in the steam turbine.

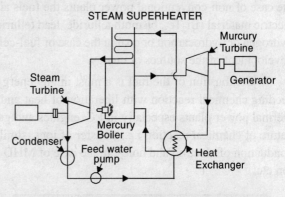

Fig. 1.7

1.13.6 REHEAT-REGENERATIVE CYCLE

In steam power plants using high steam pressure reheat regenerative cycle is used. The thermal efficiency of this cycle is higher than only reheat or regenerative cycle. Fig. 1.8 shows the flow diagram of reheat regenerative cycle. This cycle is commonly used to produce high pressure steam (90 kg/cm²) to increase the cycle efficiency.

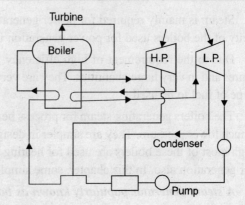

1.13.7 FORMULA SUMMARY

1. Rankine efficiency
   $$= (H_1 - H_2)/(H_1 - H_{w2})$$
2. Efficiency ratio or Relative efficiency
   $$= \text{Indicated or Brake thermal efficiency/Rankine efficiency}$$
3. Thermal efficiency = $3600/m(H_1 - H_{w2})$, m = steam flow/kw hr
4. Carnot efficiency = $(T_1 - T_2)/T_1$

1.14 FUELS AND COMBUSTION

The working substance of the energy conversion device *viz.*, prime-mover (which convert the natural resources of energy into power or electricity) is called fuel. The most common fuel is fossil fuel *viz.*, Coal, petrol, diesel or water gas in the case of steam power plants, I.C. Engines, gas turbines, and hydro-electric power plants. Uranium 235($1U^{235}$) as fissionable and $1U^{238}$ as fertile fuel in the case of fission reactors of nuclear power plant and hydrogen as fuel in the case of fusion nuclear reactor. While fission reactor is conventional fusion reactor is supposed to be non-conventional due to its uncontrolled reaction rate; and it is believed that Russian's have developed it but keeping the whole world silence. In the case of non-conventional power plants the fuels are according to their characteristics *viz.*, Thermo-electric material (Bi_2Te_3, bismuth telluride, lead telluride etc.); thermionic materials (Na, K, Cs, W etc.); hydrogen or hydrocarbon or coal in the case of fuel-cells and further water and methane etc in the recent development of the sources of energy.

Combustion of the fuel is a must in any energy conversion device. It is defined as rapidly proceeding chemical reaction with liberation of heat and light. This phenomenon incurred in the case of thermal power plants especially in I.C. engines and gas turbines. But in the case of fuel cell it is of the nature of chemical reaction *i.e.*, transfer of ions, similarly in the case of thermo-electric generator it is conduction of electron and holes, in the case of MHD power plant it is drifting of positive and negative ion etc.

1.15 STEAM GENERATORS

Steam is mainly required for power generation, process heating and pace heating purposes. The capacity of the boilers used for power generation is considerably large compared with other boilers.

Due to the requirement of high efficiency, the steam for power generation is produced at high pressures and in very large quantities. They are very large in size and are of individual design depending the type of fuel to be used.

The boilers generating steam for process heating are generally smaller in size and generate steam at a much lower pressure. They are simpler in design and are repeatedly constructed to the same design. Though most of these boilers are used for heating purposes, some, like locomotive boilers are used for power generation also. In this chapter, some simple types of boilers will be described.

A steam generator popularly known as boiler is a closed vessel made of high quality steel in which steam is generated from water by the application of heat. The water receives heat from the hot

FUNDAMENTAL OF POWER PLANT

gases though the heating surfaces of the boiler. The hot gases are formed by burning fuel, may be coal, oil or gas. Heating surface of the boiler is that part of the boiler which is exposed to hot gases on one side and water or steam on the other side. The steam which is collected over the water surface is taken from the boiler through super heater and then suitable pipes for driving engines or turbines or for some industrial heating purpose. A boiler consists of not only the steam generator but also a number of parts to help for the safe and efficient operation of the system as a whole. These parts are called mountings and accessories.

1.16 STEAM PRIME MOVERS

The prime mover convert the natural resources of energy into power or electricity.

The prime movers to be used for generating electricity could be diesel engine, steam engine, steam turbines, gas turbines, and water turbine.

Since we know that, a power plant generated a flow of mechanical or electrical energy by means of generators. When coupling runs the generator, then the generator is a prime mover.

In case of steam power plant, the prime movers is steam engine or steam turbine, which is called, steam prime movers. Presently, the steam turbine has totally replaced steam engine. The steam is generated in a boiler and is then expanded in the turbine. The output of the steam turbine is utilized to run the generator. The fuel used in the boiler is coal or oil.

1.17 STEAM CONDENSERS

Thermal efficiency of a closed cycle power developing system using steam as working fluid and working on Carnot cycle is given by an expression $(T_1 - T_2)/T_1$. This expression of efficiency shows that the efficiency increases with an increase in temperature T_1 and decrease in temperature T_2. The maximum temperature T_1 of the steam supplied to a steam prime mover is limited by material considerations. The temperature T_2 (temperature at which heat is rejected) can be reduced to the atmospheric temperature if the exhaust of the steam takes place below atmospheric pressure. If the exhaust is at atmospheric pressure, the heat rejection is at 100°C.

Low exhaust pressure is necessary to obtain low exhaust temperature. But the steam cannot be exhausted to the atmosphere if it is expanded in the engine or turbine to a pressure lower than the atmospheric pressure. Under this condition, the steam is exhausted into a vessel known as condenser where the pressure is maintained below the atmosphere by continuously condensing the steam by means of circulating cold water at atmospheric temperature.

A closed vessel in which steam is condensed by abstracting the heat and where the pressure is maintained below atmospheric pressure is known as a condenser. The efficiency of the steam plant is considerably increased by the use of a condenser. In large turbine plants, the condensate recovery becomes very important and this is also made possible by the use of condenser.

The steam condenser is one of the essential components of all modern steam power plants.

Steam condenser are of two types:

1. Surface condenser.   2. Jet condensers

1.17.1 SURFACE CONDENSERS

In surface condensers there is no direct contact between the steam and cooling water and the condensate can be re-used in the boiler: In such condenser even impure water can be used for cooling purpose whereas the cooling water must be pure in jet condensers. Although the capital cost and the space needed is more in surface condensers but it is justified by the saving in running cost and increase in efficiency of plant achieved by using this condenser. Depending upon the position of condensate extraction pump, flow of condensate and arrangement of tubes the surface condensers may be classified as follows:

(*i*) **Down flow type.** Fig. 1.9 shows a sectional view of dawn flow condenser. Steam enters at the top and flows downward. The water flowing through the tubes in one direction lower half comes out in the opposite direction in the upper half Fig. 1.10 shows a longitudinal section of a two pass down-flow condenser.

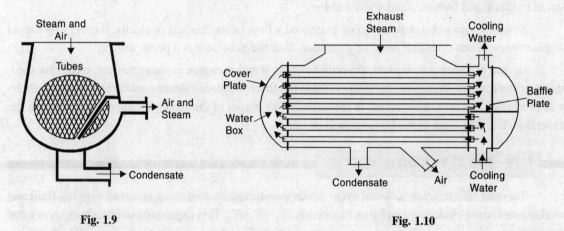

Fig. 1.9 Fig. 1.10

(*ii*) **Central flow condenser.** Fig. 1.11 shows a central flow condenser. In this condenser the steam passages are all around the periphery of the shell. Air is pumped away from the centre of the condenser. The condensate moves radially towards the centre of tube nest. Some of the exhaust steams while moving towards the centre meets the undercooled condensate and pre-heats it thus reducing undercooling.

(*iii*) **Evaporation condenser.** In this condenser (Fig. 1.12) steam to be condensed is passed through a series of tubes and the cooling waterfalls over these tubes in the form of spray. A steam of air flows over the tubes to increase evaporation of cooling water, which further increases the condensation of steam.

FUNDAMENTAL OF POWER PLANT

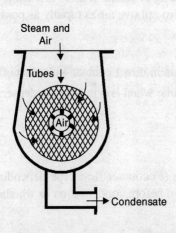

Fig. 1.11

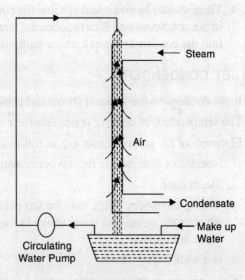

Fig. 1.12

ADVANTAGES AND DISADVANTAGES OF A SURFACE CONDENSER

The various advantages of a surface condenser are as follows:
1. The condensate can be used as boiler feed water.
2. Cooling water of even poor quality can be used because the cooling water does not come in direct contact with steam.
3. High vacuum (about 73.5 cm of Hg) can be obtained in the surface condenser. This increases the thermal efficiency of the plant.

The various disadvantages of the surface condenser are as follows:
1. The capital cost is more.
2. The maintenance cost and running cost of this condenser is high.
3. It is bulky and requires more space.

REQUIREMENTS OF A MODERN SURFACE CONDENSER

The requirements of ideal surface condenser used for power plants are as follows:
1. The steam entering the condenser should be evenly distributed over the whole cooling surface of the condenser vessel with minimum pressure loss.
2. The amount of cooling water being circulated in the condenser should be so regulated that the temperature of cooling water leaving the condenser is equivalent to saturation temperature of steam corresponding to steam pressure in the condenser.

This will help in preventing under cooling of condensate.

3. The deposition of dirt on the outer surface of tubes should be prevented.

Passing the cooling water through the tubes and allowing the steam to flow over the tubes achieve this.

4. There should be no air leakage into the condenser because presence of air destroys the vacuum in the condenser and thus reduces the work obtained per kg of steam. If there is leakage of air into the condenser air extraction pump should be used to remove air as rapidly as possible.

1.17.2 JET CONDENSERS

In jet condensers the exhaust steam and cooling water come in direct contact with each other.

The temperature of cooling water and the condensate is same when leaving the condensers.

Elements of the jet condenser are as follows:

1. Nozzles or distributors for the condensing water.
2. Steam inlet.
3. Mixing chambers: They may be (*a*) parallel flow type (*b*) counter flow type depending on whether the steam and water move in the same direction before condensation or whether the flows are opposite.
4. Hot well.

In jet condensers the condensing water is called injection water.

1.17.3 TYPES OF JET CONDENSERS

1. Low level jet condensers (Parallel flow type). In this condenser (Fig. 1.13) water is sprayed through jets and it mixes with steam. The air is removed at the top by an air pump. In counter flow type of condenser the cooling water flows in the downward direction and the steam to be condensed moves upward.

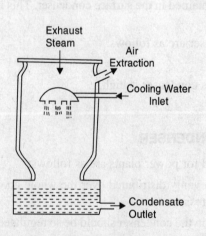

Fig. 1.13

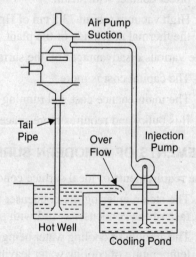

Fig. 1.14

2. High level or Barometric condenser. Fig. 1.14 shows a high-level jet condenser. The condenser shell is placed at a height of 10.33 m (barometric height) above the hot well. As compared to low level jet condenser. This condenser does not flood the engine if the water extraction pump fails. A separate air pump is used to remove the air.

3. Ejector Condenser. Fig. 1.15 shows an ejector condenser. In this condenser cold water is discharged under a head of about 5 to 6 m through a series of convergent nozzles. The steam and air enter the condenser through a non-return valve. Mixing with water condenses steam. Pressure energy is partly convert into kinetic energy at the converging cones. In the diverging come the kinetic energy is partly converted into pressure energy and a pressure higher than atmospheric pressure is achieved so as to discharge the condensate to the hot well.

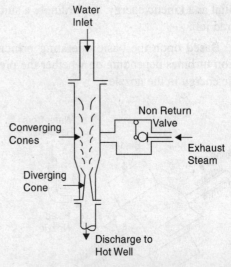

Fig. 1.15

1.18 WATER (HYDRAULIC) TURBINES

Turbine is a machine wherein rotary motion is obtained by centrifugal forces, which result from a change in the direction of high velocity fluid jet that issues from a nozzle.

Water turbine is a prime mover, which uses water as the working substance to generate power.

A water turbine uses the potential and kinetic energy of water and converts it into usable mechanical energy. The fluid energy is available in the natural or artificial high level water reservoirs, which are created by constructing dams at appropriate places in the flow path of rivers. When water from the reservoir is taken to the turbine, transfer of energy takes place in the blade passages of the unit. Hydraulic turbines in the form of water wheels have been used since ages; presently their application lies in the field of electric power generation. The mechanical energy made available at the turbine shaft is used to run an electric generator, which is directly coupled, to the turbine shaft. The power generated by utilizing the potential and kinetic energy of water has the advantages of high efficiency, operational flexibility, low wear tear, and ease of maintenance.

Despite the heavy capital cost involved in constructing dams and reservoirs, in running pipelines and in turbine installation (when compared to an equivalent thermal power plant) different countries have tried to tap all their waterpower resources. Appropriate types of water turbines have been installed for most efficient utilization. A number of hydro-electric power plants have and are being installed in India too to harness the available waterpower in the present crisis of fast idling energy resources. Hydro-electric power is a significant contributor to the world's energy sources.

Water (hydraulic) turbines have been broadly classified as,

1. Impulse   2. Reaction

1.18.1 IMPULSE AND REACTION TURBINES

Hydraulic turbines are required to transform fluid energy into usable mechanical energy as efficiently as possible. Further depending on the site, the available fluid energy may vary in its quantum of potential and kinetic energy. Accordingly a suitable type of turbine needs to be selected to perform the required job.

Based upon the basic operating principle, water turbines are categorized into impulse and reaction turbines depending on whether the pressure head available is fully or partially converted into kinetic energy in the nozzle.

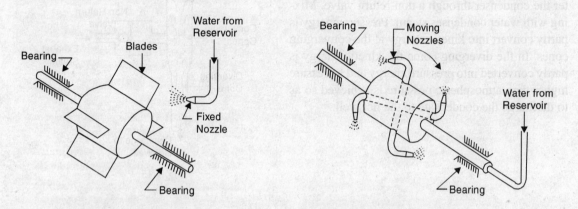

Fig. 1.16. Impulse Turbine.   **Fig. 1.17.** Reaction Turbine.

Impulse Turbine wherein the available hydraulic energy is first converted into kinetic energy by means of an efficient nozzle. The high velocity jet issuing from the nozzle then strikes a series of suitably shaped buckets fixed around the rim of a wheel (Fig. 1.16). The buckets change the direction of jet without changing its pressure. The resulting change in momentum sets buckets and wheel into rotary motion and thus mechanical energy is made available at the turbine shaft. The fluid jet leaves the runner with a reduced energy. An impulse turbine operates under atmospheric pressure, there is no change of static pressure across the turbine runner and the unit is often referred to as a free jet turbine. Important impulse turbines are: Pelton wheel, Turgo-impulse wheel, Girad turbine, Banki turbine and Jonval turbine etc., Pelton wheel is predominantly used at present.

Reaction Turbine wherein a part of the total available hydraulic energy is transformed into kinetic energy before the water is taken to the turbine runner. A substantial part remains in the form of pressure energy. Subsequently both the velocity and pressure change simultaneously as water glides along the turbine runner. The flow from inlet to outlet of the turbine is under pressure and, therefore, blades of a reaction turbine are closed passages sealed from atmospheric conditions.

Fig. 1.17 illustrates the working principle of a reaction turbine in which water from the reservoir is taken to the hollow disc through a hollow shaft. The disc has four radial openings, through tubes, which are shaped as nozzles. When the water escapes through these tubes its pressure energy decreases and there is increase in kinetic energy relative to the rotating disc. The resulting reaction force sets the disc in rotation. The disc and shaft rotate in a direction opposite to the direction of water jet. Important reaction turbines are, Fourneyron, Thomson, Francis, Kaplan and Propellor turbines Francis and Kaplan turbines are widely used at present.

The following table lists salient points of difference between the impulse and reaction turbines with regard to their operation and application.

FUNDAMENTAL OF POWER PLANT

Table 1.1

Impulse Turbine	Reaction Turbine
1. All the available energy of the fluid is converted into kinetic energy by an efficient nozzle that forms a free jet.	1. Only a portion of the fluid energy is transformed into kinetic energy before the fluid enters the turbine runner.
2. The jet is unconfined and at atmospheric pressure throughout the action of water on the runner, and during its subsequent flow to the tail race.	2. Water enters the runner with an excess pressure, and then both the velocity and pressure change as water passes through the runner.
3. Blades are only in action when they are in front of the nozzle.	3. Blades are in action all the time.
4. Water may be allowed to enter a part or whole of the wheel circumference.	4. Water is admitted over the circumference of the wheel.
5. The wheel does not run full and air has free access to the buckets.	5. Water completely fills the vane passages throughout the operation of the turbine.
6. Casing has no hydraulic function to perform; it only serves to prevent splashing and to guide the water to the tail race.	6. Pressure at inlet to the turbine is much higher than the pressure at outlet ; unit has to be sealed from atmospheric conditions and, therefore, casing is absolutely essential.
7. Unit is installed above the tail race.	7. Unit is kept entirely submerged in water below the tail race.
8. Flow regulation is possible without loss.	8. Flow regulation is always accompanied by loss.
9. When water glides over the moving blades, its relative velocity either remains constant or reduces slightly due to friction.	9. Since there is continuous drop in pressure during flow through the blade passages, the relative velocity does increase.

1.19 SCIENCE VS. TECHNOLOGY

The difference between science and technology is science is the knowing of what is going on, what is happening in nature, and to increase knowledge. Science is a lot slower than technology. Technology is to control and use of science to provide a practical use.

1.19.1 SCIENTIFIC RESEARCH

INDIA has had a long and distinguished tradition in Science from accomplishments of ancient times to great achievements during this century; the latter half, prior to Independence has been related largely to pure research. At the time of Independence, our scientific and technological infrastructure was neither strong nor organised in comparison with that of the Developed World. This had resulted in our being technologically dependent on skills and expertise available in other countries during early years of Independence. In the past four decades, an infrastructure and capability largely commensurate with meeting national needs has been created minimising our dependence on other countries. But, we still have a long way to go in this field to be self-sufficient. A range of industries from small to the most sophisticated has been established covering wide-range of utilities, services and goods. There is now a reservoir of expertise well acquainted with the most modern advances in basic and applied areas that is equipped to make choices between available technologies, to absorb readily new technologies and provide a framework for future national Development.

1.19.2 SCIENCE AND TECHNOLOGY INFRASTRUCTURE

Scientific research in India is carried out fewer than three major sectors, *viz.*, Central Government, state governments and various in-house research and development units of industrial undertakings, both under public and private sectors besides cooperative Reserved & Development associations. Bulk of research effort in the country is financed by major scientific departments/agencies such as Departments of Science and Technology, Atomic Energy, Space, Scientific and Industrial Research, Electronics, Non-Conventional Energy Sources, Environment, Ocean Development, Biotechnology Agencies *i.e.,* Indian Council of Medical Research, Council of Scientific and Industrial Research, Indian Agricultural Research Institute. etc. There are about 200 research laboratories within the purview of these major scientific agencies carrying out research in different, areas. Besides. There are a large number of scientific institutions under the Central ministries departments which carryout research programmes of practical relevance to their areas of responsibility. States supplement the efforts of Central government in areas like agriculture, animal husbandry, fisheries, public health, etc. Institutions of higher education carryout sizeable work in science and technology and are supported by the University Grants Commission and Central and state governments. They also carryout sponsored research projects financed by different agencies.

Government is providing a number of incentives to industrial establishments in private and public sectors to encourage them to undertake research and development activities. Consequently, scientific research is gaining momentum in several industrial establishments. As on January 1990, there were over 1,200 in-house research and development units in public and private sectors, reorganised by the Department of Scientific and Industrial Research. Also, recently public funded research institutions through Department of Science and Technology have introduced a 'Pass Book' Scheme for import of scientific equipment liberally.

1.20 FACTS VS. VALUES

Fact is the regulatory ideas without false ability not arguments. Values are the judgment of good and bad regulations. The Indian constitution is based on values, which are a shared set of understandings of what is good or bad. Science is above the plane of values, free from what is good and bad, because science is an objective.

1.21 ATOMIC ENERGY

India is recognized as one of few countries in the world, which have made considerable advances in the field of atomic energy. Despite the closely guarded nature of this technology at the international level, the country is self-reliant in the same and has established competence in carrying out activities over the entire nuclear fuel cycle. The executive agency for all activities pertaining to atomic energy in the country is the Department of Atomic Energy (DAE), which was set-up in 1954. The Atomic Energy Commission (AEC) lays down policies pertaining to the functioning of DAE, which was set-up in 1948. The portfolio of DAE has all along been under the charge of the Prime Minister.

The activities of DAE are primarily in the area of nuclear power generation, research and development in atomic energy and in the industries and minerals sector. These activities are carried out by its constituent units, Public Sector Units (PSUs) and by institutions which are given financial assistance by DAE. India has also been offering training facilities, fellowships, scientific visits, etc., and makes available the service of its scientists and engineers for expert assignments in several countries both through the International Atomic Energy Agency (IAEA) and through bilateral agreements.

1.22 HIGHLIGHTS OF THE NUCLEAR POWER PROGRAMME

When the country's atomic energy programme was launched in the 1940s, a three-stage nuclear energy programme was envisaged to use the available Uranium and vast Thorium Resources. The first stage was to comprise of Natural Uranium Fuelled Pressurised Heavy Water Reactors (PHWRs), which would produce power, and Plutonium as a by-product. The second stage is expected to have Plutonium Fuelled Fast Breeder Reactors (FBRs), which in addition to producing power and Plutonium, will also yield Uranium-233 from Thorium. The third stage reactors would be based on the Thorium Cycle to produce more Uranium-233 for fuelling additional breeder reactors.

The present installed capacity of nuclear power reactors in India is 1,465 MWe. The total electricity generated by nuclear power stations during 1988–89 and 1989–90 was 5,817 and 4,625 million kW hours respectively, and the target for 1990–91 has been fixed at 6850 million units. Excepting for the first two units at Tarapur, which are of the Boiling Water Reactor (BWR) type and were set-up as a turnkey by a United States of America's company, other power reactors in the country are of the PHWR Type which constitute the first stage of the programme. DAE aims at establishing about 10,000 mW of nuclear power generation capacity from PHWRs during the coming ten to fifteen years. In addition, two reactors of the Pressurised Water Reactor (PWR) type of 1000 mW each are being set-up at Kudankulam, TamilNadu, with the assistance of the USSR. Further, work on a Prototype Fast Breeder Reactor (PFBR) of 500 mW capacities is also expected to be taken up in the near future.

Important inputs for the PHWRs are heavy water and nuclear fuel, which are made available by organisations within DAE. Amongst these, there are units which carry out exploration and survey of Uranium resources and subsequently mining and processing them for production of Uranium Concentrates. Other units are responsible for production of nuclear fuel and heavy water. Facilities are also available for the back-end of the nuclear fuel cycle to reprocess spent fuel from nuclear power reactors and for management of radioactive wastes.

A significant feature of the Indian Atomic Energy Programme is that it has all long been backed-up by a comprehensive R and D programme encompassing a wide-range of multi-disciplinary activities relating to atomic energy. This includes fundamental research in basic sciences to disciplines like Nuclear Engineering, Metallurgy, Medicine, Agriculture, Isotopes, etc. Research is also being carried out in FBR technology and frontline areas like fusion, lasers and accelerators.

All the organisations of DAE which are engaged in these activities, can be considered to be one of the following categories, namely, R and D units, PSUs, Industries and Mineral (I and M) sector units, Aided Institutions or Service Sector Units.

1.23 NUCLEAR POWER CORPORATION OF INDIA LIMITED

This is the most recent and largest of the PSUs. It was set-up in 1987 to implement the nuclear power generation programme on commercial lines by converting the erstwhile Nuclear Power Board into Nuclear Power Corporation of India Limited (NPCIL). It is responsible for designing, constructing, commissioning and operating all Nuclear Power Reactors in the country. The seven operating reactors have a total installed capacity of 1435 mW and comprise of : two units of 160 mW each at Tarapur near Bombay; two units of 220 mW each at Rawatbhata near Kota in Rajasthan; two units of 235 mW each at Kalpakkam near Madras, and one unit of 235 mW at Narora in Uttar Pradesh. Excepting Tarapur units, which are of the Boiling Water Reactors (BWR) type, all others are of the PHWR Type. While the Rajasthan Reactors were set-up with the assistance of Canada, all subsequent reactors are of indigenous design and construction.

Several more 235 mW PHWRs are in various stages of construction. The second unit at Narora is nearing completion and is expected to become critical during 1990–91. Construction of two reactors at Kakrapar near Surat in Gujarat is also in an advanced stage and the first of these is also expected to be commissioned during 1991–92. Work is in progress on four more reactors two each at Kaiga in Karnataka and Rawatbhata in Rajasthan, which are expected to be completed during 1995–96.

As regards future power reactors advance action has been initiated on four more 235 mW units and four PHWR units of 500 mW each. Detailed design and engineering of the 500 mW PHWR units are also being done in house by NPCIL. To meet the growing demand for electricity in southern region, it has also been decided to set-up two 1,000 mW VVWR units (of the Pressurised Water Reactor type) in Kudankulam, Tamil Nadu, with Soviet assistance.

1.24 OCEAN ENGINEERING APPLICATIONS

Software to retrieve and analyse the raw data on heave/pitch/roll time series to obtain Directional Wave Spectra has been developed by NIO. The European Molecular Biology Laboratory (EMBL) for worldwide distribution as DNACLONE package has adopted software generated by IMTECH scientists associated with the National Facility of Distributed Information Centre on Enzyme Engineering, Immobilized Biocatalysts, Microbial Fermentation and Bioprocessing Engineering.

The computer software packages developed by SERC (M) continued to attract several user agencies in the Government and Public and Private Sectors. Fifty-four packages were licensed to twenty parties in different parts of India. An improved version of the Flosolver Parallel Computer with sixteen Intel 80386–80387 Processors (32 bit) has become operational during the year, marking a significant advance in NAL's Parallel Computer Development Programme. The new version attains a sustained speed of three-four MFLOPS.

Inherently present Josephson junctions have been exploited in making a two hole SQUID which operates at Liquid Nitrogen Temperature (77 K). It is an r.f. SQUID and is made out of bulk yttrium-barium-copper oxide (YBCO) Superconductor which remains super conducting up to about 90 K.

OCEAN DEVELOPMENT

For centuries, people of India have been using the seas around the Indian Sub-continent for transport, communication and food During the last few years, exploration and exploitation of living and non-living resources of the seas have acquired a new thrust. The new 'Ocean Regime' established by United Nations Convention on the Law of the Sea, 1982, which has been signed by 159 countries including India and ratified by 42 countries besides United Nations Council of Namibia as on 25 November, 1989, assigns much of the World Ocean to Exclusive Economic Zones where coastal states have jurisdiction over exploration and exploitation of resources and for other economic purposes.

FUNDAMENTAL OF POWER PLANT

Recognizing the importance of oceans in economic development and progress of the Nation, the Government set-up Department of Ocean Development in July, 1981, for planning and coordinating oceanographic survey, research and development, management of ocean resources, development of manpower and marine technology. The department is entrusted with the responsibility for protection of marine environment on the high seas.

The budget outlay for various schemes for ocean development during 1990–91 is Rs. 35 crore under Plan and Rs. 7.02 crore under Non-Plan. The revised estimates are Rs. 43.50 crore under Plan and Rs. 12.17 crore under Non-Plan.

The objectives of 'ocean development' have been laid down by Parliament in the Ocean Policy Statement of November 1982. The domain of our concern for development of oceanic resources and its environment extends from the coastal lands and islands lapped by Brackish Water to the wide Indian Ocean. India's Costline is more than 6000 km long and its territory include 1256 islands. Its Exclusive Zone covers an area of 2.02 million-sq.km. and the continental shelf extends up to 350 nautical miles from the coast. Briefly stated, the objectives of development of the oceanic regime are:

1. To explore and assess living and non-living resources;
2. To harness and manage its resources (materials, energy and biomass) and create additional resources such as mariculture;
3. To cope with and protect its environment (weather, waves and coastal front);
4. To develop human resources (knowledge, skill and expertise); and
5. To play our rightful role in Marine Science and Technology in the International Arena.

SOLVED EXAMPLES

Example 1. *Steam at a pressure of 15 kg/cm² (abs) and temperature of 250°C. is expanded through a turbine to a pressure of 5 kg/cm² (abs.). It is then reheated at constant pressure to a temperature of 200°C after which it completes its expansion through the turbine to an exhaust pressure of 0.1 kg/cm²(abs). Calculate theoretical efficiency.*

(a) *Taking reheating into account*

(b) *If the steam was expanded direct to exhaust pressure without reheating.*

Solution. From Mollier diagram

H_1 = Total heat of steam at 15 kg/cm² and 250°C = 698 Kcal/kg

H_2 = Total heat of steam at 5 kg/cm² = 646 Kcal/kg

Now steam is reheated to 200°C at constant pressure

H_3 = Heat in this stage = 682 Kcal/kg

This steam is expanded to 0.1 kg/cm²

H_4 = Heat in this stage = 553 Kcal/kg

H_{w4} = Total Heat of water at 0.1 kg/cm² = 45.4 Kcal/kg

Theoretical efficiency = $\dfrac{\{(H_1 - H_2) + (H_3 - H_4)\}}{\{H_1 + (H_3 - H_2) - H_{w4}\}}$

$$= \frac{\{(698-646)+(682-533)\}}{\{698+(642-646)-45.4\}}$$

$$= \mathbf{0.293 \text{ or } 29.3\% \text{ Ans.}}$$

Example 2. *Determine the thermal efficiency of the basic cycle of a steam power plant (Rankine Cycle), the specific and hourly steam consumption for a 50 mW steam turbine operating at inlet conditions: pressure 90 bar and temperature 500°C. The condenser pressure is 0.40 bar.*

Solution. From Mollier diagram

H_1 = Total heat of steam at point 1 = 3386.24 kJ/kg

H_2 = Total heat of steam at point 2 = 2006.2 kJ/kg

H_{w2} = Total Heat of water at point 2 = 121.42 kJ/kg

(a) Thermal efficiency = $\dfrac{(H_1 - H_2)}{(H_1 - H_{w2})}$

$$= \frac{(3386.24 - 2006.2)}{(3386.24 - 121.42)} = 42.27\%$$

(b) Specific steam consumption is the amount of steam in kg per kW-hr.

Now 1 kW-hr = 3600 kJ

Specific steam consumption = $\dfrac{3600}{(H_1 - H_2)} = \dfrac{3600}{(3386.24 - 2006.2)}$

$$= 2.61 \text{ kg/kW-hr}$$

(c) Hourly steam consumption = 2.61 × Kilowatts

$$= 2.61 \times 50{,}000 = \mathbf{1.305 \text{ Tonnes/hr Ans.}}$$

Example 3. *A steam power plant, operating with one regenerative feed water heating is run at the initial steam conditions of 35.0 bar and 440°C with exhaust pressure of 0.040 bar. Steam is bled from the turbine for feed water heating at a pressure of 1.226 bar. Determine*

(1) Specific heat consumption

(2) Thermal efficiency of the cycle

(3) Economy percentage compared with the cycle of a simple condensing power plant.

Solution. From Mollier diagrams and steam table,

H_1 = 3314 kJ/kg

H_2 = 2560 kJ/kg

H_3 = 2100 kJ/kg

H_{w2} = 439.43 kJ/kg

H_{w3} = 121.42 kJ/kg

From the heat balance for the feed water heater

$$m(H_2 - H_{w2}) = (1 - m)(H_{w2} - H_{w3})$$

$$m(2560 - 439.43) = (1 - m)(439.43 - 121.42)$$

On solving, we get $m = 0.1304$ kg

Total work done $= 1 \times (H_1 - H_2) + (1 - m)(H_2 - H_3)$

$= (3314 - 2460) + (1 - 0.1304)(2560 - 2100)$

$= 1154$ kJ/kg

(1) Specific steam consumption $= \dfrac{3600}{1154} = \mathbf{3.12 Kg/kW\text{-}hr.}$ **Ans.**

(2) Thermal efficiency $= \dfrac{1154}{(3314 - 439.43)} = \mathbf{40.15\%.}$ **Ans.**

(3) With out regeneration feed water heating the work done will be

$H_1 - H_2 = 3314 - 2100 = 1214$ kJ/kg

Steam consumption $= \dfrac{3600}{1214} = 2.94$ kg/kW-hr

Without regeneration heating, the thermal efficiency

$$\eta = \dfrac{(H_1 - H_3)}{(H_1 - H_{w3})}$$

Now from the steam tables

$H_{w3} = 121.42$ kJ/kg

$$\eta = \dfrac{(3314 - 2100)}{(3314 - 121.42)} = 0.38$$

Increase in thermal efficiency due to regeneration feed water heating is

$$= \dfrac{(0.4015 - 0.38)}{0.4015} = \mathbf{5.5 \%} \textbf{ Ans.}$$

THEORETICAL QUESTIONS

1. What is the concept of Power plant?
2. Define the types of energy.
3. What are the resources for power development in India?
4. What is the present position of power in India?
5. What is the future planning for power generation?
6. Define different types of power cycle.
7. Write short notes on fuel and combustion.
8. Write short notes on steam generators.
9. Write short notes on steam condenser.
10. Briefly describe water turbine.
11. Differentiate between impulse and reaction turbine.

EXERCISES

1. A simple Rankine cycle works between pressure of 30 bar and 0.04 bar, the initial condition of steam being dry saturated, calculate the cycle efficiency work ratio and specific steam consumption. **[Ans. 35%, 0.997, 3.84 kg/kWh]**

2. A steam power plant works between 40 bar and 0.05 bar. If the steam supplied is dry saturated and the cycle of operation is Rankine. Find (*a*) cycle efficiency, and (*b*) specific steam consumption. **[Ans. 35.5%, 3.8 kg/kWh]**

3. An engine operating on ideal carnot cycle uses steam at 10 bar and 90% dryness at the end of the isothermal expansion process. The pressure during isothermal compression is 1.5bar. Find the thermal efficiency of the cycle. Also find the power developed by the engine if the engine uses 0.5 kg of steam per cycle and makes 200 cycles/min. Assume that the liquid is saturated at the beginning of isothermal expansion(evaporation). **[Ans. 15.1%, 456 kW]**

4. Steam at 28 bar and 50°C superheat is passed through a turbine and expanded to a pressure where the steam is dry and saturated. It is then reheated at constant pressure to its original temperature and then expanded to the condenser pressure of 0.2 bar. The expansion being isentropic, find
   (*i*) Work done per kg of steam
   (*ii*) Thermal efficiency with and without reheats. **[Ans. 880 kJ/kg, 30.3%]**

5. The steam at a pressure of 100 bar and 500°C is supplied to a steam turbine. It comes out at 0.07 bar and 0.85 dry. One stage reheating is used and reheating is carried out upto its original temperature. Determine the theoretical thermal efficiency of the plant. Also find out the pressure at which reheating is carried out. Assume expansions at both stages are isentropic. Show the processes on h-s chart. If the net output is 1400 kJ/kg of steam find out the actual efficiency of the plant. **[Ans. 42.6%, 39.3%]**

Chapter 2

Non-Conventional Energy Resources and Utilisation

2.1 INTRODUCTION

The major sources of energy, aside from human and animal power, are petroleum resources, natural gas, coal, hydropower, biomass, geothermal, nuclear, wind and solar. Theoretically these resources can be substituted for one another in order to perform a specific task. However, in practice, substitution would be subject to technical and economic limitations that circumscribe the use of such energy resources for specific purposes at given locations and periods of time. Few energy resources are used and consumed at the same location in which they are found, and most of them require generally elaborate transportation and conversion facilities to make them useful for performing the intended task. Analysis of energy resources, therefore, should cover availability and also the production, processing and distribution facilities.

2.2 ENERGY SCIENCE

Science is a systematized body of knowledge about any department of nature, internal or external to man. The energy science deals with scientific principles, characteristics, laws, rules, units/dimensions, measurements, processes etc. about various forms of energy and energy transformations. Science involves experimentation, measurement, mathematical calculations, laws, observations, etc.

Energy science has interface with every other science. Energy science is the mother science of physics, thermodynamics, electromagnetic, nuclear science, mechanical science, chemical science, biosciences etc. Each science deals with some 'activity'. Energy is the essence of activities. Energy management has national priority.

Energy science focusses attention on the 'energy' and 'energy transformations' involved in the various other branches of science, to National Economy and Civilization.

2.3 VARIOUS ENERGY SCIENCE

PHYSICS: It is a branch of natural science dealing with properties and changes in matter and energy. Physics deals with continuous changes in matter and energy and includes mechanics, electro magnetic, heat, optics, nuclear energy etc. and laws governing the energy transformations. Physicists have developed energy science.

THERMODYNAMICS: It is a branch of physics dealing with transformation of thermal energy into other forms of energy, especially mechanical energy and laws governing the conversions. Thermodynamics plays a dominant role in Energy Technologies.

BIOLOGICAL SCIENCES: It deals with biomass and biological processes. Biosciences are concerned with the physical characteristics; life processes of living vegetation and animals on land and in water and their remains.

BIOMASS: It is the matter derived from vegetation and animals. Biomass is a natural non-conventional source of energy and is being given highest priority in recent years. (1980s onwards) Biomass is the important non-conventional energy for the 21st century.

CHEMISTRY: It is a science dealing with composition and properties of substances and their reactions to form other substances. The chemical reactions are accompanied by release of thermal energy (exothermic reactions) or absorption of thermal energy (endothermic reaction). Chemical Reactions are intermediate energy conversion processes. Many useable energy forms are obtained from chemical reactions. (*e.g.* petroleum products, synthetic gases and liquids). Natural Gas and Petroleum products are most important energy forms in the world during 20th and 21st century.

ELECTROMAGNETIC: The flow of electrons and electrical charges through a circuit produces associated electromagnetic fields and electrical power. Electromagnetic is a branch of physics dealing with electricity, magnetism and various transformations of other forms of energy (mechanical, thermal, chemical etc.) into electrical energy and vice-versa. Electrical energy is the most superior; efficient, useful form of energy which can be generated, transmitted, distributed, controlled, utilized. Electrical energy is an intermediate and secondary form of energy being used very widely all over the world.

Fig. 2.1 illustrates the various branches of science and technology concerning 'energy'.

ENERGY SCIENCE: It is the mother science dealing with motion of particles or objects (microscopic or macroscopic), associated energy transformations and effects, concerning various physical, biological and environmental sciences. The energy science correlates various branches of science from energy point of view.

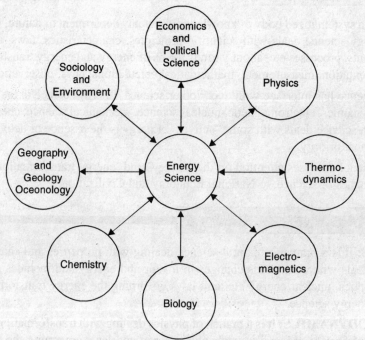

Fig. 2.1. Energy science and other sciences are correlated.

TECHNOLOGY (GREEK "TECHNOLOGIA"): It is a systematic treatment of practical or industrial arts, applied sciences resulting in technical progress by the use of plants and machinery and automation in industry, agriculture, transportation, human and social activities etc.

Technology is an applied science dealing with specific technical problems; Technology is concerned with satisfying short term, mid term and long term needs of society and its members.

2.4 ENERGY TECHNOLOGY

The applied part of energy sciences for work and processes, useful to human society, nations and individuals is called Energy Technology. Energy technologies deal with various primary energies, processing, useful energies and associated plants and processes. The coverage including exploration, transportation, conversion, utilization.

Energy Technology is concerned with 'demand' for various forms of secondary energy (usable energy) and the methods of 'supply'. Various factors affecting the 'demand' and 'supply' are in the scope of energy-technology.

Energy can be supplied via various alternative routes. *e.g.* We may burn wood or natural gas or use electrical energy to obtain heat. Energy Technology deals with various alternatives. The energy chains (routes) between various raw energies (primary energy resources) and final energy consumption are analysed and compared to decide the suitable choice.

Energy Technologies deal with plants and processes involved in the energy transformation and analysis of the useful energy (exergy) and worthless energy (anergy). Energy Technology includes study of efficiencies and environmental aspects of various processes.

Energy technology deals with the complete energy route and its steps such as:

1. Exploration of energy resources; Discovery of new sources
2. Extraction or Tapping of Non-conventional or Growing of Bio-farms
3. Processing
4. Intermediate storage
5. Transportation/Transmission
6. Reprocessing
7. Intermediate storage
8. Distribution
9. Supply
10. Utilization, Conservation, Recycling.

In every step measurements in terms of standard units are involved.

The Energy Strategies include the long-term policies, short-term and mid-term planning, economic planning, social and environmental aspects of various energy routes. These are analysed from the perspectives of the world, region, nation, states, sub-regions, various economic sectors, communities and individuals.

The energy science and technologies give a systematizea, qualitative and quantitative approach to energy studies for the entire human society, to an individual and the environment with reference to the, past, present and future.

The present and future standard of living, economy and environment are shaped by the Energy Technologies.

The subject matter of Energy Science and Energy Technology is of immense interest to the Planners, Economists, Scientists, Engineers, Professionals and Industrialists, Societies and Individuals, etc.

2.5 ENERGY TECHNOLOGY AND ENERGY SCIENCES

"Energy Science and Technology" deal with several useful natural and artificial (man-made) energy systems. The basic objectives are to extract, convert, transform, transport, distribute and reconvert different types of energy with least pollution and with highest economy.

Energy technology is a systematized knowledge of various branches of energy flow and their relationship with human society as viewed from scientific, economic, social, technological, industrial aspects for benefit of man and environment.

The science of energy is concerned with the natural rules and characteristics of energy, energy resources, energy conversion processes and various phenomena related directly or indirectly to the extraction conversion and use of energy resources essential to the economy and prosperity.

The science of energy deals with the phenomena related with energy conversion plants and processes for generating secondary energy (electricity, heat, steam, fuel, gas, etc.) by converting various kinds of primary energy sources. The energy science deals with aspects of useful energy, (exergy, work, power, efficiency and worthless energy anergy) losses etc.

Energy technology correlates various sciences and technologies. Energy technology is theoretically related with a series of physical, engineering and social sciences, which are normally, regarded as independent departments of study. Each branch of engineering and social science has specific coverage and objectives, which are rather independent of other branches. For example electrical power engineering deals with energy in terms of kW, MW, voltage, current etc. and does not deal in detail with various energy resources, energy cycle and ecological aspects. Thermodynamics deals with heat and work but does not cover control of electrical energy system obtained from thermal power plants. Social sciences may not deal with chemical formulae and chemical reactions but are concerned with the pollution and environment, demand and supply of energy and economy.

Energy science and Technology takes an overview of entire energy perspective and goes into details of interrelationships between various branches of science and technology, and Management.

National planning and economic strategies are influenced by energy technology. Industrial project planning and a plant process economy are influenced by energy technology.

The problems of energy technology were totally ignored during the earlier course of history, as the natural resources were available in plenty. Each branch of science and technology was developed rather independently and with disregard to energy resources and energy management. Such approach was partly due to availability of abundant and cheap fossil fuels, fire wood, etc. and very modest demand for useful energy.

During the twentieth century, the energy consumption increased Steeply.

After the oil price rise in 1973, and with global pollution due to energy conversion processes, following important aspects have emerged:

1. Increasing cost of energy sources is affecting individual, social and national life and economy.

2. Depleting energy reserves of fossil fuels (coal, petroleum products, oil, gas) has produced energy crisis.

3. Increasing energy consumption rate and depleting supplies all over the world is resulting in continuing inflation and energy shortage. This is called energy crisis.

4. Large fossil power plants without emission control devices are emitting flay ash SO_x, NO_x, CO etc. in atmosphere resulting in pollution, ecological disasters, global warming, acid rains etc. Energy, Environment and power plants are closely inter-linked.

Though the, nature supplies abundant non-conventional, the technologies for conversion are in early stage of development and not yet commercially successful as against conventional. However, non-conventional are on the path of slow rise all over the world.

Independent study of individual branch of science and technology does not give solution to energy problems facing a man, society, nation and the world.

These limitations of isolated studies have focussed the attention of planners, economists, sociologists, engineers, technologists, environmentalists etc. on the Integrated Energy Technology.

Energy technology integrates the energy aspects of various branches of science, with broad overview of various energy problems. Energy technology suggests alternative solutions within the framework of available science and technological base and energy resources.

Each branch of engineering science has certain theories, laws, equations, units and dimensions. Energy science co-relates the equations and analysis between various different branches of Engineering Sciences.

Energy technology takes an overview of entire energy conversion process from raw energy input to final delivery of secondary energy. This involves chemical, thermal, mechanical, electrical, bio, nuclear and other engineering sciences.

2.6. LAW OF CONSERVATION OF ENERGY

"Energy cannot be newly created. Energy cannot be destroyed. In a closed system, the total mass and energy remains unchanged. In a closed system the energy is conserved".

The law does not distinguish between useful and wasted energy. The law neglects 'losses' from the closed system into the environment.

In an energy conservation process, energy is transmitted from one or more forms to another form/forms resulting in certain work, heat transfer and energy loss.

In a closed system containing certain quantity of matter and energy, several energy transformations may take place from one form to another. The total mass and energy in the closed system remains unchanged.

2.7 FACTS AND FIGURES ABOUT ENERGY

These are the figures about energy of World Energy Production;

27% Coal
21% Natural Gas
39% Crude Oil
6% Nuclear
6% Hydro-electric Power

Fossil Fuels are 87% of the world's energy production. Wind and Solar energy are less than 1% of the world's production.

Distribution of World Oil Reserves

66.3% Middle East
6% Africa
8.4% North America
5.0% Far East/Oceania
5.9% East Europe
1.5% West Europe
6.9% Central South America

*This was the reason for the gulf war.

2.8 INDIAN AND GLOBAL ENERGY SOURCES

We do different types of work every day. Energy is required for it. Heat energy from burning fuels like wood, coal, petrol and cooking gas are widely used for cooking and other purposes. Different forms of energy are converted into electrical energy. The electrical energy thus obtained is used to operate radio, television, and lamps and to run vehicles and heavy machines in factories. Think of a situation where there is no energy. Not only our daily activities but also our life will come to a stand still. Thus energy is an inseparable part of our life. This topic deals with different sources of energy and their characteristics.

Sources of energy are those, which can supply adequate amount of energy in a suitable form for long periods. The major sources of energy, aside from human and animal power, are petroleum resources, natural gas, coal, hydropower, biomass, geothermal, nuclear, wind and solar. Theoretically these resources can be substituted for one another in order to perform a specific task. However, in practice, substitution would be subject to technical and economic limitations that circumscribe the use of such energy resources for specific purposes at given locations and periods of time. Few energy resources are used and consumed at the same location in which they are found, and most of them require generally elaborate transportation and conversion facilities to make them useful for performing the intended task. Analysis of energy resources, therefore, should cover availability and also the production, processing and distribution facilities. Some of the energy sources are:

2.8.1 THE SUN

Sun is the source of many forms of energy available to us. The most abundant element in sun is hydrogen. It is in a plasma state. This hydrogen at high temperature, high pressure and high density undergoes nuclear fusion and hence releases an enormous amount of energy. This energy is emitted as radiations of different forms in the electromagnetic spectrum. Out of these X-rays, gamma rays and most of ultraviolet rays do not pass through the earth's atmosphere. But heat energy and light energy are the main radiations that reach the earth. This energy is the basis for the existence of life on earth.

Sun is a sphere of intensely hot gaseous matter with a diameter of $1.39e^9$ m and $1.5e^{11}$ m away from earth. Sun has an effective black body temperature of 5762 K and has a temperature of $8e^6$ K to $40e^6$ K. The sun is a continuous fusion reactor in which hydrogen (4 protons) combines to form helium (one He nucleus). The mass of the He nucleus is less than that of the four protons, mass having been lost in the reaction and converted to energy. The energy received from the sun on a unit area perpendicular to the direction of propagation of radiation outside atmosphere is called solar constant, and has a value 1353 Wm^{-2}. This radiation when received on the earth has a typical value of 1100 Wm^{-2} and is variable.

The wavelength range is 0.29 to 2.5 micrometers. This energy is typically converted into usual energy form through natural and man-made processes. Natural processes include wind and biomass. Man-made processes include conversion into heat and electricity.

2.8.2 PETROLEUM

Petroleum products are by far the most versatile and useful energy resources available at present. Their low costs until 1973, ease of transportation and infinite divisibility are the three attributes that made petroleum products the most suitable and economical commercial energy resources. Petroleum products constitute 50–95 percent of commercial energy supplies and almost all the needs of transportation sector and mobile equipment are currently met by petroleum products. There are only a few possible substitutes that too on a limited scale. They also constitute the basic fuel for electric power plants while coal, natural gas and hydro resources are used in those locations where they are available. Kerosene and LPG are the favored cooking fuels and kerosene is the major lighting fuel in areas where there is no electricity. Before going further, it is interesting to know a bit on the history of oil and its economic and political implications in the past.

Petroleum was first found in Pennsylvania (USA). Petroleum is used to make gasoline, heating oil, diesel fuel, and lubricating oils. Following is brief times on energy development:

1879: Standard oil controlled 90% of refining capacity.

1870–1880: Kerosene is the largest volume manufactured good.

1882: Standard Oil Trust was established, which was a shield of legality and flexibility. Edison demonstrates electricity.

1885: 250,000 light bulbs in use.

1896: Henry Ford was the chief engineer at Edison, where he builds his first gasoline-powered engine.

1900: 8,000 automobiles, Drilling began in Texas.

1905: Oil discoveries in Louisiana and Oklahoma.

1909: Standard Oil Trust was broken up in 38 companies.

1912: 902,000 automobiles in use

1913: Ford introduced the assembly line.

1928: Texas number one oil producer.

1938: Oil discovery in Kuwait and Saudi Arabia.

1940's: United States shifts from petroleum exporting to petroleum importing.

1960: OPEC formed.

1986: Oil price collapse.

Petroleum is rock oil that exists down in the earths crust. They drill for petroleum to determine the size of the reserve and to produce oil at a controlled rate. There are three steps in recovering petroleum. The first step is the primary recovery, which is when oil flows by natural pressure or simple pumping. The maximum recovery is usually 30% of the oil available in the well. The next step is the secondary recovery, which is when water or gas is pumped into the well to force oil out. This adds an additional 10–20% to be recovered. The third step is the tertiary recovery, where hot gases and chemicals are pumped into the well to make the oil less viscous for easier pumping. Petroleum is classified according to its viscosity and sulfur contents. Pennsylvania's crude oil is low in sulfur and viscosity,

Venezuela's crude oil is high in sulfur and viscosity, and Middle East crude oil is usually low in sulfur. Petroleum refining separates different components of petroleum. It changes the chemical composition of petroleum component to produce desirable fuels and chemicals. Petroleum refining has 3 major processes. The first process is a physical process called distillation, which separates components according to their boiling points. The second step is the cracking, which breaks down long chains to make more gasoline, diesel, and jet fuel. This is a chemical process using a catalyst. The third process is the reforming process, where it converts straight chains into branched chains for better performance in gasoline engines. Petroleum is transported long distances by super tankers across oceans, and pipelines across continents. For short distances, petroleum is transported by barges, trucks, and rail cars.

Petroleum products are used in internal combustion engines, where the fuel is put right into the cylinder with the piston. A spark ignites the gasoline engines, and compression ignites diesel engines.

HISTORY OF OIL

Though oil has been known for thousands of years, the first modern commercial drilling and production of oil is usually said to have begun in 1859 in the US, when Col. Edwin L. Drake sunk a well in Pennsylvania near some natural oil seepage and within a few years it was in widespread use throughout the US. The producers, weakened by overproduction, were gradually taken over by the refining and distribution companies led by Rockefeller's Standard Oil Trust. Standard Oil dominated the oil industry in the US until, under anti-trust legislation, it was ordered in 1911 to divest itself of all its subsidiaries. Of the 38 companies in the group, three companies, Exxon, Mobil and Socal took a major role in the world oil market. Together with four other major companies Gulf, Texaco, Shell and BP, these seven companies (the 'Seven sisters') dominated the world oil scene throughout the first half this century.

During the 1920s and 30s, there was a period of intense competition, with a threat of over production aggravated by new discoveries in Mexico, Venezuela, Sumatra and Iran and a fall in demand during the economic depression. The major international oil companies led by Exxon, Shell and BP developed in 1928 a secret agreement to accept their current volumes of business, to decide jointly the shares in future increases in production. The resulting cartel continued until it was terminated by anti-trust in the 1940s in the US. Throughout this period the prices paid for crude oil were determined by negotiation between oil companies and governments in producing countries. This procedure continued into the 1960s, but by that time the continuing discovery and development of large low cost oil supplies in the Middle East had led to a post war decline in the price paid to producing countries. In an attempt to halt this decline, a group of producing countries, *viz.*, Iran, Iraq, Kuwait, Saudi Arabia and Venezuela whose GNP was substantially dependent on oil income, formed OPEC, the Organization of Oil Exporting Countries.

The foundation of OPEC in 1960 was seen as a defensive measure by the producers following a unilateral reduction by Exxon of the posted price they would pay for the supplies of Middle East crude oil, which was followed by other major oil companies. The five founder members of OPEC were at that time responsible for 80 percent of internationally traded crude oil. Intervention by governments in the activities of oil companies in their countries had begun dramatically in Mexico in 1938, when all operating companies in their countries were nationalised. Much earlier, in 1913, to ensure oil supplies for the UK Navy, Churchill has taken control of BP (then Anglo Persian) but UK rarely involved in commercial management. In 1938, under threat of nationalisation, Venezuela, then a major exporter, obliged the major companies (Exxon, Shell, Gulf) to increase their royalty payments, and ten years later in 1948, it successfully implemented a law giving the Venezuelan government a 50% share in all profits. This profit sharing arrangement was soon demanded elsewhere and in the 1950s and 1960s it was adopted in most oil producing countries.

In 1973 the world oil outlook changed dramatically, following an embargo imposed by the Arab members of OPEC on countries that they believed were providing assistance to Israel at the time of the 1973 October war between Israel and her neighbors. By coincidence, at the same time OPEC ministers decided to raise the oil price from $3 to $5.12. The day following this announcement in Oct. 73, the Arab members (OAPEC) agreed an immediate 5% reduction in oil production. Subsequently, the international oil price rose to $20 a barrel by Dec. 73. Shortly after this OPEC increased the oil price to $11.65 per barrel, giving a five-fold increase over the price two years earlier. After this the price declined gradually in real terms, due to inflation until late in 1978, when once again the spot market rose in response to local scarcities on the interruption of Iran's oil production. Following the lead from the spot market, OPEC began to move posted prices upwards again.

OPEC AND ITS MEMBERS

1. Algeria
2. Libya
3. Indonesia
4. Nigeria
5. Iran
6. Qatar
7. Iraq
8. United Arab Emirates
9. Kuwait
10. Saudi Arabia
11. Venezuela

Origins of Oil. Oil and gas are names given to a wide variety of hydrocarbons found in sedimentary basins on or under the earth's surface. Oil or petroleum is generally a complex mixture of the heavier (non-gaseous) hydrocarbons, averaging about two atoms of hydrogen to each carbon atom. Oil found in different reservoirs differs in composition, and many even vary within a single reservoir. Its properties vary from a light fluid to viscous heavy oil, grading to asphalt.

The process of oil formation started with the mixing of marine organisms with sand and salt to form sedimentary deposits, in periods ranging from tens of millions of hundreds of millions of years ago. Continued deposits of material led to burial, with a concomitant rise in pressure and temperature, resulting in compaction of the sediment into sedimentary rock, called the 'source rock', and conversion of the organic material into hydrocarbons (oil) embedded in the source rock. Increasing pressure from continued burial, together with the movement of water, with which rock below the water table is saturated, resulted in movement of the small oil globules into the porous and permeable environment of reservoir rocks. In some situations the oil became trapped in the reservoir rocks by a neighboring layer of impermeable rock, and these oil bearing reservoir rocks are the sources from which oil is now obtained.

Many types of geological structure can give rise to possible traps for oil. The first, called an anticline trap is in the form of a dome, in which gas, oil and water are held within the reservoir rock overlain by a layer of impermeable rock that prevents the oil and gas, more buoyant than the underlying water, from escaping to the surface. The second type, is called a fault trap, and may occur where impermeable rock at a fault in the strata of reservoir rocks prevents upward movement of oil. In the fourth type, the reservoir rock changes in permeability so that further movement of oil through the pores of the reservoir rock becomes possible.

To Summaries: Hydrocarbons are generated in source rocks from the remains of marine organism deposited and buried in the rocks. They are transported by surface tension, gravitational and pressure forces into reservoir rocks, where, if there are suitable traps, they accumulate in the pores of the rock and form the reservoirs of oil and gas found today. For oil to be formed within the source rocks, they must have been buried for a million years or more at depths over 1 km, to get the pressure and temperature high enough, but rarely more than about 4 km or the higher temperature at those depths would usually decompose the oil, leaving methane gas and petroleum coke.

Exploration and Production: First oil wells are only a few to hundred meters deep but most of today's producing accumulations lie in the depth range 500 to 3000 meters, but the deepest producing wells are at 6500 m for oil and 7500 meter for gas. Similarly, variations in pressure from atmosphere to 1000 atm. have been found although the pressure usually increases by 100 to 150 atm. per km in depth corresponding to the depth of the overlying column of rock pore water. Temperatures also increases with depth at a rate given by the geothermal gradient, generally in the range 15 to 40°C per km in oil producing areas, though temperatures in oil reservoirs are usually below 110°C.

Early exploration methods like geological surveys, measuring the angles of tilt of the rock strata that emerged at the surface, correlation of nearby drilling data, are augmented by seismic surveying; geomagnetic and gravitational surveys; geochemical tests; geothermal, radiation and electrical conductivity surveys, etc. These exploration methods allow the identification of structure that may be traps but they can only rarely establish the presence of oil that can be ensured only by drilling.

Oil is driven from the reservoir rocks into the borehole by the difference in pressure. Hence the rate of production from an oil well is limited. A measure of the rate of production from a reservoir is the reserves to production ratio (R/P) measured in years. R/P is high in the early years but tends to become constant in the range from 5 to about 15 years. An assessment of the amount of oil that may be recovered from a reservoir requires information on the amount of oil in place and an estimate of the recovery factor. The amount that can be extracted is related to the conditions in the reservoir, oil composition and the method of extraction. The world average recovery factor at present is about 25–30%.

The production of conventional oil depends on the reservoir fluids flowing under pressure out of the reservoir rock into the borehole. Oil recovery processes are usually considered as falling into three categories:

- **Primary Recovery :** The oil recovered by the natural displacement processes that occur as oil is produced from a reservoir;
- **Secondary Recovery :** The additional oil recovered as a result of water/gas injection into the reservoir to complement the naturally occurring drive processes;
- **Enhanced Recovery/Tertriary Recovery :** Oil recovery by processes aimed at higher displacement efficiencies than those obtained through the natural processes of gas and water drive, like use of chemicals, CO_2 and heat.

Oil Reserves. The proven reserves are defined as the quantity of oil that can be commercially produced with existing technology. At present the total world proved reserves amount to be about 1047 barrels (1047 bbl) of which 77% lies in OPEC countries (1996 estimate). The total world consumption of crude oil in 1996 was 71.7 million barrels per day. OPEC estimates that total world oil consumption could reach from the 70 million barrels a day in 1995 to around 100 million barrels per day by the year 2020. Table 2.1 gives the world's largest proven crude oil reserves and Table 2.2 gives the production rate. It is expected that oil's share of the world wide energy market will fall from almost 40% in 1995 to less than 37% in 2020. But oil will still be the world's single largest source of energy. Oil is a limited resource, so it may eventually run out, although not for many years to come. OPEC's oil reserves are sufficient to last another 80 years at the current rate of production, while non-OPEC oil producers' reserves might last less than 20 years. The worldwide demand for oil is rising and if we manage our resources well, use the oil efficiently and develop new fields, then our oil reserves should last for many more generations to come.

Table 2.1. World's largest proven crude oil reserves (1996)

Country	Reserves (millions of barrels)
Saudi Arabia	261,444
Iraq	112,000
United Arab Emirates	97,800
Kuwait	96,500
Iran	92,600

Table 2.2. World oil production (1996)

Country	Production (million barrels per day)
Saudi Arabia	8.1
Former Soviet Union	6.9
United States	6.5
Iran	3.6
China	3.2

2.8.3 NATURAL GAS

Do you know the fact that natural gas known in the short form as CNG is used in buses, trucks etc. in Delhi? Natural gas is a fossil fuel. This is usually formed in the Earth along with petroleum. Its main constituent is methane. It also contains small quantities of Ethane and Propane. Natural gas liquefied by applying high pressure is CNG (Compressed Natural Gas). In automobiles, houses and factories, CNG is used as a fuel. It is also used as a source of hydrogen required in the manufacture of fertilizers.

Natural Gas (CNG) is generally a mixture of the lighter hydrocarbons with methane (CH_4) predominating, often with varying fractions of nitrogen and impurities such as hydrogen sulfide. Natural Gas meet nearly 20% of world's energy needs. Increase in NG supplies during this century has been almost as dramatic as those of oil. However, development of NG industry has been limited to markets that could be economically connected by pipeline to natural gas reserves. The expense of constructing costly pipeline networks could only be justified where there are both large reserves and an assured demand. The future role of NG will be largely determined by transport costs and the world depends largely on large gas reserves and resources in areas further away from major markets. World proven gas reserves are estimated to be about 394 billion boe (1975 values) and estimated undiscovered resources to be about 1358 billion boe. Much will depend on when and how much of the estimated undiscovered resources in North America and Western Europe are found and developed.

Almost all of the present world production of NG is transported by pipeline. An alternative to gas pipelines is provided by transport by tankers carrying liquefied natural gas (LNG). The technology is commercially available since 1960s and the costs are still high. The gas must first be liquefied by cooling to $-161°C$, then carried in specially designed refrigerated tanks, and re-gasified at receiving terminals. Approximately 25% of the energy is lost in processing the LNG, and allowing also for transport, only about two thirds of the original supply of gas is delivered to the consumer. An LNG processing and transport system requires high capital expenditure and this will limit the rate of growth of international trade in LNG. NG provides a clean and convenient fuel and an important chemical feedstock.

2.8.4 COAL

Coal has been used as a fuel for several millennia in China. In Europe, coal was known to the Greeks and called 'anthrax' from which the name anthracite is derived. Its use was very much limited until the firewood crisis in England in the 16th century which led to wide spread use of coke when Darby developed the use of coke for reducing the iron ore. The changing rates of coal production are explained by the change in market nature. When the development of railway began, the demand for coal increased directly but also permitted it to be transported much more cheaply. The coal market in USA was disturbed by the rapid market growth for oil and later NG.

Formation. Coal is composed mainly of carbon though it also contains hydrogen and oxygen and varying small amounts of nitrogen, sulfur and other elements. It was formed by the decomposition of the remains of vegetation growing in swamps or in large river deltas undergoing intermittent subsidence. The decomposed material from plants and trees was transformed first by bacterial action into peat which become buried by later sedimentary deposits. Later under the movement of the earth's crust, the layers of peat become more deeply buried, and under the influence of heat and biochemical reactions they were transformed into various types of coal or lignite, during this coalification process, the carbon content increased as oxygen and hydrogen were released. Methane (CH_4) was formed and either escaped into atmosphere or migrated until it was captured in a geological trap so that it formed a natural gas reservoir contained by an impervious layer similar to those that contain petroleum.

Properties. Coals are ranked according to their carbon content. Under mild conditions of heat and pressure, the lowest rank coals were formed, consisting of brown coal and lignite. At higher temperatures and pressures, sub-bituminous and bituminous coals were formed, and under very high pressures, the highest rank coals, called anthracites, were formed. The anthracites contain more than 92% carbon, 2–3% hydrogen together with oxygen, volatile matter and impurities. Bituminous coal contains about 5% hydrogen and has a carbon content of 70–80%. The lowest ranks of lignite and brown coal may have less than 50% carbon content. The rank by carbon content approximates to a ranking by heat content though with some overlap between classes. Other classifications of importance include the coking qualities for mechanical strength, ash content, and volatile matter content. Sulfur is an important impurity as it appears in combustion products as oxides of sulfur (SO_2), which pollutes the environment.

Mining. Most hard coal (bituminous and anthracite) is obtained by deep mining though modern technology has led to the increasing use of open-cast methods using large excavators capable of shifting hundreds of tonnes per hour and the mines may reach depths of several hundreds of feet. Surface mining is cheaper than deep mining and rapid expansion is possible. Deep mining requires minimum of two shafts and expects to take 10 years to bring into operation. The mining shafts play a crucial role in providing ventilation to the mine, to remove methane associated with coal and to reduce heat and humidity. Two principal methods of mining in use are:

1. Longwall
2. Bord and Pillar or Room and Pillar

In Long wall method, coal is extracted in one operation from a face that may be of 600 m in length. In the older Bord and Pillar method, the area is divided into rectangles by driving a series of roadways at right angles to each other and then mining from each of these rectangles or pillars. In modern mines, over 90% of the coal is mined, loaded and transported mechanically. Transport underground is mainly by means of conveyor belts (replacing earlier tubs), which bring the coal to the main shaft for rising to the surface. At the mine head, the coal is cleaned, sorted or screened and blended.

Resources and Reserves. Geological resources include all coal that may become economic at some time in the future. Reserves include all coal that is known to be technically and economically recoverable under today's conditions. These estimates are old (1978) and not updated recently. The world total estimates for coals resources are (in billion tonnes) : Hard coal : 7725 and Brown coal : 2399. The proven reserves are 493 and 144 resp. World possesses vast resources of coal, far more than any of those of any other fossil fuel. However, almost 90% of the coal resources are concentrated in four countries: USSR (45%), US (24%), China (13%) and Australia (6%). US, USSR and China produce about 60% of coal output and the other major producers are (25% share): Poland, Germany, UK, Australia, South Africa and India. Today, most of the world's coal production is still consumed in the countries where it is mined. Only about 10% is traded internationally. The major portion of coal is used for electricity generation.

2.8.5 NUCLEAR ENERGY

Nuclear power production is based on the energy released when an atomic nucleus such as uranium undergoes fission following the absorption of a neutron to form a compound nucleus. This compound nucleus is unstable and may break into two or three smaller atomic nuclei with the simultaneous emission of several neutrons together with the release of considerable amount of energy. These neutrons may themselves be absorbed by other nuclei, and if enough of these are uranium nuclei, it is possible for a chain reaction to develop. Chain reactions form the basis of the operation of a nuclear reactor. Fission of a single atom of uranium yields 200 MeV ($= 3.2e^{-11}$ J), whereas the oxidation of one carbon atom releases only 4 eV. Natural uranium consists of 99.3% ^{238}U and only 0.7% of lighter isotope ^{235}U, but it is the latter that provides the most readily available fission energy in nuclear reactor. The maintenance of chain reaction, with exactly one neutron (on average) eventually causing another fission, is the design objective of any nuclear reactor.

Nuclear Energy Generation. If the ratio of ^{235}U to ^{238}U in a mixture is low, it is necessary to arrange that the neutrons be slowed down by a moderator (a light material like water, heavy water, helium gas, beryllium, carbon, mixed, usually in homogeneously, with the fuel) in order to take advantage of the increase in fission cross section for low energy neutrons. If the ratio is high, it is possible to design reactors that are based on fission caused by fast (high energy) neutrons. Reactors using slow neutrons are called thermal reactors in contrast to fast reactors whose design makes use of fission caused by fast neutrons. To reduce the size and increase the options for the choice of materials for a reactor, it is possible to enrich the uranium that is to enhance the fissile ^{235}U in some portion of the available natural uranium at the expense of the remainder. Higher the enrichment, easier it becomes to maintain the chain reaction, so the volume of the reactor may be reduced and a moderator with a lower moderating ratio may be used. Light water reactors use uranium enriched from 0.7 % to about 3%.

In a thermal reactor, the production of fissile isotopes is lower than burn-up of the fissile component of uranium ^{235}U in the fuel. However, in a fast reactor, using high-energy neutrons, the number of neutrons produced per fission is higher than in a thermal reactor, and some fission of ^{238}U also occurs, so that there are more spare neutrons available for absorption by the common uranium isotope ^{238}U, giving a higher rate of fissile decay products. By suitable design the conversion gain can be chosen so that more fissile material is produced than is consumed. Reactors of this type are called fast breeder reactors. Practically all power reactors in operation use ^{235}U as a fuel.

Resources. About 150 tonnes per year of natural uranium is required to meet the current demand. Proven resources are 2191000 tonnes there may be additional resources of 2177000 tonnes available. It is expected that FBR will takeover the future requirements and hence the future needs may not increase drastically.

2.8.6 LPG (LIQUEFIED PETROLEUM GAS)

LPG is a widely used fuel in homes. From where is this obtained? You have learnt earlier that petroleum gas is a constituent obtained when petroleum is subjected to fractional distillation. If high pressure is applied to this gas it will be liquefied. This liquid is LPG (Liquefied Petroleum Gas).

This is filled in strong cylinders and distributed. The main constituent of this is Butane. Small quantities of Ethane and Propane are also found.

Accident From LPG. Gases in LPG are odorless. What happens if it leaks? We will not know even if it fills the whole room. What will be the result then, if an electric switch is switched on or a matchstick is struck? Big fire or explosion will take place. Therefore to detect the leakage of the LPG another gas, Ethyl merchantman, having a special smell is mixed with it. Smell of this is sometimes felt when the gas cylinder is opened. If this smell is felt never try to light the match or to operate electrical appliances. Doors and windows must be opened and check whether there is any leak in the cylinder. When not in use it is better to have the valve of the cylinder closed.

2.8.7 ALCOHOL

Spirit lamp is used in classrooms for experiments. The spirit being used in spirit lamp is alcohol. This is a good fuel. Atmospheric pollution is much less when it is burnt. In certain countries a mixture containing alcohol and petrol is used as fuel in automobiles.

2.8.8 GASOHOL

A mixture of petrol (gasoline) and alcohol is being used as fuel in automobiles in Brazil and Zimbabwe. This fuel is gasohol. 'gaso' from gasoline and 'hol' from alcohol.

2.8.9 HYDRO POWER

Water is the only non-conventional energy source that has been exploited by man on a large scale. The technology is well established and simple. The industrial infrastructure for the manufacture of water turbines, valves, gates, generators and associated electrical equipment are well established in many countries.

Based on the capacity, it could be a micro, mini, small and big power plant. Based on the head, it is called a low head (<15 m), medium head (15-50 m) or high head (>50 m). Based on the type of load it may be base load or peak load. Based on the hydraulic features, it may be conventional, pumped storage or tidal type. Based on construction, it may be run of river, valley dam type, a diversion canal type or a high head diversion plant.

2.9 ENERGY EXPLOITED

Energy is derived from conventional and non-conventional resources and the former are in the process of depletion. These are fossil fuels-oil, coal and natural gas. It took million of years to build up these resources. Non-conventional resources are solar energy, wind energy, water energy and biomass. Approximately 80% of the world's energy is produced by fossil fuels. However, in France, the French Atomic Energy Commission established nuclear reactors, which produce enough energy to meet 70% of country's requirement.

World demand for oil (according to UN reports) rose from 436 million tonnes in 1960 to 2189 million tonnes in 1970 and to 3200 million tonnes in 1999. The corresponding figures for coal are 1043, 1635 and 2146 and for natural gas the figures are 187, 1022 and 2301. The demand will continue to grow. Of the developing countries, China has the highest per capita consumption of energy. For India, per capita consumption is lower than that of China. It may be mentioned that consumption figures represent commercial energy and do not take into account the non-commercial energy used by developing countries where poor people use wood that is acquired by gathering without any payment.

Among non-conventional resources, hydropower is the largest. Hydropower projects are in operation both in developed and developing countries - notable among the latter are China, India and Brazil. Hydropower potential is huge and at present only 15 percent of the potential in the developing world is being utilized. Wind power has also a great potential. Windmills and sails have been in use since ancient times. It is a fast growing resource. In 1980s, wind energy generation of the world was 10 megawatts. In the year 2000 it was 14000 megawatts. Green Piece International estimates that if the present trend continues wind power could supply 10% of world's electricity by 2020.

The use of solar energy is through photovoltaic cells. The photovoltaic news reported that world's photovoltaic production climbed from 0.1 megawatt to 200 megawatts in 1999. The biomass resources are various types of cultivated or uncultivated vegetation. Wood forms the chief resource and is the primary fuel for the people in Africa and Asia. Excessive use of wood has led to depletion of forests.

Coal, oil, gas and water constitute the main sources of energy in our country. The share of various energy sources in the commercial consumption of energy is mostly from coal (56%) and petroleum (32%), the other sources being nuclear, natural gas and water. Apart from commercial energy, a large amount of traditional energy sources in the form of fuel wood, agriculture waste and animal residue are used.

Commercial energy consumption has grown from 130.7 MTOE (million tonnes of oil equivalent) in 1991-92 to 176.08 MTOE in 1997-98. The main drivers of this increase are the accompanying structural change of economic growth and a rise in population together with rapid urbanisation.

Industrial sector is the largest consumer of energy consuming about 50% of the total commercial energy produced in the country followed by the transport sector. Among the most energy intensive industries which together account for nearly 80% of the total industrial energy consumption are the fertiliser, aluminum, textiles, cement, iron and steel, pulp and paper and chloro-alkali.

Transport sector is the largest consumer of petroleum products mainly in the form of high speed diesel and gasoline and accounts for nearly 50% of the total consumption.

With increase mechanisation and modernisation of its activities, the agricultural sector's consumption of commercial energy has grown considerably. The share of the farm sector in electrical energy consumption has increased from a mere 3.9% in 1950-51 to about 32.5% in 1996-97.

In the domestic sector, the consumption of natural fuel (mostly wood) energy is very high. Around 78% of rural and 30% of urban households depend on firewood. However, the mix of traditional fuels in the national energy mix is decreasing as more efficient commercial fuels are increasingly substituting these. In particular between 1970-71 and 1994-95, the annual consumption of electricity per household went up from 7 kWh to 53 kWh; of kerosene from 6.6 kg to 9.9 kg and of cooking gas from 0.33 kg to 3.8 kg. There is, however, a marked disparity in the level of energy and type of fuel consumed in rural and urban areas.

2.10 ENERGY DEMAND

Humanity today is on the verge of another catastrophe, *i.e.,* the energy crisis. Increasing industrialisation and unsuitable consumption patterns are escalating the environmental problems due to depletion of resources and energy. The unsustainable use of renewable resources and generation of toxic materials are creating problems to biodiversity, environment and human health. Energy is a primary input in any industrial operation. Energy is also a major input in sectors such as commerce, transport, telecommunication etc. besides the wide range of services required in the household and industrial sectors. Energy use is not an end in itself. Energy plays a dual role. It is an input into the productive sectors of the economy, *i.e.,* industry and agriculture as well as supporting infrastructure of transport. It is also a consumer good, *i.e.,* energy consumed in households has a direct impact on the quality of life. In India, per capita consumption is one-fourth of the world average and one-twenty fifth that of USA. Traditional fuels like animal dung, fuel wood and modern fuels are steadily replacing crop residue account to 30% of the total energy consumption in our country but these. The development of energy source is highly capital intensive and large investments are needed for meeting the demands of energy for different consuming sectors. It would be really ironic if fuel becomes more expensive than food. Gulf war and Iran-Iraq war had also brought into sharper focus the energy predicament.

Energy demand is not an exception to the economic theory of limited means and unlimited wants. The pace of exploitation of the energy resources has been growing over time, and has resulted in gradual depletion of the scarce reserves. The critical link between energy and economy has exposed the vulnerability of nations to the volatile energy situation. Energy today has become a key factor in deciding the product cost at micro level as well as indicating the inflation and the debt burden at the macro level. Energy cost is a significant factor in economic activity, at par with factors of production like capital, land and labour. The imperatives of an energy shortage situation calls for energy conservation measure, which essentially mean using less energy for the same level of activity. While on one hand the demand for energy is increasing, on the other hand the energy resources are becoming scarce and costlier. This steady increase in gap has not only compelled technocrats and decision makers in the industry to develop new measures of energy conservation but also to have systematic approach towards present trend of energy consumption through energy auditing and application of modern techniques and methods for minimizing energy wastage. Energy conservation is considered as a quick and economical way to solve the problem of power shortage as also a means of conserving the country's finite sources of energy. Energy conservation measures are cost effective, require relatively small investments and have short gestation as well as pay back periods. The studies conducted by Energy Management Centre, New Delhi have indicated that there is about 25 % potential of energy conservation W the industrial sector.

Energy technologies deal with various primary energies, processing, useful energies and associated plants and processes. The coverage including exploration, transportation, conversion, utilization. Energy Technology is concerned with 'demand' for various forms of secondary energy (usable energy) and the methods of 'supply'. Various factors affecting the 'demand' and 'supply' are in the scope of energy-technology. Energy is the key input for domestic demand, industrial and economic development. It is also a pre-requisite for sustaining industrial growth. With industrialization and urbanization, the demand of energy has continuously grown. In the past few decades, its need has multiplied many times. Presently, the demand of energy arises in a very large number of applications. Main amongst them is listed below.

- **In power plants.** to run the turbine. The rotation of turbine is then used to rotate the alternator to generate electricity.

- **In transportation sector.** to propel automobiles, trains ships, submarines, helicopters, aircrafts etc.
- **In military lists.** to propel missiles, tanks, weapons etc.
- **In industrial sectors.** for manufacturing steel, aluminium and other metals; in producing cement, plastics, chemicals, fertilizers; in oil refineries etc.
- **For domestic purposes.** in refrigerators, air-conditioners, fans lighting, television, music systems, washing machine etc.

The world energy consumption in the past four decades (between 1960 and 2000) has gone-up by more than three times, and is still growing rapidly. The energy consumption of some countries (in 1998) is given below as reference

- India 13.3×10^{18} J
- Japan 22.6×10^{18} J
- Russia 27.5×10^{18} J
- China 36×10^{18} J
- USA 100.5×10^{18} J

World 215.5×10^{18} J 3.9

2.11 ENERGY PLANNING

The Energy Strategies include the long-term policies, short-term and mid-term planning, economic planning, social and environmental aspects of various energy routes. These are analysed from the perspectives of the world, region, nation, states, sub-regions, various economic sectors, communities and individuals.

Considering the importance of power industry in the overall development of the country, power sector has been given high priority in the country's development plans. Energy sector alone accounts for about 29% of sixth plan investment. If investments in coal and oil transport and other infrastructures are taken into account, the total investment in the energy sector will account for about 40% of the plan investments. The fact alone is sufficient to exhibit the importance of power industry for the country's development. From a mere Rs.149 crores in the First Plan, the outlay for power during sixth plan period has increased to Rs. 15750 crores. The installed generating capacity has grown ten-fold from 2300 MW in 1951 to 25900 MW in 1978. Of this, 11000 MW was in hydel, 14000 MW in thermal and less than 1000 MW in nuclear power stations. The total number of power stations of 20 MW capacities and above at the end of March 1978, was 127, of which 65 were hydel, 60 thermal and 2 nuclear. Power generation rose from 7514 million kWh in 1950-51 to 103754 million kWh in 1978-79, *i.e.*, nearly 15 times. The total users of electricity have risen from 15 lakes in 1950 to 2641akhs in 1978-79. The per capita consumption of electricity rose from 18 kWh in 1950-51 to 121 kWh in 1978-79.

In spite of these measures, this industry is unable to meet the demands. Power shortages have become a recurrent feature in the country. Against an estimated requirement of 108656 million kWh in 1978-79, the actual availability was only a 97588 million kWh a deficit of about 11070 million kWh or 10.2°C. With the programme of large-scale industrialization and increased agricultural activity, the demand for power in the country is increasing at a rapid rate. If the present trend continues, the demand for power by the end of this century would be about 125 to 150 million kW. Allowing for adequate reserve margins required for scheduled maintenance, a total generating capacity of about 175 to 200 million kW would be needed by the year 2000 to meet the anticipated demands. This would mean 8 to 10 fold increase of the existing capacity.

Only proper development of hydel, thermal and nuclear resources of the country can achieve the required growth. Out of total available hydel-potential (41,000 MW), only 16% has been developed, therefore there is sufficient scope to develop this source of power in future. The major hydel potential is available in the northern region. Even if all the hydel potential is developed, it will not be possible to meet the growing demand. Therefore, it is necessary to supplement the hydel potentials with thermal. The coal deposits are rich and ample, though in terms of per capita it is hardly 176 tonnes in India which is certainly poor compared with other countries as 1170 tonnes in China, 13500 tonnes in the U.S.A. and 22000 tonnes in the former U.S.S.R. The available coal is also unevenly distributed in the country (60% only in Bihar and Bengal). This further requires the development of transportation facilities. Therefore, it is also not possible to depend wholly on thermal power development. The consideration for the use of nuclear fuel for power production in future is equally essential particularly in those states, which are far away from coal resources and poor in hydel potential. The future planning in the power development should aim at optimum exploitation of resources available so that power mix of hydel, thermal and nuclear is achieved.

Another step to be taken in the power development industry is setting up super-thermal power plants the central sector at different places in the country. The super-thermal power stations are at Farakka, Ramagundam, Korba and Singrauli and these are supplying power for the past 20 years. Presently all of them are supplying power through the national grid to deficit states.

In our country even 20 MW hydro potentials have not been developed, whereas it appears to be advantageous to develop even 20 kW units. Development of small hydro potentials as in China has, to a great extent, reduced the strain in existing plants. The development of biogas can ease the strain on oil supply to domestic users, which can otherwise diverted to power generation.

Another suggestion to face the present alarming power situation in the country is energy plantation. India receives large amount of solar radiation and photosynthesis is the process by which solar energy is converted into food and fuel by green plants. Fast growing species of trees give a yield of about 15 to 35 tonnes/hectare/year. The land, which is presently not used either for agriculture or forest, can be used for energy plantation where average rainfall is 80 to 100 cm per annum. With present Forest Technology, planned production forestry offers an unusual opportunity. If the forest area is increased from present 22 to 30%, increase in forest area is 30 million hectares of land) it can yield sufficient energy after next 20 years. The Government does not seriously think this phase of energy production but it looks a fruitful proposition.

As per the present planning of the Government, the problem of increased power demand will be solved only by proper mixed development of hydel, thermal and nuclear at least during one more decade. The severity of the power problem can be partly solved by the conservation of power. The efficiency hest thermal power plant is 35%. In India, it is hardly 25%. If auxiliary consumption and line loss are taken into account, the efficiency still goes to hardly 16%. The problem can be partly solved by proper maintenance and good quality of fuel supply.

The efficiency of the power plant operation is also defined as kWh generated per kW installed. The maximum kWh per annum per kW is 8760. The average figure in India is hardly 4000, which shows that the utilisation is only 45%. If this utilisation is increased, need for new capacity for power generation will be reduced. Increasing load factors can reduce the capacity of the power industry. The proper planning to develop hydel, thermal and nuclear resources in India in addition to measures taken to reduce outages and with proper load management will definitely go a long way in meeting the increasing power demand of the country.

2.12 INTRODUCTION TO VARIOUS SOURCES OF ENERGY

There are mainly two types of sources of energy
1. Conventional Sources of Energy (Non-Renewable Sources of Energy)
2. Non-conventional Sources of Energy (Renewable Sources of Energy).

2.12.1 CONVENTIONAL SOURCES OF ENERGY

These resources are finite and exhaustible. Once consumed, these sources cannot be replaced by others. Examples include coal, timber, petroleum, lignite, natural gas, fossil fuels, nuclear fuels etc. The examples are

(*i*) fossil fuel (*ii*) nuclear energy (*iii*) hydro energy

Have you not seen the filling of fuel in automobiles? What are the fuels that are being used in automobiles? What type of sources of energy are they? Are they non-conventional? Fossil fuel is an invaluable source of energy produced due to chemical changes taking place in the absence of oxygen, in plants and animals that have been buried deep in the earth's crust for many million years. Fossil fuels like coal, petroleum and natural gas are formed in this manner. These are conventional sources of energy. For example, energy from, Petroleum, natural gas, coal, nuclear energy, etc

THERMAL POWER

Thermal generation accounts for about 70% of power generation in India. Thermal energy generation is based on coal, furnace oil and natural gas. Steam cycle, rankin cycle or sterling cycle can be used for energy production. Now clean coal technologies (with 10% ash content) have been used in thermal power plants on commercial scale.

NATIONAL THERMAL POWER CORPORATION (NTPC)

It was incorporated in November 1975 as a public sector undertaking with the main objectives of planning, promoting and organising integrated development of thermal power. Installed capacity of NTPC projects stands at 16000 MW.

2.12.2 NON-CONVENTIONAL SOURCES OF ENERGY

These sources are being continuously produced in nature and are not exhaustible. Examples include wood, geothermal energy, wind energy, tidal energy, nuclear fusion, gobar gas, biomass, solar energy etc. The examples are

(*i*) Solar energy (*ii*) wind energy (*iii*) geothermal energy (*iv*) ocean energy such as tidal energy, wave energy (*v*) biomass energy such as gobar gas.

It is evident that all energy resources based on fossil fuels has limitations in availability and will soon exhaust. Hence the long term option for energy supply lies only with non-conventional energy sources. These resources are in exhaustible for the next hundreds of thousands of years.

The sources which are perennial and give energy continuously and which do not deplete with use are the Non conventional sources of energy.

For example, energy from, solar energy, bio-energy, wind energy, geothermal energy, wave, tidal and OTEC.

2.13 INTRODUCTION TO VARIOUS NON CONVENTIONAL (RENEWABLE) SOURCES OF ENERGY

Renewable energy development programme is gaining momentum in India. It has emerged as a viable option to achieve the goal of sustainable development. However, Indian renewable energy programme need more thrust at this stage. India has now the world's largest programme for deployment of renewable energy products and systems, the spread of various renewable energy technologies in the country has been supported by a variety of incentives and policy measures.

Power generation from non-conventional renewable sources has assumed significance in the context of environmental hazards posed by the excessive use of conventional fossil fuels. Renewable energy technologies have provied viable for power generation not so much as a substitute, but as supplement to conventional power generation. Currently renewables contribute over 3500 MW, which represents almost 3.5 percent of the total installed generating capacity of one lakh MW from all sources. Of this, wind power alone accounts for 1617 MW, while biomass power accounts for 450 MW and small hydros 1438 MW. An additional 4000 MW of power from renewable sources is to be added during the Tenth Five Year Plan period (2002–07) mainly through wind, biomass, small hydros, waste energy and solar energy system. Further, India has set a goal elevating the share of renewable energy sources in power generation up to 10 percent share of new capacity addition or 10,000 MW to come from renewables by 2012.

Today, India has the largest decentralised solar energy programme, the second largest biogas and improved stove programmes and the fifth largest wind energy programme in the world. A substantial manufacturing base, has been created in a variety of renewable energy technologies placing India in a positron not only to export technologies; but also offer technical expertise to other countries.

Table 2.3

NON-CONVENTIONAL ENERGY POTENTIAL AND ACHIEVEMENTS		
SOURCE/SYSTEM	POTENTIAL	ACHIEVEMENTS (As on 31-03-2002)
Biogas Plants	120 lakh	33 lakh
Improved Chulhas	1,200 lakh	350 lakh
Wind	45,000 MW	1,617 MW
Small hydro	15,000 MW	1,438 MW
Biomass Power	19,500 MW	391 MW
Biomass Gasifieres		51 MW
Solar PV	20 MW/Sq Km	85 MW
Waste-to-energy	1,700 MW	22 MW
Solar Water Heating	1,400 lakh sqm collector area	6 lakh sqm collector area

2.14 BIO-GAS

Biogas is a good fuel. Have you thought how this is fomed? Biomass like animal excreta, vegetable wastes and weeds undergo decomposition in the absence of oxygen in a biogas plant and form a mixture of gases. This mixture is the biogas. Its main constituent is methane. This is used as a fuel for cooking and Lighting.

2.14.1 AEROBIC AND ANAEROBIC BIO-CONVERSION PROCESS

There are mainly three aerobic and anaerobic bio-conversion process for the biomass energy applications: There are:

Bioproducts: Converting biomass into chemicals for making products that typically are made from petroleum.

Biofuels: Converting biomass into liquid fuels for transportation.

Biopower: Burning biomass directly, or converting it into a gaseous fuel or oil, to generate electricity.

Bioproducts. Whatever products we can make from fossil fuels, we can make using biomass. These bioproducts, or biobased products, are not only made from renewable sources, they also often require less energy to produce than petroleum-based products.

Researchers have discovered that the process for making biofuels releasing the sugars that make up starch and cellulose in plants also can be used to make antifreeze, plastics, glues, artificial sweeteners, and gel for toothpaste.

Other important building blocks for bioproducts include carbon monoxide and hydrogen. When biomass is heated with a small amount of oxygen present, these two gases are produced in abundance. Scientists call this mixture biosynthesis gas. Biosynthesis gas can be used to make plastics and acids, which can be used in making photographic films, textiles, and synthetic fabrics.

When biomass is heated in the absence of oxygen, it forms pyrolysis oil. A chemical called phenol can be extracted from pyrolysis oil. Phenol is used to make wood adhesives, molded plastic, and foam insulation.

Biofuels. Unlike other renewable energy sources, biomass can be converted directly into Liquid fuels, biofuels. For our transportation needs (cars, trucks, buses, airplanes, and trains). The two most common types of biofuels are ethanol and biodiesel.

Ethanol is an alcohol, the same found in beer and wine. It is made by fermenting any biomass high in carbohydrates (starches, sugars, or celluloses) through a process similar to brewing beer. Ethanol is mostly used as a fuel additive to cut down a vehicle's carbon monoxide and other smog-causing emissions. But flexible fuel vehicles, which run on mixtures of gasoline and up to 85% ethanol, are now available.

Biodiesel is made by combining alcohol (usually methanol) with vegetable oil, animal fat, or recycled cooking greases. It can be used as an additive to reduce vehicle emissions (typically 20%) or in its pure form as a renewable alternative fuel for diesel engines.

Other biofuels include methanol and reformulated gasoline components. Methanol, commonly called wood alcohol, is currently produced from natural gas, but could also be produced from biomass.

There are a number of ways to convert biomass to methanol, but the most likely approach is gasification. Gasification involves vaporizing the biomass at high temperatures, then removing impurities from the hot gas and passing it through a catalyst, which converts it into methanol.

Most reformulated gasoline components produced from biomass are pollution reducing fuel additives, such as methyl tertiary butyl ether (MTBE) and ethyl tertiary butyl ether (ETBE).

Biopower. Biopower, or biomass power, is the use of biomass to generate electricity. There are six major types of biopower systems: direct fired, cofiring, gasification, anaerobic digestion, pyrolysis, and small, modular.

Most of the biopower plants in the world use direct fired systems. They burn bioenergy feedstocks directly to produce steam. This steam is usually captured by a turbine, and a generator then converts it into electricity. In some industries, the steam from the power plant is also used for manufacturing processes or to heat buildings. These are known as combined heat and power facilities. For instance, wood waste is often used to produce both electricity and steam at paper mills.

Many coal fired power plants can use cofiring systems to significantly reduce emissions, especially sulfur dioxide emissions. Coal firing involves using bioenergy feedstocks as a supplementary energy source in high efficiency boilers.

Gasification systems use high temperatures and an oxygen starved environment to convert biomass into a gas (a mixture of hydrogen, carbon monoxide, and methane). The gas fuels what's called a gas turbine, which is very much like a jet engine, only it turns an electric generator instead of propelling a jet.

The decay of biomass produces a gas methane that can be used as an energy source. In landfills, wells can be drilled to release the methane from the decaying organic matter. Then pipes from each well carry the gas to a central point where it is filtered and cleaned before burning. Methane also can be produced from biomass through a process called anaerobic digestion. Anaerobic digestion involves using bacteria to decompose organic matter in the absence of oxygen.

Methane can be used as an energy source in many ways. Most facilities burn it in a boiler to produce steam for electricity generation or for industrial processes. Two new ways include the use of microturbines and fuel cells. Microturbines have outputs of 25 to 500 kilowatts. About the size of a refrigerator, they can be used where there are space limitations for power production. Methane can also be used as the "fuel" in a fuel cell. Fuel cells work much like batteries but never need recharging, producing electricity as long as there's fuel.

In addition to gas, liquid fuels can be produced from biomass through a process called pyrolysis. Pyrolysis occurs when biomass is heated in the absence of oxygen. The biomass then turns into a liquid called pyrolysis oil, which can be burned like petroleum to generate electricity. A biopower system that uses pyrolysis oil is being commercialized.

Several biopower technologies can be used in small, modular systems. A small, modular system generates electricity at a capacity of 5 megawatts or less. This system is designed for use at the small town level or even at the consumer level. For example, some farmers use the waste from their livestock to provide their farms with electricity. Not only do these systems provide renewable energy, they also help farmers and ranchers meet environmental regulations.

Small, modular systems also have potential as distributed energy resources. Distributed energy resources refer to a variety of small, modular power generating technologies that can be combined to improve the operation of the electricity delivery system.

2.14.2 RAW MATERIALS

All types of organic wastes which can form slurry are suitable for producing biogas by the process of anaerobic digestion in a biogas plant. Wood and sugar biogases are difficult and time consuming with this process and incineration may be preferred. The choice of raw material (in feed) is based on availability of the waste. The biogas plant is designed to suit particular type of in feed.

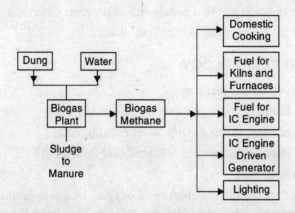

Fig. 2.2. Energy Route of Biogas (Gobar Gas).

Biogas production taken different time period depending upon raw material; temperature; process adopted etc.

The biomass used as a raw material can be classified into the following categories.

Waste	Cultivated and Harvested
Agricultural wastes	Agricultural energy crops
Rural animal wastes	Aquatic crops
Poultry waste	
Butchary waste	
Urban waste (garbage)	Forest crops
Aquatic wastes	
Forest wastes	
coconut husk waste	
Industrial wastes	

Others are poultry waste, piggery waste, sheep, goat, cow, horse dung, Slaughter house waste, coconut shell, husk ,waste garbage, fruit skins and leftovers.

The waste is generated periodically and can be converted into useful biogas. The problem of waste disposal is solved as the sludge is used as manure.

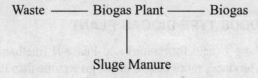

The cultivated or harvested biomass is specially grown on land or in sea/lake for obtaining raw materials for biogas production.

2.14.3 PROPERTIES OF BIO GAS

Main properties of bio gas are:
1. Comparatively simple and can be produced easily.
2. Burns without smoke and without leaving ash as residues.
3. Household wastes and bio-wastes can be disposed of usefully and in a healthy manner.
4. Reduces the use of wood and to a certain extent prevents deforestation.
5. The slurry from the biogas plant is excellent manure.

2.14.4 BIO GAS PLANT TECHNOLOGY

The important parts of biogas plant are
1. The tank where biomass undergoes decomposition (digester)
2. The tank where biomass is mixed with water (mixing tank)
3. The tank where slurry of biomass is collected (out flow tank)
4. Arrangement to store gas.

Due to the action of bacteria in the absence of oxygen, biogas is produced in the plant. This is collected in the tank. In the gasholder type plant, the cylinder rises up as the gas fills the tank and the storage capacity increases. The gas storage capacity of dome type will be less than that of gasholder type. Residue of biomass (slurry) can be used as good manure.

Biogas plants are built in several sizes, small (0.5 m^3/day) to very large 2500 m^3/day). Accordingly, the configurations are simpler to complex.

Biogas plants are classified into following main types.
—Continuous type or batch type.
—Drum type and dome type.

There are various configurations within these types.

CONTINUOUS TYPE

Continuous type biogas plant delivers the biogas con-tinuously and is fed with the biomass regularly. Continuous type biogas plant is of two types.

(A) SINGLE STAGE CONTINUOUS TYPE BIOGAS PLANT

In such a plant Phase-I (acid formation) and Phase-II (methanation) are carried out in the same chamber without barrier. Such plants are simple, economical, easy to operate and control. These plants are generally preferred for small and medium size biogas plants. Single stage plants have lesser rate of gas production than the two stage plant.

(B) TWO STATE CONTINUOUS TYPE BIOGAS PLANT

In such a plant the Phase-I (acid formation) and Phase-II (methane formation) take place in separate chambers. The plant produces more biogas in the given time than the single stage plant. However, the process is complex and the plant is costlier, difficult to operate and maintain. Two stage plant is preferred for larger biogas plant systems.

BATCH TYPE BIOGAS PLANT

The infeed biomass is fed in batches with large time interval between two consecutive batches. One batch of biomass infeed is given sufficient **retention** time in the digester (30 to 50 days). After completion of the digestion, the residue is emptied and the fresh charge is fed. The fresh biomass charge may be subjected to aeration or nitrogenation after feeding and then the digester covers are closed for the digestion process. Thereafter, the Biogas is derived from the digester after 10 to 15 days. Fermentation continues for 30 to 50 days.

Salient Features:

1. Batch type biogas plant delivers gas intermittently and dis-continuously.
2. Batch type biogas plant may have several digesters (reacters) which are fed in a sequential manner and discharged in a sequential manner to obtain the output biogas continuously.
3. Batch type biogas plants have longer digestion time and are therefore more suitable for materials which are difficult for anaerobic digestion (*e.g.* harder, fibrous biomass).
4. Batch type biogas plant needs initial seeding to start the anaerobic fermentation.
5. Batch type biogas plant needs larger volume of the digester to accommodate large volume of the batch. Hence initial cost is higher.
6. Operation and maintenance is relatively more complex. Batch type biomass plants need well organised and planned feeding. Such plants are preferred by European farmers. Such plants are not yet popular in India.

FIXED DOME TYPE DIGESTER

In the fixed dome type digester biogas plant, the digester and gas-collector (gas dome) are enclosed in the same chamber. This type of construction is suitable for batch type biogas plant. The digester is conveniently built at or below ground level in comparatively cooler zone. The construction of the digester is with locally available materials like, bricks, tera-cota. The pressure inside the digester increases as the biogas is liberated. The biogas gets collected in the upper portion of the digester in a dome shaped cavity. The outlet pipe is provided at the tope of the fixed dome. Alternatively the gas collector (gas holder) is a separately installed chamber. The digester tank and gas collector chamber are separated by a water seal tank.

The arrangement of a separate gas collector is preferred as the tapping of gas from the gas holder does not affect the pressure and the digestion process in the main digester. The water seal tank prevents the return of the gas from the gas collector to the digester chamber.

An additional displacement chamber may be provided for provid-ing space to the displacement slurry in the digester due to gas pressure in the upper dome of the fixed type digester. The fixed dome type digester can be fed on daily basis with small quantities of the slurry. The excess slurry in the digester gets accommodated in the displacement chamber. The level of the slurry in the main digester and the displacement collector can vary in accordance with the pressure and volume of the biogas in the fixed type of dome. The pressure in the fixed dome and the displacement gas collector are almost the same as they are connected by the outlet from the main digester.

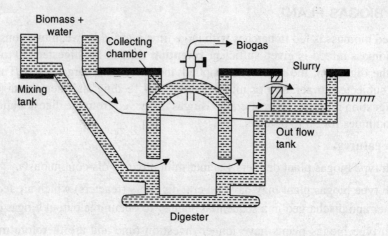

Fig 2.3. Fixed Dome type.

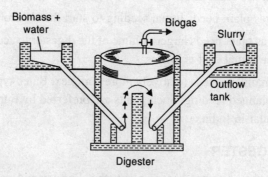

Fig 2.4. Floating gas holder type.

Floating Gas Holder Type. In this design a dome made floats above the slurry in the disaster. In the Fig. 2.4, The disaster tank is of cylindrical masonry construction. The floating dome is of fabricated steel construction. The dome guide shaft provides the axial guide to the floating dome. As the gas is collected in it. The sliding bearing provides smooth sliding surface and guide to the floating dome. The gas generated in the slurry gets collected in the dome and the dome arises. The water seal tank provides separation between the gas in the dome and the outlet gas.

2.15 WIND ENERGY

Wind energy is another potential source of energy. Winds are the motion of air caused by uneven heating of the earth's surface by the sun and rotation of the earth. It generates due to various global phenomena such as 'air-temperature difference' associated with different rates of solar heating. Since the earth's surface is made up of land, desert, water, and forest areas, the surface absorbs the sun's radiation differently. Locally, the strong winds are created by sharp temperature difference between the land and the sea. Wind resources in India are tremendous. They are mainly located near the sea coasts. Its potential in India is estimated to be of 25×10^3 mW. According to a news release from American Wind Energy Association the installed wind capacity in India in the year 2000 was 1167 mW and the wind energy production was 2.33×10^6 mWh. This is 0.6% of the total electricity production.

During the day, air above the land heats more quickly than air above water. The hot air over the land expands and rises, and the heavier, cooler air over a body of water rushes in to take its place, creating local winds. At night, the winds are reversed because air-cools more rapidly over land than over water. Similarly, the large atmospheric winds that circle the earth are created because land near the equator is heated more by the sun than land near the North and South Poles.

Today people can use wind energy to produce electricity. Wind is called a renewable energy source because we will never run out of it. Winds are natural phenomena in the atmosphere and have two different origins.

(1) Planetary winds are caused by daily rotation of earth around its polar axis and unequal temperature between Polar Regions and equatorial regions.

(2) Local Winds are caused by unequal and heating and cooling of ground surface of ocean 1 lake surfaces during day and night.

2.15.1 WIND MACHINE FUNDAMENTALS

Throughout history people have harnessed the wind. Over 5,000 years ago, the ancient Egyptians used wind power to sail their ships on the Nile River. Later people built windmills to grind their grain. The earliest known windmills were in Persia (the area now occupied by Iran). The early windmills looked like large paddle wheels. Centuries later, the people in Holland improved the windmill. They gave it propeller type blades and made it so it could be turned to face the wind. They have been used for pumping water or grinding grain. Windmills helped Holland become one of the world's most industrialized countries by the 17th century. Today, the windmill's modern equivalent — a *wind turbine* — can use the wind's energy to generate electricity.

American colonists used windmills to grind wheat and corn, to pump water, and to cut wood at sawmills.

In this century, people used windmills to generate electricity in rural areas that did not have electric service. When power lines began to transport electricity to rural areas in the 1930s, the electric windmills were used less and less.

Then in the early 1970s, oil shortages created an environment eager for alternative energy sources, paving the way for the re-entry of the electric windmill on the American landscape.

Today's wind machine is very different from yesterday's windmill. Along with the change in name have come changes in the use and technology of the windmill. While yesterday's machines were used primarily to convert the wind's kinetic energy into mechanical power to grind grain or pump water, today's wind machines are used primarily to generate electricity. Like old-fashioned windmills, today's wind machines still use blades to collect the wind's kinetic energy. Windmills work because they stow down the speed of the wind. The wind flows over the airfoil shaped blades causing lift, like the effect on airplane wings, causing them to turn. The blades are connected to a drive shaft that turns an electric generator to produce electricity.

Modern wind machines are still wrestling with the problem of what to do when the wind isn't blowing. Large turbines are connected to the utility power network-some other type of generator picks up the load when there is no wind. Small turbines are often connected to diesel/electric generators or sometimes have a battery to store the extra energy they collect when the wind is blowing hard.

2.15.2 AEROFOIL DESIGN

A wind turbine changes the kinetic energy of the wind into rotary motion (or torque) that can do work. It could power a water pump, or turn a generator.

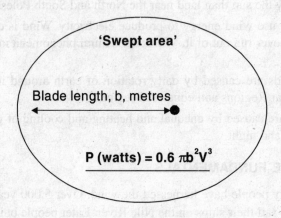

Fig. 2.5

Winds are the motion of air caused by uneven heating of the earth's surface by the sun and rotation of the earth. There is a direct relationship between the swept area of the turbine blades and the turbine's power output (see above). The estimated total power capacity of the winds passing over the land is about $1e^{15}$ W. But the total exploitable wind power is only $2e^{13}$ W.

The theoretical wind power can be estimated as:

Power density = $0.6 \ k. \ \rho.v^3 = 0.6\pi b^2 v^3$

where; k = Energy pattern factor (depends on type of wind)

ρ = Wind density

v = The average wind velocity.

Of the theoretical quantity energy that can be extracted from the wind, large commercial wind turbines are unlikely to get more than 25% of this. Small and less high-tech. designs might only get 15%. But the effect of this equation is that if the wind speed doubles, the power output increases eight-fold. So small increases in wind velocity can create large increases in power output.

The large amounts of energy that are produced at very high wind speeds means that most wind turbines have a pre-designed maximum power output to prevent the machinery ripping itself apart. Large wind turbines rated 150 kW and above are very complex machines. All wind turbines must 'feather' the blades turning then slightly out of the wind as wind speed increases in order to prevent the turbine running away. If this didn't happen, the centripetal force could rip the blades off. But large turbines also have complex automatic gearboxes that keep the generator turning at the optimum speed for power generation. Rarely does the wind blow constantly. This means that the rated output of the turbine will never be achieved as a constant output. On average turbines produce about 30% of their rated capacity as continuous power. So, to compare wind turbines to continuous power sources, you have to multiply the continuous capacity by 3.33 to get the amount of wind turbine capacity required to produce the same power output. For example, a 1,000 mW coal-fired power station would require 3,333 mW of wind capacity to provide the same average power output. As yet there is no efficient form of large scale power storage that would allow the variations in wind turbine output to be evened out over time. Reversible hydrogen fuel cells may be an option in the future — they can produce hydrogen to store energy, and then use the hydrogen to produce power at other times when the wind turbine output is low.

The larger the wind turbine, the higher the tower that supports the blades must be. This isn't just to keep the blades a safe distance off the ground. The higher you go from ground level, the faster and more uniform the wind so you get more power. So, as the power output of turbines increases, so does their size. The only restriction on the size of a wind turbine is the strength of the materials from which it is built. This is a serious engineering problem because whilst the turbine must be light enough to turn in a light breeze, the structure must be strong enough to withstand storm force winds.

Forces on the Blades. There are two types of forces operating on the blades of a propeller-type wind turbine. They are the circumferential forces in the direction of wheel rotation that provide the torque and the axial forces in the direction of the wind stream that provide an axial thrust that must be counteracted by proper mechanical design.

The circumferential force, or torque, T is obtained from

$$T = \frac{P}{\omega} = \frac{P}{\pi DN} \qquad \ldots(1)$$

where
T = torque, N or lb,
ω = angular velocity of turbine wheel, m/s
D = diameter of turbine wheel = $\sqrt{4\,A/\pi}$, m
N = wheel revolutions per unit time, s^{-1}

For a turbine operating at power P, the torque is given by

$$T = \eta \frac{1}{8g_c} \frac{\rho D V_1^3}{N} \qquad \ldots(2)$$

For a turbine operating at maximum efficiency $\eta_{max} = 16/27$, the torque is given by T_{max},

$$T_{max} = \frac{2}{27 g_c} \frac{\rho D V_1^3}{N}$$

The axial force, or axial thrust, is

$$F_a = \frac{1}{2g_c} \rho A (V_1^2 - V_2^2) = \frac{\pi}{8g_c} \rho D^2 (V_1^2 - V_2^2)$$

The axial force on a turbine wheel operating at maximum efficiency where $V_e = 1/3; V_i$ is given by

$$F_{a,\,max} = \frac{4}{9g_c} \rho A V_1^2 = \frac{\pi}{9g_c} \rho D^2 V_1^2$$

The axial forces are proportional to the square of the diameter of the turbine wheel which makes them difficult to cope with in extremely large-diameter machines. There is thus an upper limit of diameter that must be determined by design and economical considerations.

The performance of a wind mill rotor stated as coefficient of performance is expressed as:

$$C_p = A/P_{max}$$
$$= A/(1/2\, \rho V^3)$$

where
ρ = Density of air
A = Swept area
V = Velocity of the wind

Further the tip speed ratio being the function of speed at the tip of the rotor to the wind speed, *i.e.* U/V and in most of the parts of India, the wind velocity being low (through the wind energy average around 3 kWh/m² day) The exploitation of wind mills in India is feasible. Depending upon the survey of velocity in a region the appropriate value of design parameter may be computed.

2.15.3 WIND POWER SYSTEMS

You usually stand in an open space to enjoy the wind. You know how wind originates. Moving air is wind. Since the wind has velocity it has kinetic energy. This is the energy of the wind. We shall see how the kinetic energy of the wind can be used to produce electricity. For that, we can use windmills. Windmills are devices, which work on wind. How the kinetic energy of the wind is made use of in windmills shall be looked into. We shall examine the working of a windmill. The important part of a windmill is a structure with large leaves, fixed at the top of a high tower.

What will happen when wind blows on these leaves? You may have seen paper fans available at festival places, rotating when the wind blows. In a similar manner the speed of leaves changes with the speed of the wind. What happens if the rotation of the windmill is given to the rotor of a generator? Rotor also rotates. Then electricity is obtained from the generator. What happens if the windmill is connected to a water pump? As the leaves of the windmill rotate pump works pumping out water.

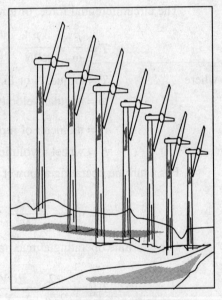

Fig. 2.6. Windmill farm.

Wind machines are just as efficient as coal plants. Wind plants convert 30 percent of the wind's kinetic energy into electricity. A coal-fired power plant converts about 30-35 percent of the heat energy in coal into electricity. It is the capacity factor of wind plants that puts them a step behind other power plants. Capacity factor refers to the capability of a plant to produce energy. A plant with a 100 percent capacity rating would run all day, every day at full power. There would be no down time for repairs or refueling, an impossible dream for any plant. Wind plants have about a 25 percent capacity rating because wind machines only run when the wind is blowing around nine mph or more. In comparison, coal plants typically have a 75 percent capacity rating since they can run day or night, during any season of the year.

One wind machine can produce 275-500 thousand kilowatt-hours (kWh) of electricity a year. That is enough electricity for about 50 homes per year.

In this country, wind machines produce about three billion kWh of energy a year. Wind energy provides 0.12% of the nation's electricity, a very small amount. Still, that is enough electricity to serve more than 300,000 households, as many as in a city the size of San Francisco or Washington, D.C. California produces more electricity from the wind than any other state of USA. It produces 98 percent of the electricity generated from the wind in the United States. Some 16,000 wind machines produce more than one percent of California's electricity. (This is about half as much electricity as is produced by one nuclear power plant.) In the next 15 years, wind machines could produce five percent of California's electricity. The United States is the world's leading wind energy producer. The U.S. produces about half of the world's wind power. Other countries that have invested heavily in wind power research are Denmark, Japan, Germany, Sweden, The Netherlands, United Kingdom, and Italy. The American Wind Energy Association (AWEA) estimates wind energy could produce more than 10 percent of the nation's electricity within the next 30 years.

So, wind energy may be an important alternative energy source in the future, but it will not be the sole answer to our energy problems. We will still need other energy sources to meet our growing demand for electricity.

2.15.4 ECONOMIC ISSUES

On the economic front, there is a lot of good news for wind energy. First, a wind plant is far less expensive to construct than a conventional energy plant. Wind plants can simply add wind machines as electricity demand increases. Second, the cost of producing electricity from the wind has dropped dramatically in the last two decades. Electricity generated by the wind cost 30 cents per kWh in 1975, but now costs less than five cents per kWh. In comparison, new coal plants produce electricity at four cents per kWh. In the 1970s and 1980s, oil shocks and shortages pushed the development of alternative energy sources. In the 1990s, the push may come from something else, a renewed concern for the earth's environment.

We will use two terms to describe wind energy production: efficiency and capacity factor. Efficiency refers to how much useful energy (electricity, for example) we can get from an energy source. A 100 percent energy efficient machine would change all the energy put into the machine into useful energy. It would not waste any energy. (You should know there is no such thing as a 100 percent energy efficient machine. Some energy is always "lost" or wasted when one form of energy is converted to another. The "lost" energy is usually in the form of heat.)

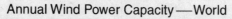
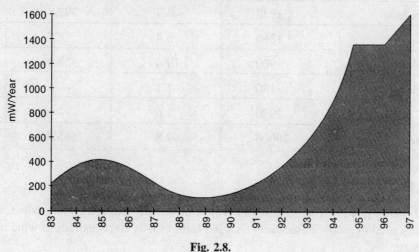

Fig. 2.8.

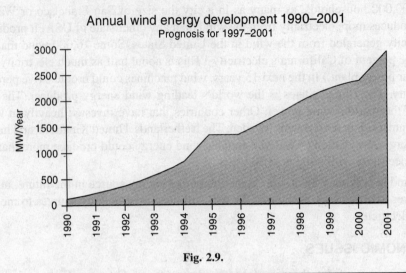

Fig. 2.9.

State-wise Wind Power Installed Capacity in India				
State	As on 31.03.2002			
	Gross Potential (MW)	Demonstration Projects (MW)	Private Sector Projects (MW)	Total Capacity (MW)
Andhra Pradesh	2200	5.4	87.2	92.6
Gujarat	3100	17.3	149.6	166.9
Karnataka	4120	2.6	66.0	68.6
Kerala	380	2.0	—	2.0
Madhya Pradesh	3000	0.6	22.0	22.6
Maharashtra	1920	6.4	392.8	399.2
Rajasthan	1210	6.4	9.7	16.1
Tamil Nadu	900	19.4	838.1	857.5
West Bengal	180	1.1	—	1.1
Others	2990	1.6	—	1.6
Total (All India)	**20000**	**62.8**	**1565.4**	**1628.2**

2.15.5 SELECTION OF WIND MILL

Wind power plants, or wind farms or wind mill as they are sometimes called, are clusters of wind machines used to produce electricity. A wind farm usually has hundreds of wind machines in all shapes and sizes.

Unlike coal or nuclear plants, public utility companies do not own most wind plants. Instead they are owned and operated by business people who sell the electricity produced on the wind farm to electric utilities. These private companies are known as Independent Power Producers.

Operating a wind power plant is not as simple as plunking down machines on a grassy field. Wind plant owners must carefully plan where to locate their machines. They must consider wind availability (how much the wind blows), local weather conditions, nearness to electrical transmission lines, and local zoning codes.

Wind plants also need a lot of land. One wind machine needs about two acres of land to call its own. A wind power plant takes up hundreds of acres. On the plus side, farmers can grow crops around the machines once they have been installed.

After a plant has been built, there are still maintenance costs. In some states, maintenance costs are offset by tax breaks given to power plants that use renewable energy sources. The Public Utility Regulatory Policies Act, or PURPA; also requires utility companies to purchase electricity from independent power producers at rates that are fair and nondiscriminatory.

2.15.6 RECENT DEVELOPMENTS

The present windmill technology is inadequate for the low wind speed regions in the plains. Special development projects in the following areas must be taken up so that wind energy can also be used in the low wind speed regions.

Artificial Winds. Generation of artificial winds to drive windmills by heating large surfaces with favorable thermodynamic properties is technically feasible. A project report has been prepared to heat a large surface in which case the resulting current (artificial wind) can drive turbines. The efforts needed to pursue the project in the form of money, manpower and time are huge.

Aeroelectric Plant. The low wind velocity in the plains can be augmented by the use of diffuser at intake to wind mills. Besides the propellers, Madaras and Darrieus, there has been a plethora of designs for wind machines. One intriguing power plant design, called the aeroelectric plant, uses the flow up a tower that looks like a cooling tower as shown in Fig. 2.10. Its walls are heated by solar radiation. Since the walls are circular, the sun's rays need not be tracked as it changes position in the sky during the day. The heated walls, in turn, heat the inside air and a flow up the tower is established. This air flow is made to drive a number of air turbines located near the top of the tower. The driving pressure causing air flow is given by the well-known chimney effect.

P. Carlson of Californica has proposed a slightly modified form in which, the interior air in a very tall tower would be cooled by pumping water to the top. The water evaporates in the low pressure air there, causing a downward flow of cooled air. The driving pressure can be calculated in a manner similar to that for wet cooling towers. A conceptual design of such a plant called for 2.4 km high, 300 m diameter tower located in a hot desert and 10 wind turbines surrounding the tower periphery at the bottom producing 2500 MW.

Low Wind Speed Turbines. The turbines available in India and abroad are suitable for a rated wind speed of 3.5 m/s or more whereas low wind speed turbines for rated values of 1.5 – 2 m/s are needed for plain areas. Special efforts are, therefore, needed to develop cheap and simple rotors, which can cut in at low wind speeds available in the plains. The Savonious rotor and American multi-blade type windmills have opti-mum power coefficients at a very low tip-speed and can therefore be used as starting point to develop windmills suitable for low wind speeds.

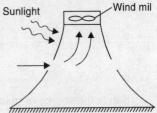

Fig. 2.10. Aeroelectric Plant

2.16 SOLAR ENERGY

The sun is the source of the vast majority of the energy we use on earth. Most of the energy we use has undergone various transformations before it is finally utilized, but it is also possible to tap this source of solar energy as it arrives on the earth's surface. There are many applications for the direct use of solar thermal energy, space heating and cooling, water heating, crop drying and solar cooking. It is a technology, which is well understood and widely used in many countries throughout the world. Most solar thermal technologies have been in existence in one form or another for centuries and have a well-established manufacturing base in most sun-rich developed countries.

The most common use for solar thermal technology is for domestic water heating. Hundreds of thousands of domestic hot water systems are in use throughout the world, especially in areas such as the Mediterranean and Australia where there is high solar insulation (the total energy per unit area received from the sun). As world oil prices vary, it is a technology, which is rapidly gaining acceptance as an energy saving measure in both domestic and commercial water heating applications. Presently, domestic water heaters are usually only found amongst wealthier sections of the community in developing countries. Other technologies exist which take advantage of the free energy provided by the sun. Water heating technologies are usually referred to as **active solar** technologies, whereas other technologies, such as space heating or cooling, which passively absorb the energy of the sun and have no moving components, are referred to as **passive solar** technologies. More sophisticated solar technologies exist for providing power for electricity generation. We will look at these briefly later in this fact sheet.

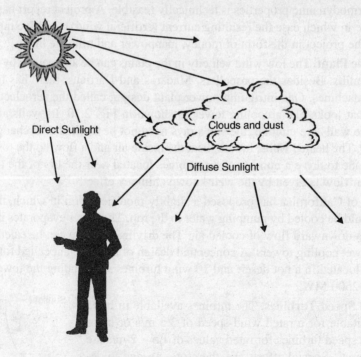

Fig. 2.11. Direct and Diffuse Solar Radiation.

Sun is the source of many forms of energy available to us. Do you know how energy is obtained from the sun? The most abundant element in sun is hydrogen. It is in a plasma state. This hydrogen at high temperature, high pressure and high density undergoes nuclear fusion and hence releases an enormous amount of energy. This energy is emitted as radiations of different forms in the electromagnetic spectrum.

Out of these X-rays, gamma rays and most of ultraviolet rays do not pass through the earth's atmosphere. But heat energy and light energy are the main radiations that reach the earth. This energy is the basis for the existence of life on earth.

Sun is a sphere of intensely hot gaseous matter with a diameter of $1.39e^9$ m and $1.5e^{11}$ m away from earth. Sun has an effective black body temperature of 5762 K and has a temperature of $8e^6$ K to $40e^6$ K. The sun is a continuous fusion reactor in which hydrogen (4 protons) combines to form helium (one He nucleus). The mass of the He nucleus is less than that of the four protons, mass having been lost in the reaction and converted to energy. The energy received from the sun on a unit area perpendicular to the direction of propagation of radiation outside atmosphere is called solar constant, and has a value 1353 Wm^{-2}. This radiation when received on the earth has a typical value of 1100 Wm^{-2} and is variable. The wavelength range is 0.29 to 2.5 micro meters. This energy is typically converted into usual energy form through natural and man-made processes. Natural processes include wind and biomass. Man-made processes include conversion into heat and electricity.

2.16.1 SOLAR RADIATIONS

Radiation from sun on entering the earth's atmosphere gets scattered by the atmospheric gas molecules and dust particles and received on earth from all directions and is called diffuse radiation. The portion of radiation received on earth from sun without change in original quality is called beam or direct radiation.

The earth revolves about the sun in an approximately circular path, with the sun located slightly off center of the circle. The earth's axis of rotation is tilted 23.5 degrees with respect to its pane of revolution about the sun, the position of the earth relative to the sun's rays at the time of winter solstice when the North Pole is inclined 23.5 degree away from the sun. All points on the earth's surface north of 66.5 N latitude are in total darkness while all regions within 23.5 degree of the South Pole receive continuous sunlight. At the time of the summer solstice, the situation is reversed. At the time of the two equinoxes, both poles are equidistant from the sun and all points on the earth's surface have 12 hours of daylight and 12 hours of darkness. The sun's ray passing through the center of the earth lies in the equatorial plane at the time of equinoxes. From vernal equinox to autumnal equinox, the rays lie north of the equatorial plane. From autumnal equinox to vernal equinox, the rays lie south of the equatorial plane. The average direction of the sun's rays for the entire year lies in the equatorial plane. Accordingly to intercept maximum amount of solar energy over the whole year, a solar collector in the northern hemisphere should be tilted and face due south.

The Nature and Availability of Solar Radiation. Solar radiation arrives on the surface of the earth at a maximum power density of approximately 1 kilowatt per metre squared (kWm^{-2}). The actual usable radiation component varies depending on geographical location, cloud cover, hours of sunlight each day, etc. In reality, the solar flux density (same as power density) varies between 250 and 2500 kilowatt hours per metre squared per year ($kWhm^{-2}$ per year). As might be expected the total solar radiation is highest at the equator, especially in sunny, desert areas. Solar radiation arrives at the earth's outer atmosphere in the form of a direct beam. This light is then partially scattered by cloud, smog, dust or other atmospheric phenomenon. We therefore receive solar radiation either as direct radiation or scattered or diffuse radiation, the ratio depending on the atmospheric conditions. Both direct and diffuse components of radiation are useful, the only distinction between the two being that diffuse radiation cannot be concentrated for use.

Solar radiation arriving from the sun reaches the earth's surface as short wave radiation. All of the energy arriving from the sun is eventually re-radiated into deep space otherwise the temperature of the earth would be constantly increasing. This heat is radiated away from the earth as long-wave radiation. The art of extracting the power from the solar energy source is based around the principle of capturing the short wave radiation and preventing it from being reradiated directly to the atmosphere. Glass and other selective surfaces are used to achieve this. Glass has the ability to allow the passage of short wave radiation whilst preventing heat from being radiated in the form of long wave radiation. For storage of this trapped heat, a liquid or solid with a high thermal mass is employed. In a water heating system this will be the fluid that runs through the collector, whereas in a building the walls will act as the thermal mass. Pools or lakes are sometimes used for seasonal storage of heat.

2.16.2 SOLAR THERMAL POWER PLANT

In the solar power plant, solar energy is used to generate electricity. Sunrays are focused using concave reflectors on to copper tubes filled with water and painted black outside. The water in the tubes then boils and become steam. This steam is used to drive steam turbine, which in turn causes the generator to work. A plant using this principle is working on experimental basis in Gurgaon in Haryana. Its capacity is 500 kilowatt. Another plant of similar type is being constructed in Jodhpur in Rajastan.

Many power plants today use fossil fuels as a heat source to boil water. The steam from the boiling water rotates a large turbine, which activates a generator that produces electricity. However, a new generation of power plants, with concentrating solar power systems, uses the sun as a heat source. There are three main types of concentrating solar power systems: parabolic-trough, dish/engine, and power tower.

Parabolic-trough systems concentrate the sun's energy through long rectangular, curved (U-shaped) mirrors. The mirrors are tilted toward the sun, focusing sunlight on a pipe that runs down the center of the trough. This heats the oil flowing through the pipe. The hot oil then is used to boil water in a conventional steam generator to produce electricity.

A dish/engine system uses a mirrored dish (similar to a very large satellite dish). The dish-shaped surface collects and concentrates the sun's heat onto a receiver, which absorbs the heat and transfers it to fluid within the engine. The heat causes the fluid to expand against a piston or turbine to produce mechanical power. The mechanical power is then used to run a generator or alternator to produce electricity.

A power tower system uses a large field of mirrors to concentrate sunlight onto the top of a tower, where a receiver sits. This heats molten salt flowing through the receiver. Then, the salt's heat is used to generate electricity through a conventional steam generator. Molten salt retains heat efficiently, so it can be stored for days before being converted into electricity. That means electricity can be produced on cloudy days or even several hours after sunset.

'Solar Power Tower' Power Plant The first is the 'Solar Power Tower' design which uses thousands of sun-tracking reflectors or heliostats to direct and concentrate solar radiation onto a boiler located atop a tower. The temperature in the boiler rises to 500 – 7000°C and the steam raised can be used to drive a turbine, which in turn drives an electricity producing turbine. There are also called central Receiver Solar Power Plants.

It can be divided into solar plant and conventional steam power plant. The flow diagram is given in Fig. 2.12.

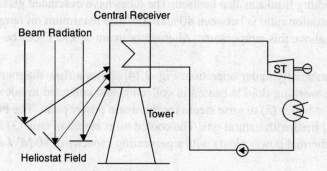

Fig. 2.12. Central Receiver Solar Power Plant.

A heliostat field consists of a large number of flat mirrors of 25 to 150 m² area which reflects the beam radiations onto a central receiver mounted on a tower. Each mirror is tracked on two axis. The absorber surface temperature may be 400 to 1000°C. The concentration ratio (total mirror area divided by receiver area) may be 1500. Steam, air or liquid metal may be used as working fluid. Steam is raised for the conventional steam power plant.

'Distributed (Parabolic) Collector System' Power Plant. The second type is the distributed collector system. It is also called solar farm power plant as a number of solar modules consisting of parabolic trough solar collectors are interconnected. This system uses a series of specially designed 'Trough' collectors which have an absorber tube running along their length. Large arrays of these collectors are coupled to provide high temperature water for driving a steam turbine. Such power stations can produce many megawatts (mW) of electricity, but are confined to areas where there is ample solar insulation.

Every module consists of a collector as shown in Figs. 2.13 and 2.14. It is rotated about one axis by a sun tracking mechanism. Thermo-oil is mostly used as heating fluid as it has very high boiling

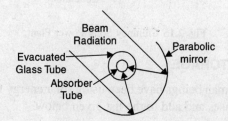

Fig. 2.13. Distributed (Parabolic) Solar Collector.

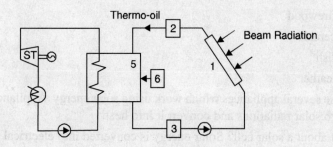

Fig. 2.14. Distributed (Parabolic) Trough Solar Power Plant.

point. Water/steam working fluid can also be used. The tubes have evacuated glass enclosure to reduce the losses. The concentration ratio is between 40 and 100. The maximum oil temperature is limited to 400°C as oil degrades above this temperature. Alternately steam at 550°C can be directly generated in the absorber tube.

These are commercially under operation. Fig. 2.14. shows a flow diagram of parabolic trough solar power plant. The working fluid is heated in collectors and collected in hot storage tank (2). The hot thermo-oil is used in boiler (5) to raise steam for the steam power plant. The boiler also is provided with a back-up unit (6) fired with natural gas. The cooled oil is stored in tank (3) and pumped (4) back to collector (1). Solar thermal power plants with a generating capacity of 80 MW are functioning in the USA.

Solar Chimney Power Plant. The air stream is heated by solar radiation absorbed by the ground and covered by a transparent cover. The hot air flow through or chimney which gives the air a certain velocity due to pressure drop caused by the chimney effect. The hot air flows through an air turbine to generate power.

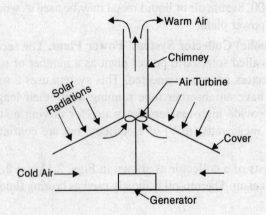

Fig. 2.15 Chimney Solar Power Plant.

2.16.3 SOLAR ENERGY STORAGE

It is well known that human beings have been using solar energy for different uses, from ancient days. Find examples of these uses and add to the list given below.

1. To get salt from sea water.
2. To dry wet clothes
3. To dry firewood
4. To dry cereals
5. To dry fish
6. To dry leather

We now use several appliances which work using solar energy. Appliances like solar cooker and solar heater absorb solar radiations and convert it into heat.

Then what about a solar cell? Solar energy is converted into electrical energy and it is directly used or stored in a battery.

There are eight possible pathways for conversion of solar radiation to useful energy. Solar thermal conversion method converts radiation to heat using solar flat collectors. Solar thermo chemical conversion method converts radiation to heat and produce steam then to kinetic energy using a pump or turbine. Solar thermal electric conversion method converts radiation to steam and to kinetic and electrical energy through a turbine and generator to electrical energy. The above route through a further electrolysis process gives chemical energy (H_2 fuel). A high temperature catalytic conversion process produces chemical energy (H_2 fuel) directly. Photovoltaic conversion of solar radiation gives direct electrical energy. Photosynthesis process produces chemical energy directly from radiation. Chemical energy (H_2 fuel) is directly produced from solar radiation using the electricity produced by the photovoltaic method. A few of these methods are dealt in detail further.

Commercial and industrial buildings may use the same solar technologies photovoltaic, passive heating, day lighting, and water heating that are used for residential buildings. These nonresidential buildings can also use solar energy technologies that would be impractical for a home. These technologies include ventilation air preheating, solar process heating and solar cooling.

Many large buildings need ventilated air to maintain indoor air quality. In cold climates, heating this air can use large amounts of energy. A solar ventilation system can preheat the air, saving both energy and money. This type of system typically uses a transpired collector, which consists of a thin, black metal panel mounted on a south-facing wall to absorb the sun's heat. Air passes through the many small holes in the panel. A space behind the perforated wall allows the air streams from the holes to mix together. The heated air is then sucked out from the top of the space into the ventilation system.

Solar process heating systems are designed to provide large quantities of hot water or space heating for nonresidential buildings. A typical system includes solar collectors that work along with a pump, a heat exchanger, and/or one or more large storage tanks. The two main types of solar collectors used an evacuated tube collector and a parabolic trough collector can operate at high temperatures with high efficiency. An evacuated-tube collector is a shallow box full of many glass, double-walled tubes and reflectors to heat the fluid inside the tubes. A vacuum between the two walls insulates the inner tube, holding in the heat. Parabolic troughs are long, rectangular, curved (U-shaped) mirrors tilted to focus sunlight on a tube, which runs down the center of the trough. This heats the fluid within the tube.

The heat from a solar collector can also be used to cool a building. It may seem impossible to use heat to cool a building, but it makes more sense if you just think of the solar heat as an energy source. Your familiar home air conditioner uses an energy source, electricity, to create cool air. Solar absorption coolers use a similar approach, combined with some very complex chemistry tricks, to create cool air from solar energy. Solar energy can also be used with evaporative coolers (also called "swamp coolers") to extend their usefulness to more humid climates, using another chemistry trick called desiccant cooling.

SPACE HEATING

In colder areas of the world (including high altitude areas within the tropics) space heating is often required during the winter months. Vast quantities of energy can be used to achieve this. If buildings are carefully designed to take full advantage of the solar insolation which they receive then much of the heating requirement can be met by solar gain alone. By incorporating certain simple design principles a new dwelling can be made to be fuel efficient and comfortable for habitation. The bulk of these technologies are architecture based and passive in nature. The use of building materials with a high thermal mass (which stores heat), good insulation and large glazed areas can increase a buildings capacity to capture and store heat from the sun. Many technologies exist to assist with diurnal heating needs but seasonal storage is more difficult and costly.

For passive solar design to be effective certain guidelines should be followed:
1. A building should have large areas of glazing facing the sun to maximise solar gain
2. Features should be included to regulate heat intake to prevent the building from overheating
3. A building should be of sufficient mass to allow heat storage for the required period
4. Contain features which promote the even distribution of heat throughout the building.

One example of a simple passive space heating technology is the Trombe wall. A massive black painted wall has a double glazed skin to prevent captured heat from escaping. The wall is vented to allow the warm air to enter the room at high level and cool air to enter the cavity between the wall and the glazing. Heat stored during the wall during the day is radiated into the room during the night. This type of technology is useful in areas where the nights are cold but the days are warm and sunny.

SPACE COOLING

The majority of the worlds developing countries, however, lie within the tropics and have little need of space heating. There is a demand, however, for space cooling. The majority of the worlds warm-climate cultures have again developed traditional, simple, elegant techniques for cooling their dwellings, often using effects promoted by passive solar phenomenon. There are many methods for minimising heat gain. These include siting a building in shade or near water, using vegetation or landscaping to direct wind into the building, good town planning to optimise the prevailing wind and available shade. Buildings can be designed for a given climate domed roofs and thermally massive structures in hot arid climates, shuttered and shaded windows to prevent heat gain, open structure bamboo housing in warm, humid areas. In some countries dwellings are constructed underground and take advantage of the relatively low and stable temperature of the surrounding ground. There are as many options as there are people.

2.16.4 RECENT DEVELOPMENTS IN SOLAR POWER PLANTS

Solar Thermal Applications. The applications include water heating for domestic, commercial and industrial use, space heating and drying, solar distillation, solar cooling through absorption & adsorption cycles, solar water pumping and solar power generation.

Solar Photovoltaics. Photovoltaic (PV) or solar cells refers to the creation of voltage from light. A solar cell is a converter; it changes the light energy into electrical energy. A cell does not store any energy, so when the source of light (typically the sun) is removed, there is no electrical current from the cell. If electricity is needed in the night, a battery must be included in the circuit. There are many materials that can be used to make solar cells, but the most common is the element silicon. A typical solar cell is 3-6 inches in diameter and are now available in various shapes like circular, square, etc. The conversion processes occurs instantly whenever there is light falling on the surface of a cell. And the output of the cell is proportional to the input light.

2.17 ELECTROCHEMICAL EFFECTS AND FUEL CELLS

Fuel cells produce electricity from an electrochemical reaction between hydrogen and oxygen. Fuel cells are efficient, environmentally benign and reliable for power production. The use of fuel cells has been demonstrated for stationary/portable power generation and other applications.

Fuel cell is device that converts the chemical energy stored in a fuel *directly* to electrical energy. Fuel cell is an *open system*.

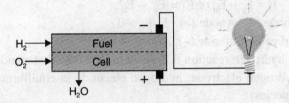

Fig. 2.16

Some similarity to a battery except that energy must be stored or built into a battery. Batteries are *closed systems*.

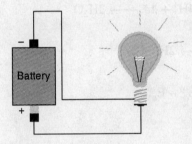

Fig. 2.17

There are two Principles of the fuel cell

1. Chemo electricity
   — Chemistry must occur before energy flows
   — F/C system like an entire chemical plant
2. Match Energy Source to Application
   — Stationary/Vehicle/Portable
   — Sometimes F/Cs won't work.

2.17.1 REVERSIBLE CELLS

Since oxidation and reduction are physically separate, two things must happen to complete the reaction in a fuel cell:

1. Ions travel through electrolyte:
   — acidic f/c: cations to cathode
   — alkaline f/c: anions to anode
2. Electrons travel anode to cathode electrons fall through "potential gradient" and thus do

Cathode: electrode to which cations migrate

Anode: electrode to which anions migrate

Mnemonic

— Reduction occurs at the cathode (redcats)

— Oxidation occurs at the anode

Cell Potential: Cell potential E or E_o is the difference between the cathode potential E_c and the anode potential E_a.

$$E = E_c - E_a \text{ or } E^\circ = E_c^\circ - E_a^\circ$$

In Fuel cells: Cathode (+); Anode (–) ; E, $E_o > 0$

In Electrochemical cells: Cathode (–); Anode (+); E, $E^\circ < 0$

By definition, the hydrogen reaction is defined to be 0.000 V at standard conditions

she = standard hydrogen electrode; hydrogen electrode in equilibrium at standard conditions (298 K, unit activity of species)

In Reversible Cell

For H_2/O_2 fuel cell:

Cathode (reduction) : $\quad E^\circ_{she}/v$

$$O_2 + 4H^+ + 4e^- \longrightarrow 2H_2O$$

Anode (oxidation):

$$H_2 \longrightarrow 2H^+ + 2e^-$$

Reversible cell potential

$$E^\circ = E_c^\circ - E_a^\circ$$

2.17.2 IDEAL FUEL CELLS

Over Potentials. In a real process the electrodes cannot operate at their equilibrium potentials. Nonidealities in real processes lead to efficiency losses or resistances to the process.

Electrodes must shift to potentials more favorable for oxidation or reduction to overcome efficiency losses. These shifted potentials are called overpotentials.

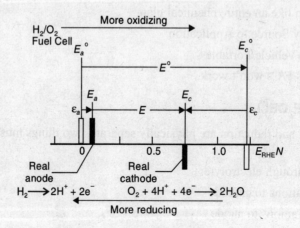

Fig. 2.18

In a fuel cell the chemical energy of the fuel drives the overpotential, which in turn drives the reaction.

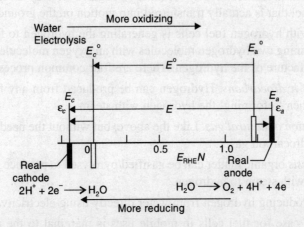

Fig. 2.19

Ideal Fuel Cell Potentials. Overpotentials always reduce fuel cell potentials, so that less voltage is delivered per electron transferred.

Conversely, over potentials always increase electrochemical cell potentials so that more voltage is required per electron transferred.

Ideal Cell Potentials

$$E = E_c - E_a = (E_{eq, c} - e_c) = (E_{eq, a} + e_a)$$

Ideal fuel cell Energy Conversion

1. Reactant/product transport
3. Ion transport through e-lyte
4. Electron transport
2. Reaction at electrocatalyst.

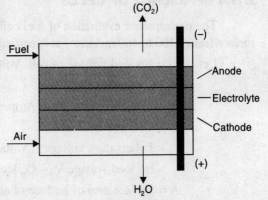

Fig. 2.20

2.17.3 OTHER TYPES OF FUEL CELLS

Fuel cells are a means of converting a fuel to electrical energy using an electrochemical membrane. The most popular to date has been the proton exchange hydrogen fuel cell. It takes two molecules of hydrogen and one molecule of oxygen and produces two molecules of water leaving behind four spare electrons to generate an electric current. In terms of the energy value of the hydrogen, the conversion process is around 75% to 80% efficient.

Some fuel cells use other chemical fuels as a source of hydrogen such as methanol, which is processed into hydrogen for the use by the fuel cell.

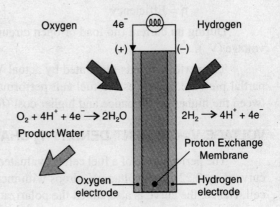

Fig. 2.21

Although this means that the system doesn't have to store large quantities of highly explosive hydrogen, it does reduce the efficiency of the electricity generation process to 30% or 40%. This is still more efficient than burning the methanol directly combustion engines are only around 20% efficient in terms of the energy of the fuel that is actually transferred into motion on the ground.

The problem with hydrogen fuel cells is generating the hydrogen to fuel them. To get those 4 electrons out by combining two hydrogen molecules with an oxygen molecule, you have to put them in at the point you manufacture of the hydrogen. There are four common processes:

Reformation of hydrocarbons. Hydrogen can be produced from any fossil fuel, such as oil or coal, by heating and then 'reforming' the hydrogen with steam.

Steam reformation of natural gas. Like the above, but without the need to initially turn the solid hydrocarbons into hydrocarbon gases.

Biomass pyrolysis organic matter can be gasified/pyrolysed to produce hydrogen rich gases than can then be reformed with steam to hydrogen.

Electrolysis. Producing hydrogen from water directly using electricity.

In general, the case for fuel cells in mobile uses is marginal to the use of other fuels. They potentially have an application in balancing out the variations from certain forms of renewable energy such as wind or tidal power.

2.17.4 EFFICIENCY OF CELLS

The performance evaluation of fuel cells is represented in terms of current density at electrode surface (range 100 to 400 mA/cm^2) at specified temperature and reactant partial pressures and voltage.

Let, V_o = No load voltage of cell, Volts, DC

V_c = Cell voltage on load

I_c = Cell current on load, Ampere

P_c, = Cell power, Watts

V_p = Polarization voltage = Voltage drop in the cell?

= No load voltage V_o – O_n load voltage V,

A = Surface area of on face of an electrode, m

I_d = Current density of cell, = I_c/A.... A/m....

η = Efficiency

During no current (no load or open circuit), the cell voltage is maximum and is called no load voltage (V_o).

The performance is illustrated by actual V_c vs. I_d curve. Increase in operating temperature and partial pressure, improves the fuel cell performance (increase in V_c and P_c). There is a trade-off between the higher performance and higher cost (for high temperature, pressure design).

VOLTAGE V_c-CURRENT DENSITY I_d CHARACTERISTIC (POLARIZATION CURVE)

The performance of a fuel cell is evaluated by the cell voltage V_c vs. electrode current density Id curve (Fig. 2.22). Cell voltage V_c drops with increase in current density due to polarization within the cell. Hence, the curve is also called the polarization curve of the fuel cell.

Polarization is internal chemical, electrical, thermal effect within the fuel cell resulting in inefficiencies. Polarization the cause of internal energy loss and is measured by in terms of polarization voltage V_p.

$V_p = v_o - v_c$

V_P = Polarization voltage of the cell = Voltage Drop

= No load voltage V_o – On load voltage V.

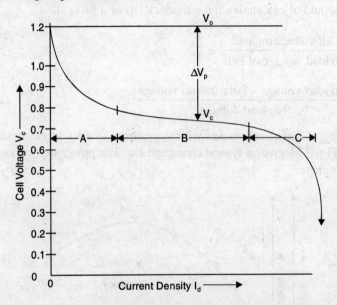

Fig. 2.22. Polarisation curve.

As the load on the cell ($I_s \times A = I$) increases, the internal electro-chemical reactions trying to oppose the cause also increase. The internal losses increase and the terminal voltage V_c drops. These internal losses and inefficiencies increasing with current are called Polarization. The drop in voltage V_p is called Polarization Voltage VP.

Power per Cell P_c,

Power = Voltage × Current

$P_c = V_c \times 4$

The power of a cell increase with the increase in current density, and reaches a saturation point at due to polarization effects.

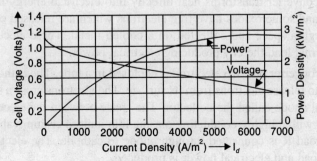

Fig. 2.23. Power-Current Density Curve.

Input power – Polarization losses = Output power

and, Output power/Input power = Efficiency(η)

Efficiency of Cells. In terms of the final energy output these options are only 40% to 60% efficient — the exception being electrolysis which can be up to 80% efficient. The argument of fuel cell advocates is that this cycle still represents an improvement for cars. The overall efficiency is something like 40% if running fuel rather than 20% if running an internal combustion engine.

The simple method of calculating the efficiency (η) of a fuel cell is

$$\eta = \frac{\text{Cell voltage on load}}{\text{No load voltage of cell}} = \frac{V_c}{V_o}$$

$$\eta = \frac{\text{No load voltage} - \text{Polarization voltage}}{\text{No load voltage}}$$

The efficiency of a fuel cell varies with the current density at electrode surface due to the Polarization Effect. Fig. 2.24 gives a typical characteristic. The power loss is converted to waste heat and released to atmosphere.

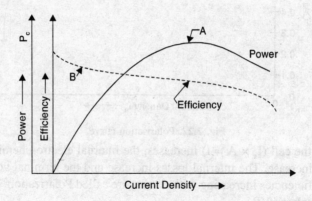

Fig. 2.24. Power and Efficiency Curve.

After reaching saturation level, the and power per cell starts decreasing. The losses increase and are converted to waste heat.

2.18 THERMIONIC SYSTEMS AND THENNIONIC EMISSION

A thermionic converter transforms heat directly into electrical energy by utilizing thermionic emission. All metals and some oxides have free electrons which are released on heating. These electrons can travel through a space and collected on a cooled metal. These electrons can return to hot metal through an external load thereby producing electrical power.

A thermionic converter has two elec-trodes enclosed in a tube. The cathode is called an emitter and is heated enough to release electrons from its surface. The electrons cross a small gap and accumulate on a cooled metal anode called the collector. The space between the electrodes is maintained at high vacuum or filled with a highly conducting plasma like ionised cesium vapour to minimal energy losses. The external load R is connected through anode to cathode. The electrons return to cathode through the external load and electrical power is produced.

Thermionic conversions is a sealed and evacuated device comprising of
1. A heated cathode (electron emitter)
2. An anode (electron collector)
3. Vacuum gap between 1 and 2 (with ionised vapour to neutralise space charge).

The gap is only about a mm. External electrical circuit is connected between anode and cathode. (Ref. Fig. 2.25). The thermionic converter converts thermal energy directly to electrical energy by virtue of flow of electrons through the vacuum gap.

Heat is supplied to emitter. Electrons released from emitter flow through small vacuum gap seeded with ionising substance. Heat is rejected from collector. Electrical energy is tapped from the terminals.

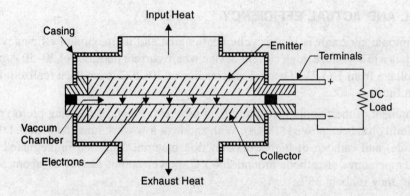

Fig. 2.25. Principle of Thermoionic Converter (Generator).

2.18.1 THERMOIONIC CONVERSION

The emission of an electron from a metal surface is opposed by a potential barrier equal to the difference between the energies of an electron outside and inside the metal. Therefore, a certain amount of energy has to be spent to release the electron from the surface. This energy is called surface work function (Φ).

The maximum electron current per unit area emitted from the surface is given by the following Richardson Dushman equation:

$$J = A_1 T^2 e^{\left(-\frac{\sigma}{kT}\right)}$$

where
J = current density, [A/m^2]
T = Temperature, [K]
Φ = work function, [eV]
$K = 1.38 \times 10^{-23}$ J/molecule K
= Boltzmann constant
$A_1 = 120$ A/cm^2-K^2) = Emission constant.

The kinetic energy of the free electrons at absolute zero would occupy discrete energy levels from zero upto some maximum value defined by the Fermi energy level, ε_f. Each energy level contains a limited number of free electrons.

Above absolute zero temperature, some electrons may have energies higher than the Fermi level. The energy that must be supplied to overcome the weak attrac-tive force on the outermost orbital electrons is the work function, Φ, so that the electron leaving the emitter has an energy level $\Phi + \varepsilon_f$. When emitter is heated, some high energy free electrons at the Fermi level receive energy equal to emitter work function Φ_c, and escape the emitter surface. They move through the gap and strike the collector. The K.E. (ε_{fa}) plus the energy equal to collector work function Φ_a is given up and this energy is rejected as heat from the low temperature collector.

The electron energy is reduced to the Fermi energy level of the anode ε_{fa} This energy state is higher than that of the electron at the Fermi energy level of cathode ε_{fc}. Therefore, the electron is able to pass through the external load from anode to cathode. The cathode materials are selected with low Fermi levels as comprised to anode materials which must have higher Fermi level.

2.18.2 IDEAL AND ACTUAL EFFICIENCY

A thermionic generator is like a cyclic heat engine and its maximum efficiency is limited by Carnot's law. It is a low-voltage, high current device where current densities of 20–50 A/cm^2 have been achieved at voltage from 1 to 2V. Thermal efficiencies of 10–20% have been realized. Higher values are possible in future.

Development of thermionic generators is in progress. Practical working prototypes have been built and feasibility has been proved (1980s). With anode of low work function material (barium oxide, stron-tium oxide) and cathode of high work-function material (tungsten impregnated barium compound), and temperatures at cathode around 2000°C, power output density of about 6 W/m can be achieved. Efficiency is about 35%.

The positively charged cathode tends to pull the emitted electrons back. The electrons already in the gas exert a retarding force on the electrons trying to cross the gap. This pro-duces a space charge barrier Fig. 2.26 shows the characteristic curve of a thermionic generator with an interspace retarding potential equivalent to S volts above the anode work function Φ_a.

Potential barrier,

$$V_c > \Phi_c \text{ and } V_c > \Phi_a$$

The current densities are :

$$J_c = A_1 T_c^2 e^{\left(-\frac{V_c}{kT_c}\right)} \text{ [A/cm}^2\text{]}$$

$$J_a = A_1 T_c^2 e^{\left(-\frac{V_c}{kT_2}\right)} \text{ [A/cm}^2\text{]}$$

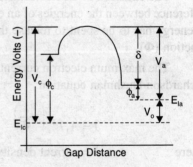

Fig. 2.26. Characteristic Curve.

The output voltage across the electrical resistance R,

$$V_o = V_c - V_a = \phi_c - \phi_a = \frac{1}{e}(\varepsilon_{fa} - \varepsilon_{fc})$$

Each electron has to overcome the interspace potential ($V_c - \Phi_c$) and work function Φ_c when it leaves the cathode. The net energy carried,

$$Q_{1c} = J_c (V_c - \phi_c + \phi_c) = J_c V_c \text{ [W/cm}^2\text{]}$$

Each electron also carries away its K.E. which is equal to $2KT_c$, i.e.,

$$Q_{2c} = J_c \frac{2KT_c}{e} \quad [\text{W/cm}^2]$$

The back emission from the anode must similarly carry energy to the cath-ode. The net rate of energy supply to the cathode,

$$Q_1 = J_c \left(V_c + \frac{2KT_c}{e}\right) - J_a \left(V_a + \frac{2KT_c}{e}\right)$$

where $e = 1.602 \times 10^{-19}$ coulomb.

The power output of the generator

$$W = V_o (J_c - J_a)$$

The thermal efficiency of the thermionic generator

$$\eta = \frac{V_o(J_c - J_a)}{J_c \left(V_c + \frac{2KT_c}{e}\right) - J_a \left(V_a + \frac{2KT_e}{e}\right)}$$

Now $V_o = V_c - V_a$ substituting the following values.

$$\frac{V_c}{KT_c} = \beta_c \; ; \quad \frac{V_a}{KT_a} = \beta_a \text{ and } \frac{T_a}{T_c} = \theta.$$

$$\eta = \frac{(\beta_c KT_c - \beta_a KT_a)(J_c - J_a)}{J_c \left(\beta_c KT_c + \frac{2KT_c}{e}\right) - J_a \left(\beta_a KT_a + \frac{2KT_e}{e}\right)}$$

$$= \frac{(\beta_c - \theta\beta_a)[1 - \theta^2 e^{(\beta_c - \beta_a)}]}{(\beta_c + 2) - \theta^2 (\beta_2 + 2\theta) e^{(\beta_c - \beta_e)}}$$

It is found that for all values of θ, the efficiency curve peaks are very near to the value of $\beta_a = \beta_c$

If $\beta_a = \beta_c$

$$\eta_{max} = [1 - \theta] \frac{\beta}{\beta + 2} \left[\frac{1 - \theta^2}{1 - \frac{\theta^2 (\beta + 2\theta)}{\beta + 2}}\right]$$

$$\frac{1 - \theta^2}{1 - \frac{\theta^2 (\beta + 2\theta)}{\beta + 2}} = 1$$

$$\eta_{max} = (1 - \theta) \frac{\beta}{\beta + 2}$$

Carnot efficiency

$$\eta_c = 1 - \frac{T_a}{T_c} = (1 - \theta)$$

If $\beta = 18$,

$$\eta_{max} = 0.9(1 - \theta)$$

η_{max} occurs when $\beta_a = \beta_c$

or

$$\frac{V_c}{T_c} = \frac{V_a}{T_a}.$$

2.19 THERMOELECTRIC SYSTEMS

A loop of two dissimilar metals develops an e.m.f. when the two junctions of the loop are kept at different temperatures. This is called Seebeck effect. This effect is used in a thermocouple to measure temperature.

Thermoelectric generator is a device which directly converts heat energy into electrical energy using the Seebeck thermoelectric effect. The device is very simple but thermal efficiency is very low of the order of 3%. Efficiency of thermoelectric generator depends upon the temperature of hot and cold junctions and the material properties. The semiconductor materials have more favourable properties which can withstand high temperatures and can give reasonable efficiency. The probability of developing peak load power stations of the order of 100 mW working at 20 percent thermal efficiency is high. Where cheap fuels are available thermoelectric generators can be developed for base load and standby power generation also. Another important application is the use of radioactive decay heat to generate power in space and other remote locations. The use of solar energy to supply heat for generating electricity can be an attractive application of thermoelectric devices if high efficiency materials can be developed.

2.19.1 PRINCIPLE OF WORKING

The operation of a thermoelectric generator is shown in Fig. 2.27. The net useful power output is given by

$$W = I^2 R \ [W]$$

where I = current [A]

R = External load resistance [Ω]

The current in the circuit is given by

$$I = \frac{\alpha \Delta T}{(R_i + R)} \ [A]$$

where α = Seeback coefficient (V/K)

ΔT = Temperature difference between hot and cold junctions [K]

R_i = Internal resistance of thermoelectric generator [Ω]

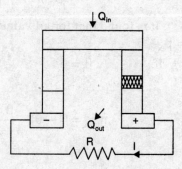

Fig. 2.27. Thermoelectric Generator.

The magnitude of potential difference depends on the pair of conductor materials and on the temperature difference between the junctions.

For a loop made of copper and constant wires, the value of Seebeck coefficient a is 0.04 mV/K. For a temperature difference of 600 K between the junctions, a voltage of 24 mV will be developed. In order to achieve higher potential difference many generators have to be connected in parallel.

For increasing the useful power output, parallel and series connections are used.

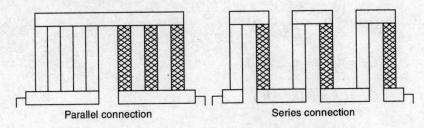

Fig. 2.28. Cascading of Thermoelectric Generators.

2.19.2 PERFORMANCE

The thermo-elements of a thermoelectric generator are made up of semiconduc-tors p and n type. Heat is supplied to the hot junction and from the cold junction heat is removed. Both the junctions are made of copper, see Fig. 2.29.

Let T_1 = source temperature [K]

T_o = sink temperature [K]

L_p, L_n = length of semiconductor elements [m]

A_p, A_n = cross-sectional area of thermoelectric elements [m^2]

k_p, k_n = thermal conductivity of elements [W/mK]

ρ_p, ρ_n = electric resistivity of elements [Ω-m]

k_p, k_n = thermal conductivity of elements [W/K] = $\dfrac{kA}{L}$

R_p, R_n = electrical resistance of elements [Ω] = $\dfrac{\rho L}{A}$

α_p, α_n = Seebeck coefficient [V/K]

π_p, π_n = Peltier coefficient [V]

Seebeck coefficient, $\alpha_{p,n} = \underset{\Delta T \to 0}{\text{Lt}} \dfrac{\Delta V}{\Delta T}$

Peltier heat, $\alpha_{p,n} = \pi_{p,n} I$

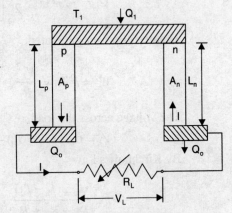

Fig 2.29. Circuit Diagram of Thermo-electric Power Generator.

When a current (I) flows through the junction of two elements, Peltier heat is produced. This is called Peltier effect. The Peltier coefficient.

$\pi_{p,n} = a_{p,n} \cdot I$

From Ist law of thermodynamics as applied to upper plate (as control volume), the temperature difference $(T_1 - T_o)$ will generate a Seebeck voltage, $a_{pn}(T_1 - T_o)$. There will be an electrical current I which will flow through the external load R_L.

The heat Q_1 will flow into hot junction and conducted into the two legs, Q_k. The Peltier heat Q_p will be produced at the junction due to current flowing through the circuit.

Joule heat $Q_1/2$ will flow into the junction. It is assumed that half Joulean heat appears at each junction. Heat balance at the junction will give,

$$\dot{Q}_1 = \frac{1}{2}\dot{Q}_j = \dot{Q}_p + \dot{Q}_k$$

where
$$\dot{Q}_p = \pi_{p,n} I = \alpha_{p,n} = I.T_1$$

$$\dot{Q}_j = I^2 R = I^2(R_p + R_n)$$

$$= I^2 \left[\frac{\rho_p L_p}{A_p} + \frac{\rho_n L_n}{A_n} \right]$$

$$\dot{Q}_k = k\Delta T = (k_p + k_n)(T_1 - T_0)$$

$$= \left(\frac{k_p A_p}{L_p} + \frac{k_n A_n}{L_n} \right)(T_1 - T_0)$$

Substituting the values into above equation

$$\dot{Q}_1 = \alpha_{p,n} I T_1 - \frac{1}{2}\left[\frac{\rho_p L_p}{A_p} + \frac{\rho_n L_n}{A_n}\right] I^2 + \left[\frac{k_p A_p}{L_p} + \frac{k_n A_n}{L_n}\right](T_1 - T_2)$$

The useful power generated,

$$W_L = I^2 R_L = \frac{V_L^2}{R_L}$$

The voltage across the load,

$$V_L = \alpha_{p,n}(T_1 = T_0) - I(R_p - R_n)$$

By Kirchhoff's Law

$$I = \frac{\alpha_{p,n}(T_1 - T_0)}{R_L + R_p + R_n}$$

Let $\quad m = \dfrac{R_L}{R_p + R_n}$ = Resistance ratio

$$\therefore \quad m + 1 = \frac{R_p + R_n + R_L}{R_p + R_n}$$

Now $\quad I = \dfrac{\alpha_{p,n}(T_1 - T_0)}{(R_p + R_n)(1 + m)}$

$$W_L = \frac{\alpha_{p,n}^2 (T_1 - T_0)^2}{(R_p + R_n)^2 (1+m)^2} (R_p + R_n)m$$

$$= \frac{m}{(1+m)^2} \cdot \frac{\alpha_{p,n}^2 (T_1 - T_0)^2}{R_p + R_n}$$

The required heat input,

$$\dot{Q}_1 = \alpha_{p,n}^2 \frac{T_1 (T_1 - T_0)}{(R_p + R_n)(1+m)} - \frac{1}{2} \frac{\alpha_{p,n}^2 (T_1 - T_0)^2}{(1+m)^2 (R_p + R_n)} + (R_p + k_n)(T_1 - T_0)$$

The efficiency of thermoelectric generator,

$$\eta = \frac{W_L}{\dot{Q}_1} = \frac{\dfrac{m}{(1+m)^2} \cdot \dfrac{\alpha_{p,n}^2}{R_p + R_n}(T_1 - T_0)^2}{\alpha_{p,n}^2 \dfrac{T_1(T_1 - T_0)}{(R_p + R_p)(1+m)} - \dfrac{1}{2} \dfrac{(T_1 - T_0)^2 \alpha_{p,n}^2}{(1+m)^2 (R_p + R_n)} + (k_p + k_n)(T_1 - T_0)}$$

$$= \frac{T_1 - T_0}{T_1} \cdot \frac{m}{(1+m)\dfrac{T_1 - T_0}{T_1} - \dfrac{1}{2} \dfrac{T_1 - T_0}{T_1} + \dfrac{(k_p + k_n)(R_n + R_p)(1+m)^2}{\alpha_{p,n}^2 T_1}}$$

Let the figure of merit,

$$Z = \frac{\alpha_{p,n}^2}{(k_n + k_p)(R_n + R_p)}$$

The figure of merit as defined above consists of the material properties of the two semiconductors. The n will increase with increase of Z.

Let $\qquad R = R_n + R_p$

and $\qquad K = k_n + k_p$

When the product (R, K) is minimum Z is maximum and therefore I is maximum.

$$(R, K) = \left(\frac{\rho_p L_p}{A_p} + \frac{\rho_n L_n}{A_n} \right) \left(\frac{k_p A_p}{L_p} + \frac{k_n A_n}{L_n} \right)$$

For a given pair of elements,

$$\frac{d\dot{W}_L}{d_m} = 0 \text{ which gives } m = 1.$$

$$\therefore \quad \dot{W}_L = \frac{1}{4} \frac{\alpha_{p,n}^2 (T_1 - T_0)^2}{R_p + R_n}$$

Maximum efficiency corresponding to max power,

$$\eta_{\text{max, power}} = \frac{T_1 - T_0}{T_1} \cdot \frac{1}{2 - \frac{1}{2}\frac{T_1 - T_0}{T_1} + \frac{4}{ZT_1}}$$

2.20 GEO THERMAL ENERGY

We live between two great sources of energy, the hot rocks beneath the surface of the earth and the sun in the sky. Our ancestors knew the value of geothermal energy; they bathed and cooked in hot springs. Today we have recognized that this resource has potential for much broader application.

The term geothermal comes geo meaning earth and thermal meaning heat. Heat from the Earth, or geothermal from the Greek — Geo (Earth) + thermal (heat) — energy can be and already is accessed by drilling water or steam wells in a process similar to drilling for oil.

The core of the earth is very hot and it is possible to make use of this geothermal energy (in Greek it means heat from the earth). These are areas where there are volcanoes, hot springs, and geysers, and methane under the water in the oceans and seas. In some countries, such as in the USA water is pumped from underground hot water deposits and used to heat people's houses.

Geothermal energy is an enormous, underused heat and power resource that is **clean** (emits little or no greenhouse gases), **reliable** (average system availability of 95%), and **homegrown** (making us less dependent on foreign oil). The geothermal fields were first discovered in 1847 by William Bell Elliot, an explorer surveyor who was hiking in the mountains between Cloverdale and Calistoga, California, in search of grizzly bears. He discovered steam seeping out of the ground along a quarter of a mile on the steep slope of a canyon near colb Mountain, an extinct volcano, now known as the Geysers. The first application of geothermal energy was for space heating, cooking, and medicinal purposes.

The center of the earth is estimated at temperature up to 10,000 K due to decay process of radioactive isotopes. The total steady geothermal energy flow towards earth's surface is 4.2×10^{10} kW. But the average flow energy is only 0.063 W/m^2.

The utilization of geothermal energy for the production of electricity dates back to the early part of the twentieth century. For 50 years the generation of electricity from geothermal energy was confined to Italy and interest in this technology was slow to spread elsewhere. In 1943 the use of geothermal hot water was pioneered in Iceland.

The following general objectives of geothermal energy:

(1) Reduction of dependence on nonrenewable energy and stimulation of the state's economy through development of geothermal energy.

(2) Mitigation of the social, economic, and environmental impacts of geothermal development.

(3) Financial assistance to counties to offset the costs of providing public services and facilities necessitated by the development of geothermal resources within their jurisdictions.

(4) Maintenance of the productivity of renewable resources through the investment of proceeds from these resources.

2.20.1 HOT SPRINGS

Earth tremors in the early Cenozoic period caused the magma to come close to the earth's surface in certain places and crust fissures to open up. The hot magma near the surface thus causes active volcanoes and hot springs and geysers where water exists. It also causes steam to vent through the fissures.

The hot magma near the surface solidifies into igneous rock. The heat of the magma is conducted upward to this igneous rock. Ground water that finds its way down to this rock through fissures in it will be heated by the heat of the rock or by mixing with hot gases and steam emanating from the magma. The heated water will then rise convectively upward and into a porous and permeable reservoir above the igneous rock. A layer of impermeable solid rock that traps the hot water in the reservoir caps this reservoir. The solid rock, however, has fissures that act as vents of the giant underground boiler. The vents show up at the surface as geysers, fumaroles, or hot springs. The natural heat in the earth has manifested itself for thousands of years in the form of hot springs. A well taps steam from the fissure for use in a geothermal power plant.

Fig. 2.30. Hot Springs in Steamboat Springs, Nevada.

Geothermal power stations have been installed at a number of places around the world, where geothermal steam is available.

The share of geothermal produced electricity in the year 2000 is 0.3% of the total electricity produced in the world. In India, there are more than 300 hot water springs.

2.20.2 STEAM EJECTION

Hot water geothermal energy deposits are present in several locations around the earth. Underground water collects heat from surrounding hot rocks. Such hot water reserves are with small con-tent of steam. Rain water collected over the land areas of several hundreds of square kilometers percolates through the ground to the depths of 1 to 6 km where it is heated by thermal conduction from the surrounding hot rocks. The hot water moves upwards through the defects of restricted areas in the rocks. The 'defects' are of fractures and highly permeable portions in the rock. The hot water moves upwards to the surface with relatively little or no storage in between. If however a zone of geothermal energy deposits is covered by a impermeable rock with a few fractures or defects, the energy deposits will be stored under ground readily available for extraction.

The energy available in such deposits can be extracted by means of production wells drilled through the impermeable rocks. In hydro-geothermal energy deposits, the geothermal fluid are in form of geothermal brine, hot mineral water and steam. Steam deposits are very few in number.

However, largest geothermal energy reserve of petro-geothermal type and are called Hot Dry Rock type that is without underground water. HDR deposits have largest geothermal energy potential in the world.

Extraction of geothermal energy through these hot dry rocks requires injection of water into artificially created fractured rock cavities in hot dry rock and extraction of hot water and steam by means of production wells. Cold water is injected into the well by means of injection wells and hot water and steam is extracted by the production wells. Water injected into the well acts as a heat collecting and heat transporting medium. Cavity in hot dry rock acts like a boiler steam generator. Cold water is injected and hot water/steam is obtained.

2.20.3 SITE SELECTION

India has about 150 known geothermal sites having geothermal fluid of moderate and low temperature (< 160°C). The geothermal fields in India are in the form of hot water springs (40 to 98°C) and shallow water reservoir temperatures are less than 160°C.

The important hydro-geothermal resource locations are

Puga Hydro-Geothermal Field, Jammu and Kashmir.

West-Coast Hydro-Geothermal Field, Maharashtra, Gujarat.

Tattapani-Hydro-Geothermal Field, Madhya Pradesh.

Due to moderate and low temperatures of geothermal fluids, the prospects of geothermally electrical power plants in India are very low. However, Geothermal Hydrothermal Energy is likely to have several applications in the temperature range of 30°C to 190°C.

For site selection of geothermal energy, the following factor may be considered.

(1) Borax deposits present

(2) Some locations with water at 120°C at 200 to 500 m depths

(3) Na-Ca-Cl-SO_4 contents

(4) Some shallow depth reservoir with water at 80 to 110°C.

2.20.4 GEOTHERMAL POWER PLANTS

The first mechanical conversion was in 1897 when the steam of the field at Larderello, Italy, was used to heat a boiler producing steam which drove a small steam engine. The first attempt to produce electricity also took place at Larderello in 1904 with an electric generator that powered four light bulbs

Fig. 2.31. First Geothermal Power Plant, 1904 Lardarello, Italy.

Fig. 2.32. Modern Geothermal Power Plant.

(as shown in Fig. 2.31). This was followed in 1912 by a condensing turbine; and by 1914, 8.5 mW of electricity was being produced. By 1944 Larderello was producing 127 mW. The plant was destroyed near the end of World War II, but was fortunately rebuilt and expanded and eventually reached 360 mW in 1981.

Fig. 2.33. A Geothermal Power Plant at the Geysers.

In the United States, the first attempt at developing the geysers field was made in 1922. Steam was successfully tapped, but the pipes and turbines of the time were unable to cope with the corrosive and abrasive steam. The effort was not revived until 1956 when two companies, Magma Power and Thermal Power, tapped the area for steam and sold it to Pacific Gas and Electric Company. By that time stainless steel alloys were developed that could withstand the corrosive steam, and the first electric-generating unit of 11 mW capacity began operation in 1960. Since then 13 generally progressively larger units have been added to the system. The latest is a 109 mW unit that began operation in September 1982 and which brought the Geysers total capacity to 909 mW. Two more units are under construction and four more are planned, which will bring the total capacity to 1514 mW by the late 1980s.

Other electric-generating fields of note are in New Zealand (where the main activity at Wairakei dates back to 1958), Japan, Mexico (at Cerro Prieto), the Phillipines, the Soviet Union, and Iceland (a large space-heating program).

Future world projections for geothermal electric production, based on the decade of the 1970s, are 7 percent per year. In the last four years of that decade, however, the growth rate was 19 percent per year. In the United States, the projections are for growth between 13.5 and 22 percent per year through the 1980s, which is 2.5 to 4 times the 5.3 percent per year growth rate of the total electric-generating capacity. This includes the steam field at the Geysers and other fields of different types.

The U.S. Geological Survey predicts a U.S. potential from currently iden-tified sources to be around 23,000 mW of electric power and around 42×10 SkJ of space and process heat for 30 years with existing technology, and 72,000 to 127,000 mW of electricity and 144 to 294×10^{15} Btu of heat from unidentified sources. Areas of geothermal potential in the North American continent, Geysers Region in Northern California, the Imperial Valley in Southern California, and the Yellowstone Region in Idaho, Montana, and Wyoming.

Most power plants need steam to generate electricity. The steam rotates a turbine that activates a generator, which produces electricity. Many power plants still use fossil fuels to boil water for steam. Geothermal power plants, however, use steam produced from reservoirs of hot water found a couple of miles or more below the Earth's surface. There are three types of geothermal power plants: *dry steam, flash steam, and binary cycle.*

Let
$$\gamma_p = \frac{A_p}{L_p}$$

and
$$\gamma_n = \frac{A_n}{L_n}$$

$$\therefore \quad (R, K) = \left(\frac{\rho_p}{\gamma_p} + \frac{\rho_n}{\gamma_n}\right)(k_p \gamma_p + k_n \gamma_n)$$

$$= k_p \rho_p + k_n \rho_n \frac{\gamma_p}{\gamma_n} + k_n \rho_n \frac{\gamma_n}{\gamma_p} + k_p \rho_n$$

When materials are fixed, the values of $\gamma_n, \gamma_p, A_p/L_n$ are fixed. In order to minimize the product (R, K),

$$\frac{d(R, k)}{d\left(\frac{\gamma_n}{\gamma_p}\right)} = 0 = k_n \rho_p - k_p \rho_n \left(\frac{\gamma_n}{\gamma_p}\right)^{-2}$$

$$\frac{\gamma_n}{\gamma_p} = \sqrt{\frac{k_p \rho_n}{k_n \rho_p}}$$

$$(R, K)_{min} = k_p \rho_p + k_p \rho_n \left[\frac{k_n \rho_p}{k_p \rho_n}\right]^{1/2} + k_n \rho_p \left[\frac{k_p \rho_n}{k_n \rho_p}\right]^{1/2} + k_n \rho_n$$

$$= [\sqrt{\rho_p k_p} + \sqrt{\rho_n k_n}]^2$$

$$Z_{max} = \left[\frac{\alpha_{p,n}}{\sqrt{\rho_p k_p} + \sqrt{\rho_n k_n}}\right]^2$$

$$\eta = \frac{T_1 - T_0}{T_1} \cdot \frac{m}{(1+m) - \frac{1}{2}\frac{T_1 - T_0}{T_1} + \frac{(1+m)^2}{ZT_1}}$$

For a given pair of materials with Z maximum and T_1, T_o are fixed, n will depend upon optimum value of m.

$$\frac{d\eta}{dm} = 0$$

$$m_{opt} = \left[1 + Z\frac{T_1 + T_o}{2}\right]^{1/2} = m_o$$

$$m_o = 1 = \frac{Z}{m_o - 1} \cdot \frac{T_1 + T_o}{2}$$

$$\eta = \frac{T_1 + T_o}{T_1} \cdot \frac{m_o Z T_1}{(1+m_0)^2} \left[\frac{1}{1 - \frac{1}{2}\frac{ZT_1}{(1+m_o)^2} \cdot \frac{T_1 + T_o}{T_1} + \frac{ZT_1(1+m_o)}{(1+m_o)^2}} \right]$$

$$= \frac{(T_1 - T_o) m_0 Z}{(m_o + 1)^2 + (m_o + 1) ZT_1 - \frac{1}{2} Z(T_1 - T_o)}$$

$$= \frac{(T_1 - T_o) m_0 Z}{\frac{m_o + 1}{m_o - 1} + \frac{Z}{2}(T_1 + T_o) + (m_0 + 1)ZT_1 - \frac{1}{2} Z(T_1 - T_o)}$$

$$= \frac{(T_1 - T_o) m_o (m_o - 1)}{(m_o + 1)(T_1 + T_o) + 2(m_o^2 - 1)T_1 - (T_1 - T_o)(m_o - 1)}$$

$$= \frac{(T_1 - T_o) 2m_o (m_o + 1)}{2m_o T_1 \left[m_o + \frac{T_o}{1} \right]}$$

$$\eta = \frac{T_1 - T_0}{T_1} \cdot \frac{(m_0 - 1)}{m_0 + \frac{T_0}{T_1}}$$

$$\dot{W}_L = \frac{m}{(1+m)^2} \cdot \frac{\alpha_{p,n}^2 (T_1 - T_0)^2}{R_p + R_n}$$

Small-scale geothermal power plants (under 5 megawatts) have the potential for widespread application in rural areas, possibly even as distributed energy resources. Distributed energy resources refer to a variety of small, modular power-generating technologies that can be combined to improve the operation of the electricity delivery system.

Dry Steam Power Plant (Vapour Dominated System). Dry steam power plants draw from underground resources of steam. The steam is piped directly from underground wells to the power plant, where it is directed into a turbine/generator unit. There are only two known underground resources of steam in the United States: The Geysers in northern California and Yellowstone National Park in Wyoming, where there's a well-known geyser called Old Faithful. Since Yellowstone is protected from development, the only dry steam plants in the country are at The Geysers.

Power plants using dry steam systems were the first type of geothermal power generation plants built. They use the steam from the geothermal reservoir as it comes from wells, and route it directly through turbine/generator units to produce electricity. It is the rarest form of geothermal energy but the most suitable generation and the most developed of all geothermal resources or system.

Fig. 2.34 shows a schematic diagram of a dry steam power system also called vapour dominated system. Dry steam from the turbine at perhaps 200°C is used. It is near saturated at the bottom of the wall and may have a shut-off pressure up to 35 bar. Pressure drops through the well cause it to slightly superheated at the well head. An example of a dry steam generation operation is at the Geysers in northern California.

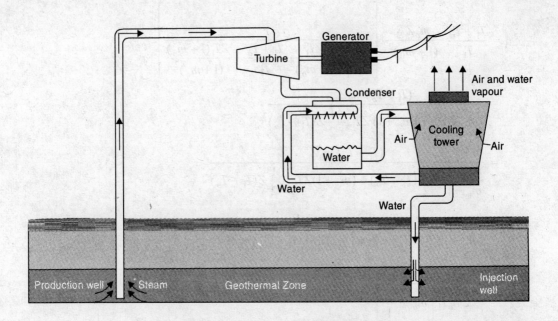

Fig. 2.34. Schematic of the Dry Steam Power Plant.

Flash Steam Power Plant (Liquid Domain System). Flash steam power plants are the most common. They use geothermal reservoirs of water with temperatures greater than 182°C. This very hot water flows up through wells in the ground under its own pressure. As it flows upward, the pressure decreases and some of the hot water boils into steam. The steam is then separated from the water and used to power a turbine/generator. Any leftover water and condensed steam are injected back into the reservoir, making this a sustainable resource.

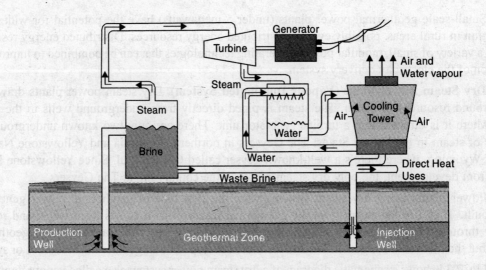

Fig. 2.35. Schematic of the Flash Steam Power Plant.

Flash-steam power plants built in the 1980s tapped into reservoirs of water with temperatures greater than 182°C. The hot water flows up through wells in the ground under its own pressure. As it flows upward, the pressure decreases and some of the hot water "flashes" into steam. The Geysers in northern California, which uses steam piped directly from wells, produces the world's largest single source of geothermal power.

Flash steam plants are the most common type of geothermal power generation plants in operation today. They use water at temperatures greater than 182°C that is pumped under high pressure to the generation equipment at the surface. Upon reaching the generation equipment the pressure is suddenly reduced, allowing some of the hot water to convert or "flash" into steam. This steam is then used to power the turbine/generator units to produce electricity. The remaining hot water not flashed into steam, and the water condensed from the steam is generally pumped back into the reservoir. An example of an area using the flash steam operation is the Cal Energy Navy I flash geothermal power plant at the Coso geothermal field.

Fig. 2.36. The Cal Energy Navy I Flash Geothermal Power plant at the Coso Geothermal Field.

Binary Cycle Power Plant (Liquid Dominatd Systems). Binary cycle power plants operate on water at lower temperatures of about 107°–182°C. These plants use the heat from the hot water to boil a working fluid, usually an organic compound with a low boiling point. The working fluid is vaporized in a heat exchanger and used to turn a turbine. The water is then injected back into the ground to be reheated. The water and the working fluid are kept separated during the whole process, so there are little or no air emissions.

Binary cycle power plant operates on water at lower temperatures of about 107 degrees Celsius to 182 degrees Celsius. These plants use the heat from the hot water to boil a fluid, usually an organic compound with a low boiling point.

Binary cycle geothermal power generation plants differ from Dry Steam and Flash Steam systems in that the water or steam from the geothermal reservoir never comes in contact with the turbine/generator units. In the Binary system, the water from the geothermal reservoir is used to heat another "working fluid" which is vaporized and used to turn the turbine/generator units. The geothermal water, and the "working fluid" are each confined in separate circulating systems or "closed loops" and never come in contact with each other. The advantage of the binary cycle plant is that they can operate with lower temperature waters (225°F–360°F), by using working fluids that have an even lower boiling point than water. They also produce no air.

Hybrid Geothermal Power Plant-Fossil System. The concept of hybrid geothermal-fossil-fuel systems utilizes the relatively low-tem-perature heat of geothermal sources in the low-temperature end of a conventional cycle and the high-temperature heat from fossil-fuel combustion in the high-temperature end of that cycle. The concept thus combines the high-efficiency of a high-temperature cycle with a natural source of heat for part of the heat addition, thus reducing the consumption of the expensive and nonrenewable fossil fuel.

There are two possible arrangements for hybrid plants. These are

(1) Geothermal preheat, suitable for low-temperature liquid-dominated systems, and

(2) Fossil superheat, suitable for vapor-dominated and high-temperature liquid-dominated systems.

Geothermal-Preheat Hydrid Systems. In these systems the low-temperature geothermal energy is used for feed water heating of an otherwise conventional fossil-fueled steam plant. Geothermal heat replaces some, or all, of the feed water heaters, depending upon its temperature. A cycle operating on this principle is illustrated in Fig. 2.37. As shown, geothermal heat heats the feed water throughout the low-temperature end prior to an open-type deaerating heater. The DA is followed by a boiler feed pump and three closed-type feed water heaters with drains cascaded backward. These receive heat from steam bled from higher-pressure stages of the turbine. No steam is bled from the lower-pressure stages because geothermal brine fulfills this function.

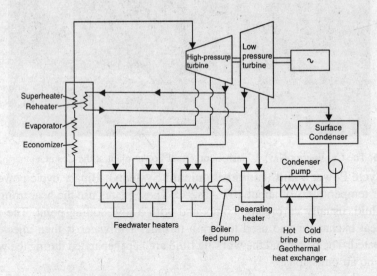

Fig 2.37. Schematic of a Geothermal-Preheat Hybrid System.

Fossil-Superheat Hybrid Systems. In these systems, the vapor-dominated steam, or the vapor obtained from a flash separator in a high-temperature liquid-dominated system, is superheated in a fossil fired super heater.

Fig. 2.38 show schematic flow. It comprises a double-flash geothermal steam system. Steam pro-duced at 4 in the first-stage flash separator is preheated from 4 to 5 in a regenerator by exhaust steam from the high-pressure turbine at 7. It is then superheated by a fossil fuel fired super heater to 6 and expands in the high pressure turbine to 7 at a pressure near that of the second stage steam separator. It than enters the regenerator, leaves it at 8, where it mixes with the lower pressure steam produces in the second stage flash separator at 15, and produces steam at 9, which expands in the lower pressure turbine to 10. The condensate at 11 is pumped and reinjected into ground at 12. The spent brine from the second stage evaporator is also reinjected in to ground at 16.

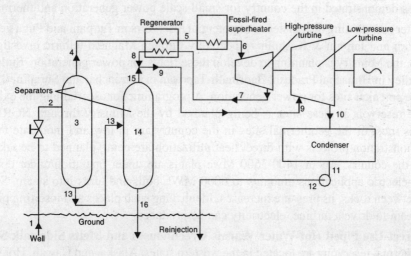

Fig. 2.38. Schematic of a fossil-superheat hybrid system with two-stage flash evaporation, regenerator, and fossil-fired super heater.

Fig. 2.39. Crops surround this plant in Imperial Valley, Calif. High mineral contents of some southern California geothermal reservoirs provide salable by-products like silica and zinc.

2.20.5 ADVANCED CONCEPTS

Geothermal energy technologies use the heat of the earth for direct-use applications, geothermal heat pumps, and electrical power production. Research in all areas of geothermal development is helping to lower costs and expand its use.

Advanced technologies will help manage geothermal resources for maximum power production, improve plant-operating efficiencies, and develop new resources such as hot dry rock, geopressured brines, and magma.

Encouraged by the findings of nearly 340 hot springs in the country, a systematic collaborative, research, development and demonstration programme was undertaken with different organizations viz.

IIT, Delhi, National Aeronautic Limited, Bangalore, Geological Survey of India, National Geophysical Research Institute (NGRI), Hyderabad, Oil & Natural Gas Corporation etc. and the use of geothermal energy was demonstrated in the country for small scale power generation and thermal applications.

After ascertaining the existence of potential reservoirs in Tattpani and Puga geothermal fields in Chhattisgarh and Jammu & Kashmir respectively through Magneto-telluric investigations by NGRI, Hydrabad, the Ministry is planning to develop these fields for power generation. Sutluj-Spiti, Bcas and Parbati valley in Himachal Pradesh, Badrinath-Tapovan in Uttranchal and Surajkund in Jharkhand also have some potenital sites for power generation. A programme for ascertaining the existence and potential of the reservoir at these sites is being planned by the Ministry through NGRI, Hyderabad and NHPC. As most of the geothermal sites in the country are in low and moderate temperature range, some demonstration projects with direct heat utilization are being planned to be taken up at different places in the country. About 1400-3600 Mwe plants are under operation/under construction. World wide non-electric applications amounts to 6000 MWt. boils and turns into steam. Since this steam is trapped between rocks, its pressure increases. Identifying such places and inserting pipes there to bring out the steam to drive a turbine, electricity can be produced.

Direct-Use Piped Hot Water Warms Greenhouses and Melts Sidewalk Snow. In the U.S., most geothermal reservoirs are located in the western states, Alaska, and Hawaii. Hot water near Earth's surface can be piped directly into facilities and used to heat buildings, grow plants in greenhouses, dehydrate onions and garlic, heat water for fish farming, and pasteurize milk. Some cities pipe the hot water under roads and sidewalks to melt snow. District heating applications use networks of piped hot water to heat buildings in whole communities. For more information on direct use of geothermal energy.

Geothermal Heat Pumps use Shallow Ground Energy to Heat and Cool Buildings. Almost everywhere, the upper 10 feet of Earth's surface maintains a nearly constant temperature between 50 and 60 degrees F (10 and 16 degrees C). A geothermal heat pump system consists of pipes buried in the shallow ground near the building, a heat exchanger, and ductwork into the building. In winter, heat from the relatively warmer ground goes through the heat exchanger into the house. In summer, hot air from the house is pulled through the heat exchanger into the relatively cooler ground. Heat removed during the summer can be used as no-cost energy to heat water.

Fig. 2.40. Snow Melting on Sidewalks in Klamath Falls.

The Future of Geothermal Energy. The three technologies discussed above use only a tiny fraction of the total geothermal resource. Several miles everywhere beneath Earth's surface is hot, dry rock being heated by the molten magma directly below it. Technology is being developed to drill into this rock, inject cold water down one well, circulate it through the hot, fractured rock, and draw off the heated water from another well. One day, we might also be able to recover heat directly from the magma.

Fig. 2.41. World's Largest Heat Pump System in Louisville, KY.

Fig. 2.42. This 3,000 sq. ft. house in Oklahoma City has a verified average electric bill of $60 per month using a geothermal heat pump.

2.21 OCEAN ENERGY

The ocean can produce two types of energy: thermal energy from the sun's heat, and mechanical energy from the tides and waves. Oceans cover more than 70% of Earth's surface, making them the world's largest solar collectors. The sun's heat warms the surface water a lot more than the deep ocean water, and this temperature difference creates thermal energy. Just a small portion of the heat trapped in the ocean could power the world.

Fig. 2.43. Ocean Wave.

Ocean mechanical energy is quite different from ocean thermal energy. Even though the sun affects all ocean activity, tides are driven primarily by the gravitational pull of the moon, and waves are driven primarily by the winds. As a result, tides and waves are intermittent sources of energy, while ocean thermal energy is fairly constant. Also, unlike thermal energy, the electricity conversion of both tidal and wave energy usually involves mechanical devices.

A barrage (dam) is typically used to convert tidal energy into electricity by forcing the water through turbines, activating a generator. For wave energy conversion, there are three basic systems:

Channel systems that funnel the waves into reservoirs;

Float systems that drive hydraulic pumps; and

Oscillating water column systems that use the waves to compress air within a container.

The mechanical power created from these systems either directly activates a generator or transfers to a working fluid, water, or air, which then drives a turbine/generator.

The availability of ocean thermal energy on the earth can be calculated as;

Let,

E_t = Total terrestrial ocean thermal energy incidence

E_r = Total extraterrestrial solar energy received by the earth = 5.457×10^{18} MJ/year

C_r = Average clearness index = 0.5

F_A = fraction of the area of ocean = 0.7

$$E_t = E_r \times C_r \times F_A$$
$$= 5.457 \times 1018 \times 0.5 \times 0.7$$
$$= 1.9 \times 1018 \text{ MJ/year}$$

This corresponds to an average terrestrial incidence on the waters of the solar constant
$S = 1353 \text{ W/m}^2 \times 0.5 = 676 \text{ W/m}^2$.

This energy is not totally absorbed by the water because some of it is reflected back to the sky. A good estimate of the amount absorbed is obtained from the annual evaporation of water,

Annual evaporation (EV) = 1.20 m

Average water surface temperature (T) = 20°C

The latent heat of vaporization(C) = 2454 kJ/kg

Seawater density (Ω) = 1000 kg/m^3.

The annual energy absorbed (E_{ab}) = $E_V \times \rho \times C$
$$= 1.20 \times 1000 \times 2454$$
$$= 3 \times 106 \text{kJ/m}^2 = 95 \text{ W/m}^2.$$

This amount is, of course, replenished by rainfall back on the water and by runoff from land.

Absorbed energy as % of incident energy = $E_{ab}/S = (95/676) \times 100 = 14\%$

2.21.1 POWER PLANTS BASED ON OCEAN ENERGY

Ocean thermal energy is used for many applications, including electricity generation. There are three types of electricity conversion systems: closed-cycle, open-cycle, and hybrid.

Closed-cycle systems use the ocean's warm surface water to vaporize a working fluid, which has a low-boiling point, such as ammonia. The vapor expands and turns a turbine. The turbine then activates a generator to produce electricity. Open-cycle systems actually boil the seawater by operating at low pressures. This produces steam that passes through a turbine/generator. Hybrid systems combine both closed-cycle and open-cycle systems.

Depending Upon these electricity conversion systems the Ocean power plant can be divided mainly in to two groups.

The Open or Claude OTEC Cycle Power Plant. The Frenchman Georges Claude constructed the first OTEC plant in 1929 on the Mantanzas Bay in Cuba.

The Claude plant used an open cycle in which seawater itself plays the multiple role of heat source, working fluid, coolant, and heat sink.

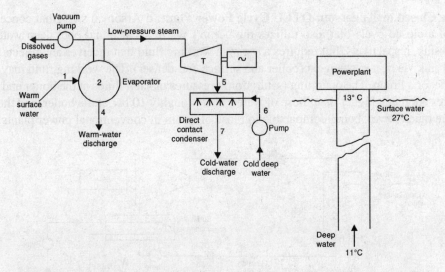

Fig. 2.44. Flow diagram and schematic of a Claude (open-cycle) OTEC power plant.

In the cycle warm surface water at 27°C is admitted into an evaporator in which the pressure is maintained at a value slightly below the saturation pressure corresponding to that water temperature. Water entering the evaporator, there four, finds itself "superheated" at the new pressure.

This temporarily superheated water undergoes volume boiling causing that water to partially flash to steam to an equilibrium two-phase condition at the new pressure and temperature. The low pressure in the evaporator is maintained by a vacuum pump that also removes the dissolved noncondensable gases from the evaporator.

The evaporator now contains a mixture of water and steam of very low quality at 2. The steam is separated from the water as saturated vapor at 3. The remaining water is saturated at 4 and is discharged as brine back to the ocean. The steam at 3 is, by conventional power plant standards, a very low-pressure, very high specific-volume working fluid (0.0317 bar, 43.40 m^3/kg, compared to about 160 bar, 0.021 m^3/kg for modern fossil power plants). It expands in a specially designed turbine that can handle such conditions to 5. Since the turbine exhaust system will be discharged back to the ocean in the open cycle, a direct-contact condenser is used, in which the exhaust at 5 is mixed with cold water from the deep cold-water pipe at 6, which results in a near-saturated water at 7. That water is now discharged to the ocean.

The cooling water reaching the condenser at 13°C is obtained from deep water at 11°C (51.8°F). This rise in temperature is caused by heat transfer between the pro-gressively warmer outside water and the cooling water inside the pipe as it ascends the cold water pipe.

There are thus three temperature differences, all about 2°C: one between warm surface water and working steam, one between exhaust steam and cooling water, and one between cooling water reaching the condenser and deep water. These represent external irreversibility's that reduce the overall temperature difference between heat source and sink from 27 − 11 = 16°C (28.8°F) to 25 − 15 = 10°C (18°F) as the temperature difference available for cycle work. It is obvious that because of the very low temperature differences available to produce work, the external differences must be kept to absolute minimum to realize as high efficiency as possible. Such a necessary approach, unfortunately, also results in very large warm and cold water flows and hence pumping power, as well as large heavy cold water pipes.

The Closed or Anderson, OTEC Cycle Power Plant. d'Arsonval's original concept in 1881 was that of a closed cycle that also utilizes the ocean's warm surface and cool deep waters as heat source and sink, respectively, but requires a separate working fluid that receives and rejects heat to the source and sink via heat exchangers (boiler and surface con-denser). The working fluid may be ammonia, propane, or a Freon. The operating (saturation) pressures of such fluids at the boiler and condenser temperatures are much higher than those of water, being roughly 10 bar at the boiler, and their specific volumes are much lower, being comparable to those of steam in conventional power plants.

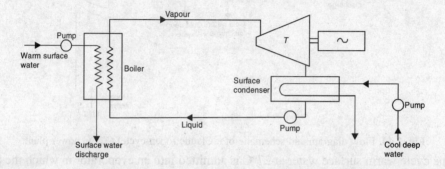

Fig. 2.45. Schematic of a closed-cycle OTEC power plant.

Such pressures and specific volumes result in turbines that are much smaller and hence less costly than those that use the low-pressure steam of the open cycle. The closed cycle also avoids the problems of the evaporator. It, however, requires the use of very large heat exchangers (boiler and condenser) because, for an efficiency of about 2 percent, the amounts of heat added and rejected are 50 times the output of the plant.

In addition, the temperature differences in the boiler and condenser must be kept as low as possible to allow for the maximum possible temperature difference across the turbine, which also contributes to the large surfaces of these units.

Barjot first proposed the closed-cycle approach in 1926, but the most recent design was by Anderson and Anderson in the 1960s. The closed cycle is sometimes referred to as the **Anderson cycle**. The Andersons chose propane as the working fluid with a 20°C temperature difference between warm surface and cool water, the latter some 600 m deep. Propane is vaporized in the boiler at 10 bars or more and exhausted in the condenser at about 5 bars.

In order to minimize the mass and the amount of material (and hence cost) used to manufacture the immensely large heat exchangers, the Anderson OTEC system employs thin plate-type heat exchangers instead of the usual heavier and more expensive shell-and-tube heat exchangers. To help reduce the thickness of the plates, the heat exchangers are placed at depths where the static pressure of the water in either ex-changer roughly equals the pressure of the working fluid. Thus if propane is the working fluid in the boiler at 26.7°C and 9.9 bar the boiler.

2.22 OTHER TECHNOLOGY

2.22.1 LIQUID FUEL

The recent controversies of tariff of IPPs in operation have been an eye opener to the issue of fuel selection. Several experts in hindsight have also suggested that liquid fuel was a wrong selection for projects such as DPC. However, this reasoning is too narrow and skewed in its interpretation of liquid fuels. Naphtha is definitely the least favorite and the most prohibitive fuel world-wide and by definition and characteristics, it does belong to the liquid fuel category. There are other veritably successful alternative liquid fuels like furnace oil, LSHS, etc. where power plants are viable in the liquid fuel mode.

Many liquid fuels based IPPs are now without ambiguity opting for safest bet and the least price volatile liquid fuel that is furnace oil. Furnace oil has its tremendous advantages. With an average gross calorific value of 10,200 Kcal/Kg furnace oil is a potent low grade fuel for energy generation. In comparison to more refined and primary fuels like naphtha and natural gas, furnace oil has no other commercial value except being combusted for energy generation.

Crude oil prices has been floating between $55 to $65 per barrel at the current international oil pool (Sept.—2005). It proffers excellent potential to have low generation cost of power. With an effective and time-tested technology of power generation with heavy fuel operated diesel engine based power plants, furnace oil will indeed be the preferred alternative for liquid fuel IPPs. Further, establishing captive power plant and IPPs on DG technology has very low gestation period in comparison to gas turbine or steam turbine based combined cycle power plants. The detractors of DG technology and using heavy fuel may trump up the bogey of high sulphur content. But that is not a very contentious issue as government is now clearing IPPs and allowing them to use furnace oil as fuel with sulphur £ 2% by weight.

For that matter, even LSHS or its Indian version LSWR if used will account for even much lower sulphur content » 0.5% to 0.8%. The pollution control norms can be met for combusting both of the above heavy fuels in the diesel engine power plants by constructing chimneys of adequate height for exhaust gases' exit. The furnace oil as well as LSHS has no problem of availability. Fuel storage, handling and transportation are far simpler and easy to establish. It is also expected that over a period of time, the most predictable and stable price levels will be for the furnace oil.

It also gives the opportunity for having lower levels of import costs on account of fuel while using furnace oil. All these will help the government to keep the fuel import bills for the future at lower levels, which is desirable. Under the circumstances, all medium and small IPPs can get their projects started if they decide to switch over to furnace oil based DG power plants. Many promoters of IPPs have already embarked on this course. It is a good sign at least a few, if not all, have realised that it is more prudent and time saving to use most technically and commercially viable fuel for liquid fuel based power generation. The IPPs can also freely source their liquid fuel requirements and no fuel linkage is prerogative for sanctioning the project.

The most contentious issue was Naphtha. This fuel was primarily being imported to fulfil the needs of the fertiliser and petrochemical industries as feedstock. Why at all it was considered for energy generation remains an enigma. Government experts thought that locally produced Naphtha, which was in surplus, could serve the needs of the liquid fuel based power plants. But the grade produced locally is HAN (High Aromatic Naphtha), which could not be used for combustion in gas turbines which needed LAN (Low Aromatic Naphtha).

This reflected a cruel lack of understanding of the technology of gas turbines used as power generation machines and their fuel application part. This policy continued till it reached a dead end.

Finally the government allowed alternate fuels and heavy fuel based Diesel Engine technology for the medium and small sized liquid fuel based power plants became popular.

Heavy fuel, especially furnace oil grades is ideally used in diesel power stations. This type of technology to burn furnace oil using four stroke engines is the most reliable type of power generation system. The concept of using diesel engine being extended to IPPs due to fuel being furnace oil makessense. The gestation for such projects could be as low as 14 months for a 35 MW power plant. Furnace oil and LSHS being residual fuels have no other commercial use than combusting for energy generation purpose. Naphtha as well as Natural gas is of use as feedstock to the fertiliser and the petro-chemical industry. The concept of residual fuels is therefore limited to the power generation industry and for marine propulsions only.

Generation cost from fuel oil power plants could be pegged at a low figure of about Rs. 3.50 per KW hr. of energy. Such generation costs is considering into effect all variables like furnace oil cost, lube oil cost, operation and maintenance cost, interest on capital and borrowings, depreciation etc.

The crude oil prices, the world over had touched the lowest levels of the decade during the past two years. One of the fundamental reasons for the liquid fuel captive power plants in India moving to diesel engine technology using heavy fuel was the comparatively less volatile nature of pricing of the heavy fuel like furnace oil, LSHS and residual fuels over the past decade. The past couple of years are also witness to weakening of the crude oil cartel mainly via the OPEC nations thus removing albeit by default monopolistic nature of the oil pool or the cartel of the select countries that control 80% of universal oil reserves.

The international benchmark Brent rests at $27–$28 per barrel. It will be strategically worth-while for countries like India to adopt heavy fuel diesel engine technology for power generation for the medium capacity (up to 150MW range) of the power stations. All the small and medium IPPs ideally should look to this route to come to best operational economics in terms of low generation cost.

Supply and demand mechanics cannot be achieved by extraneous machinations. They are in fact a function of free play of the market forces. To move the oil market positively, a better proposition would be to increase cash inflow by enhanced sales and a better market share. Crude oil producers working towards such a stratagem will be able to profit in the long run and fulfil the aspirations of their own populace in terms of improved GDP, per capita incomes and technological upgradation etc.

Lack of consensus on oil pricing and stock mobilisation generates parallel monopolistic alliances of oil exporters, bordering on opportunism. The aim of this informal consortium was to bulldoze OPEC and reduce the crude oil production so as to enable it to fetch higher price. A host of reasons can be assigned to explain this market pricing structure. A few of these are:

(1) Stock surplus in the global crude oil market.

(2) Low levels of industrial outputs worldwide.

(3) Energy markets effected by the global economic slumps and accompanied by consumption drops.

It is in this perspective that India's bilateral negotiation for a part barter deal with Iraq for supply of crude at $7 per barrel should be seen as a positive progress. The rest of the payment will go as counter trading of wheat of equivalent value to compensate for international pricing for the crude. It is such pricing arrangements that we should welcome in India to control the ballooning oil pool deficit and eke out positive technology and fuel options.

2.22.2 FUEL CELL TECHNOLOGY

Fuel cells produce electricity from an electrochemical reaction between hydrogen and oxygen. Fuel cells are efficient, environmentally benign and reliable for power production. The use of fuel cells has been demonstrated for stationary/portable power generation and other applications. Ministry of Non-Conventional Energy Sources (MNES) has taken up projects on different types of fuel cells through various organizations. These projects have led to the development of prototypes of fuel cells, materials/catalysts and components for fuel cell systems. Phosphoric Acid Fuel Cell (PAFC) stacks have been developed and demonstrated for decentralized power production. Under a project funded by MNES, a 50 kW (2 × 25 kW) PAFC power plant has been developed and tested by BHEL, Hyderabad for distributed power generation.

As per part of an R&D project funded by the Government, the SPIC Science Foundation, Chennai, had developed an improved version of 5 kW Proton Exchange Membrane (PERM) fuel cell module and successfully demonstrated its use for on-site power generation and vehicular propulsion. Efforts were on to develop durable ion-exchange membranes and establish performance and reliability of systems.

The Central electrochemical Research Institute (CECRI), Karaikudi has developed a small Molten Carbonate Fuel Cell (MCFC) stack. The Central Glass and Ceramic Research Institute (CGCRI), kolkatta is developing a 1 kW solid Oxide fuel Cell (SOFC) power pack. Under an R&D project funded by MNES, the Indian Institute of Science (IISc), Bangalore, will construct a 100-watt liquid-feed solid polymer electrolyte direct methanol fuel cell (DMFC). The Indian Institute of Technology (IIT), Madras, Chennai in collaboration with SPIC Science Foundation is also working on a project of Chemical Technology. BHEL and Indian Institute Of Chemical Technology (IICT), Hyderabad have developed catalysts and reformers for reformation methanol into hydrogen for fuel cells.

A PEMFC-based uninterrupted power supply (UPS) system to deliver single-phase AC power at 220 volts, 50 Hertz has been developed and demonstrated by SPIC Science Foundation, Chennai. This system has been sent to Indian Institute of Technology, Madras, Chennai for testing/demonstration.

SPIC Science Foundation has identified a number of polymers under an R&D project and studied their suitability as electrolytes for fuel cells. Modified nation membranes have been developed for high temperature applications. By incorporating a suitable reinforcing agent, the mechanical strength of films of block polymers f polystyrene has been improved. Membranes of up to 100 sq. cm area have been made.

The National Chemical Laboratory (NCL), Pune, has carried out literature survey on various types of polymeric membranes for PEMFC and procured important patents and reprints of relevant work. They have selected suitable monomers and synthetic strategies on the basis of anticipated transport behavior of protons. NCL has also synthesized different proton conducting polymers such as polyamides, polybenzimidazoles and surface fictionalized polymers using surface fictionalization. Membrane electrode assemblies (MEAs) for fuel cell stack have been prepared. The fabrication of a prototype using MEAs is in progress.

The Indian Institute of Technology, Madras, Chennai and SPIC Science Foundation, Chennai are jointly implementing a project for optimization of Proton Exchange Membrane Fuel Cell Stack design using advanced computational techniques. Preliminary modeling work has commenced.

The application of fuel cells has already been demonstrated for small-scale power generation and for operating an electric vehicle. It is proposed to take up projects and activities related to demonstration and testing of fuel cell systems in field conditions. Information/date/experience generated on the performance of fuel cells in field conditions will help in improving the performance of components and systems for greater reliability. MNES proposes to initiate Technology Mission of Fuel Cells during the Tenth Plan.

2.22.3 HYDROGEN ENERGY

Hydrogen is the simplest element. An atom of hydrogen consists of only one proton and one electron. It's also the most plentiful element in the universe. Despite its simplicity and abundance, hydrogen doesn't occur naturally as a gas on the Earth-it's always combined with other elements. Water, for example, is a combination of hydrogen and oxygen (H_2O). Hydrogen is also found in many organic compounds, notably the *hydrocarbons* that make up many of our fuels, such as gasoline, natural gas, methanol, and propane.

Hydrogen can be separated from hydrocarbons through the application of heat-a process known as *reforming*. Currently, most hydrogen is made this way from natural gas. An electrical current can also be used to separate water into its components of oxygen and hydrogen. This process is known as *electrolysis*. Some algae and bacteria, using sunlight as their energy source, even give off hydrogen under certain conditions.

Hydrogen is high in energy, yet an engine that burns pure hydrogen produces almost no pollution. **NASA** has used liquid hydrogen since the 1970s to propel the space shuttle and other rockets into orbit. Hydrogen fuel cells power the shuttle's electrical systems, producing a clean byproduct-pure water, which the crew drinks.

A fuel cell combines hydrogen and oxygen to produce electricity, heat, and water. Fuel cells are often compared to batteries. Both convert the energy produced by a chemical reaction into usable electric power. However, the fuel cell will produce electricity as long as fuel (hydrogen) is supplied, never losing its charge.

Fuel cells are a promising technology for use as a source of heat and electricity for buildings, and as an electrical power source for electric motors propelling vehicles. Fuel cells operate best on pure hydrogen. But fuels like natural gas, methanol, or even gasoline can be reformed to produce the hydrogen required for fuel cells. Some fuel cells even can be fueled directly with methanol, without using a reformer.

In the future, hydrogen could also join electricity as an important energy carrier. An energy carrier moves and delivers energy in a usable form to consumers. Renewable energy sources, like the sun and wind, can't produce energy all the time. But they could, for example, produce electric energy and hydrogen, which can be stored until it's needed. Hydrogen can also be transported (like electricity) to locations where it is needed.

Fig. 2.46

2.22.4 HYDROGEN ENERGY TECHNOLOGY (AS A FUEL)

You have learnt in chemistry that hydrogen is a combustible gas. When it burns it releases lots of heat. Water vapour alone is produced. No poisonous gas is produced when hydrogen is burnt. Then is it not a good fuel? Why then hydrogen is not used as a fuel in our daily life?

There is every chance for explosion when hydrogen is burnt.

Moreover, it is difficult to store hydrogen safely. Attempts are being made to burn hydrogen in small measures so as to reduce the chance of accident. We can expect that, in future, hydrogen will become a fuel which can be used by anybody. Hydrogen is being used even now as a fuel in rockets.

Hydrogen is a clean fuel and efficient energy medium for fuel cells and other devices. Hydrogen can be produced from water, non-conventional energy sources and from other fuels. Hydrogen could be used for a broad range of applications to supplement or substitute the consumption of hydrocarbon fuels and fossil fuels in an environment friendly manner. The large-scale introduction of hydrogen as a fuel would reduce the consumption of fossil fuels and keep the air clean and free from pollution. This Ministry is supporting research, development and demonstration projects on various aspects of hydrogen energy including production, storage and utilization of hydrogen as fuel at various research, scientific and educational institutions, laboratories, universities, and industries.

Hydrogen can be produced from non-conventional energy sources by various methods. Electrolytic, photolytic/photo biological, photo-electrolysis and thermos-chemical hydrogen production technologies are currently under development and use. The selection of production processes/technologies will depend on the availability of resource, expertise, infrastructure and economical aspect. The research group at Banaras Hindu University (BHU), Varanasi, carried out studies on semiconductor-septum solar cells for pilot scale production of hydrogen by photo catalytic decomposition of water.

This Government has sanctioned a project Chettiar Research Centre (MCRC), Chennai, for the production of hydrogen from organic effluents at a pre-commercial level and optimise various parameters for optimal hydrogen production. Photo bioreactors of 0.125 m and 1.25 m have been fabricated at Nillikuppam. One more reactor of 12.5m capacity is being fabricated by MCRC. For hydrogen production, 12 heterotopy bacteria and two phototropic bacteria were isolated from different sources. The project seeks to demonstrate sustained biological hydrogen production at a pre-commercial level, study and optimize various parameters and prepare documentation for commercial exploitation of the technology for treatment of industrial biological effluents. The research group at BHU, Varanasi, is also developing a laboratory scale bio-hydrogen production plant for producting hydrogen form bagasse.

The Indian Institute of Technology (IIT), Madras, (Chennai) is engaged in developing a hydrogen storage device based on indigenous Mischmetal-based alloys. A hydrogen storage device using special SS tubes, filter and 100 g of AB alloy has been designed and developed and its working performance has been studied. Design aspects of a larger hydrogen storage device using special SS tubes, heat exchanger and flow meter with alloys have been studied. Four Mischmetal-based AB and Ab alloys, which have reasonable plateau pressure at room temperature, have been used in the hydrogen storage device.

Another research group at IIT, Madras, Chennai has studied the design aspects of the novel metal hydride reactors for environment-friendly energy conversion devices. A concept has been devised for preliminary screening and selection of metal hydrides for specific energy conversion devices. Various energy conversion systems with suitable metal hydrides such as Zr-based hydrides and carbon nanotubes are being analyzed. Transient heat and mass transfer analyses are being done. Design aspect of the reactor bed are also being studied.

It is proposed to demonstrate and field test 5 hydrogen fueled two-wheelers at BHU, Varanasi, under the MNES-funded project. Each vehicle will require about 20–25 kg of hydrogen storage material to cover a distance of up to 100 km. BHU has synthesized new composite materials for storing hydrogen under this project. Procurement orders have already been placed for the purchase of material and equipment including 5 motorcycles. IIT, Kharagpur, is to design and develop a compressor driven metal hydride system for cooling and heating applications. The design optimisation of a working prototype of 1 kW space coolong system based on compressor driven metal hydride systems, using hydrogen as the working fluid. has been completed. Different components including compressors have also been selected.

With MNES support for the development of low polluting hydrogen-diesel dual-fuel engine. IIT, Delhi has successfully operated an engine (125 KVA) in the hydrogen-diesel dual fuel mode. Different parameters have been studied including the engine performance and exhaust emission characteristics of this system. MNES propose to initiate Technology Mission on Hydrogen Energy during the Tenth Plan.

2.22.5 BATTERY OPERATED VEHICLES

The Ministry of Non-Conventional Energy Sources (MNES) of the Government of India is implementing a programme on Alternative Fuel for Surface Transportation, which focuses on development and deployment of battery operated vehicles (BOVs). BOVs are environmentally benign, noise-free and consume no oil.

The Central Electrochemical Research Institute (CECRI), Karaikudi is developing high-energy lithium polymer batteries of 1 ah capacity with a life cycle of 350 for vehicular traction. CECRI has already synthesized and characterized LiCoO cathode active material and completed the optimisation of polymer electrolyte films and basic cell studies. The charge-discharge studies indicated cell efficiencies of more than 60%.

The project sanctioned at the Center for Materials for Electronics Technology (C-MET), Pune, envisages the development of novel route synthesis, characterisation and electrochemical studies on high quality cathode materials for rechargeable lithium batteries for electric vehicle use. Lithium manganese oxide is one of the cathode materials for lithium batteries. C-MET has developed cathode materials and characterised by using different characterization techniques. Work is in progress for the development of prototype lithium cells and for optimisation of various parameters for the cathode materials developed so far under this project.

Indian Institute of Science (IISc), Bangalore, has assembled and characterised laboratory scale lithium ion secondary cells in non-aqueous electrolyte with aluminum as a negative electrode and lithium manganese oxide as the positive electrode. Several carbon samples are used for separation of negative electrodes and electrochemical characterisation. Commercial lithium cobalt oxide is used for preparation of positive electrodes and electrochemical characterisation. Discharge capacity of 60–80 mAh/g has been achieved during a long cycle life.

NCL, Pune has synthesized carbonaceous materials based on coconut shell carbon for super capacitor electrode using various activation methods. The laboratory has prepared activated carbon using different processing procedures with KOH, ZnCl LiOH and CsOH etc. The BET surface area after gas phase activation is in the range of 800–1000 m/g. The research group has also prepared carbon composite electrodes using Ru and Ir metal oxides.

IISc, Bangalore, is implementing a project entitled 'Development of solid electrolyte materials for electrochemical double layer super capacitors.' The Institute has developed solid polymer electrolytes of silicate-salt composites based on sol-gel process for ultra capacitor applications. Capacitance of 300–400 Farad per gram of the material developed has been achieved. Polyacrilonitrile-based solid electrolytes have been developed and capacitances of the order of a few hundreds of Farads achieved. A number of solid electrolyte systems would be investigated for super capacitor applications. The CECRI at Karaikudi is implementing a project to develop conducting polymer based super capacitors. The Institute has fabricated and characterized n-type and p-type conducting polymer composite electrodes for super capacitor use. It is in the process of assembling and analyzing a model super capacitor to achieve higher performance.

The Nimbkar Agricultural Research Institute (NARI) at Phaltan, under another project, developed and demonstrated the operation of 20 battery-assisted cycle rickshaws in Maharasthra. Encouraged with the performance, NARI plans to develop and deploy more such passenger rickshaws. The Ministry sanctioned a pilot project to M/s Scooters India Limited (SIL), Lucknow and M/s Mahindra

Eco Mobiles Limited, Mumbai, for demonstration of 300 numbers of battery operated three wheelers (BOTWs) in Agra and other cities. 250 BOTWs are already operating in Agra, Allahabad, Ahmedabad, Delhi, Kolkata, Lucknow and Pune for demonstration, awareness, promotion and generating performance data of the vehicles in field conditions.

The Ministry broadened the scope for the demonstration programme on BOVs in the year 2002–2003 to cover battery operated passenger three wheelers and battery operated passenger cars, besides the battery operated buses/minibuses. This programme provided subsidy on the purchase of these types of indigenously manufactured vehicles through the Nodal Agencies and Departments in the States and Union Territories. MNES propose to initiate Technology Mission on Battery Operated Electric Vehicles during the Tenth Plan.

2.22.6 BIO FUEL TECHNOLOGY

Conservation of imported petroleum products and environmental pollution are the two important issues of concern today in the country. Thus, there is a need to search for alternate fuels to petrol and diesel for use in automobiles and diesel engines. Ethanol, currently used mainly as a raw material for chemical industries, in medicines and for potable purposes, is being increasingly looked upon as a potential fuel for powering automobiles. When used in blends with gasoline, ehanol enhances the combustion of gasoline due to oxygen molecules resulting in a more efficient burn and reduced emissions. Other potential bio fuels are edible and non-edible oils such as Jatropha curcas, Karanje, honge, etc. Recent developments taken place world over have made the use of ethanol petrol blend and biodiesel interesting new alternatives for conventional, unmodified diesel vehicles.

Development of technology for the production of ethanol from different routes, converting different non-edible oils to bio-diesel, developing kits/modified engines capable of using biofuels with 10% and more blends are some of the attempts being made by the Ministry to reduce the use of imported petroleum products in automobiles. Recently a policy analytical study has been carried out by the Ministry for drafting a long-term policy on biofuels. A shot loan scheme based on interest subsidies has been introduced during the current financial year for the producers of ethanol and other biofuels and the manufacturers of modified engines and kits enable to use biofuels. A research and development project has been sanctioned to Andhra University to develop aqua-porthole and develop the specifications of the engines enable to use the fuel thus developed. Efforts are being made to take up a demonstration project on the trial run of diesel vehicles with bio-diesel in collaboration with some oil company and manufacturers of diesel vehicles followed by launching of a Technology Mission on biofuels.

2.22.7 HYDROELECTRIC POWER

Flowing water creates energy that can be captured and turned into electricity. This is called *hydroelectric power* or *hydropower*.

The most common type of hydroelectric power plant uses a dam on a river to store water in a reservoir. Water released from the reservoir flows through a turbine, spinning it, which in turn activates a generator to produce electricity. But hydroelectric power doesn't necessarily require a large dam. Some hydroelectric power plants just use a small canal to channel the river water through a turbine.

Fig. 2.47

Another type of hydroelectric power plant- called a *pumped storage plant* can even store power. The power is sent from a power grid into the electric generators. The generators then spin the turbines backward, which causes the turbines to pump water from a river or lower reservoir to an upper reservoir, where the power is stored. To use the power, the water is released from the upper reservoir back down into the river or lower reservoir. This spins the turbines forward, activating the generators to produce electricity. A small or micro-hydroelectric power system can produce enough electricity for a home, farm, or ranch.

2.22.8 INNOVATIVE HEAT EXCHANGER TO SAVE ENERGY

Energy saves is Energy produced is not just a cliché but in-thing in most of the industries of the world. Heat Exchanger is one such device that conserves the use of energy resources like coal, oil, gas etc. It has the potential to bring the Indian fertilizer industry at par with the most efficient fertilizer industries of the world.

State-of-the-art-production developed by the Department of Chemical Engineering, IIT, Delhi, the pilot plant facility of the Innovative Heat Exchanger. The cost of the pilot project is Rs. 2.46 crores which took three years of research and design to materialize.

Heat exchangers contribute to about 25 percent of the equipment installed in fertilizer industry. Presently, shell and tube heat exchangers, plate type exchangers and helical pipe exchangers are in practice in the process industries. Fertilizer plants are highly energy intensive and 70–80 percent of the total production cost of fertilizer is spent on energy alone.

The fertilizer industry made a lankmark 11.20 million tons production during 1996-97, comprising of about 8.38 million tons of nitrogenous fertilizer and 2.82 million tons of phosphatic fertilier Today, the industry is not only an essential link in the food chain, but also has made its impact on the national economy. Plant designers continue to design and build ammonia plants with lower energy consumption. This has resulted in reduction of energy consumption from the earlier levels of 16-18 Giga Calories/Matric Tonnes to the present levels of 7.5-8.0 G Cal/MT of ammonia for Naptha based plants.

Energy conservation measures in Fertilizer industries have gained importance in the recent past and all new plants are constructed with the latest concept of low energy consumption. The theoretical thermodynamic heat requirement for ammonia production is about 4.47 GCal/MT, as against the current average consumption of 8 GCal/MT is lot in cooling water or ambient air through stack or radiation losses., though all processes will have some losses of energy from the system, the potential to reduce the energy consumption is sustantial.

The innovative heat exchanger consists of flatter velocity profiles and lower temperature gradient, which improves its performance, reduces residence time and thermal time distributions can be obtained by increasing the mixing between the fluid elements of different age groups and temperatures. Innovative Heat Exchanger also finds extensive use owing to the cross-sectional mixing induced by centrifugal force. Uniform thermal environment is an extremely desirous factor for the improved performance of any heat exchanger.

The idea of the innovative heat exchanger is based on the concept of centrifugal force. In the present device technique has been innovated for the effective utilization of the centrifugal force to advantage. The flow generated in this device due to curvature of a stationary surface bounding the flow changes direction continuously causing a local deflection of the velocity vector. This results in complex secondary flows, which is one of the principal features of fluid flow in this device. The new, flow geometry is capable of rotating the plane of vortex formation by any angle thereby exploiting the advantage of centrifugal force. The occurrence of this phenomenon increases mixing between the fluid elements of different age groups and temperatures. This leads to considerable increase in the heat transfer coefficient.

After the technical discussion with the management and technical group in fertilizer industry, various potential areas where this innovative heat exchanger can replace the existing heat exchangers were identified. Some of which are: In Ammonia plants: Methanator feed preheater, CO_2 strip reboil/shift effluent coolers feed gas, CO_2 stripper overhead trim cooler, Lean-solution Cooler (Air cooler), CO_2 stripper condenser air cooler, CO_2 ejector steam generator, CO_2 ejector steam reboiler, NH_3 refrigeration condenser, Lean solution/BFW exchanger and in the Urea plants: Distillation pre-heater, HP hydrolyser preheater and Distillation tower reboiler.

There is 15–20 percent improvement in heat transfer with 60–70 percent reduction in the exchanger area as compared to shell and tube heat exchanger. This device has two-fold advantage of intensifying the convective transfer processes (*i.e.*, increase heat and mass transfer coefficients) and also provide increased transfer area per unit volume of space. It offers higher film-coefficient (*i.e.*, the rate at which heat is transferred through a wall from one fluid to another) and more effective use of available pressure drop result in efficient and less expensive designs. The Innovative Heat Exchanger geometry permits handling of high temperatures and extreme temperature differentials without high-induced stresses or costly expansion joints. The compact size provides a distinct benefit and ease of fabrication and its performance is substantially closer to plug flow system.

It can, not only work as a heat exchanger but also as inline mixer, separation devices and in chemical reactors. It has a variety of applications: in coiled membranes blood oxygenators, kidney dialysis devices due to their effectiveness in reducing concentration polarization, chemical reactors due to increased residence time and minimized axial dispersion, heat exchangers, cryogenic systems, bio-sensors, clean steam generators, natural gas heaters, freeze condensers, chromato graphic columns, sample coolers and room heaters.

SOLVED EXAMPLES

Example 1. *A nuclear fission reaction power plant converts energy in matter to electrical energy by following energy chain*

$$\text{Energy in Matter} \rightarrow \text{Thermal Energy} \rightarrow \text{Mechanical Energy} \rightarrow \text{Electrical Energy}$$

Neglecting losses, how much matter is converted into electrical energy per day by a 10 mW power plant ?

Solution. Matter converted = Electrical Energy delivered

Energy delivered = Power × Time

$$E_e = (10 \text{ mW} \times 10^6) \times (24 \text{ hr} \times 3600)$$
$$= (10^7 \text{ W}) \times (8.64 \times 10^4 \text{ s})$$
$$E_e = 8.64 \times 10^{11} \text{ J}$$

Electrical energy delivered = Matter converted into energy

$$E_0 = m_0 C^2$$
$$m_0 = E/C^2$$

$$= \frac{8.64 \times 10^{11} \text{ J}}{(3 \times 10^8 \text{ m/s})^2} = 9.6 \times 10^{-4} \text{ kg}$$

The dimensional relationship of $[E/C^2]$ is

$$\frac{J}{(m/s)^2} = \frac{N.m.s^2}{m^2} = \frac{N}{m} s^2 = kg.$$

Example 2. *A turbine generator unit has output of 150 mW and efficiency of 0.80. Calculate energy supplied per hours by steam generator.*

Solution. Output of turbine generator = 150 mW

Input to turbine = $\dfrac{150}{0.8}$ = 187.5 mW

Energy input for 1 hour operation

= 187.5 (mW) × 1 (hr) = 187.5 mW-hr

1 mW hr = 3.6×10^9 J

187.5 mW hr = $\mathbf{675 \times 10^9}$ **J.**

Example 3. *A 100 mW geothermal power plant is operated for 11 months in a year. 1 month is for maintenance shutdown. The cost of electrical energy supplied is Rs. 2.5/- per kW-hr. Calculate the total earning by the power plant neglecting losses.*

Solution. Total hours = (24 hr) × (30 days) × (11 months) = 7920 hrs

Total energy generated = MW × hr

= 100 × 7920 = 7920 × 100 mW hr = 792×10^3 kW hr

Total cost of energy sold

= kW hr × Rs. × kW hr = $792 \times 10^3 \times 2.5$ Rs.

= **Rs. 1980000.00.**

Example 4. *Calculate annual requirement of lignite (fuel) for a thermal power plant rated 2000 mW under following conditions.*

Plant rating 2000 mW,

Annual load factor = $\dfrac{\text{Energy Delivered}}{\text{Rated power} \times \text{Total hours in year}}$ = 0.5

Plant efficiency = $\dfrac{\text{Electrical power output}}{\text{Thermal power input}}$ = 0.25

Utility factor of fuel = $\dfrac{\text{Useful energy in fuel}}{\text{Available energy in fuel}}$ = 0.7

Available energy density in coal = 14 mJ/kg.

Solution. Total energy delivered in year

(mW hr)$_e$ = Rated power × Hours × Load factor hours in year

= 365 × 24 = 8760 hrs

(mW hr)$_e$ = 2000 × 8760 × 0.5 = 8760000

$$\text{Thermal energy input} = \frac{\text{Electrical energy output}}{\text{Plant efficiency}}$$

$$= \frac{8760000}{0.25} = 35040000 \text{ (mW hr)}_t$$

Thermal energy input to plant for one year = 35040000 (mW hr)$_t$

Lignite input to plant for one year

$$= \frac{\text{(mW hr)}_t}{\text{(Useful mW/kg) in lignite}}$$

Useful energy in fuel per kg = Available energy in fuel × Utility factor

$$= (14 \text{ mJ/kg}) \times 0.7 = 9.8 \text{ mJ/kg}$$

Lignite input per year = 35040000 (mW hr)$_t$

Conversion factor:

1 mW hr = 1 mW × 3600 s = 3600 mW.s = 3600 mJ

Lignite required per year = $35040000 \times \dfrac{3600}{9.8}$

$$= 1.287 \times 10^{10} \text{ kg} = 12870 \times 10^6 \text{ kg}$$

Since (1 MT = 1000 kg) = **12870 × 10³ MT.**

Example 5. *The incident beam of sunlight has power density of 0.9 kW/m² in the direction of the beam. The angle of incidence θ is 60°. Calculate power collected by the surface having total flat area of 100 m².*

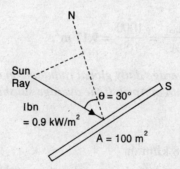

Fig. 2.48.

Solution. Equivalent solar power falling on the surface S

P_N (Watts) = I_N (W/m²) × A (m²)

I_N (kW/m²) = Ibn . cos θ

$$= (0.9 \times 10^3) \times 0.5 \text{ W/m}^2 = 4.5 \times 10^3 \text{ W/m}^2$$

P_N (kW) = $(0.45 \times 10^3 \text{ W/m}^2) \times 100 \text{ m}^2 = 0.45 \times 10^5$ W = **0.045 mW.**

Example 6. *When a photovoltaic cell is exposed to solar insulation of 950 W/m², the short circuit current is 220 A/m² both based on a unit area of the exposed junction. The open circuit voltage is 0.60 V and the temperature is 300 K. Calculate*

(a) reversed saturation current
(b) the voltage that maximizes the power
(c) the load current that maximizes the power
(d) the maximum power
(e) the maximum conversion efficiency
(f) the cell area for an output of 1 kW at the condition of maximum power.

Solution. $\dfrac{I_s/A}{I_o/A} = \exp\left(\dfrac{eV_{ac}}{kT}\right) - 1 = 1.193 \times 10^{10} \quad \dfrac{I_o}{A} = \dfrac{220}{1.193 \times 10^{10}} = 1.8 \times 10^{-8} \text{ A/m}^2$

The voltage Vm_p that maximizes the power is given by

$$\exp\left(\dfrac{eV_{mp}}{kT}\right)\left(1 + \dfrac{eV_{mp}}{kT}\right) = 1 + \dfrac{I_g/A}{I_o/A}$$

$$V_{mp} = 0.52 \text{ V}$$

$$\dfrac{I_{mp}}{A} = \dfrac{eV_{mp}/kT}{1 - eV_{mp}/kT}\left(\dfrac{I_s}{A} - \dfrac{I_o}{A}\right) = 210 \text{ A/m}^2$$

$$\dfrac{p_{max}}{A} = \left(\dfrac{I_{mp}}{A}\right) V_{mp} = 109 \text{ W/m}^2$$

$$\eta_{max} = \dfrac{P_{max}/A}{P_{in}/A} = \dfrac{109}{950} = 11.5\%$$

$$A = \dfrac{P_{out\ required}}{P_{max}/A} = \dfrac{1000}{109} = 9.17 \text{ m}^2.$$

Example 7. *Estimate the average daily global radiation on a horizontal surface at Ahmedabad (22°00′ N 73°10′E) during the month of April. If the average sunshine hours per day are 10. Assume a = 0.28 and b=0.48.*

Solution. Let

I_{sc} = Solar constant = 4870.8 kJ/m².hr

H_o = Daily global radiation on a m² horizontal surface at the location on a clear sky day in the month H_o is calculated from as

$$H_o = \dfrac{24}{\pi} I_{sc}\left[1 + 0.033 \cos\dfrac{360}{365} n\,(\sin\alpha.\sin s + \cos\alpha.\cos s.\sin s)\right]$$

H_g = Daily global radiation (monthly average) for a horizontal surface at the location is calculated by using value of H_o, a, b as

$$H_g = H_o a + b\left(\dfrac{L_h}{L_m}\right) \text{ kJ/m}^2.\text{ day}$$

The L_h are hours per day (average of the month)

L_m are maximum day hours in the month.

a, b are Angstrom's constants.

Angle of declination δ, from above Eqn.

$$\delta = 23.45 \sin\left[\frac{360}{365}(284+n)\right]$$

For April 15, n is calculated as :

Jan.		Feb.		March		April	$= n$
31	+	28	+	31	+	15	$= 105$

$$= 23.45 \sin\left[\frac{360}{365}(389)\right]$$

$$= 23.45 \sin 383.67 = 23.45 \times 0.4 = 9.41°$$

From above Eqn. sunshine hour angle ω_s

$\omega_s = \cos^{-1}(-\tan\phi \cdot \tan\delta)$
$= \cos^{-1}(-\tan 22° \cdot \tan 9.41°)$
$= \cos^{-1}(0.40 \times 0.61) = \cos^{-1}(0.064)$
$= 86.33°$

ω_s is converted from degrees to radians. $180° = \pi$ radians

$$86.33° = \frac{86.33}{180} \times \pi = 1.507 \text{ rad}$$

Maximum length $L_m = \dfrac{2}{15} \times 86.33°$ hours $= 11.51$ hours

$L_h = 10$ hours (given). Now H_o as

$$H_o = \frac{24}{\pi} I_{sc}\left(1 + 0.033 \cos\frac{360}{365}n\right) \times \int_{-\omega_s}^{+\omega_s} \omega_s \cdot \sin\phi \cdot \sin\delta + \cos\phi \cdot \cos\delta \cdot \cos\omega_s)$$

Substituting above calculated values, we get

$= 37210 [1 + 0.033 \cos(360/365) \times 105] \times (1.507 \sin 22° \cdot \sin 9.41 + \cos 22° \cdot \cos 9.41 \cdot \sin 86.3)$
$= 37210 [1 + 0.033 \times (-0.23) \times 1.507 \times 0.375 \times 0.16 + 0.93 \times 0.987 \times 0.998]$
$= 37210 [1.0076 \times (0.09 + 0.916)]$
$H_0 = 37210(1.09) = 40559$ kJ/m² day

$$H_g = H_o\left[a + b\left(\frac{L_h}{L_m}\right)\right]$$

$= 40559 [0.28 + 0.48 (10/11.51)]$
$= 40559 \times 0.697 = \textbf{28270 kJ/m day}$

Average global radiation per day in April = **28270 kg/m² day.**

Example 8. *Calculate the day length on a horizontal surface at New Delhi (28°35′ N, 77°12′ E) on December 1.*

Solution. Day length $= (2/15) \cos^{-1}(-\tan \Phi \cdot \tan \delta)$

Day length $= \dfrac{2}{15} \cos^{-1}[-\tan 28.58° \tan(-22.11°)]$

$= \mathbf{10.30 \text{ h}}.$

Example 9. *Determine solar time corresponding to 1430 h (IST) at Bombay (190°07′ N, 72°51′ E) on July 1. In India standard time is based on 82.50° E.*

Solution. Solar time $= 1430 \text{ h} - 4(82.50 - 72.85) \text{ minutes} + (-4 \text{ minutes})$

$= 1430 \text{ h} - 38.6 \text{ minutes} - 4 \text{ minutes}$

$= \mathbf{1347 \text{ h}}.$

Example 10. *A 100 mW vapour-dominated system uses saturated steam from a well with a shut-off pressure of 28 bar. Steam enters the turbine at 5.5 bar and condenses at 0.15 bar. The turbine polytropic efficiency is 0.82 and the turbine-generator combined mechanical efficiency is 0.9. The cooling tower exist is at 20°C. Calculate the necessary steam flow, the cooling water flow and the plant efficiency and heat rate if reinjection occurs prior to cooling tower.*

Solution.

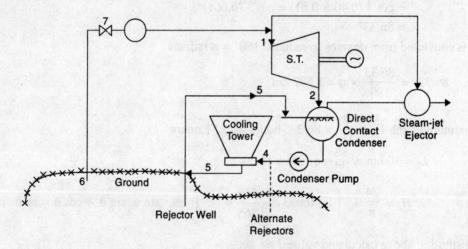

Fig. 2.49. Vapour-dominated Power Plant.

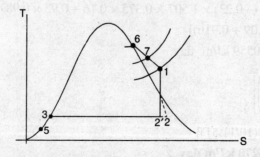

Fig. 2.50. T-s Diagram

$h_6 = h_g$ at 28 bar = 2801.7 kJ/kg

h_1 at 5.5 bar = 2801.7 kJ/kg

$t_1 = 176°C$ (20°C superheat)

$s_1 = 6.897$ kJ/kg-K

$v_1 = 0.356$ m³/kg

s_{2s} at 0.15 bar = $s_1 = 6.897 = 0.7327 + x_{4s}$ (7.306)

$x_{2s} = 0.8437$

$h_{2s} = 218.7 + 0.8437 (2377.24)$

$= 2224.38$ kJ/kg

Isentropic work = $h_1 - h_{2,s}$

$= 2801.7 - 2224.38 = 577.32$ kJ/kg

Actual turbine work = $0.82 \times 577.32 = 473.4$ kJ/kg.

$h_2 = h_1 - W_T$

$= 2801.7 - 473.4 = 2328.3$ kJ/kg

Ignoring pump work

$h_3 = h_4 = 218.7$ kJ/kg

$h_5 = h_f$ at 20°C = 88.5 kJ/kg

Turbine steam flow = $\dfrac{100 \times 10^3}{(473.4 \times 0.9)} = 234.7$ kg/s

$= 0.845 \times 10^6$ kg/hr

Turbine volume flow = $(0.845 \times 10^6 \times v_1)/60$

$= (0.845 \times 10^6 \times 0.356)/60$

$= 5013.4$ m³/min

Cooling water flow to condenser, m_5

$m_5 (h_3 - h_5) = m_2 (h_2 - h_3)$

$m_5 = (2328.3 - 218.7) \, m^2/(218.7 - 88.5)$

$= 2109.6 \times 0.845 \times \dfrac{10^6}{130.2}$

$= 13.7 \times 10^6$ kg/hr

Heat added = $h_6 - h_4$

$= 2801.7 - 218.7 = 2583$ kJ/kg

Plant efficiency = $W_T \times 0.9 / (h_6 - h_4) = (473.4 \times 0.9)/2583 = 16.49\%$

Plant heat rate = $\dfrac{3600}{0.1649}$ = **21831.4 kJ/kWh.**

Example 11. *A horizontal shaft, propeller type wind-turbine is located in area having following wind characteristics :*

Speed of wind 10 m/s at 1 atm and 15°C. Calculate the following:

1. Air density ρ

2. Total power density in wind stream, W/m²

3. Maximum possible obtainable power density, W/m²

4. Actual obtainable power density, W/m²

5. Total power from a wind-turbine of 120 m dia.

6. Torque and axial thrust (Na) on the wind-turbine operating at 40 rpm and at maximum efficiency of 42%.

Solution. 1. Air-density of wind

We use the equation $\rho = \dfrac{P}{RT}$ ...(1)

where
P = pressure of air, Pa
T = temperature of air, K
R = gas constant 287 J/kg.k 1 atm

Pressure 1 atm = 1.01325×10^5 Pa

Temperature 15°C = 15 + 273.15 = 378.15 K

Substituting in Eqn. (1)

Air density $\rho = \dfrac{P}{RT}$

$1.01325 \times \dfrac{10^5}{287} \times 378.15 = \mathbf{1.226 \text{ kg/m}^3}$.

2. For total power P_t

$\dfrac{P_t}{A} = 0.5\rho V_i^3 \ 0.5 \times 1.226 \times 10^3 = \mathbf{613 \text{ W/m}^2}$.

3. Maximum possible power

$\dfrac{P_{max}}{A} = (8/27)\rho V_i^3 = (8/27 \times 1) \times 1.226 \times 10^3 = \mathbf{363 \text{ W/m}^2}$.

4. Assuming $\eta = 42\%$

$P/A = 0.42 \ (P_{tot}/A) = 0.42 \times 613 = \mathbf{257 \text{ W/m}^2}$

5. $P = 0.257 \times \pi D^2/4 = 0.257 \times \pi 120^2/4 = \mathbf{2906 \text{ kW}}$.

6. $T_{max} = \dfrac{2}{27}(\rho D V_i^3/N_s) = \dfrac{2}{27}[1.20 \times 1.226 \times 10^3/(40/60)]$

$= \mathbf{16,347 \text{ N}}$.

Axial thrust (F_x) in newtons F_x max = $(\pi/9) \rho D^2 V_i^2 = \pi/9 (1.226 \times 120^2 \times 10^2)$
$$= 616{,}255 \text{ N}.$$

Example 12. *A 10 m/s wind is at 1 standard atmosphere and 15°C. Calculate:*
1. *The total power density in the wind stream.*
2. *The maximum obtainable power density.*
3. *A reasonably obtainable power density.*
4. *Total power produced if the turbine diameter is 120 m.*

Solution. The air density,

$$\rho = \frac{P}{RT}$$

$$= (1.01325 \times 10^5)/[287 (15 + 273)] = \mathbf{1.226 \text{ kg/m}^3}.$$

1. Total power density

$$\frac{\rho_{total}}{A} = \frac{\rho V_i^3}{2}$$

$$= 1.226 \times (10)^3/2 = \mathbf{613 \text{ W/m}^3}$$

2. Maximum power density

$$\frac{\rho_{max}}{A} = \frac{8}{27} \rho V_i^3$$

$$= (8/27) \times 1.226 \times (10)^3 = \mathbf{363 \text{ W/m}^3}.$$

3. Assuming $\eta = 40\%$

Actual power density,

$$\frac{P}{A} = 0.4 \left(\frac{P_{wt}}{A}\right)$$

$$= 0.4 \times 613 = \mathbf{245 \text{ W/m}^3}.$$

4. Total power produced,

$$P = \left(\frac{P}{A}\right) \frac{\pi D^2}{4}$$

$$= 0.245 \times \pi (120)^2/4 = \mathbf{2770 \text{ kW}}.$$

Example 13. *A 10 metre diameter rotor has 30 blades, each 0.25 metre wide. Calculate its solidity.*

Solution. Solidity = [(No. of blade × width)/(π × dia of rotor)] × 100 in %

Given that

Number of blade = 30

Width = 0.25 m

Diameter of rotor = 10 m

Putting all the values, we get

Solidity = $[(30 \times 0.25)/(\pi \times 10)] \times 100$ in %

= **24%**.

Example 14. *A 5 metre diameter rotor is rotating at 15 revolutions per minute (rpm) and the wind speed is 3 m/s, calculate tip speed ratio of the rotor.*

Solution. Tip speed ratio = $[(\pi \times$ dia of rotor $\times$ revolution per sec)/ Wind speed]

Given that;

Diameter of rotor = 5 m

Revolution = 15 RPM = 15/60 = 0.25 revolution per second

Wind speed = 3 m/sec

Tip speed ratio = $[(\pi \times 5 \times 0.25)/ 3] = 1.3$

Tip speed ratio = **1.3**.

Example 15. *Ocean wave on the coast of Tamil Nadu, India were with following data :*

Amplitude 1 m, Period 6 s. Calculate the following:

Wavelength, velocity, energy density, density, power extracted from a wave of 10.0 m with a power density, energy in 100 m wide wave. Assume density of ocean water as 1000 kg/m³.

Solution. For ocean wave;

Wavelength (λ) = $1.56\ T^2$ m

In this example $\lambda = 1.56\ T^2 = 1.56 \times 6^2 =$ **56.16 m**

Wave velocity, c = Wavelength(λ)/Period (T) = 56.6/6 = **9.36 m/s**

Frequency, $f = 1/T = 1/6\ s^{-1} = 0.1667$

Surface energy density

$$E/A = 1/2.\ D.\ a^2.g\ J/m^2 = 1.2 \times 1000 \times 1 \times 9.81\ J/m^2$$
$$= 4.905 \times 1000\ J/m^2.$$

Energy in 100 m wide wave

Area A = Wavelength (λ) × Width (W) = **56.16 × 100 m.**

Energy = $E/A \times A$

= $(4.905 \times 1000) \times (56.16 \times 100)$

= 275.46×100000 J = **27.546 J.**

Power = $E \times f$ W

= $27.546 \times 10^3 \times (1/6) =$ **4.56 × 10 W.**

Power density = P/A

= $(4.56 \times 10^3)/(56.16 \times 100)$

= 256.08×10^5 W/m

= **25.60 kW/m².**

THEORETICAL QUESTIONS

1. List various type of source of energy.
2. List the advantage of liquid fuel.
3. Write short notes on non-conventional energy.
4. Write short notes on petroleum.
5. Describe biofuels technology.
6. Write short notes on geothermal energy.
7. What is tidal energy and write the source of tidal energy ?
8. Briefly describe about ocean energy.
9. What is Wind energy and wind mill system ?
10. Write short note on biogas plant.
11. What are the features of biomass energy and describe what is the source of biomass?
12. What is solar cell technologies ?
13. Write down working of solar cells.
14. Write short notes on water heater.
15. Write short notes on box type solar cooker.
16. Write short notes on nuclear energy generation.
17. Write down properties and formation of coal.

Chapter 3

Power Plant Economics and Variable Load Problem

3.1 TERMS AND FACTORS

The main terms and factors are as follows:

1. Load Factor

It is defined as the ratio of the average load to the peak load during a certain prescribed period of time. The load factor of a power plant should be high so that the total capacity of the plant is utilized for the maximum period that will result in lower cost of the electricity being generated. It is always less than unity.

High load factor is a desirable quality. Higher load factor means greater average load, resulting in greater number of power units generated for a given maximum demand. Thus, the fixed cost, which is proportional to the maximum demand, can be distributed over a greater number of units (kWh) supplied. This will lower the overall cost of the supply of electric energy.

2. Utility Factor

It is the ratio of the units of electricity generated per year to the capacity of the plant installed in the station. It can also be defined as the ratio of maximum demand of a plant to the rated capacity of the plant. Supposing the rated capacity of a plant is 200 mW. The maximum load on the plant is 100 mW at load factor of 80 per cent, then the utility will be

$$= (100 \times 0.8)/(200) = 40\%$$

3. Plant Operating Factor

It is the ratio of the duration during which the plant is in actual service, to the total duration of the period of time considered.

4. Plant Capacity Factor

It is the ratio of the average loads on a machine or equipment to the rating of the machine or equipment, for a certain period of time considered.

Since the load and diversity factors are not involved with 'reserve capacity' of the power plant, a factor is needed which will measure the reserve, likewise the degree of utilization of the installed equipment. For this, the factor "Plant factor, Capacity factor or Plant Capacity factor" is defined as,

Plant Capacity Factor = (Actual kWh Produced)/(Maximum Possible Energy that might have produced during the same period)

Thus the annual plant capacity factor will be,

= (Annual kWh produced)/[Plant capacity (kW) × hours of the year]

The difference between load and capacity factors is an indication of reserve capacity.

5. Demand Factor

The actual maximum demand of a consumer is always less than his connected load since all the appliances in his residence will not be in operation at the same time or to their fullest extent. This ratio of the maximum demand of a system to its connected load is termed as demand factor. It is always less than unity.

6. Diversity Factor

Supposing there is a group of consumers. It is known from experience that the maximum demands of the individual consumers will not occur at one time. The ratio of the sum of the individual maximum demands to the maximum demand of the total group is known as diversity factor. It is always greater than unity.

High diversity factor (which is always greater than unity) is also a desirable quality. With a given number of consumers, higher the value of diversity factor, lower will be the maximum demand on the plant, since,

Diversity factor = Sum of the individual maximum Demands/Maximum demand of the total group

So, the capacity of the plant will be smaller, resulting in fixed charges.

7. Load Curve

It is a curve showing the variation of power with time. It shows the value of a specific load for each unit of the period covered. The unit of time considered may be hour, days, weeks, months or years.

8. Load Duration Curve

It is the curve for a plant showing the total time within a specified period, during which the load equaled or exceeded the values shown.

9. Dump Power

This term is used in hydro plants and it shows the power in excess of the load requirements and it is made available by surplus water.

10. Firm Power

It is the power, which should always be available even under emergency conditions.

11. Prime Power

It is power, may be mechanical, hydraulic or thermal that is always available for conversion into electric power.

12. Cold Reserve

It is that reserve generating capacity which is not in operation but can be made available for service.

13. Hot Reserve

It is that reserve generating capacity which is in operation but not in service.

14. Spinning Reserve

It is that reserve generating capacity which is connected to the bus and is ready to take the load.

15. Plant Use Factor

This is a modification of Plant Capacity factor in that only the actual number of hours that the plant was in operation is used. Thus Annual Plant Use factor is,

= (Annual kWh produced) / [Plant capacity (kW) × number of hours of plant operation]

3.2 FACTOR EFFECTING POWER PLANT DESIGN

Following are the factor effecting while designing a power plant.
(1) Location of power plant
(2) Availability of water in power plant
(3) Availability of labour nearer to power plant
(4) Land cost of power plant
(5) Low operating cost
(6) Low maintenance cost
(7) Low cost of energy generation
(8) Low capital cost

3.3 EFFECT OF POWER PLANT TYPE ON COSTS

The cost of a power plant depends upon, when a new power plant is to set up or an existing plant is to be replaced or plant to be extended. The cost analysis includes

1. Fixed Cost

It includes Initial cost of the plant, Rate of interest, Depreciation cost, Taxes, and Insurance.

2. Operational Cost

It includes Fuel cost, Operating labour cost, Maintenance cost, Supplies, Supervision, Operating taxes.

3.3.1 INITIAL COST

The initial cost of a power station includes the following:
1. Land cost
2. Building cost
3. Equipment cost
4. Installation cost
5. Overhead charges, which will include the transportation cost, stores and storekeeping charges, interest during construction etc.

POWER PLANT ECONOMICS AND VARIABLE LOAD PROBLEM

To reduce the cost of building, it is desirable to eliminate the superstructure over the boiler house and as far as possible on turbine house also.

Adopting unit system where one boiler is used for one turbogenerator can reduce the cost on equipment. Also by simplifying the piping system and elimination of duplicate system such as steam headers and boiler feed headers. Eliminating duplicate or stand-by auxiliaries can further reduce the cost.

When the power plant is not situated in the proximity to the load served, the cost of a primary distribution system will be a part of the initial investment.

3.3.2 RATE OF INTEREST

All enterprises need investment of money and this money may be obtained as loan, through bonds and shares or from owners of personal funds. Interest is the difference between money borrowed and money returned. It may be charged at a simple rate expressed as % per annum or may be compounded, in which case the interest is reinvested and adds to the principal, thereby earning more interest in subsequent years. Even if the owner invests his own capital the charge of interest is necessary to cover the income that he would have derived from it through an alternative investment or fixed deposit with a bank. Amortization in the periodic repayment of the principal as a uniform annual expense.

3.3.3 DEPRECIATION

Depreciation accounts for the deterioration of the equipment and decrease in its value due to corrosion, weathering and wear and tear with use. It also covers the decrease in value of equipment due to obsolescence. With rapid improvements in design and construction of plants, obsolescence factor is of enormous importance. Availability of better models with lesser overall cost of generation makes it imperative to replace the old equipment earlier than its useful life is spent. The actual life span of the plant has, therefore, to be taken as shorter than what would be normally expected out of it.

The following methods are used to calculate the depreciation cost:
(1) Straight line method
(2) Percentage method
(3) Sinking fund method
(4) Unit method.

Straight Line Method. It is the simplest and commonly used method. The life of the equipment or the enterprise is first assessed as also the residual or salvage value of the same after the estimated life span. This salvage value is deducted from the initial capital cost and the balance is divided by the life as assessed in years. Thus, the annual value of decrease in cost of equipment is found and is set aside as depreciation annually from the income. Thus, the rate of depreciation is uniform throughout the life of the equipment. By the time the equipment has lived out its useful life, an amount equivalent to its net cost is accumulated which can be utilized for replacement of the plant.

Percentage Method. In this method the deterioration in value of equipment from year to year is taken into account and the amount of depreciation calculated upon actual residual value for each year. It thus, reduces for successive years.

Sinking Fund Method. This method is based on the conception that the annual uniform deduction from income for depreciation will accumulate to the capital value of the plant at the end of life of the plant or equipment. In this method, the amount set aside per year consists of annual installments and the interest earned on all the installments.

Let,

A = Amount set aside at the end of each year for n years.

n = Life of plant in years.

S = Salvage value at the end of plant life.

i = Annual rate of compound interest on the invested capital.

P = Initial investment to install the plant.

Then, amount set aside at the end of first year = A

Amount at the end of second year

= A + interest on A = $A + Ai = A(1 + i)$

Amount at the end of third year

= $A(1 + i)$ + interest on $A(1 + i)$

= $A(1 + i) + A(1 + i)i$

= $A(1 + i)^2$

Amount at the end of nth year = $A(1 + i)^{n-1}$

Total amount accumulated in n years (say x)

= sum of the amounts accumulated in n years

i.e., $x = A + A(1 + i) + A(1 + i)^2 + \ldots\ldots + A(1 + i)^{n-1}$

$= A[1 + (1 + i) + (1 + i)^2 + \ldots\ldots + (1 + i)^{n-1}]$ ...(1)

Multiplying the above equation by $(1 + i)$, we get

$x(1 + i) = A[(1 + i) + (1 + i)^2 + (1 + i)^3 + \ldots\ldots + (1 + i)^n]$ ...(2)

Subtracting equation (1) from (2), we get

$x.i = [(1 + i)^n - 1] A$

$x = [\{(1 + i)^n - 1\}/i]A$, where $x = (P - S)$

$P - S = [\{(1 + i)^n - 1\}/i]A$

$A = (P - S)[i/\{(1 + i)^n - 1\}]A$

Unit Method. In this method some factor is taken as a standard one and, depreciation is measured by that standard. In place of years equipment will last, the number of hours that equipment will last is calculated. This total number of hours is then divided by the capital value of the equipment. This constant is then multiplied by the number of actual working hours each year to get the value of depreciation for that year. In place of number of hours, the number of units of production is taken as the measuring standard.

3.3.4 OPERATIONAL COSTS

The elements that make up the operating expenditure of a power plant include the following

(1) Cost of fuels.

(2) Labour cost.

(3) Cost of maintenance and repairs

(4) Cost of stores (other than fuel).
(5) Supervision.
(6) Taxes.

3.3.5 COST OF FUELS

In a thermal station fuel is the heaviest item of operating cost. The selection of the fuel and the maximum economy in its use are, therefore, very important considerations in thermal plant design. It is desirable to achieve the highest thermal efficiency for the plant so that fuel charges are reduced. The cost of fuel includes not only its price at the site of purchase but its transportation and handling costs also. In the hydro plants the absence of fuel factor in cost is responsible for lowering the operating cost. Plant heat rate can be improved by the use of better quality of fuel or by employing better thermodynamic conditions in the plant design.

The cost of fuel varies with the following:
(1) Unit price of the fuel.
(2) Amount of energy produced.
(3) Efficiency of the plant.

3.3.6 LABOUR COST

For plant operation labour cost is another item of operating cost. Maximum labour is needed in a thermal power plant using. Coal as a fuel. A hydraulic power plant or a diesel power plant of equal capacity requires a lesser number of persons. In case of automatic power station the cost of labour is reduced to a great extent. However labour cost cannot be completely eliminated even with fully automatic station, as they will still require some manpower for periodic inspection etc.

3.3.7 COST OF MAINTENANCE AND REPAIRS

In order to avoid plant breakdowns maintenance is necessary. Maintenance includes periodic cleaning, greasing, adjustments and overhauling of equipment. The material used for maintenance is also charged under this head. Sometimes an arbitrary percentage is assumed as maintenance cost. A good plan of maintenance would keep the sets in dependable condition and avoid the necessity of too many stand-by plants.

Repairs are necessitated when the plant breaks down or stops due to faults developing in the mechanism. The repairs may be minor, major or periodic overhauls and are charged to the depreciation fund of the equipment. This item of cost is higher for thermal plants than for hydro-plants due to complex nature of principal equipment and auxiliaries in the former.

3.3.8 COST OF STORES

The items of consumable stores other than fuel include such articles as lubricating oil and greases, cotton waste, small tools, chemicals, paints and such other things. The incidence of this cost is also higher in thermal stations than in hydro-electric power stations.

3.3.9 SUPERVISION

In this head the salary of supervising staff is included. A good supervision is reflected in lesser breakdowns and extended plant life. The supervising staff includes the station superintendent, chief engineer, chemist, engineers, supervisors, stores incharges, purchase officer and other establishment. Again, thermal stations, particularly coal fed, have a greater incidence of this cost than the hydro-electric power stations.

3.3.10 TAXES

The taxes under operating head includes the following:

(*i*) Income tax

(*ii*) Sales tax

(*iii*) Social security and employee's security etc.

3.4 EFFECT OF PLANT TYPE ON RATES (TARIFFS OR ENERGY ELEMENT)

Rates are the different methods of charging the consumers for the consumption of electricity. It is desirable to charge the consumer according to his maximum demand (kW) and the energy consumed (kWh). The tariff chosen should recover the fixed cost, operating cost and profit etc. incurred in generating the electrical energy.

3.4.1 REQUIREMENTS OF A TARIFF

Tariff should satisfy the following requirements:

(1) It should be easier to understand.

(2) It should provide low rates for high consumption.

(3) It should encourage the consumers having high load factors.

(4) It should take into account maximum demand charges and energy charges.

(5) It should provide less charges for power connections than for lighting.

(6) It should avoid the complication of separate wiring and metering connections.

3.4.2 TYPES OF TARIFFS

The various types of tariffs are as follows,

(1) Flat demand rate

(2) Straight line meter rate

(3) Step meter rate

(4) Block rate tariff

(5) Two part tariff

(6) Three part tariff.

The various types of tariffs can be derived from the following general equation:

$$Y = DX + EZ + C$$

POWER PLANT ECONOMICS AND VARIABLE LOAD PROBLEM

where

Y = Total amount of bill for the period considered.
D = Rate per kW of maximum demand.
X = Maximum demand in kW.
E = Energy rate per kW.
Z = Energy consumed in kWh during the given period.
C = Constant amount to be charged from the consumer during each billing period.

Various type of tariffs are as follows:

(1) Flat Demand Rate. It is based on the number of lamps installed and a fixed number of hours of use per month or per year. The rate is expressed as a certain price per lamp or per unit of demand (kW) of the consumer. This energy rate eliminates the use of metering equipment. It is expressed by the expression.

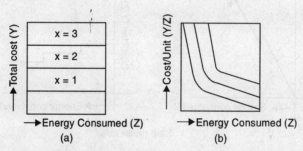

Fig. 3.1

(2) Straight Line Meter Rate. According to this energy rate the amount to be charged from the consumer depends upon the energy consumed in kWh which is recorded by a means of a kilowatt hour meter. It is expressed in the form

$$Y = EZ$$

This rate suffers from a drawback that a consumer using no energy will not pay any amount although he has incurred some expense to the power station due to its readiness to serve him. Secondly since the rate per kWh is fixed, this tariff does not encourage the consumer to use more power.

(3) Step Meter Rate. According to this tariff the charge for energy consumption goes down as the energy consumption becomes more. This tariff is expressed as follows.

$Y = EZ$ If $0 \leq Z \leq A$
$Y = E_1 Z_1$ If $A \leq Z_1 \leq B$
$Y = E_2 Z_2$ If $B \leq Z_2 \leq C$

And so on. Where E, E_1, E_2 are the energy rate per kWh and A, B and C, are the limits of energy consumption.

(4) Block Rate Tariff. According to this tariff a certain price per units (kWh) is charged for all or any part of block of each unit and for succeeding blocks of energy the corresponding unit charges decrease.

It is expressed by the expression

$$Y = E_1 Z_1 + E_2 Z_2 + E_3 Z_3 + E_4 Z_4 + \ldots$$

where E_1, E_2, E_3.... are unit energy charges for energy blocks of magnitude Z_1, Z_2, Z_g,.... respectively.

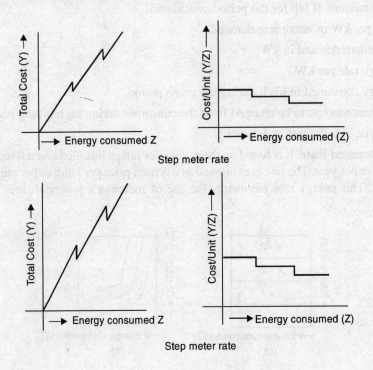

Fig. 3.2

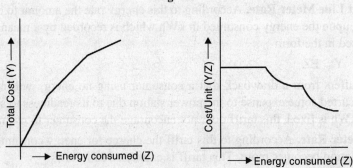

Fig. 3.3

(5) Two Part Tariff (Hopkinson Demand Rate). In this tariff the total charges are based on the maximum demand and energy consumed. It is expressed as

$$Y = D \cdot X + EZ$$

A separate meter is required to record the maximum demand. This tariff is used for industrial loads.

(6) Three-Part Tariff (Doherty Rate). According to this tariff the customer pays some fixed amount in addition to the charges for maximum demand and energy consumed. The fixed amount to be charged depends upon the occasional increase in fuel price, rise in wages of labour etc. It is expressed by the expression

$$Y = DX + EZ + C$$

POWER PLANT ECONOMICS AND VARIABLE LOAD PROBLEM   129

3.5 EFFECT OF PLANT TYPE ON FIXED ELEMENTS

Various types of fixed element are :
(1) Land
(2) Building
(3) Equipment
(4) Installation of Machine
(5) Design and planning

The fixed element means which are not movable, and for any types of power plant, the fixed elements play a major role. Since each cost is added to the final cost of our product (electricity in case of Power plant). So when a power plant is established, the first selection is fixed element.

Effect of plant on land is as cost of land.

3.6 EFFECT OF PLANT TYPE ON CUSTOMER ELEMENTS

The costs included in these charges depend upon the number of customers. The various costs to be considered are as follows:

(1) Capital cost of secondary distribution system and depreciation cost, taxes and interest on this capital cost.

(2) Cost of inspection and maintenance of distribution lines and the transformers.

(3) Cost of labour required for meter reading and office work.

(4) Cost of publicity.

3.7 INVESTOR'S PROFIT

If the power plant is the public property, as is the case in India, then the customers will be the taxpayers to share the burden of the government. For this purpose, there is an item in the rates to cover taxes in place of the investor's profit. The consumers in the form of electric consumption bills will pay these taxes. This amount is collected in twelve installments per year or six installments per year.

The investor expects a satisfactory return on the capital investment. The rate of profit varies according to the business conditions prevailing in different localities.

Adopting the following economical measures can reduce cost of power generation:
(1) By reducing initial investment in the power plant.
(2) By selecting generating units of adequate capacity.
(3) By running the power plant at maximum possible load factor.
(4) By increasing efficiency of fuel burning devices so that cost of fuel used is reduced.
(5) By simplifying the operation of the power plant so that fewer power-operating men are required.
(6) By installing the power plant as near the load centre as possible.
(7) By reducing transmission and distribution losses.

3.8 ECONOMICS IN PLANT SELECTION

A power plant should be reliable. The capacity of a power plant depends upon the power demand. The capacity of a power plant should be more than predicted maximum demand. It is desirable that the number of generating units should be two or more than two. The number of generating units should be so chosen that the plant capacity is used efficiently. Generating cost for large size units running at high load factor is substantially low. However, the unit has to be operated near its point of maximum economy for most of the time through a proper load sharing programme. Too many stand bys increase the capital investment and raise the overall cost of generation.

The thermal efficiency and operating cost of a steam power plant depend upon the steam conditions such as throttle pressure and temperature.

The efficiency of a boiler is maximum at rated capacity. Boiler fitted with heat recovering devices like air preheater, economiser etc. gives efficiency of the order of 90%. But the cost of additional equipment (air preheater economiser) has to be balanced against gain in operating cost.

Power can be produced at low cost from a hydropower plant provided water is available in large quantities. The capital cost per unit installed is higher if the quantity of water available is small. While installing a hydropower plant cost of land, cost of water rights, and civil engineering works cost should be properly considered as they involve large capital expenditure.

The other factor, which influences the choice of hydropower plant, is the cost of power transmission lines and the loss of energy in transmission. The planning, design and construction of a hydro plant is difficult and takes sufficient time.

The nuclear power plant should be installed in an area having limited conventional power resources. Further a nuclear power plant should be located in a remote or unpopulated are to avoid damage due to radioactive leakage during an accident and also the disposal of radioactive waste should be easy and a large quantity of water should be available at the site selected. Nuclear power becomes competitive with conventional coal fired steam power plant above the unit size of 500 mW.

The capital cost of a nuclear power plant is more than a steam power plant of comparable size. Nuclear power plants require less space as compared to any other plant of equivalent size. The cost of maintenance of the plant is high.

The diesel power plant can be easily located at the load centre. The choice of the diesel power plant depends upon thermodynamic considerations. The engine efficiency improves with compression ratio but higher pressure necessitates heavier construction of equipment with increased cost. Diesel power plants are quite suitable for smaller outputs. The gas turbine power plant is also suitable for smaller outputs. The cost of a gas turbine plant is relatively low. The cost of gas turbine increases as the sample plant is modified by the inclusion of equipment like regenerator, reheater, and intercooler although there is an improvement in efficiency of the plant by the above equipment. This plant is quite useful for regions where gaseous fuel is available in large quantities.

In order to meet the variable load the prime movers and generators have to act fairly quickly to take up or shed load without variation of the voltage or frequency of the system. This requires that supply of fuel to the prime mover should be carried out by the action of a governor. Diesel and hydropower plants are quick to respond to load variation as the control supply is only for the prime mover. In a steam power plant control is required for the boilers as well as turbine. Boiler control may be manual or automatic for feeding air, feed water fuel etc. Boiler control takes time to act and therefore, steam powers plants cannot take up the variable load quickly. Further to cope with variable load operation it is necessary for the power station to keep reserve plant ready to maintain reliability and continuity of power supply at all times. To supply variable load combined working of power stations is also economical.

For example to supply a load the base load may be supplied by a steam power plant and peak load may be supplied by a hydropower plant or diesel power plant.

The size and number of generating units should be so chosen that each will operate on about full load or the load at which it gives maximum efficiency. The reserve required would only be one unit of the largest size. In a power station neither there should be only one generating unit nor should there be a large number of small sets of different sizes. In steam power plant generating sets of 80 to 500 mW are quite commonly used whereas the maximum size of diesel power plant generating sets is about 4000 kW. Hydro-electric generating sets up to a capacity of 200 mW are in use in U.S.A.

3.9 ECONOMIC OF POWER GENERATION

Economy is the main principle of design of a power plant. Power plant economics is important in controlling the total power costs to the consumer. Power should be supplied to the consumer at the lowest possible cost per kWh. The total cost of power generation is made up of fixed cost and operating cost. Fixed cost consists of interest on capital, taxes, insurance and management cost. Operating cost consists of cost of fuel labour, repairs, stores and supervision. The cost of power generation can be reduced by,

(*i*) Selecting equipment of longer life and proper capacities.

(*ii*) Running the power station at high load factor.

(*iii*) Increasing the efficiency of the power plant.

(*iv*) Carrying out proper maintenance of power plant equipment to avoid plant breakdowns.

(*v*) Keeping proper supervision as a good supervision is reflected in lesser breakdowns and extended plant life.

(*vi*) Using a plant of simple design that does not need highly skilled personnel.

Power plant selection depends upon the fixed cost and operating cost. The fuel costs are relatively low and fixed cost and operation and maintenance charges are quite high in a case of a nuclear power plant. The fuel cost in quite high in a diesel power plant and for hydro power plant the fixed charges are high of the order of 70 to 80% of the cost of generation. Fuel is the heaviest items of operating cost in a steam power station. A typical proportion of generating cost for a steam power station is as follows :

Fuel cost = 30 to 40%
Fixed charges for the plant = 50 to 60%
Operation and maintenance cost = 5 to 10%

The power generating units should be run at about full load or the load at which they can give maximum efficiency. The way of deciding the size and number of generating units in the power station is to choose the number of sets to fit the load curve as closely at possible. It is necessary for a power station to maintain reliability and continuity of power supply at all times. In an electric power plant the capital cost of the generating equipment's increases with an increase in efficiency. The benefit of such increase in the capital investment will be realised in lower fuel costs as the consumption of fuel decreases with an increase in cycle efficiency.

Fig. 3.4 shows the variation of fixed cost and operation cost with investment.

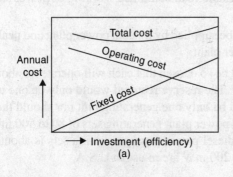

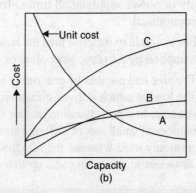

Fig. 3.4   **Fig. 3.5**

Fig. 3.5 shows the variation of various costs of power plant versus its capacity.

3.10 INDUSTRIAL PRODUCTION AND POWER GENERATION COMPARED

Industrial production is directly related to the power generation. Since in India, the major problem is of electricity. It is not possible to give 24 hr electricity to the industries. And each industrial production is based on power generation, each machine is runs by electricity, so if there is a problem of electricity in any industry, then it is directly suffer the total production.

So for run a plant for 24 hr, there is necessary to a power generation unit. And the power generation unit is of any type (*i.e.*, diesel, steam, gas turbine, etc.), which we learn in next chapters.

3.11 LOAD CURVES

The load demand on a power system is governed by the consumers and for a system supplying industrial and domestic consumers, it varies within wide limits. This variation of load can be considered as daily, weekly, monthly or yearly. Typical load curves for a large power system are shown in Fig. 3.6. These curves are for a day and for a year and these show the load demanded by the consumers at any particular time. Such load curves are termed as "Chronological load Curves". If the ordinates of the chronological load curves are arranged in the descending order of magnitude with the highest ordinates on left, a new type of load curve known as "load duration curve" is obtained. Fig. 3.6 shows such a curve. If any point is taken on this curve then the abscissa of this point will show the number of hours per year during which the load exceeds the value denoted by its ordinate. Another type of curve is known as "energy load curve" or the "integrated duration curve". This curve is plotted between the load in kW or MW and the total energy generated in kWh. If any point is taken on this curve, abscissa of this point show the total energy in kWh generated at or below the load given by the ordinate of this point. Such a curve is shown in Fig. 3.6. In Fig. 3.6(*b*), the lower part of the curve consisting of the loads which are to be supplied for almost the whole number of hours in a year, represents the "Base Load", while the upper part, comprising loads which are required for relatively few hours per year, represents the "Peak Load".

POWER PLANT ECONOMICS AND VARIABLE LOAD PROBLEM 133

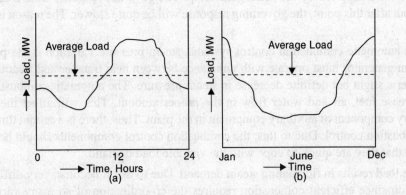

Fig. 3.6. Chronological Load Curves (*a*) Daily Load Curve (*b*) Yearly Load Curve.

3.12 IDEAL AND REALIZED LOAD CURVES

From the standpoint of equipment needed and operating routine, the ideal load on a power plant would be one of constant magnitude and steady duration. However, the shape of the actual load curve (more frequently realized) departs far from this ideal, Fig. 3.7. The cost to produce one unit of electric power in the former case would be from 1/2 to 3/4 of that for the latter case, when the load does not remain constant or steady but varies with time. This is because of the lower first cost of the equipment due to simplified control and the elimination of various auxiliaries and regulating devices.

Also, the ideal load curve will result in the -improved operating conditions with the various plant machines (for example turbine and generators etc.) operating at their best efficiency. The reason behind the shape of the actual realized load curve is that the various users of electric power (industrial, domestic etc.) impose highly variable demands upon the capacity of the plant.

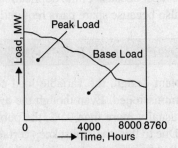

Fig. 3.7. Load Duration Curve.

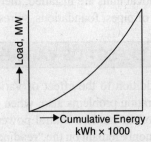

Fig. 3.8. Energy Load Curve.

3.14 EFFECT OF VARIABLE LOAD ON POWER PLANT DESIGN

The characteristics and method of use of power plant equipment is largely influenced by the extent of variable load on the plant. Supposing the load on the plant increases. This will reduce the rotational speed of the turbo-generator. The governor will come into action operating a steam valve and admitting more steam and increasing the turbine speed to its normal value. This increased amount of steam will have to be supplied by the seam generation. The governor response from load to turbine is quite prompt, but after this point, the governing response will be quite slower. The reason is explained as given below:

In most automatic combustion control systems, steam pressure variation is the primary signal used. The steam generator must operate with unbalance between heat transfer and steam demand long enough to suffer a slight but definite decrease in steam pressure. The automatic combustion controller must then increase fuel, air and water flow in the proper amount. This will affect the operation of practically every component of auxiliary equipment in the plant. Thus, there is a certain time lag element present in combustion control. Due to this, the combustion control components should be of most efficient design so that they are quick to cope with the variable load demand.

Variable load results in fluctuating steam demand. Due to this it become, very difficult to secure good combustion since efficient combustion requires the co-ordination of so many various services. Efficient combustion is readily attained under steady steaming conditions. In diesel and hydro power plants, the total governing response is prompt since control is needed only for the prime mover.

The variable load requirements also modify the operating characteristics built into equipment. Due to non-steady load on the plant, the equipment cannot operate at the designed load points. Hence for the equipment, a flat-topped load efficiency curve is more desirable than a peaked one.

Regarding the plant units, if their number and sizes have been selected to fit a known or a correctly predicted load curve, then, it may be possible to operate them at or near the point of maximum efficiency. However, to follow the variable load curve very closely, the total plant capacity has usually to be sub-divided into several power units of different sizes. Sometimes, the total plant capacity would more nearly coincide with the variable load curve, if more units of smaller unit size are employed than a few units of bigger unit size. Also, it will be possible to load the smaller units somewhere near their most efficient operating points. However, it must be kept in mind that as the unit size decreases, the initial cost per kW of capacity increases.

Again, duplicate units may not fit the load curve as closely as units of unequal capacities. However, if identical units are installed, there is a saving in the first cost because of the duplication of sizes, dimensions of pipes, foundations, wires insulations etc. and also because spare parts required are less.

3.14 EFFECT OF VARIABLE LOAD ON POWER PLANT OPERATION

In addition to the effect of variable load on power plant design, the variable load conditions impose operation problems also, when the power plant is commissioned. Even though the availability for service of the modern central power plants is very high, usually more than 95%, the public utility plants commonly remain on the "readiness-to-service" bases. Due to this, they must keep certain of their reserve capacity in "readiness-to-service". This capacity is called "spinning reserve" and represents the equipment standby at normal operating conditions of pressure, speed etc. Normally, the spinning reserve should be at least equal to the least unit actively carrying load. This will increase the cost of electric generation per unit (kWh).

In a steam power plant, the variable load on electric generation ultimately gets reflected on the variable steam demand on the steam generator and on various other equipments. The operation characteristics of such equipments are not linear with load, so, their operation becomes quite complicated. As the load on electrical supply systems grow, a number of power plants are interconnected to meet the load. The load is divided among various power plants to achieve the utmost economy in the whole system. When the system consists of one base load plant and one or more peak load plants, the load in excess of base load plant capacity is dispatched to the best peak system, all of which are nearly equally efficient, the best load distribution needs thorough study and full knowledge of the system.

SOLVED EXAMPLES

Example 1. *Determine the thermal efficiency of a steam power plant and its coal bill per annum using the following data.*

Maximum demand = 24000 kW
Load factor = 40%
Boiler efficiency = 90%
Turbine efficiency = 92%
Coal consumption = 0.87 kg/Unit
Price of coal = Rs. 280 per tonne

Solution.

η = Thermal efficiency
  = Boiler efficiency × Turbing efficiency
  = 0.9 × 0.92 = 0.83

Load factor = Average Load/Maximum Demand
Average Load = 0.4 × 24000 = 9600 kW
E = Energy generated in a year = 9600 × 8760 = 841 × 10^5 kWh
Cost of coal per year = (E × 0.87 × 280)/1000
  = (841 × 10^5 × 0.87 × 280)/1000
  = **Rs. 205 × 10^5.** Ans.

Example 2. *The maximum (peak) load on a thermal power plant of 60 mW capacity is 50 mW at an annual load factor of 50%. The loads having maximum demands of 25 mW, 20 mW, 8 mW and, 5 mW are connected to the power station.*

Determine: (a) Average load on power station (b) Energy generated per year (c) Demand factor (d) Diversity factor.

Solution.

(a) Load factor = Average load/Maximum demand
  Average load = 0.5 × 50 = 25 mW

(b) E = Energy generated per year
  = Average load × 8760
  = 219 × 10^6 kWh.

(c) Demand factor = Maximum demand/Connected load
  = 50/(25 + 20 + 8 + 5) = 0.86

(d) Diversity factor = $\dfrac{M_1}{M_2}$

where M_1 = Sum of individual maximum demands = 25 + 20 + 8 + 5 = 58 mW

M_2 = Simultaneous maximum demand = 50 mW

Diversity factor = $\dfrac{58}{50}$ = **1.16. Ans.**

Example 3. *In a steam power plant the capital cost of power generation equipment is Rs. 25×10^5. The useful life of the plant is 30 years and salvage value of the plant to Rs. 1×10^5. Determine by sinking fund method the amount to be saved annually for replacement if the rate of annual compound interest is 6%.*

Solution.
P = Capital cost = Rs. 20×10^5

S = Salvage value = Rs. 1×10^5

n = Useful life = 30 years

r = Compound interest

A = Amount to be saved per year for replacement

$$A = \dfrac{[(P-S)r]}{\{(1+r)^n - 1\}} = \dfrac{[(20 \times 10^5 - 1 \times 10^5)0.06]}{\{(1+0.06)^{30} - 1\}}$$

= **Rs. 24,000. Ans.**

Example 4. *A hydro power plant is to be used as peak load plant at an annual load factor of 30%. The electrical energy obtained during the year is 750×10^5 kWh. Determine the maximum demand. If the plant capacity factor is 24% find reserve capacity of the plant.*

Solution.
E = Energy generated = 750×10^5 kWh

Average load = $\dfrac{(750 \times 10^5)}{8760}$ = 8560 kW

where 8760 is the number of hours in year.

Load factor = 30%

M = Maximum demand

Load factor = Average load/Maximum demand

$M = \dfrac{85,600}{0.3}$ = 28,530 kW

C = Capacity of plant

Capacity factor = $\dfrac{E}{(C \times 8760)}$

$0.24 = \dfrac{(750 \times 10^5)}{(C \times 8760)}$

C = 35,667 kW

Reserve capacity = C − M = 35,667 − 28,530

= **7137 kW. Ans.**

Example 5. *A diesel power station has fuel consumption 0.2 kg per kWh. If the calorific value of the oil is 11,000 kcal per kg determine the overall efficiency of the power station.*

Solution. For 1 kWh output

Heat input = $11,000 \times 0.2 = 2200$ kcal.

Now 1 kWh = 862 kcal.

Overall efficiency = $\dfrac{\text{Output}}{\text{Input}} = \dfrac{866}{2200} = 39.2\%$. **Ans.**

Example 6. *A steam power station has an installed capacity of 120 MW and a maximum demand of 100 MW. The coal consumption is 0.4 kg per kWh and cost of coal is Rs. 80 per tonne. The annual expenses on salary bill of staff and other overhead charges excluding cost of coal are Rs.50×10^5. The power station works at a load factor of 0.5 and the capital cost of the power station is Rs. 4×10^5. If the rate of interest and depreciation is 10% determine the cost of generating per kWh.*

Solution. Maximum demand = 100 mW

Load factor = 0.5

Average load = $100 \times 0.5 = 50$ MW = $50 \times 1000 = 50,000$ kW.

Energy produced per year = $50,000 \times 8760 = 438 \times 10^6$ kWh.

Coal consumption = $438 \times 10^6 \times (0.4/1000) = 1752 \times 10^6$ tonnes.

Annual Cost

(1) Cost of coal = $1752 \times 10^2 \times 80 = $ Rs. $14,016 \times 10^2$

(2) Salaries = Rs. 50×10^5

(3) Interest and depreciation = $(10/100) \times 4 \times 10^5 = $ Rs. 4×10^4

Total cost = Rs. $14,016 \times 10^3$ + Rs. 50×10^5 + Rs. 4×10^4

= Rs. $19,056 \times 10^3$

Cost of generation per kWh = $\left\{ \dfrac{(19,056 \times 10^3)}{(438 \times 10^6)} \right\} \times 100$

= **4.35 paise.   Ans.**

Example 7. *Any undertaking consumes 6×10^6 kWh per year and its maximum demand is 2000 kW. It is offered two tariffs.*

(a) Rs. 80 per kW of maximum demand plus 3 paise per kWh.

(b) A flat rate of 6 paise per kWh.

Calculate the annual cost of energy.

Solution.

(a) (According to first tariff the cost of energy)

$$= 2000 \times 80 + \left(\dfrac{3}{100}\right) \times 6 \times 10^6$$

$= 160,000 + 180,000 = $ **Rs. 340,000.   Ans.**

(b) Cost of energy according to flat rate

$$= \left(\frac{6}{100}\right) \times 6 \times 10^6 = \textbf{Rs. 360,000.} \quad \textbf{Ans.}$$

Example 8. *Two lamps are to be compared:*

(a) Cost of first lamp is Re. 1 and it takes 100 watts.

(b) Cost of second lamp is Rs. 4 and it takes 60 watts.

Both lamps are of equal candlepower and each has a useful life of 100 hours. Which lamp will prove economical if the energy is charged at Rs. 70 per kW of maximum demand per year plus 5 paise per kWh? At what load factor both the lamps will be equally advantageous?

Solution. (a) **First Lamp**

$$\text{Cost of lamp per hour} = \frac{(1 \times 100)}{1000} = 0.1 \text{ paise}$$

$$\text{Maximum demand per hour} = \frac{100}{1000} = 0.1 \text{ kW}$$

Maximum demand charge per hour

$$= \frac{0.1 \times (70 \times 100)}{7860} = 0.08 \text{ paise}$$

Energy consumed per hour = $0.1 \times 1 = 0.1$ kWh

Energy charge per hour = $0.1 \times 5 = 0.5$ paise

Total cost per hour = $0.1 + 0.08 + 0.5 = 0.68$ paise.

(b) **Second Lamp**

$$\text{Cost of lamp per hour} = \frac{(4 \times 100)}{1000} = 0.4 \text{ paise}$$

$$\text{Maximum demand per hour} = \frac{60}{1000} = 0.06 \text{ kW}$$

$$\text{Maximum demand charge per hour} = \frac{0.06 \times (70 \times 100)}{8760} = 0.048 \text{ paise}$$

Energy consumed per hour = $0.06 \times 1 = 0.06$ kWh

Energy charge per hour = $0.06 \times 5 = 0.3$ paise

Total cost per hour = $0.4 + 0.048 + 0.3 = 0.748$ paise

Therefore the first lamp is economical

Let x be the load factor at which both lamps become equally advantageous. Only maximum demand charge changes with load factor.

$$0.1 + \frac{0.08}{x} + 0.5 = 0.4 + \frac{0.048}{x} + 0.3$$

$$x = 0.32$$

32% Ans.

POWER PLANT ECONOMICS AND VARIABLE LOAD PROBLEM

Example 9. *A new factory having a minimum demand of 100 kW and a load factor of 25% is comparing two power supply agencies.*

(a) *Public supply tariff is Rs. 40 per kW of maximum demand plus 2 paise per kWh.*
Capital cost = Rs. 70,000
Interest and depreciation = 10%

(b) *Private oil engine generating station.*
Capital Cost = Rs. 250,000
Fuel consumption = 0.3 kg per kWh
Cost of fuel = Rs. 70 per tonne
Wages = 0.4 paise per kWh
Maintenance cost = 0.3 paise per kWh
Interest and depreciation = 15%.

Solution. Load factor = Average load/Maximum demand

Average load = Load factor × Maximum demand
$$= 0.25 \times 700 = 175 \text{ kW}.$$

Energy consumed per year = $175 \times 8760 = 153.3 \times 10^4$ kWh.

(a) **Public Supply**

Maximum demand charges per year = $40 \times 700 =$ Rs. 28,000.

Energy charge per year = $\left(\dfrac{2}{100}\right) \times 153.3 \times 10^4 = 30{,}660$

Interest and depreciation = $\left(\dfrac{10}{100}\right) \times 70{,}000 =$ Rs. 7,000.

Total cost = Rs. [28,000 + 30,660 + 7,000] = Rs. 65,660

Energy cost per kWh = $\left(\dfrac{65{,}660}{153.3 \times 10^4}\right) \times 100 = 429$ paise

(b) **Private oil engine generating station**

Fuel consumption = $\dfrac{(0.3 \times 153.3 \times 10^4)}{1000} = 460$ tonnes

Cost of fuel = $460 \times 70 =$ Rs. 32,000

Cost of wages and maintenance
$$= \{(0.4 + 0.3)100\} \times 153.3 \times 10^4 = \text{Rs. } 10{,}731.$$

Interest and depreciation
$$= \left(\dfrac{15}{100}\right) \times 250{,}000 = \text{Rs. } 37{,}500$$

Total cost = Rs. [33,203 + 10,731 + 37,500]
= Rs. 80,431

Energy cost per kWh

$$\left(\frac{80,431}{153.3 \times 10^4}\right) \times 100 = 5.2 \text{ paise.} \quad \textbf{Ans.}$$

THEORETICAL PROBLEMS

1. Define: load factor, utility factor, plant operating factor, capacity factor, demand factor and diversity factor.
2. What is the difference between demand factor and diversity factor?
3. What is 'diversity factor' ? List its advantages in a power system.
4. Prove that the load factor of a power system is improved by an increase in diversity of load.
5. What is meant by load curve? Explain its importance in power generation.
6. Differentiate 'dump power', 'firm power' and 'prime power'.
7. Define 'depreciation' and explain its significance.
8. Explain the sinking fund method of calculating the depreciation.
9. Discuss the factors to be considered for, 'plant selection' for a
10. How 'load duration curve' is obtained from 'load' curve ?
11. What are the principal factors involved in fixing of a tariff?

NUMERICAL PROBLEMS

12. The following data is available for a steam power station:

    Maximum demand = 25,000 kW; Load factor = 0.4; Coal consumption = 0.86 kg/kWh; Boiler efficiency = 85%; Turbine efficiency = 90%; Price of coal = Rs. 55 per tonne.

    Determine the following:

    (i) Thermal efficiency of the station.

    (ii) Coal bill of the plant for one year. **[Ans.** (i) 76.5% (ii) Rs. 41,43,480]

13. The annual peak load on a 30 mW power station is 25 mW. The power station supplies load having maximum demands of 10 mW, 8.5 mW, 5 mW and 4.5 mW. The annual load factor is 0.45. Find:

    (i) Average load (ii) Energy supplied per year
    (iii) Diversity factor (iv) Demand factor.

    [Ans. (i) 11.25 mW (ii) 98.55 × 10^6 kWh (iii) 1.12 (iv) 0.9]

14. A power station has a maximum demand of 15 mW, a load factor of 0.7, a plant capacity factor of 0.525 and a plant use factor of 0.85. Find:

    (i) The daily energy produced.
    (ii) The reserve capacity of the plant.

POWER PLANT ECONOMICS AND VARIABLE LOAD PROBLEM   141

   (*iii*) The maximum energy that could be produced daily if the plant operating schedule is fully loaded when in operation.

   [**Ans.**   (*i*) 252,000 kWh (*ii*) 5,000 kW (*iii*) 296,470 kWh]

15. Determine the annual cost of a feed water softener from the following data: Cost = Rs. 80,000; Salvage value = 5%, Life = 10 years; Annual repair and maintenance cost = Rs. 2500; Annual cost of chemicals = Rs. 5000; Labour cost per month = Rs. 300; Interest on sinking fund = 5%.   [**Ans.** Rs. 17,140]

16. Calculate the unit cost of production of electric energy for a power station for which data are supplied as follows :

   Capacity   = 50 MW
   Cost per kW   = Rs. 600
   Load factor   = 40%
   Interest and depreciation = 10%

   Cost of fuel, taxation and salaries = Rs. 36×10^{11}.   [**Ans.** 3.71 paise]

17. Estimate the generating cost per unit supplied from a power plant having the following data :

   Plant capacity = 120 MW

   Capital cost = Rs. 600×10^6

   Annual load factor = 40%

   Annual cost of fuel, taxation, oil and salaries = Rs. 600,000 Interest and depreciation = 10%

   [**Ans.** 1.33 paise]

18. Estimate the generating cost per unit. supplied from a power plant having data :

   Output per year = 4×10^8 kWh

   Load factor = 50%

   Annual fixed charges = Rs. 40 per kW

   Annual running charges = 4 paise per kWh

19. A 50 MW generating station has the following data:

   Capital cost = Rs. 15×10^5

   Annual taxation = Rs. 0.4×10^5

   Annual salaries and wages = Rs. 1.2×10^6

   Cost of coal = Rs. 65 per tonne

   Calorific value of coal = 5500 kcal/kg.

   Rate of interest and depreciation = 12%,

   Plant heat rate = 33,000 kcal/kWh

   at 100% capacity and 40000 kcal/kWh at 60%.

   Calculate the generating cost/kWh at 100% and 60% capacity factor.

Chapter 4

Steam Power Plant

4.1 INTRODUCTION

Steam is an important medium of producing mechanical energy. Steam has the advantage that, it can be raised from water which is available in abundance it does not react much with the materials of the equipment of power plant and is stable at the temperature required in the plant. Steam is used to drive steam engines, steam turbines etc. Steam power station is most suitable where coal is available in abundance. Thermal electrical power generation is one of the major method. Out of total power developed in India about 60% is thermal. For a thermal power plant the range of pressure may vary from 10 kg/cm² to super critical pressures and the range of temperature may be from 250°C to 650°C.

The average all India Plant load factor (P.L.F.) of thermal power plants in 1987-88 has been worked out to be 56.4% which is the highest P.L.F. recorded by thermal sector so far.

4.2 ESSENTIALS OF STEAM POWER PLANT EQUIPMENT

A steam power plant must have following equipments :

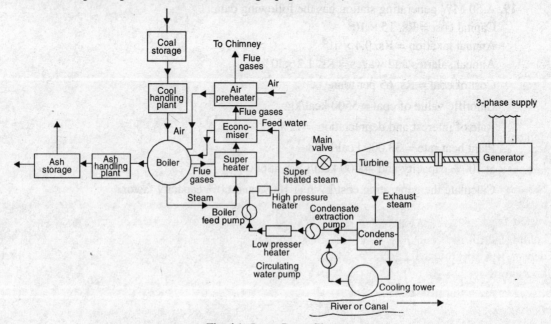

Fig. 4.1. Steam Power Plant.

STEAM POWER PLANT

1. A furnace to burn the fuel.
2. Steam generator or boiler containing water. Heat generated in the furnace is utilized to convert water in steam.
3. Main power unit such as an engine or turbine to use the heat energy of steam and perform work.
4. Piping system to convey steam and water.

In addition to the above equipment the plant requires various auxiliaries and accessories depending upon the availability of water, fuel and the service for which the plant is intended.

The flow sheet of a thermal power plant consists of the following four main circuits :

(*i*) Feed water and steam flow circuit
(*ii*) Coal and ash circuit
(*iii*) Air and gas circuit
(*iv*) Cooling water circuit.

A steam power plant using steam as working substance works basically on Rankine cycle.

Steam is generated in a boiler, expanded in the prime mover and condensed in the condenser and fed into the boiler again.

The different types of systems and components used in steam power plant are as follows :

(*i*) High pressure boiler
(*ii*) Prime mover
(*iii*) Condensers and cooling towers
(*iv*) Coal handling system
(*v*) Ash and dust handling system
(*vi*) Draught system
(*vii*) Feed water purification plant
(*viii*) Pumping system
(*ix*) Air preheater, economizer, super heater, feed heaters.

Fig. 4.1 shows a schematic arrangement of equipment of a steam power station. Coal received in coal storage yard of power station is transferred in the furnace by coal handling unit. Heat produced due to burning of coal is utilized in converting water contained in boiler drum into steam at suitable pressure and temperature. The steam generated is passed through the superheater. Superheated steam then flows through the turbine. After doing work in the turbine die pressure of steam is reduced. Steam leaving the turbine passes through the condenser which maintain the low pressure of steam at the exhaust of turbine. Steam pressure in the condenser depends upon flow rate and temperature of cooling water and on effectiveness of air removal equipment. Water circulating through the condenser may be taken from the various sources such as river, lake or sea. If sufficient quantity of water is not available the hot water coming out of the condenser may be cooled in cooling towers and circulated again through the condenser. Bled steam taken from the turbine at suitable extraction points is sent to low pressure and high pressure water heaters.

Air taken from the atmosphere is first passed through the air pre-heater, where it is heated by flue gases. The hot air then passes through the furnace. The flue gases after passing over boiler and superheater tubes, flow through the dust collector and then through economiser, air pre-heater and finally they are exhausted to the atmosphere through the chimney.

Steam condensing system consists of the following:

    (*i*) Condenser

    (*ii*) Cooling water

    (*iii*) Cooling tower

    (*iv*) Hot well

    (*v*) Condenser cooling water pump

    (*vi*) Condensate air extraction pump

    (*vii*) Air extraction pump

    (*viii*) Boiler feed pump

    (*ix*) Make up water pump.

4.2.1. POWER STATION DESIGN

Power station design requires wide experience. A satisfactory design consists of the following steps :

    (*i*) Selection of site

    (*ii*) Estimation of capacity of power station.

    (*iii*) Selection of turbines and their auxiliaries.

    (*iv*) Selection of boilers, and their auxiliaries.

    (*v*) Design of fuel handling system.

    (*vi*) Selection of condensers.

    (*vii*) Design of cooling system.

    (*viii*) Design of piping system to carry steam and water.

    (*ix*) Selection of electrical generator.

    (*x*) Design and control of instruments.

    (*xi*) Design of layout of power station. Quality of coal used in steam power station plays an important role in the design of power plant. The various factors to be considered while designing the boilers and coal handling units are as follows :

        (*a*) Slagging and erosion properties of ash.

        (*b*) Moisture in the coal. Excessive moisture creates additional problems particularly in case of pulverized fuel power plants.

        (*c*) Burning characteristic of coal.

        (*d*) Corrosive nature of ash.

4.2.2. CHARACTERISTICS OF STEAM POWER PLANT

The desirable characteristic for a steam power plant are as follows :

    (*i*) Higher efficiency.

    (*ii*) Lower cost.

STEAM POWER PLANT

(iii) Ability to burn coal especially of high ash content, and inferior coals.

(iv) Reduced environmental impact in terms of air pollution.

(v) Reduced water requirement.

(vi) Higher reliability and availability.

4.3 COAL HANDLING

Coal delivery equipment is one of the major components of plant cost. The various steps involved in coal handling are as follows : (Fig. 4.2)

(i) Coal delivery
(ii) Unloading
(iii) Preparation
(iv) Transfer
(v) Outdoor storage
(vi) Covered storage
(vii) In plant handling
(viii) Weighing and measuring
(ix) Feeding the coal into furnace.

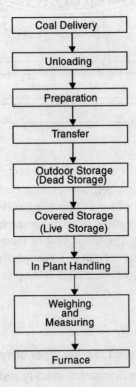

Fig. 4.2. Steps in Coal Handling.

(i) **Coal Delivery.** The coal from supply points is delivered by ships or boats to power stations situated near to sea or river whereas coal is supplied by rail or trucks to the power stations which are situated away from sea or river. The transportation of coal by trucks is used if the railway facilities are not available.

(ii) **Unloading.** The type of equipment to be used for unloading the coal received at the power station depends on how coal is received at the power station. If coal is delivered by trucks, there is no need of unloading device as the trucks may dump the coal to the outdoor storage. Coal is easily handled if the lift trucks with scoop are used. In case the coal is brought by railway wagons, ships or boats, the unloading may be done by car shakes, rotary car dumpers, cranes, grab buckets and coal accelerators. Rotary car dumpers although costly are quite efficient for unloading closed wagons.

(*iii*) **Preparation.** When the coal delivered is in the form of big lumps and it is not of proper size, the preparation (sizing) of coal can be achieved by crushers, breakers, sizers driers and magnetic separators.

(*iv*) **Transfer.** After preparation coal is transferred to the dead storage by means of the following systems :

1. Belt conveyors.
2. Screw conveyors.
3. Bucket elevators.
4. Grab bucket elevators.
5. Skip hoists.
6. Flight conveyor.

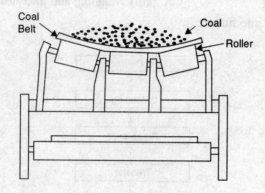

Fig. 4.3. Belt Conveyor.

1. Belt conveyor. Fig. 4.3 shows a belt conveyor. It consists of an endless belt. moving over a pair of end drums (rollers). At some distance a supporting roller is provided at the center. The belt is made, up of rubber or canvas. Belt conveyor is suitable for the transfer of coal over long distances. It is used in medium and large power plants. The initial cost of the system is not high and power consumption is also low. The inclination at which coal can be successfully elevated by belt conveyor is about 20. Average speed of belt conveyors varies between 200-300 r.p.m. This conveyor is preferred than other types.

Advantages of belt conveyor

1. Its operation is smooth and clean.
2. It requires less power as compared to other types of systems.
3. Large quantities of coal can be discharged quickly and continuously.
4. Material can be transported on moderates inclines.

2. Screw conveyor. It consists of an endless helicoid screw fitted to a shaft (Fig. 4.4). The screw while rotating in a trough transfers the coal from feeding end to the discharge end.

This system is suitable, where coal is to be transferred over shorter distance and space limitations exist. The initial cost of the system is low. It suffers from the drawbacks that the power consumption is high and there is considerable wear of screw. Rotation of screw varies between 75-125 r.p.m.

3. Bucket elevator. It consists of buckets fixed to a chain (Fig. 4.5). The chain moves over two wheels. The coal is carried by the buckets from bottom and discharged at the top.

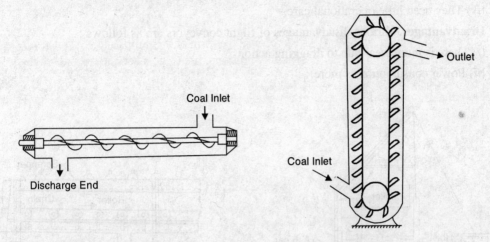

Fig. 4.4. Screw Conveyor.   Fig. 4.5. Bucket Elevator.

4. Grab bucket elevator. It lifts and transfers coal on a single rail or track from one point to the other. The coal lifted by grab buckets is transferred to overhead bunker or storage. This system requires less power for operation and requires minimum maintenance.

The grab bucket conveyor can be used with crane or tower as shown in Fig. 4.6. Although the initial cost of this system is high but operating cost is less.

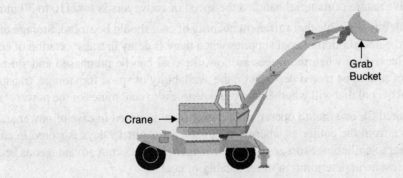

Fig. 4.6. Grab Bucket Elevator.

5. Skip hoist. It consists of a vertical or inclined hoistway a bucket or a car guided by a frame and a cable for hoisting the bucket. The bucket is held in up right position. It is simple and compact method of elevating coal or ash. Fig. 4.7 shows a skip hoist.

6. Flight conveyor. It consists of one or two strands of chain to which steel scraper or flights are attached'. which scrap the coal through a trough having identical shape. This coal is discharged in the bottom of trough. It is low in first cost but has large energy consumption. There is considerable wear.

Skip hoist and bucket elevators lift the coal vertically while Belts and flight conveyors move the coal horizontally or on inclines.

Fig. 4.8 shows a flight conveyor. Flight conveyors possess the following **advantages.**

(*i*) They can be used to transfer coal as well as ash.

(ii) The speed of conveyor can be regulated easily.

(iii) They have a rugged construction.

(iv) They need little operational care.

Disadvantages. Various disadvantages of flight conveyors are as follows :

(i) There is more wear due to dragging action.

(ii) Power consumption is more.

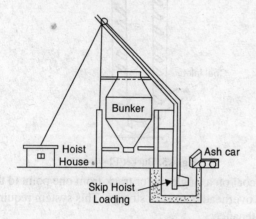

Fig. 4.7. Skip Hoist.   Fig. 4.8. Flight Conveyor.

(iii) Maintenance cost is high.

(iv) Due to abrasive nature of material handled the speed of conveyors is low (10 to 30 m/min).

(v) *Storage of coal.* It is desirable that sufficient quantity of coal should be stored. Storage of coal gives protection against the interruption of coal supplies when there is delay in transportation of coal or due to strikes in coal mines. Also when the prices are low, the coal can be purchased and stored for future use. The amount of coal to be stored depends on the availability of space for storage, transportation facilities, the amount of coal that will whether away and nearness to coal mines of the power station.

Usually coal required for one month operation of power plant is stored in case of power stations situated at longer distance from the collieries whereas coal need for about 15 days is stored in case of power station situated near to collieries. Storage of coal for longer periods is not advantageous because it blocks the capital and results in deterioration of the quality of coal.

The coal received at the power station is stored in dead storage in the form of piles laid directly on the ground.

The coal stored has the tendency to whether (to combine with oxygen of air) and during this process coal loss some of its heating value and ignition quality. Due to low oxidation the coal may ignite spontaneously. This is avoided by storing coal in the form of piles which consist of thick and compact layers of coal so that air cannot pass through the coal piles. This will minimize the reaction between coal and oxygen. The other alternative is to allow the air to pass through layers of coal so that air may remove the heat of reaction and avoid burning. In case the coal is to be stored for longer periods the outer surface of piles may be sealed with asphalt or fine coal.

The coal is stored by the following methods :

(i) *Stocking the coal in heats.* The coal is piled on the ground up to 10-12 m height. The pile top should be given a slope in the direction in which the rain may be drained off.

STEAM POWER PLANT

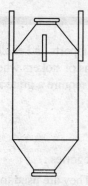

The sealing of stored pile is desirable in order to avoid the oxidation of coal after packing an air tight layer of coal.

Asphalt, fine coal dust and bituminous coating are the materials commonly used for this purpose.

(*ii*) *Under water storage.* The possibility of slow oxidation and spontaneous combustion can be completely eliminated by storing the coal under water.

Coal should be stored at a site located on solid ground, well drained, free of standing water preferably on high ground not subjected to flooding.

Fig. 4.9. Cylindrical Bucket.

(*vi*) In Plant Handling. From the dead storage the coal is brought to covered storage (Live storage) (bins or bunkers). A cylindrical bunker shown in Fig. 4.9. In plant handling may include the equipment such as belt conveyors, screw conveyors, bucket elevators etc. to transfer the coal. Weigh lorries hoppers and automatic scales are used to record the quantity of coal delivered to the furnace.

(*vii*) Coal weighing methods. Weigh lorries, hoppers and automatic scales are used to weigh the quantity coal. The commonly used methods to weigh the coal are as follows:

(*i*) Mechanical   (*ii*) Pneumatic   (*iii*) Electronic.

The Mechanical method works on a suitable lever system mounted on knife edges and bearings connected to a resistance in the form of a spring of pendulum. The pneumatic weighters use a pneumatic transmitter weight head and the corresponding air pressure determined by the load applied. The electronic weighing machines make use of load cells that produce voltage signals proportional to the load applied.

The important factor considered in selecting fuel handling systems are as follows:

(*i*) Plant flue rate

(*ii*) Plant location in respect to fuel shipping

(*iii*) Storage area available.

4.3.1 DEWATERING OF COAL

Excessive surface moisture of coal reduces and heating value of coal and creates handling problems. The coal should therefore be dewatered to produce clean coal. Cleaning of coal has the following advantages:

(*i*) Improved heating value.
(*ii*) Easier crushing and pulverising
(*iii*) Improved boiler performance
(*iv*) Less ash to handle.
(*v*) Easier handling.
(*vi*) Reduced transportation cost.

4.4 FUEL BURNING FURNACES

Fuel is burnt in a confined space called furnace. The furnace provides supports and enclosure for burning equipment. Solid fuels such as coal, coke, wood etc. are burnt by means of stokers where as burners are used to burn powdered (Pulverized) coal and liquid fuels. Solid fuels require a grate in the furnace to hold the bed of fuel.

4.4.1 TYPES OF FURNACES

According to the method of firing fuel furnaces are classified into two categories :

(i) Grate fired furnaces (ii) Chamber fired furnaces. Grate fired furnaces. They are used to burn solid fuels. They may have a stationary or a movable bed of fuel.

These furnaces are classified as under depending upon the method used to fire the fuel and remove ash and slag.

(i) Hand fired (ii) Semi-mechanized (iii) Stocker fired.

Hand fired and semi-mechanized furnaces are designed with stationary fire grates and stoker furnaces with traveling grates or stokers.

Chamber fired furnaces. They are used to burn pulverized fuel, liquid and gaseous fuels.

Furnace shape and size depends upon the following factors:

  (i) Type of fuel to be burnt.
  (ii) Type of firing to be used.
  (iii) Amount of heat to be recovered.
  (iv) Amount of steam to be produced and its conditions.
  (v) Pressure and temperature desired.
  (vi) Grate area required.
  (vii) Ash fusion temperature.
  (viii) Flame length.
  (ix) Amount of excess air to be used.

Simply furnace walls consists of an interior face of refractory material such as fireclay, silica, alumina, kaolin and diaspore, an intermediate layer of insulating materials such as magnesia with the exterior casing made up of steel sheet. Insulating materials reduce the heat loss from furnace but raise the refractory temperature. Smaller boilers used solid refractory walls but they are air cooled. In larger units, bigger boilers use water cooled furnaces.

To burn fuels completely, the burning equipment should fulfill the following conditions :

1. The flame temperature in the furnace should be high enough to ignite the incoming fuel and air. Continuous and reliable ignition of fuel is desirable.

2. For complete combustion the fuel and air should be thoroughly mixed by it.

3. The fuel burning equipment should be capable to regulate the rate of fuel feed.

4. To complete the burning process the fuel should remain in the furnace for sufficient time.

5. The fuel and air supply should be regulated to achieve the optimum air fuel ratios.

6. Coal firing equipment should have means to hold and discharge the ash.

STEAM POWER PLANT

Following factors should be considered while selecting a suitable combustion equipment for a particular type of fuel :

(i) Grate area required over which the fuel burns.

(ii) Mixing arrangement for air and fuel.

(iii) Amount of primary and secondary air required.

(iv) Arrangement to counter the effects of caping in fuel or of low ash fusion temperature.

(v) Dependability and easier operation.

(vi) Operating and maintenance cost.

4.5 METHOD OF FUEL FIRING

The solid fuels are fired into the furnace by the following methods :

1. Hand firing. 2. Mechanical firing.

4.5.1 HAND FIRING

This is a simple method of firing coal into the furnace. It requires no capital investment. It is used for smaller plants. This method of fuel firing is discontinuous process, and there is a limit to the size of furnace which can be efficiently fired by this method. Adjustments are to be made every time for the supply of air when fresh coal is fed into furnace.

Hand Fired Grates. A hand fired grate is used to support the fuel bed and admit air for combustion. While burning coal the total area of air openings varies from 30 to 50% of the total grate area. The grate area required for an installation depends upon various factors such as its heating surface, the rating at which it is to be operated and the type of fuel burnt by it. The width of air openings varies from 3 to 12 mm.

The construction of the grate should be such that it is kept uniformly cool by incoming air. It should allow ash to pass freely. Hand fired grates are made up of cast iron. The various types of hand fired grates are shown in Fig. 4.10. In large furnaces vertical shaking grates of circular type are used.

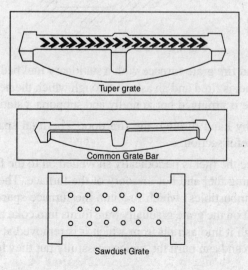

Fig. 4.10. Various Types of Hand Fired Grates.

The main characteristic of a grate fired furnaces are the heat liberation per unit of grate area and per unit of volume. The heat liberation per unit area of fire grate area is calculated as follows:

$$H = (W \times C)/A$$

where
H = Heat liberation per unit of fire grate area
W = Rate of fuel consumption (kg/sec)
C = Lower heating value of fuel (kcal/kg)
A = Fire grate area (m^2)

The heat liberation per unit of furnace volume is given by the following expression:

$$H = (W \times C)/V$$

where
H = Heat liberation per unit volume
W = Rate of fuel consumption (kg/sec)
C = Lower heating value of fuel (kcal/kg)
V = Volume of furnace (m^3).

These two characteristics depend on the following factors :

(*i*) Grade of fuel
(*ii*) Design of furnace
(*iii*) Method of combustion.

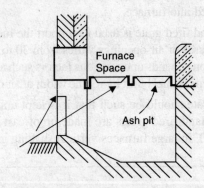

Fig. 4.11. Hand Fire Grate Furnace.

Fig. 4.11 shows a hand fire grate furnace with a stationary fuel bed. The grate divides it into the furnace space in which the fuel is fired and an ash pit through which the necessary air required for combustion is supplied. The grate is arranged horizontally and supports a stationary bed of burning fuel.

The fuel is charged by hand through the fire door. The total space in the grate used for the passage of air is called its useful section.

In a hand fired furnace the fuel is periodically shovelled on to the fuel bed burning on the grate, and is heated up by the burning fuel and hot masonry of the furnace. The fuel dries, and then evolves gaseous matter (volatiles combustibles) which rise into the furnace space and mix with air and burn forming a flame. The fuel left on the grate gradually transforms into coke and burns-up. Ash remains on the grate which drops through it into ash pit from which it is removed at regular intervals. Hand fired furnaces are simple in design and can burn the fuel successfully but they have some disadvantages also mentioned below:

STEAM POWER PLANT

(*i*) The efficiency of a hand fired furnace is low.

(*ii*) Attending to furnace requires hard manual labour.

(*iii*) Study process of fuel feed is not maintained.

Cleaning of hand fired furnaces may be mechanized by use of rocking grate bars as shown in Fig. 4.12. The grate bars loosen the slag and cause some of it to drop together with the ash into the bunker without disturbing the process of combustion.

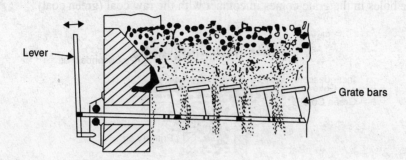

Fig. 4.12. Rocking Grate Bars.

4.5.2 MECHANICAL FIRING (STOKERS)

Mechanical stokers are commonly used to feed solid fuels into the furnace in medium and large size power plants.

The various advantages of stoker firing are as follows :

(*i*) Large quantities of fuel can be fed into the furnace. Thus greater combustion capacity is achieved.

(*ii*) Poorer grades of fuel can be burnt easily.

(*iii*) Stoker save labour of handling ash and are self-cleaning.

(*iv*) By using stokers better furnace conditions can be maintained by feeding coal at a uniform rate.

(*v*) Stokers save coal and increase the efficiency of coal firing. The main disadvantages of stokers are their more costs of operation and repairing resulting from high furnace temperatures.

Principles of Stokers. The working of various types of stokers is based on the following two principles:

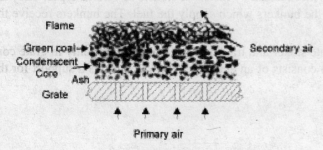

Fig. 4.13. Stokers

1. Overfeed Principle. According to this principle (Fig. 4.13) the primary air enters the grate from the bottom. The air while moving through the grate openings gets heated up and air while moving through the grate openings gets heated up and the grate is cooled.

The hot air that moves through a layer of ash and picks up additional energy. The air then passes through a layer of incandescent coke where oxygen reacts with coke to form-C02 and water vapours accompanying the air react with incandescent coke to form CO_2, CO and free H_2. The gases leaving the surface of fuel bed contain volatile matter of raw fuel and gases like CO_2, CO, H_2, N_2 and H_2O. Then additional air known as secondary air is supplied to burn the combustible gases. The combustion gases entering the boiler consist of N_2, CO_2, O_2 and H_2O and also CO if the combustion is not complete.

2. Underfeed Principle. Fig. 4.14 shows underfeed principle. In underfeed principle air entering through the holes in the grate comes in contact with the raw coal (green coal).

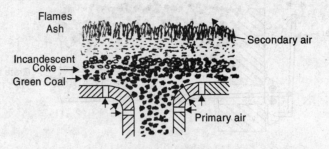

Fig. 4.14. Underfeed Principle.

Then it passes through the incandescent coke where reactions similar to overfeed system take place. The gases produced then passes through a layer of ash. The secondary air is supplied to burn the combustible gases. Underfeed principle is suitable for burning the semi-bituminous and bituminous coals.

Types of Stokers. The various types of stokers are as follows:

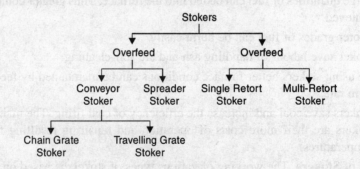

Fig. 4.15. Various Tyles of Stokers.

Charging of fuel into the furnace is mechanized by means of stokers of various types. They are installed above the fire doors underneath the bunkers which supply the fuel. The bunkers receive the fuel from a conveyor.

(i) **Chain Grate Stoker.** Chain grate stoker and traveling grate stoker differ only in grate construction. A chain grate stoker (Fig. 4.16) consists of an endless chain which forms a support for the fuel bed.

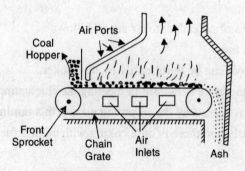

Fig. 4.16. Chain Grate Stoker.

The chain travels over two sprocket wheels, one at the front and one at the rear of furnace. The traveling chain receives coal at its front end through a hopper and carries it into the furnace. The ash is tipped from the rear end of chain. The speed of grate (chain) can be adjusted to suit the firing condition. The air required for combustion enters through the air inlets situated below the grate. Stokers are used for burning non-coking free burning high volatile high ash coals. Although initial cost of this stoker is high but operation and maintenance cost is low.

The traveling grate stoker also uses an endless chain but differs in that it carries small grate bars which actually support the fuel fed. It is used to burn lignite, very small sizes of anthracites coke breeze etc.

The stokers are suitable for low ratings because the fuel must be burnt before it reaches the rear of the furnace. With forced draught, rate of combustion is nearly 30 to 50 lb of coal per square foot of grate area per hour, for bituminous 20 to 35 pounds per square foot per hour for anthracite.

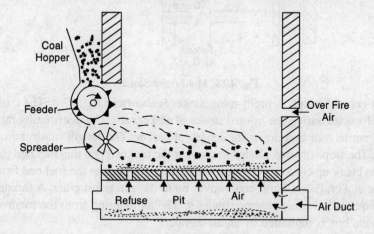

Fig. 4.17. Spreader Stoker.

(*ii*) **Spreader Stoker.** A spreader stoker is shown in Fig. 4.17. In this stoker the coal from the hopper is fed on to a feeder which measures the coal in accordance to the requirements. Feeder is a rotating drum fitted with blades. Feeders can be reciprocating rams, endless belts, spiral worms etc. From the feeder the coal drops on to spreader distributor which spread the coal over the furnace. The spreader system should distribute the coal evenly over the entire grate area. The spreader speed depends on the size of coal.

Advantages

The various advantages of spreader stoker are as follows :

1. Its operation cost is low.

2. A wide variety of coal can be burnt easily by this stoker.

3. A thin fuel bed on the grate is helpful in meeting the fluctuating loads.

4. Ash under the fire is cooled by the incoming air and this minimizes clinkering.

5. The fuel burns rapidly and there is little coking with coking fuels.

Disadvantages

1. The spreader does not work satisfactorily with varying size of coal.

2. In this stoker the coal burns in suspension and due to this fly ash is discharged with flue gases which requires an efficient dust collecting equipment.

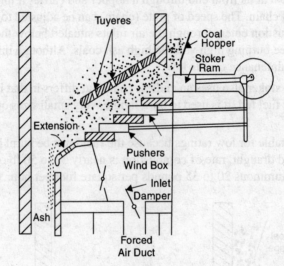

Fig. 4.18. Multi-retort Stoker.

(*iii*) **Multi-retort Stoker.** A multi-retort stoker is shown in Fig. 4.18. The coal falling from the hopper is pushed forward during the inward stroke of stoker ram. The distributing rams (pushers) then slowly move the entire coal bed down the length of stoker. The length of stroke of pushers can be varied as desired. The slope of stroke helps in moving the fuel bed and this fuel bed movement keeps it slightly agitated to break up clinker formation. The primary air enters the fuel bed from main wind box situated below the stoker. Partly burnt coal moves on to the extension grate. A thinner fuel bed on the extension grate requires lower air pressure under it. The air entering from the main wind box into the extension grate wind box is regulated by an air damper.

As sufficient amount of coal always remains on the grate, this stoker can be used under large boilers (upto 500,000 lb per hr capacity) to obtain high rates of combustion. Due to thick fuel bed the air supplied from the main wind box should be at higher pressure.

4.6 AUTOMATIC BOILER CONTROL

By means of automatic combustion control it becomes easy to maintain a constant steam pressure and uniform furnace draught and supply of air or fuel can be regulated to meet the changes in steam demand. The boiler operation becomes more flexible and better efficiency of combustion is achieved. This saves manual labour also.

Hagan system of automatic combustion control is shown in Fig. 4.19. Master relay R_1, is sensitive to small variations in steam pressure and is connected to steam pressure gauge.

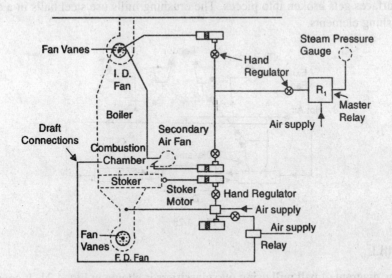

Fig. 4.19. Hagan System of Automatic Combustion Control.

A fall in pressure operates the master relay R_1 which in turn operates the servomotor coupled to the vanes of the induced draught (I.D.) fan to open them slightly and simultaneously the secondary air fan damper gets opened proportionately. By this readjustment of induced draught takes place and stabilized conditions in the combustion chamber get changed. These changes operate relay R_2 to alter the position of forced draught fan servo-motor to adjust the position of forced draught fan vanes so that stable conditions in combustion chamber are maintained. This change causes more air to flow through passage which in turn operates relay R_3. This causes stoker motor to supply extra fuel into the furnace. In case of an increase of pressure of steam the above process is reversed. Hand regulators are provided to servo motors and master relay for manual control of system.

4.7 PULVERIZED COAL

Coal is pulverized (powdered) to increase its surface exposure thus permitting rapid combustion. Efficient use of coal depends greatly on the combustion process employed.

For large scale generation of energy the efficient method of burning coal is confined still to pulverized coal combustion. The pulverized coal is obtained by grinding the raw coal in pulverising mills. The various pulverising mills used are as follows:

(*i*) Ball mill   (*ii*) Hammer mill
(*iii*) Ball and race mill   (*iv*) Bowl mill.

The essential functions of pulverising mills are as follows: (*i*) Drying of the coal (*ii*) Grinding

(*iii*) Separation of particles of the desired size.

Proper drying of raw coal which may contain moisture is necessary for effective grinding.

The coal pulverising mills reduce coal to powder form by three actions as follows:

(i) Impact   (ii) Attrition (abrasion)   (iii) Crushing.

Most of the mills use all the above mentioned all the three actions in varying degrees. In impact type mills hammers break the coal into smaller pieces whereas in attrition type the coal pieces which rub against each other or metal surfaces to disintegrate. In crushing type mills coal caught between metal rolling surfaces gets broken into pieces. The crushing mills use steel balls in a container. These balls act as crushing elements.

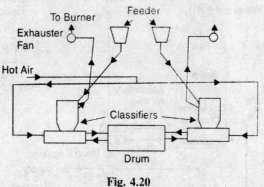

Fig. 4.20

4.7.1 BALL MILL

A line diagram of ball mill using two classifiers is shown in Fig. 4.21. It consists of a slowly rotating drum which is partly filled with steel balls. Raw coal from feeders is supplied to the classifiers from where it moves to the drum by means of a screw conveyor.

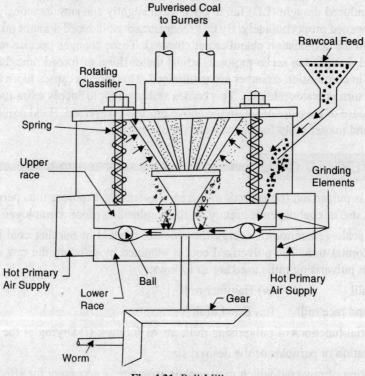

Fig. 4.21. Ball Mill.

As the drum rotates the coal gets pulverized due to the combined impact between coal and steel balls. Hot air is introduced into the drum. The powdered coal is picked up by the air and the coal air mixture enters the classifiers, where sharp changes in the direction of the mixture throw out the oversized coal particles. The over-sized particles are returned to the drum. The coal air mixture from the classifier moves to the exhauster fan and then it is supplied to the burners.

4.7.2 BALL AND RACE MILL

Fig. 4.22 shows a ball and race mill. In this mill the coal passes between the rotating elements again and again until it has been pulverized to desired degree of fineness. The coal is crushed between two moving surfaces namely balls and races. The upper stationary race and lower rotating race driven by a worm and gear hold the balls between them. The raw coal supplied falls on the inner side of the races. The moving balls and races catch coal between them to crush it to a powder. The necessary force needed for crushing is applied with the help of springs. The hot air supplied picks up the coal dust as it flows between the balls and races, and then enters the classifier. Where oversized coal particles are returned for further grinding, where as the coal particles of required size are discharged from the top of classifier.

In this mill coal is pulverized by a combination of crushing, impact and attrition between the grinding surfaces. The advantages of this mill are as follows :

(i) Lower capital cost (ii) Lower power consumption
(iii) Lower space required (iv) Lower weight.

However in this mill there is greater wear as compared to other pulverizes.

The use of pulverized coal has now become the standard method of firing in the large boilers. The pulverized coal burns with some advantages that result in economic and flexible operation of steam boilers.

Preparation of pulverized fuel with an intermediate bunker is shown in Fig. 4.22. The fuel moves to the automatic balance and then to the feeder and ball mill through which hot air is blown. It dries the pulverized coal and carries it from the mill to separator.

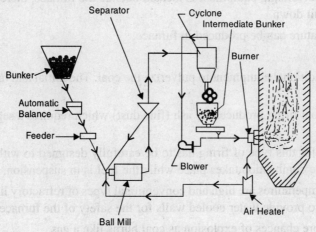

Fig. 4.22. Ball and Race Mill.

The air fed to the ball mill is heated in the air heater. In the separator dust (fine pulverized coal) is separated from large coal particles which are returned to the ball mill for regrinding. The dust moves to the cyclone. Most of the dust (about 90%) from cyclone moves to bunker. The remaining dust is mixed with air and fed to the burner.

Coal is generally ground in low speed ball tube mill. It is filled to 20-35% of its volume. With steel balls having diameter varying from 30-60 mm. The steel balls crush and ground the lumps of coal. The average speed of rotation of tube or drum is about 18-20 r.p.m. [Fig. 4.23].

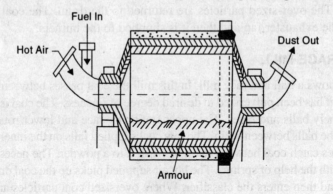

Fig. 4.23

Advantages

The advantages of using pulverized coal are as follows :

1. It becomes easy to burn wide variety of coal. Low grade coal can be burnt easily.

2. Powdered coal has more heating surface area. They permits rapids and high rates of combustion.

3. Pulverized coal firing requires low percentage of excess air.

4. By using pulverized coal, rate of combustion can be adjusted easily to meet the varying load.

5. The system is free from clinker troubles.

6. It can utilize highly preheated air (of the order of 700°F) successfully which promotes rapid flame propagation.

7. As the fuel pulverising equipment is located outside the furnace, therefore it can be repaired without cooling the unit down.

8. High temperature can be produced in furnace.

Disadvantages

1. It requires additional equipment to pulverize the coal. The initial and maintenance cost of the equipment is high.

2. Pulverized coal firing produces fly ash (fine dust) which requires a separate fly ash removal equipment.

3. The furnace for this type of firing has to be carefully designed to withstand for burning the pulverized fuel because combustion takes place while the fuel is in suspension.

4. The flame temperatures are high and conventional types of refractory lined furnaces are inadequate. It is desirable to provide water cooled walls for the safety of the furnaces.

5. There are more chances of explosion as coal burns like a gas.

6. Pulverized fuel fired furnaces designed to burn a particular type of coal can not be used to any other type of coal with same efficiency.

7. The size of coal is limited. The particle size of coal used in pulverized coal furnace is limited to 70 to 100 microns.

STEAM POWER PLANT

4.7.3 SHAFT MILL

Fig. 4.24 shows fuel pulverization with a shaft mill. The fuel from bunker is moved to feeder via automatic balance. Then from duct fuel goes to mill where it is crushed by beaters secured on the spindle of the mill rotor.

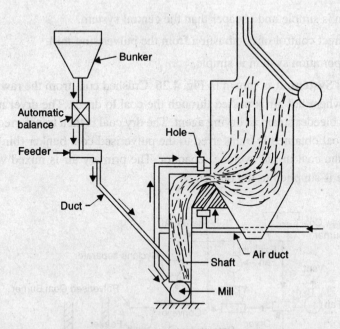

Fig. 4.24. Pulverization with a Shaft Mill.

The pulverised fuel is dried up and then blown into shaft by hot air. Secondary air is delivered into the furnace through holes to burn the fuel completely.

4.8 PULVERISED COAL FIRING

Pulverised coal firing is done by two system :

(i) Unit System or Direct System.

(ii) Bin or Central System.

Unit System. In this system (Fig. 4.25) the raw coal from the coal bunker drops on to the feeder.

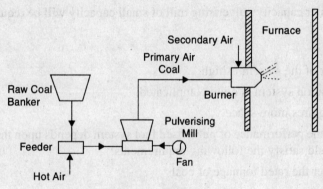

Fig. 4.25. Unit or Direct System.

Hot air is passed through coal in the feeder to dry the coal. The coal is then transferred to the pulverising mill where it is pulverised. Primary air is supplied to the mill, by the fan. The mixture of pulverised coal and primary air then flows to burner where secondary air is added. The unit system is so called from the fact that each burner or a burner group and pulveriser constitute a unit.

Advantages

(*i*) The system is simple and cheaper than the central system.

(*ii*) There is direct control of combustion from the pulverising mill.

(*iii*) Coal transportation system is simple.

Bin or Central System. It is shown in Fig. 4.26. Crushed coal from the raw coal bunker is fed by gravity to a dryer where hot air is passed through the coal to dry it. The dryer may use waste flue gases, preheated air or bleeder steam as drying agent. The dry coal is then transferred to the pulverising mill. The pulverised coal obtained is transferred to the pulverised coal bunker (bin). The transporting air is separated from the coal in the cyclone separator. The primary air is mixed with the coal at the feeder and the mixture is supplied to the burner.

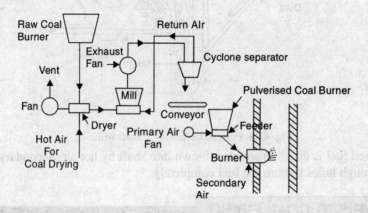

Fig. 4.26. Bin or Central System.

Advantages

1. The pulverising mill grinds the coal at a steady rate irrespective of boiler feed.

2. There is always some coal in reserve. Thus any occasional breakdown in the coal supply will not effect the coal feed to the burner.

3. For a given boiler capacity pulverising mill of small capacity will be required as compared to unit system.

Disadvantages

1. The initial cost of the system is high.

2. Coal transportation system is quite complicated.

3. The system requires more space.

To a large extent the performance of pulverised fuel system depends upon the mill performance. The pulverised mill should satisfy the following requirements:

1. It should deliver the rated tonnage of coal.

2. Pulverised coal produced by it should be of satisfactory fineness over a wide range of capacities.

3. It should be quiet in operation.

4. Its power consumption should be low.

5. Maintenance cost of the mill should be low.

Fig. 4.27 shows the equipments for unit and central system of pulverised coal handling plant.

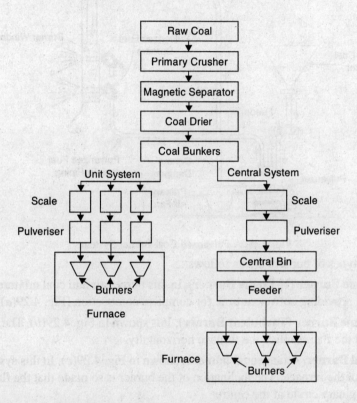

Fig. 4.27. Equipments for Central and Unit System.

4.9 PULVERISED COAL BURNERS

Burners are used to burn the pulverised coal. The main difference between the various burners lies in the rapidity of air-coal mixing *i.e.*, turbulence. For bituminous coals the turbulent type of burner is used whereas for low volatile coals the burners with long flame should be used. A pulverised coal burner should satisfy the following requirements:

(*i*) It should mix the coal and primary air thoroughly and should bring this mixture before it enters the furnace in contact with additional air known as secondary air to create sufficient turbulence.

(*ii*) It should deliver and air to the furnace in right proportions and should maintain stable ignition of coal air mixture and control flame shape and travel in the furnace. The flame shape is controlled by the secondary air vanes and other control adjustments incorporated into the burner. Secondary air if supplied in too much quantity may cool the mixture and prevent its heating to ignition temperature.

(*iii*) Coal air mixture should move away from the burner at a rate equal to flame front travel in order to avoid flash back into the burner.

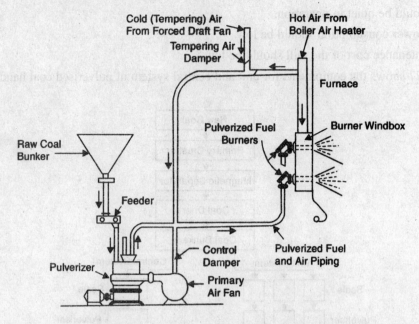

Fig. 4.28. A Pulverised Coal Burner System.

The various types of burners are as follows :

1. Long Flame Burner (U-Flame Burner). In this burner air and coal mixture travels a considerable distance thus providing sufficient time for complete combustion [Fig. 4.29(*a*)].

2. Short Flame Burner (Turbulent Burner). It is shown in Fig. 4.29(*b*). The burner is fitted in the furnace will and the flame enters the furnace horizontally.

3. Tangential Burner. A tangential burner is shown in Fig. 4.29(*c*). In this system one burner is fitted attach corner of the furnace. The inclination of the burner is so made that the flame produced are tangential to an imaginary circle at the centre.

4. Cyclone Burner. It is shown in Fig. 4.29(*d*). This burner uses crushed coal intend of pulverised coal. Its advantages are as follows :

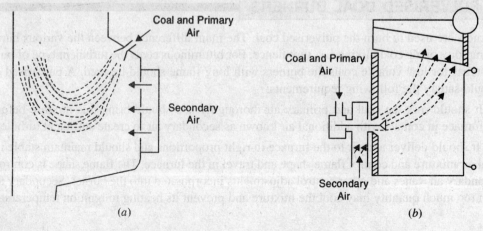

(*a*)   (*b*)

STEAM POWER PLANT

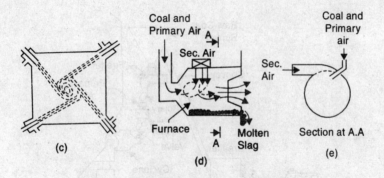

Fig. 4.29. Various Types of Burners.

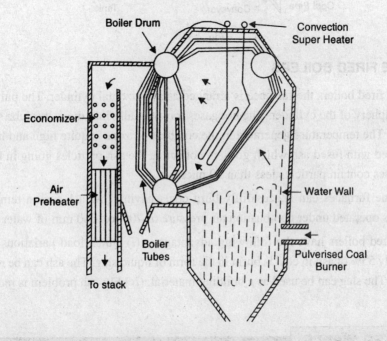

Fig. 4.30. Pulverised Coal-fired Boiler.

(*i*) It saves the cost of pulverisation because of a crusher needs less power than a pulveriser.

(*ii*) Problem of fly ash is reduced. Ash produced is in the molten form and due to inclination of furnace it flows to an appropriate disposal system.

Fig. 4.30 shows a pulverised coal-fired boiler.

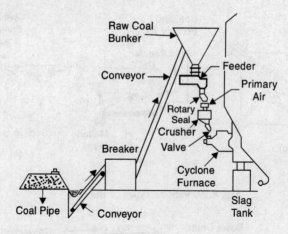

Fig. 4.31

4.9.1 CYCLONE FIRED BOILERS

In cyclone fired boilers the fur-nace is arranged as a horizontal cylinder. The pulverised fuel is bed along the periphery of the cylinder. The hot gases travel axially into the water tube section having a tight helix path. The temperature generated in the combustion zone is quite high and because of this the tubes are coated with fused ash which goes on collecting the ash particles going in the flue gases. The out going gases contain particles less than 20 microns.

The cyclone furnaces can successfully burn coals having low ash fusion temperature. The cyclone furnace is operated under combustion air pressure of 700 to 1000 mm of water gauge.

Cyclone fired boilers have the following advantages : (*i*) Quick load variations can be easily handled. (*ii*) Nearly 55% of ash in coal is burnt in the form of liquid slag. The ash can be removed in the molten form. (*iii*) The slag can be used as a building material. (*iv*) Fly-ash problem is reduced to much lower limits.

4.10 WATER WALLS

Larger central station type boilers have water cooled furnaces. The combustion space of a furnace is shielded wholly or partially by small diameter tubes placed side by side. Water from the boiler is made to circulate through these tubes which connect lower and upper headers of boiler.

The provision of water walls is advantageous due to following reasons: (1) These walls provide a protection to the furnace against high temperatures. (2) They avoid the erosion of the refractory material and insulation. (3) The evaporation capacity of the boiler is increased.

STEAM POWER PLANT

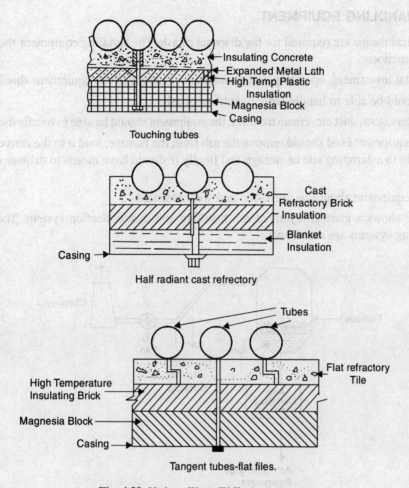

Fig. 4.32. Various Water Walls Arrangement.

The tubes are attached with the refractory materials on the inside or partially embedded into it. Fig. 4.32 shows the various water walls arrangement.

4.11 ASH DISPOSAL

A large quantity of ash is, produced in steam power plants using coal. Ash produced in about 10 to 20% of the total coal burnt in the furnace. Handling of ash is a problem because ash coming out of the furnace is too hot, it is dusty and irritating to handle and is accompanied by some poisonous gases. It is desirable to quench the ash before handling due to following reasons:

1. Quenching reduces the temperature of ash.
2. It reduces the corrosive action of ash.
3. Ash forms clinkers by fusing in large lumps and by quenching clinkers will disintegrate.
4. Quenching reduces the dust accompanying the ash.

Handling of ash includes its removal from the furnace, loading on the conveyors and delivered to the fill from where it can be disposed off.

4.11.1 ASH HANDLING EQUIPMENT

Mechanical means are required for the disposal of ash. The handling equipment should perform the following functions:

(1) Capital investment, operating and maintenance charges of the equipment should be low.

(2) It should be able to handle large quantities of ash.

(3) Clinkers, soot, dust etc. create troubles, the equipment should be able to handle them smoothly.

(4) The equipment used should remove the ash from the furnace, load it to the conveying system to deliver the ash to a dumping site or storage and finally it should have means to dispose of the stored ash.

(5) The equipment should be corrosion and wear resistant.

Fig. 4.33 shows a general layout of ash handling and dust collection system. The commonly used ash handling systems are as follows :

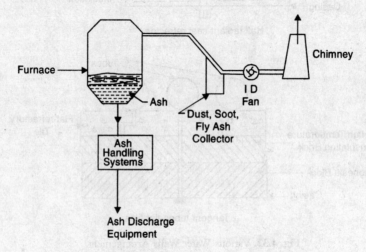

Fig. 4.33. Ash Handling and Dust Collections System.

(i) Hydraulic system

(ii) pneumatic system

(iii) Mechanical system.

The commonly used ash discharge equipment is as follows:

(i) Rail road cars

(ii) Motor truck

(iii) Barge.

The various methods used for the disposal of ash are as follows :

(i) **Hydraulic System.** In this system, ash from the furnace grate falls into a system of water possessing high velocity and is carried to the sumps. It is generally used in large power plants. Hydraulic system is of two types namely low pressure hydraulic system used for continuous removal of ash and high pressure system which is used for intermittent ash disposal. Fig. 4.34 shows hydraulic system.

STEAM POWER PLANT

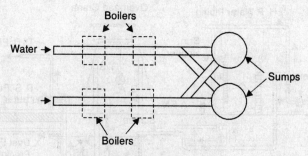

Fig. 4.34. Hydraulic System.

In this method water at sufficient pressure is used to take away the ash to sump. Where water and ash are separated. The ash is then transferred to the dump site in wagons, rail cars or trucks. The loading of ash may be through a belt conveyor, grab buckets. If there is an ash basement with ash hopper the ash can fall, directly in ash car or conveying system.

(*ii*) **Water Jetting.** Water jetting of ash is shown in Fig. 4.35. In this method a low pressure jet of water coming out of the quenching nozzle is used to cool the ash. The ash falls into a trough and is then removed.

(*iii*) **Ash Sluice Ways and Ash Sump System.** This system shown diagrammatically in Fig. 4.36 used high pressure (H.P.) pump to supply high pressure (H.P.) water-jets which carry ash from the furnace bottom through ash sluices (channels) constructed in basement floor to ash sump fitted with screen. The screen divides the ash sump into compartments for coarse and fine ash. The fine ash passes through the screen and moves into the dust sump (D.S.). Dust slurry pump (D.S. pump) carries the dust through dust pump (D.P), suction pipe and dust delivery (D.D.) pipe to the disposal site. Overhead crane having grab bucket is used to remove coarse ash. A.F.N represents ash feeding nozzle and S.B.N. represents sub way booster nozzle and D.A. means draining apron.

(*iv*) **Pneumatic system.** In this system (Fig. 4.37) ash from the boiler furnace outlet falls into a crusher where larger ash particles are crushed to small sizes. The ash is then carried by a high velocity air or steam to the point of delivery. Air leaving the ash separator is passed through filter to remove dust etc. so that the exhauster handles clean air which will protect the blades of the exhauster.

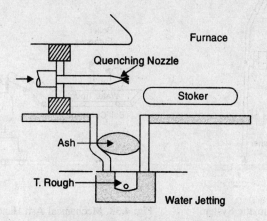

Fig. 4.35. Water Jetting of Ash.

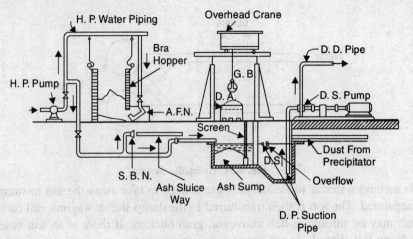

Fig. 4.36. Ash Sump System.

(v) **Mechanical ash handling system.** Fig. 4.38 shows a mechanical ash handling system. In this system ash cooled by water seal falls on the belt conveyor and is carried out continuously to the bunker. The ash is then removed to the dumping site from the ash bunker with the help of trucks.

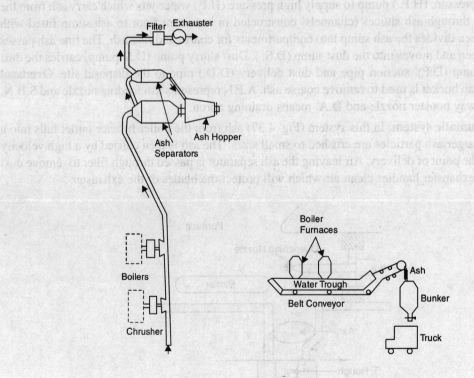

Fig. 4.37. Pneumatic System.   Fig. 4.38. Mechanical Ash Handling.

Efficient Combustion of Coal. The factors which affect the efficient combustion of coal are as follows :

1. Type of coal. The important factors which are considered for the selection of coal are as follows :

   (*i*) Sizing
   (*ii*) Caking
   (*iii*) Swelling properties
   (*iv*) Ash fusion temperature.

The characteristics which control the selection of coal for a particular combustion equipment are as follows :

   (*i*) Size of coal
   (*ii*) Ultimate and proximate analysis
   (*iii*) Resistance of degradation
   (*iv*) Grindability
   (*v*) Caking characteristics
   (*vi*) Slagging characteristics
   (*vii*) Deterioration during storage
   (*viii*) Corrosive characteristics
   (*ix*) Ash Content.

The average ash content in Indian coal is about 20%. It is therefore desirable to design the furnace in such a way as to burn the coal of high ash content. The high ash content in coal has the following:

   (*i*) It reduces thermal efficiency of the boiler as loss of heat through unburnt carbon, excessive clinker formation and heat in ashes is considerably high.
   (*ii*) There is difficulty of hot ash disposal.
   (*iii*) It increases size of plant.
   (*iv*) It increases transportation cost of fuel per unit of heat produced.
   (*v*) It makes the control difficult due to irregular combustion. High as content fuels can be used more economically in pulverised form. Pulverised fuel burning increases the thermal efficiency as high as 90% and controls can be simplified by just adjusting the position of burners in pulverised fuel boilers. The recent steam power plants in India are generally designed to use the pulverised coal.

2. Type of Combustion equipment. It includes the following:

   (*i*) Type of furnace
   (*ii*) Method of coal firing such as :
      (*a*) Hand firing
      (*b*) Stoker firing
      (*c*) Pulverised fuel firing.
   (*iii*) Method of air supply to the furnace. It is necessary to provide adequate quantity of secondary air with sufficient turbulence.

(*iv*) Type of burners used.

(*v*) Mixing arrangement of fuel and air.

The flames over the bed are due to the burning of volatile gases, lower the volatile content in the coal, shorter will be the flame. If the volatiles burn up intensely high temperature is generated over the furnace bed and helps to burn the carbon completely and *vice versa*.

For complete burning of volatiles and prevent unburnt carbon going with ash adequate quantity of secondary air with sufficient turbulence should be provided.

4.12 SMOKE AND DUST REMOVAL

In coal fed furnaces the products of combustion contain particles of solid matter floating in suspension. This may be smoke or dust. The production of smoke indicates that combustion conditions are faulty and amount of smoke produced can be reduced by improving the furnace design.

In spreader stokers and pulverised coal fired furnaces the coal is burnt in suspension and due to this dust in the form of fly ash is produced. The size of dust particles is designated in microns (1 μ = 0.001 mm). Dust particles are mainly ash particles called fly ash intermixed with some quantity of carbon ash material called cinders. Gas borne particles larger than 1μ in diameter are called dust and when such particles become greater in size than 100p they are called cinders. Smoke is produced due to the incomplete combustion of fuels, smoke particles are less than 10p in size.

The disposal smoke to the atmosphere is not desirable due to the following reasons :

1. A smoky atmosphere is less healthful than smoke free air.

2. Smoke is produced due to incomplete combustion of coal. This will create a big economic loss due to loss of heating value of coal.

3. In a smoky atmosphere lower standards of cleanliness are prevalent. Buildings, clothings, furniture etc. becomes dirty due to smoke. Smoke corrodes the metals and darkens the paints.

To avoid smoke nuisance the coal should be completely burnt in the furnace.

The presence of dense smoke indicates poor furnace conditions and a loss in efficiency and capacity of a boiler plant. A small amount of smoke leaving chimney shows good furnace conditions whereas smokeless chimney does not necessarily mean a better efficiency in the boiler room.

To avoid the atmospheric pollution the fly ash must be removed from the gaseous products of combustion before they leaves the chimney.

The removal of dust and cinders from the flue gas is usually effected by commercial dust collectors which are installed between the boiler outlet and chimney usually in the chimney side of air preheater.

4.13 TYPES OF DUST COLLECTORS

The various types of dust collectors are as follows :

1. Mechanical dust collectors.

2. Electrical dust collectors.

Mechanical dust collectors. Mechanical dust collectors are sub-divided into wet and dry types. In wet type collectors also known as scrubbers water sprays are used to wash dust from the air. The basic principles of mechanical dust collectors are shown in Fig. 4.38. As shown in Fig. 4.39(*a*) by increasing the cross-sectional area of duct through which dust laden gases are passing, the velocity of gases is reduced and causes heavier dust particles to fall down. Changing the direction of flow [Fig. 4.39(b)] of flue gases causes the heavier particles of settle out. Sometime baffles are provided as shown in Fig. 4.39(c) to separate the heavier particles.

Mechanical dust collectors may be wet type or dry type. Wet type dust collectors called scrubbers make use of water sprays to wash the dust from flue gases.

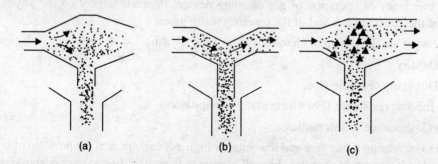

Fig. 4.39. Mechanical Dust Collector.

Dry type dust collectors include gravitational, cyclone, louvred and baffle dust collectors.

A cyclone dust collector uses a downward flowing vortex for dust laden gases along the inner walls. The clean gas leaves from an inner upward flowing vortex. The dust particles fall to the bottom due to centrifuging action.

Electrostatic Precipitators. It has two sets of electrodes, insulated from each other that maintain an electrostatic field between them at high voltage. The flue gases are made to pass between these two sets of electrodes. The electric field ionises the dust particle; that pass through it attracting them to the electrode of opposite charge. The other electrode is maintained at a negative potential of 30,000 to 60,000 volts. The dust particles are removed from the collecting electrode by rapping the electrode periodically. The electrostatic precipitator is costly but has low maintenance cost and is frequently employed with pulverised coal fired power stations for its effectiveness on very fine ash particles and is superior to that of any other type.

The principal characteristics of an ash collector is the degree of collection.

$$\eta = \text{Degree of collection}$$
$$= \frac{(G_1 - G_2)}{G_1}$$
$$= \frac{(C_1 - C_2)}{C_1}$$

where

G_1 = Quantity of ash entering an ash collector per unit time (kg/s)

G_2 = Quantity of uncollected ash passing through the collector per unit time (kg/s)

C_1 = Concentration of ash in the gases at the inlet to the ash collector (kg/m^3)

C_2 = Ash concentration at the exist (kg/m^3).

Depending on the type of fuel and the power of bailer the ash collection in industrial boilers and thermal power stations can be effected by mechanical ash collectors, fly ash scrubbers and electrostatic precipitators.

For fly ash scrubbers of large importance is the content of free lime (CaO) in the ash. With a high concentration of CaO the ash can be cemented and impair the operation of a scrubber.

The efficiency of operation of gas cleaning devices depends largely on the physico-chemical properties of the collected ash and of the entering waste gases.

Following are the principal characteristics of the fly ash:

(*i*) Density

(*ii*) Dispersity (Particle size)

(*iii*) Electric resistance (For electrostatic precipitators)

(*iv*) Coalescence of ash particles.

Due to increasing boiler size and low sulphur high ash content coal the problem of collecting fly ash is becoming increasingly complex. Fly ash can range from very fine to very coarse size depending on the source. Particles colour varies from light tan to grey to black. Tan colour indicates presence of ion oxide while dark shades indicate presence of unburnt carbon. Fly ash particles size varies between 1. micron (1 μ) to 300 μ. Fly ash concentration in flue gases depends upon mainly the following factors :

(*i*) Coal composition.

(*ii*) Boiler design and capacity.

Percentage of ash in coal directly contributes to fly ash emission while boiler design and operation determine the percentage retained in the furnace as bottom ash and fly ash carried away by flue gas. Fly ash concentration widely varies around 20-90 g/mm^3 depending on coal and boiler design. Fly ash particle size distribution depends primarily on the type of boiler such as pulverised coal fired boiler typically produces coarser particles then cyclone type boilers. Electrostatic precipitator (ESP) is quite commonly used for removal of fly ash from flue gases.

4.13.1 FLY ASH SCRUBBER

Fig. 4.40 shows a fly wash centrifugal scrubber. It is similar to a mechanical ash collector but has a flowing water film on its inner walls. Due to this film, the collected ash is removed more rapidly from the apparatus to the bin and there is less possibility for secondary. Capture of collected dust particles by the gas flow. The degree of ash collection in scrubbers varies from 0.82 to 0.90. The dust laden gas enters through the inlet pipe.

Cinder Catcher. Cinder catcher is used to remove dust and cinders from the gas. In this catcher the dust laden gas is made to strike a series of vertical baffles that change its direction and reduce its velocity. The separated dust and cinders fall to the hopper for removal. Cinder catchers are ordinarily used with stoker firing.

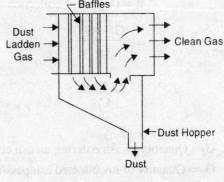

Fig. 4.40

4.13.2 FLUIDISED BED COMBUSTION (FBC)

Burning of pulverised coal has some problems such as particle size of coal used in pulverised firing is limited to 70-100 microns, the pulverised fuel fired furnances designed to burn a particular can not be used other type of coal with same efficiency, the generation of high temp. about (1650 C) in the furnace creates number of problems like slag formation on super heater, evaporation of alkali metals in ash and its deposition on heat transfer surfaces, formation of SO_2 and NO_X in large amount.

Fluidised Bed combustion system can burn any fuel including low grade coals (even containing 70% ash), oil, gas or municipal waste. Improved desulphurisation and low NO_X emission are its main characteristics. Fig. 4.41 shows basic principle of Fluidised bed combustion (FBC) system. The fuel and inert material dolomite are fed on a distribution plate and air is supplied from the bottom of distribution plate. The air is supplied at high velocity so that solid feed material remains in suspension condition during burning. The heat produced is used to heat water flowing through the tube and convert water into steam: During burning SO_2 formed is absorbed by the dolomite and thus prevents its escape with the exhaust gases. The molten slag is tapped from the top surface of the bed. The bed temperature is nearly 800-900'C which is ideal for sulphur retention addition of limestone or dolomite to the bed brings down SO_2 emission level to about 15% of that in conventional firing methods.

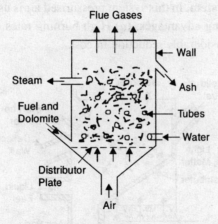

Fig. 4.41

The amount of NO_X is produced is also reduced because of low temperature of bed and low excess air as compared to pulverised fuel firing.

The inert material should be resistant to heat and disintegra-tion and should have similar density as that of coal. Limestone, or dolomite, fused alumina, sintered ash are commonly used as inert materials.

Various advantages of FBC system are as follows:

(i) FBC system can use any type of low grade fuel including municipal wastes and therefore is a cheaper method of power generation.

(ii) It is easier to control the amount of SO_2 and NO_X, formed during burning. Low emission of SO_2 and NO_X. will help in controlling the undesirable effects of SO_2 and NO_X. during combustion. SO_2 emission is nearly 15% of that in conventional firing methods.

(iii) There is a saving of about 10% in operating cost and 15% in the capital cost of the power plant.

(*iv*) The size of coal used has pronounced effect on the operation and performance of FBC system. The particle size preferred is 6 to 13 mm but even 50 mm size coal can also be used in this system.

4.13.3 TYPES OF FBC SYSTEMS

FBC systems are of following types :

(*i*) **Atmospheric FBC system :**

(*a*) Over feed system

(*b*) Under feed system.

In this system the pressure inside the bed is atmospheric.

Fig. 4.42 shows commercial circulation FBC system. The solid fuel is made to enter the furnace from the side of walls. The Low Velocity (LV), Medium Velocity (MV) and High Velocity (HV) air is supplied at different points along the sloping surface of the distribution ash is collected from the ash port. The burning is efficient because of high lateral turbulence.

(*ii*) **Pressurised FBC system.** In this system pressurised air is used for fluidisation and combustion. This system : the following advantages: (*a*) High burning rates. (*b*) Improved desulphurisation and low NO, emission. (*c*) Considerable reduction in cost.

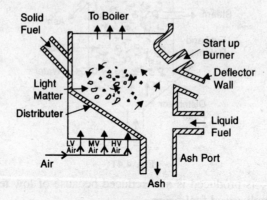

Fig. 4.42

1. (*a*) What is meant by 'over feed' and 'under feed' principles of firing coal ?

    (*b*) What are the different methods of firing coal ? Discuss the advantages of mechanical methods of firing coal.

    (*c*) Make neat sketch and explain the working of: (*i*) Chain grate stoker (*ii*) Spreader stoker. (*iii*) Multi retort stoker.

    (*d*) What is Fluidised Bed Combustion system ? Sketch and describe a Fluidised Bed Combustion (FBC) system. State the advantages of FBC system.

2. Describe the various types of grates used with hand fired furnaces.
3. Name the various methods of ash handling. Describe the pneumatic system of ash handling. Why it is essential to quench the ash before handling ?
4. (a) Describe the various methods used to fire pulverised coal.
   (b) Make a neat sketch of ball and Race mill and explain its working.
   (c) State the advantages of pulverised fuel firing.
5. Name the different types of coal-pulverising mills. Describe Ball-Mill.
6. Describe the various types of burners used to burn pulverised coal.
7. Name various draught systems. Describe the operation of a balanced draught system.
8. What is the cause of smoke ? State the factors necessary for its prevention.
9. Name the different types of chimneys used. State the advantages of steel chimney. Derive an expression for the height of chimney.
10. Describe the various methods used to control the degree of superheat. Name the advantages gained by using super-heat steam.
11. What is condenser ? Name the different types of condenser. Describe the operation of (i) Surface condenser (ii) Jet condenser.
12. What is a steam trap ? Where it is located ? Describe Ball Float steam trap.
13. What are the requirement of a well designed pipe line in a steam power plant. Name and describe the various expansion bends used in piping steam.
14. What are the advantages of using large capacity boilers ? Describe the operation of:
    (i) Velox Boiler (ii) Benson Boiler (iii) Loeffler Boiler.
15. State the advantages and disadvantages of a steam power station as compared to hydro-electric power station and nuclear power station.
16. Describe the various factors which determine the location of a steam power station.
17. Write short notes on the following: (a) Cyclone and collector (b) Industrial steam turbines (c) Hydraulic test of boiler (d) Draught fans (e) Steam separator (f) Economiser (g) Cyclone fired boilers (h) Pressure Filter. (i) Air preheater (j) Pipe fittings (k) Heat flow in steam plant.
18. What is the difference between water-tube and fire tube boilers ? Describe the working principle of Cochran Boiler or Lancashire Boiler.
19. (a) How will you classify various types of boilers ? (b) Write short notes on the following:
    (i) Efficiency of boiler
    (ii) Maintenance of boiler
    (iii) Accessories of a boiler
    (iv) Overall efficiency of steam power plant
    (v) Steam turbine specifications
    (vi) Causes of heat loss in boiler.

20. Explain the methods used to increase thermal efficiency of a steam power plant.
21. Write short notes on the following:
    (a) pH value of water.
    (b) Power plant pumps.
    (c) Steam turbine capacity.
    (d) Comparison of forced and induced draft system for boiler.
    (e) Principles of steam power plant design.
    (f) Korba super thermal power station.
    (g) Singrauli super thermal power plant.
22. Determine the quantity of air per kg of coal burnt in a furnace if the stack height is 58 m and draught produced is 35 mm of water. The temperature of flue gases is 380 C.

Chapter 5

Steam Generator

5.1 INTRODUCTION

Boiler is an apparatus to produce steam. Thermal energy released by combustion of fuel is transferred to water, which vaporizes and gets converted into steam at the desired temperature and pressure. The steam produced is used for:

(i) Producing mechanical work by expanding it in steam engine or steam turbine.

(ii) Heating the residential and industrial buildings

(iii) Performing certain processes in the sugar mills, chemical and textile industries.

Boiler is a closed vessel in which water is converted into steam by the application of heat. Usually boilers are coal or oil fired. A boiler should fulfill the following requirements

(i) **Safety.** The boiler should be safe under operating conditions.

(ii) **Accessibility.** The various parts of the boiler should be accessible for repair and maintenance.

(iii) **Capacity.** The boiler should be capable of supplying steam according to the requirements.

(iv) **Efficiency.** To permit efficient operation, the boiler should be able to absorb a maximum amount of heat produced due to burning of fuel in the furnace.

(v) It should be simple in construction and its maintenance cost should be low.

(vi) Its initial cost should be low.

(vii) The boiler should have no joints exposed to flames.

(viii) The boiler should be capable of quick starting and loading.

The performance of a boiler may be measured in terms of its evaporative capacity also called power of a boiler. It is defined as the amount of water evaporated or steam produced in kg per hour. It may also be expressed in kg per kg of fuel burnt or kg/hr/m^2 of heating surface.

5.2 TYPES OF BOILERS

The boilers can be classified according to the following criteria.

According to flow of water and hot gases.

1. Water tube.
2. Fire tube.

In water tube boilers, water circulates through the tubes and hot products of combustion flow over these tubes. In fire tube boiler the hot products of combustion pass through the tubes, which are surrounded, by water. Fire tube boilers have low initial cost, and are more compacts. But they are more likely to explosion, water volume is large and due to poor circulation they cannot meet quickly the change in steam demand. For the same output the outer shell of fire tube boilers is much larger than the shell of water-tube boiler. Water tube boilers require less weight of metal for a given size, are less liable to explosion, produce higher pressure, are accessible and can response quickly to change in steam demand. Tubes and drums of water-tube boilers are smaller than that of fire-tube boilers and due to smaller size of drum higher pressure can be used easily. Water-tube boilers require lesser floor space. The efficiency of water-tube boilers is more.

Water tube boilers are classified as follows.

1. Horizontal straight tube boilers

   (*a*) Longitudinal drum   (*b*) Cross-drum.

2. Bent tube boilers

   (*a*) Two drum   (*b*) Three drum

   (*c*) Low head three drum   (*d*) Four drum.

3. Cyclone fired boilers

Various advantages of water tube boilers are as follows.

(*i*) High pressure of the order of 140 kg/cm^2 can be obtained.

(*ii*) Heating surface is large. Therefore steam can be generated easily.

(*iii*) Large heating surface can be obtained by use of large number of tubes.

(*iv*) Because of high movement of water in the tubes the rate of heat transfer becomes large resulting into a greater efficiency.

Fire tube boilers are classified as follows.

1. External furnace:

   (*i*) Horizontal return tubular

   (*ii*) Short fire box

   (*iii*) Compact.

2. Internal furnace:

   (*i*) Horizontal tubular

      (*a*) Short firebox   (*b*) Locomotive   (*c*) Compact   (*d*) Scotch.

   (*ii*) Vertical tubular.

      (*a*) Straight vertical shell, vertical tube

      (*b*) Cochran (vertical shell) horizontal tube.

Various advantages of fire tube boilers are as follows.

(*i*) Low cost

(*ii*) Fluctuations of steam demand can be met easily

(*iii*) It is compact in size.

According to position of furnace.

(*i*) Internally fired   (*ii*) Externally fired

In internally fired boilers the grate combustion chamber are enclosed within the boiler shell whereas in case of extremely fired boilers and furnace and grate are separated from the boiler shell.

According to the position of principle axis.

(*i*) Vertical   (*ii*) Horizontal   (*iii*) Inclined.

According to application.

(*i*) Stationary   (*ii*) Mobile, (Marine, Locomotive).

According to the circulating water.

(*i*) Natural circulation   (*ii*) Forced circulation.

According to steam pressure.

(*i*) Low pressure   (*ii*) Medium pressure   (*iii*) Higher pressure.

5.3 COCHRAN BOILER

This boiler consists of a cylindrical shell with its crown having a spherical shape. The furnace is also hemispherical in shape. The grate is also placed at the bottom of the furnace and the ash-pit is located below the grate. The coal is fed into the grate through the fire door and ash formed is collected in the ash-pit located just below the grate and it is removed manually. The furnace and the combustion chamber are connected through a pipe. The back of the combustion chamber is lined with firebricks. The hot gases from the combustion chamber flow through the nest of horizontal fire tubes (generally 6.25 cm in external diameter and 165 to 170 in number). The passing through the fire tubes transfers a large portion of the heat to the water by convection. The flue gases coming out of fire tubes are finally discharged to the atmosphere through chimney (Fig. 5.1).

The spherical top and spherical shape of firebox are the special features of this boiler. These shapes require least material for the volume. The hemi spherical crown of the boiler shell gives maximum strength to withstand the pressure of the steam inside the boiler. The hemi-spherical crown of the fire box is advantageous for resisting intense heat. This shape is also advantageous for the absorption of the radiant heat from the furnace.

Coal or oil can be used as fuel in this boiler. If oil is used as fuel, no grate is provided but the bottom of the furnace is lined with firebricks. Oil burners are fitted at a suitable location below the fire door. A manhole near the top of the crown of shell is provided for cleaning. In addition to this, a number of hand-holes are provided around the outer shell for cleaning purposes. The smoke box is provided with doors for cleaning of the interior of the fire tubes.

The airflow through the grate is caused by means of the draught produced by the chimney. A damper is placed inside the chimney (not shown) to control the discharge of hot gases from the chimney and thereby the supply of air to the grate is controlled. The chimney may also be provided with a steam nozzle (not shown; to discharge the flue gases faster through the chimney. The steam to the nozzle is supplied from the boiler.

The outstanding features of this boiler are listed below:

1. It is very compact and requires minimum floor area.
2. Any type of fuel can be used with this boiler.
3. It is well suited for small capacity requirements.

4. It gives about 70% thermal efficiency with coal firing and about 75% with oil firing.

5. The ratio of grate area to the heating surface area varies from 10: 1 to 25: 1.

It is provided with all required mountings. The function of each is briefly described below:

1. Pressure Gauge. This indicates the pressure of the steam in the boiler.

2. Water Level Indicator. This indicates the water level in the boiler The water level in the boiler should not fall below a particular level otherwise the boiler will be overheated and the tubes may burn out.

3. Safety Valve. The function of the safety valve is to prevent the increase of steam pressure in the holler above its design pressure. When the pressure increases above design pressure, the valve opens and discharges the steam to the atmosphere. When this pressure falls just below design pressure, the valve closes automatically. Usually the valve is spring controlled.

4. Fusible Plug. If the water level in the boiler falls below a predetermined level, the boiler shell and tubes will be overheated. And if it is continued, the tubes may burn, as the water cover will be removed. It can he prevented by stopping the burning of fuel on the grate. When the temperature of the shell increases above a particular level, the fusible plug, which is mounted over the grate as shown in the Fig. 4.1, melts and forms an opening. The high-pressure steam pushes the remaining water through this hole on the grate and the fire is *extinguished.*

5. Blow-off Cock. The water supplied to the boiler always contains impurities like mud, sand and, salt Due to heating, these are deposited at the bottom of the boiler, and if they are not removed, they are accumulated at the bottom of the boiler and reduces its capacity and heat transfer rates. Also the salt content will goes on increasing due to evaporation of water. These deposited salts are removed with the help of blow off cock. The blow-off cock is located at the bottom of the boiler as shown in the figure and is operated only when the boiler is running. When the blow-off cock is opened during the running of the boiler, the high-pressure steam pushes the water and the collected material at the bottom is blown out. Blowing some water out also reduces the concentration of the salt. The blow-off cock is operated after every 5 to 6 hours of working for few minutes. This keeps the boiler clean.

6. Steam Stop Valve. It regulates the flow of steam supply outside. The steam from the boiler first enters into an ant-priming pipe where most of the water particles associated with steam are removed.

7. Feed Check Valve. The high pressure feed water is supplied to the boiler through this valve. This valve opens towards the boiler only and feeds the water to the boiler. If the feed water pressure is less than the boiler steam pressure then this valve remains closed and prevents the back flow of steam through the valve.

STEAM GENERATOR

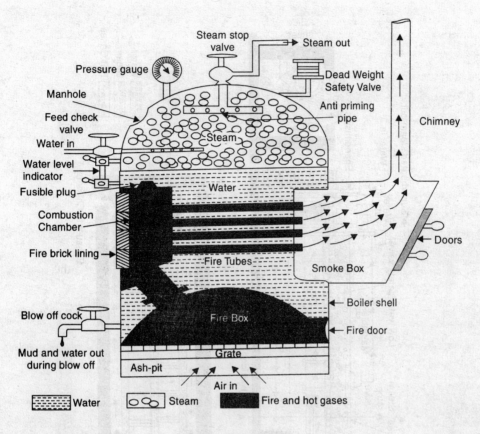

Fig. 5.1. Cochran Boiler.

5.4 LANCASHIRE BOILER

It is stationary fire tube, internally fired, horizontal, natural circulation boiler. This is a widely used boiler because of its good steaming quality and its ability to burn coal of inferior quality. These boilers have a cylindrical shell 2 m in diameters and its length varies from 8 m to 10 m. It has two large internal flue tubes having diameter between 80 cm to 100 cm in which the grate is situated. This boiler is set in brickwork forming external flue so that the external part of the shell forms part of the heating surface.

The main features of the Lancashire boiler with its brickwork shelling are shown in figure. The boiler consists of a cylindrical shell and two big furnace tubes pass right through this. The brick setting forms one bottom flue and two side flues. Both the flue tubes, which carry hot gases, lay below the water level as shown in the Fig. 5.2.

The grates are provided at the front end of the main flue tubes of the boiler and the coal is fed to the grates through the fire doors. A low firebrick bridge is provided at the end of the grate, as shown in the Fig. 5.2, to prevent the flow of coal and ash particles into the interior of the furnace tubes. Otherwise, the ash and coal particles carried with gases form deposits on the interior of the tubes and prevent the heat transfer to the water. The firebrick bridge also helps in deflecting the hot gases upward to provide better heat transfer:

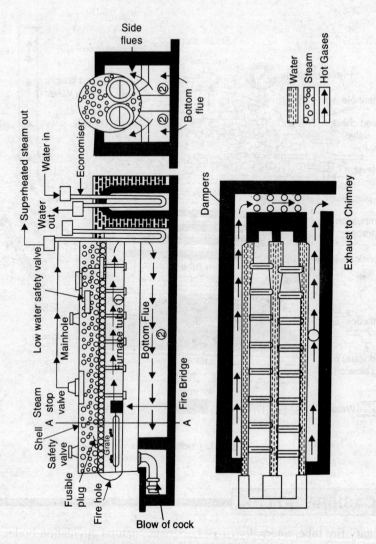

Fig. 5.2. Lancashire Boiler.

The hot gases leaving the grate pass up to the back end of the tubes and then in the downward direction. They move through the bottom flue to the front of the boiler where they are divided into two and pass to the side flues as shown in the figure. Then they move along the two-side flues and come to the chimney as shown in the figure.

With the help of this arrangement of the flow passages of the gases, the bottom of the shell is first heated and then its sides. The heat is transferred to the water through surfaces of the two flue tubes (which remain in water) and bottom part and sides of the main shell. This arrangement increases the heating surface to a large extent.

Dampers in the form of sliding doors are placed at the end of side flues to control the flow of gases. This regulates the combustion rate as well as steam generation rate. These dampers are operated by chains passing over a pulley at the front of the boiler. This boiler is fitted with usual mountings. The pressure gauge and water level indicator are provided at the front whereas steam stop valve, safety valve, low water and high steam safety valve and manhole are provided on the top of the shell.

The blow-off cock is situated beneath the front portion of the boiler shell for the removal of sediments and mud. It is also used to empty the water in the boiler whenever required for inspection.

The fusible plugs are mounted on the top of the main flues just over the grates as shown in the figure to prevent the overheating of boiler tubes by extinguishing the fire when the water level falls below a particular level. A low water level alarm is usually mounted in the boiler to give a warning in case the water level going below the precast value.

A feed check valve with a feed pipe is fitted on the front end plate. The feed pipe projecting into the boiler is perforated so that the water is uniformly distributed into the shell.

The outstanding features of this boiler are listed below:

1. Its heating surface area per unit volume at the boiler is considerably large.

2. Its maintenance is easy.

3. It is suitable where a large reserve of hot water is needed. This boiler due to the large reserve capacity can easily meet load fluctuations.

4. Super-heater and economizer can be easily incorporated into the system, therefore; overall efficiency of the boiler can be considerably increased (80-85%).

The super-heater is placed at the end of the main flue tubes. The hot gases before entering the bottom flue are passed over the super-heater tubes as shown in the figure and the steam drawn through the steam stop-valve are passed through the super-heater. The steam passing through the super-heater absorbs heat from hot gases and becomes superheated.

The economizer is placed at the end of side flues before exhausting the hot gases to the chimney. The water before being fed into the boiler through the feed check valve is passed through the economizer. The feed water is heated by absorbing the heat from the exhaust gases, thus leading to better boiler efficiency. Generally, a chimney is used to provide the draught.

5.5 LOCOMOTIVE BOILER

Locomotive boiler is a horizontal fire tube type mobile boiler. The main requirement of this boiler is that it should produce steam at a very high rate. Therefore, this boiler requires a large amount of heating surface and large grate area to burn coal at a rapid rate. Providing provides the large heating surface area a large number of fire tubes and heat transfer rate is increased by creating strong draught by means of steam jet.

A modern locomotive boiler is shown in Fig. 5.3. It consists of a shell or barrel of 1.5 meter in diameter and 4 meters in length. The cylindrical shell is fitted to a rectangular firebox at one end and smoke box at the other end. The coal is manually fed on to the grates through the fire door. A brick arch as shown in the figure deflects the hot gases, which are generated due to the burning of coal. The firebox is entirely surrounded by narrow water spaces except for the fire hole and the ash-pit. The deflection of hot gases with the help of brick arch prevents the flow of ash and coal particles with the gases and it also helps for heating the walls of the firebox properly and uniformly. It also helps in igniting the volatile matter from coal. The walls of the firebox work like an economizer. The ash-pit, which is situated below the firebox, is fitted with dampers at its front and back end shown in the figure to control the flow of air to the grate.

The hot gases from the firebox are passed through the fire tubes to the smoke box as shown in the figure. The gases coming to smoke box are discharged to the atmosphere through a short chimney with the help of a steam jet. All the fire tubes are fitted in the main shell. Some of these tubes (24 in number) are of larger diameter (13 cm diameter) fitted at the upper part of the shell and others (nearly 160 tubes) of 4.75 cm in diameter are fitted into the lower part of the shell. The shell contains water surrounding all the tubes. The top tubes are made of larger diameter to accommodate the super-heater tubes. Absorbing heat from the hot gases flowing over the tubes superheats the steam passing through the super-heater tubes. The steam generated in the shell is collected over the water surface. A dome-shaped chamber, known as steam dome, is fitted on the top of the shell. The dome helps to reduce the priming as the distance of the steam entering into the dome and water level is increased. The steam in the shell flows through a pipe mounted in the steam dome as shown in the figure into the steam header which is divided into two parts. One part of the steam header is known as saturated steam header and the other part is known as superheated steam header. The saturated wet steam through the steam pipe enters into the saturated steam header and then it is passed through the super-heater tubes as shown in the figure. The superheated steam coming out of super-heater tubes is collected in the superheated header and then fed to the steam engines. A stop valve serving also as a regulator for steam flow is provided inside a cylindrical steam dome as shown in the figure. This is operated by the driver through a regulator shaft passing from the front of the boiler.

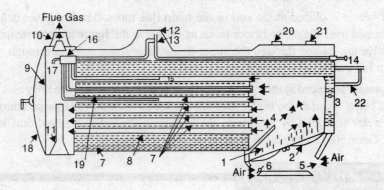

1. Fire box	2. Grate	3. Fire hole	4. Fire bride arch
5. Ash pit	6. Damper	7. Fine tubes	8. Barrel or shell
9. Smoke box	10. Chimney (short)	11. Exhaust steam pipe	12. Steam dome
13. Regulator	14. Lever	15. Superheater tubes	16. Superheater header
17. Superheater exist pipe	18. Smoke box door	19. Feed check valve	20. Safety valve
21. Whistle	22. Water gauge		

Fig. 5.3. Locomotive Boiler.

The supply of air to the grate is obtained by discharging the exhaust steam from the engine through a blast pipe which is placed below the chimney. The air-flow caused by this method is known as induced draught. A large door at the front end of the smoke box is provided which can be opened for cleaning the smoke box and fire tubes.

The height of the chimney must be low to facilitate the locomotive to pass through tunnels and bridges. Because of the short chimney, artificial draught has to be created to drive out the hot gases. The draught is created with the help of exhaust steam when locomotive is moving and with the help of live steam when the locomotive is stationary. The motion of the locomotive helps not only to increase the draught, but also to increase the heat transfer rate.

The pressure gauge and water level indicators are located m the driver's cabin at the front of the fire box as shown in the figure. The spring loaded safety valve and fusible plug are located as shown in the figure. Blow-off cock is provided at the bottom of the water wall to remove the debris and mud.

The outstanding features of this boiler are listed below :

1. Large rate of steam generation per square metre of heating surface. To some extent this is due to the vibration caused by the motion.

2. It is free from brickwork, special foundation and chimney. This reduces the cost of installation.

3. It is very compact.

The pressure of the steam is limited to about 20 bar. The details of W.G.Type Locomotive

Diameter and length of shell Ordinary tubes

Large size tubes

Pressure and temperature of steam Grate area

Heating surface area = 270 m².

The capacity of this boiler under normal load is 8500 kg/hr at 14.76 bar and 370°C burning 158.5 kg of coal per hour/m² of grate area.

Boiler manufactured at Chittaranjan are listed below : = 208.5 cm and 520.7 cm

= 116 and 57.15 mm in diameter = 38 and 114.3 mm in diameter = 14.76 bar and 370°C

= 4.27′11²

5.6 BABCOCK WILCOX BOILER

As classified earlier, in a water tube boiler, the water is inside the tubes and hot gases flow over the tubes. Babcock and Wilcox original model is a straight water tube boiler. A simple stationary boiler of this type is described here.

The boiler with its parts is shown in Fig. 5.4. The boiler shell known as water and steam drum is made of high quantity steel. It is connected by short tubes with the uptake header or riser and by longer tubes to the down take header. The water level in the drum is slightly above the center. The water tubes are connected to the top and bottom header and are kept inclined at an angle of 15° to the horizontal. The headers are provided with hand holes in the front of the tubes and are covered with caps. This arrangement helps in cleaning of the tubes. The inclined position helps the flow of water.

The furnace is arranged below the uptake header. Coal is fed to the grate through the fire door. Two firebrick baffles are arranged in such a manner that the hot gases from the grate are compelled to move in the upward and downward directions. First the hot gases rise upward and then go down and then rise up again and finally escape to the chimney through the smoke chamber.

The outer surface of the water tubes and half of the bottom cylindrical surface of the drum form the heating surface through which heat is transferred from the hot gases to the water.

The front portion of the water tubes come in contact with the hot gases at higher temperature. So the water from this portion rises in the upper direction due to decreased density and passed into the drum through the uptake header. Here the steam and water are separated and the steam being lighter is collected in the upper part of the drum. From the back portion of the drum, the water enters into the water tubes through the down take header. Thus, a continuous circulation of water from the drum to the water tubes and water tubes to the drum is maintained. The circulation of water is maintained by convective currents and is known as natural circulation.

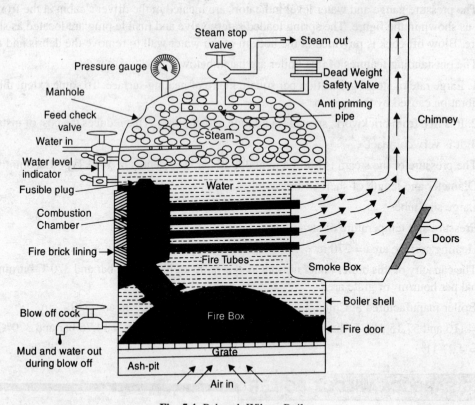

Fig. 5.4. Babcock Wilcox Boiler.

A super-heater is placed between the drum and water-tubes as shown in the figure. During the first turn of the hot gases, the gases are passed over the super-heater tubes and the steam is passed through the super-heater and becomes superheated steam. The steam is taken into the super-heater from the steam space of the drum through a tube as shown in the figure. The superheated steam coming out through super-heater is supplied through steam-pipe and steam stop valve to the turbine. When the steam is being raised from cold boiler, the super-heater is filled with water to the drum water level. This is essential to prevent the overheating of the super-heater tubes. The super-heater remains flooded with water until the steam reaches the working pressure. Once the rated pressure of steam is achieved in the boiler, then the water from the super-heater is drained and steam is fed to it for superheating purposes.

A mud box is fitted to the down header as shown in the figure. The impurities and mud particles from the water are collected in the mud box and they are blown-off from time to time by means of a blow off valve as shown in the figure.

The access to the interior of the boiler is provided by the doors. This is necessary for cleaning the tubes and removing the soots from their surfaces. The draught is regulated by a damper which is provided in the back chamber as shown in the figure. The damper position is controlled with the help of chain connected to it from the pulley as shown in figure.

The outstanding features of this boiler are listed below :

1. The evaporative capacity of this boilers is high compared with other boilers (20,000 to 40,000 kg/hr). The operating pressure lies between 11.5 to 17.5 bar.

2. The draught loss is minimum compared with other boilers.

3. The defective tubes can be replaced easily.

4. The entire boiler rests over an iron structure, independent of brick work, so that the boiler may expand or contract freely. The brick walls which form the surroundings of the boiler are only to enclose the furnace and the hot gases.

5.7 INDUSTRIAL BOILERS

The boilers are generally required in chemical industries, paper industries, pharmaceutical industries and many others. Efficiency, reliability and cost are major factors in the design of industrial boilers similar to central stations. Boiler's capacity varies from 100 to 400 tons of steam per hour. Industrial companies in foreign countries with large steam demands have considerable interest in cogeneration, the simultaneous production of steam and electricity because of federal legislation. High temperature and high pressure boilers 350°C and 75 ata) are now-a-days used even though high pressure and temperature are rarely, needed to. process requirement but they are used to generate electricity to surging prices of the oil, most of the industrial boilers are designed to use wood, municipal - pulverized coal, industrial solid waste and refinery gas few industrial boilers which are in common use are discussed below.

Packaged Water-tube Boilers. The boilers having a capacity of 50 tons/hr are generally designed with water cooled furnaces. Advantages of this design include minimum weight and maintenance as well rigidity and safety. Presently the boilers are also designed to burn coal, wood and process waste also. The much larger furnace volumes required in units designed for solid fuels restrict the capacity of packaged units to about 40 tons/hr or about one-third of a oil-gas fired unit that can be shipped by railroad.

5.8 MERITS AND DEMERITS OF WATER TUBE BOILERS OVER FIRE TUBE BOILERS MERITS

1. Generation of steam is much quicker due to small ratio of water content to steam content. This also helps in reaching the steaming temperature in short time.
2. Its evaporative capacity is considerably larger and the steam pressure range is also high-200 bar.
3. Heating surfaces are more effective as the hot gases travel at right angles to the direction of water flow.
4. The combustion efficiency is higher because complete combustion of fuel is possible as the combustion space is much larger.
5. The thermal stresses in the boiler parts are less as different parts of the boiler remain at uniform temperature due to quick circulation of water.
6. The boiler can be easily transported and erected as its different parts can be separated.
7. Damage due to the bursting of water tube is less serious. Therefore, water tube boilers are sometimes called safety boilers.
8. All parts of the water tube boilers are easily accessible for cleaning, inspecting and repairing.
9. The water tube boiler's furnace area can be easily altered to meet the fuel requirements.

Demerits :

1. It is less suitable for impure and sedimentary water, as a small deposit of scale may cause the overheating and bursting of tube. Therefore, use of pure feed water is essential.
2. They require careful attention. The maintenance costs are higher.
3. Failure in feed water supply even for short period is liable to make the boiler over-heated.

5.9 REQUIREMENTS OF A GOOD BOILER

A good boiler must possess the following qualities :

1. The boiler should be capable to generate steam at the required pressure and quantity as quickly as possible with minimum fuel consumption.
2. The initial cost, installation cost and the maintenance cost should be as low as possible.
3. The boiler should be light in weight, and should occupy small floor area.
4. The boiler must be able to meet the fluctuating demands without pressure fluctuations.
5. All the parts of the boiler should be easily approachable for cleaning and inspection.
6. The boiler should have a minimum of joints to avoid leaks which may occur due to expansion and contraction.
7. The boiler should be erected at site within a reasonable time and with minimum labour.
8. The water and flue gas velocities should be high for high heat transfer rates with minimum pressure drop through the system.
9. There should be no deposition of mud and foreign materials on the inside surface and soot deposition on the outer surface of the heat transferring parts.
10. The boiler should conform to the safety regulations as laid down in the *Boiler Act*.

5.10 HIGH PRESSURE BOILERS

In all modern power plants, high pressure boilers (> 100 bar) are universally used as they offer the following advantages.

In order to obtain efficient operation and high capacity, forced circulation of water through boiler tubes is found helpful. Some special types of boilers operating at super critical pressures and using forced circulations are described in this chapter.

1. The efficiency and the capacity of the plant can be increased as reduced quantity of steam is required for the same power generation if high pressure steam is used.
2. The forced circulation of water through boiler tubes provides freedom in the arrangement of furnace and water walls, in addition to the reduction in the heat exchange area.
3. The tendency of scale formation is reduced due to high velocity of water.
4. The danger of overheating is reduced as all the parts are uniformly heated.
5. The differential expansion is reduced due to uniform temperature and this reduces the possibility of gas and air leakages.
6. Some special types of high pressure supercritical boilers are described in this chapter.

STEAM GENERATOR

5.10.1 LA MONT BOILER

A forced circulation boiler was first introduced in 1925 by La Mont. The arrangement of water circulation and different components are shown in Fig. 5.5.

The feed water from hot well is supplied to a storage and separating drum (boiler) through the economizer. Most of the sensible heat is supplied to the feed water passing through the economizer. A pump circulates the water at a rate 8 to 10 times the mass of steam evaporated. This water is circulated through the evaporator tubes and the part of the vapour is separated in the separator drum. The large quantity of water circulated (10 times that of evaporation) prevents the tubes from being overheated.

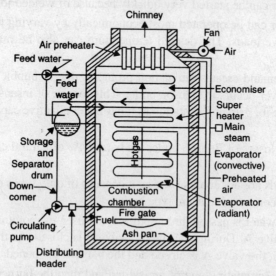

Fig. 5.5. La Mont Boiler.

The centrifugal pump delivers the water to the headers at a pressure of 2.5 bar above the drum pressure. The distribution headers distribute the water through the nozzle into the evaporator.

The steam separated in the boiler is further passed through the super-heater.

Secure a uniform flow of feed water through each of the parallel boiler circuits a choke is fitted entrance to each circuit.

These boilers have been built to generate 45 to 50 tonnes of superheated steam at a pressure of 120 bar and temperature of 500°C. Recently forced circulation has been introduced in large capacity power ?

5.10.2 BENSON BOILER

The main difficulty experienced in the La Mont boiler is the formation and attachment of bubbles on the inner surfaces of the heating tubes. The attached bubbles reduce the heat flow and steam generation as it offers higher thermal resistance compared to water film

1. Benson in 1922 argued that if the boiler pressure was raised to critical pressure (225 atm.), the steam and water would have the same density and therefore the danger of bubble formation can be completely

2. Natural circulation boilers require expansion joints but these are not required for Benson as the pipes are welded. The erection of Benson boiler is easier and quicker as all the parts are welded at site and workshop job of tube expansion is altogether avoided.

3. The transport of Benson boiler parts is easy as no drums are required and majority of the parts are carried to the site without pre-assembly.

4. The Benson boiler can be erected in a comparatively smaller floor area. The space problem does not control the size of Benson boiler used.

5. The furnace walls of the boiler can be more efficiently protected by using small diameter and close pitched tubes.

6. The superheater in the Benson boiler is an integral part of forced circulation system, therefore no special starting arrangement for superheater is required.

7. The Benson boiler can be started very quickly because of welded joints.

8. The Benson boiler can be operated most economically by varying the temperature and pressure at partial loads and overloads. The desired temperature can also be maintained constant at any pressure.

9. Sudden fall of demand creates circulation problems due to bubble formation in the natural circulation boiler which never occurs in Benson boiler. This feature of insensitiveness to load fluctuations makes it more suitable for grid power station as it has better adaptive capacity to meet sudden load fluctuations.

10. The blow-down losses of Benson boiler are hardly 4% of natural circulation boilers of same capacity.

11. Explosion hazards are not at all severe as it consists of only tubes of small diameter and has very little storage capacity compared to drum type boiler.

During starting, the water is passed through the economiser, evaporator, superheater and back to the feed line via starting valve A. During starting the valve B is closed. As the steam generation starts and it becomes superheated, the valve A is closed and the valve B is opened.

During starting, first circulating pumps are started and then the burners are started to avoid the overheating of evaporator and superheater tubes.

5.10.3. LOEFFLER BOILER

The major difficulty experienced in Benson boiler is the deposition of salt and sediment on the inner surfaces of the water tubes. The deposition reduced the heat transfer and ultimately the generating capacity. This further increased the danger of overheating the tubes due to salt deposition as it has high thermal resistance.

The difficulty was solved in Loeffler boiler by preventing the flow of water into the boiler tubes. Most of the steam is generated outside from the feedwater using part of the superheated steam coming out from the boiler.

The pressure feed pump draws the water through the economiser and delivers it into the evaporator drum as shown in the figure. About 65% of the steam coming out of superheater is passed through the evaporator drum in order to evaporate the feed water coming from economiser.

The steam circulating pump draws the saturated steam from the evaporator drum and is passed through the radiant superheater and then connective superheater. About 35% of the steam coming out from the superheater is supplied to the H.P. steam turbine. The steam coming out from H.P. turbine is passed through reheater before supplying to L.P. turbine as shown in the figure.

The amount of steam generated in the evaporator drum is equal to the steam tapped (65%) from the superheater. The nozzles which distribute the superheated steam through the water into the evaporator drum are of special design to avoid priming and noise.

This boiler can carry higher salt concentration than any other type and is more compact than indirectly heated boilers having natural circulation. These qualities fit it for land or sea transport power generation. Loeffler boilers with generating capacity of 94.5 tonnes/hr and operating at 140 bar have already been commissioned.

5.10.4. SCHMIDT-HARTMANN BOILER

The operation of the boiler is similar to an electric transformer. Two pressures are used to effect an interchange of energy.

In the primary circuit, the steam at 100 bar is produced from distilled water. This steam is passed through a submerged heating coil which is located in an evaporator drum as shown in the figure. The high pressure steam in this coil possesses sufficient thermal potential and steam at 60 bar with a heat transfer rate of 2.5 kW/m^2-°C is generated in the evaporator drum.

The steam produced in the evporator drums from impure water is further passed through 'the superheater and then supplied to the prime-mover. The high pressure condensate formed in the submerged heating coil is circulated through a low pressure feed heater on its way to raise the feed water temperature to its saturation temperature. Therefore, only latent heat is supplied in the evaporator drum.

Natural circulation is used in the primary circuit and this is sufficient to effect the desired rate of heat transfer and to overcome the thermo-siphon head of about 2 m to 10 m.

In normal circumstances, the replenishment of distilled water in the primary circuit is not required as every care is taken in design and construction to prevent leakage. But as a safeguard against leakage, a pressure gauge and safety valve are fitted in the circuit.

Advantages

1. There is rare chance of overheating or burning the highly heated components of the primary circuit as there is no danger of salt deposition as well as there is no chance of interruption to the circulation either by rust or any other material. The highly heated parts run very safe throughout the life of the boiler.

2. The salt deposited in the evaporator drum due to the circulation of impure water can be easily brushed off just by removing the submerged coil from the drum or by blowing off the water.

3. The wide fluctuations of load are easily taken by this boiler without undue priming or abnormal increase in the primary pressure due to high thermal and water capacity of the boiler.

4. The absence of water risers in the drum, and moderate temperature difference across the heating coil allow evaporation to proceed without priming.

5.10.5. VELOX-BOILER

Now, it is known fact that when the gas velocity exceeds the sound-velocity, the heat is transferred from the gas at a much higher rate than rates achieved with sub-sonic flow. The advantages of this theory are taken to effect the large heat transfer from a smaller surface area in this boiler.

Air is compressed to 2.5 bar with an help of a compressor run by gas turbine before supplying to the combusion chamber to get the supersonic velocity of the gases passing through the combustion chamber and gas tubes and high heat release rates (40 MW/m^3). The burned gases in the combustion chamber are passed through the annulus of the tubes as shown in figure. The heat is transferred from gases to water while passing through the annulus to generate the steam. The mixture of water and steam thus formed then passes into a separator which is so designed that the mixture enters with a spiral flow. The centrifugal force thus produced causes the heavier water particles to be thrown outward on the walls. This effect separates the steam from water. The separated steam is further passed to superheater and then supplied to the prime-mover. The water removed from steam in the separator is again passed into the water tubes with the help of a pump.

The gases coming out from the annulus at the top are further passed over the superheater where its heat is used for superheating the steam. The gases coming out of superheater are used to run a gas turbine as they carry sufficient kinetic energy. The power output of the gas turbine is used to run the air-compressor. The exhaust gases coming out from the gas turbine are passed through the economiser to utilise the remaining heat of the gases. The extra power required to run the compressor is supplied with the help of electric motor. Feed water of 10 to 20 times the weight of steam generated is circulated through the tubes with the help of water circulating pump. This prevents the overheating of metal walls.

The size of the velox boiler is limited to 100 tons per hour because 400 KW is required to run the air compressor at this output. The power developed by the gas turbine is not sufficient to run the compressor and therefore some power from external source must be supplied as mentioned above.

Advantages

1. Very high combustion rates are possible as 40 MJ/m^3 of combustion chamber volume.
2. Low excess air is required as the pressurised air is used and the problem of draught is simplified.
3. It is very compact generating unit and has greater flexibility.
4. It can be quickly started even though the separator has a storage capacity of about 10% of the maximum hourly output.

EXERCISES

1. State how the boilers are classified ?
2. Explain the principle of fire tube and water tube boilers.
3. Describe with a neat sketch the working of Cochran boiler. Show the position of different mountings and explain the function of each.
4. Describe, giving neat sketches, the construction and working of a Lancashire boiler. Show the positions of different mountings and accessories.
5. Sketch and describe the working of a Locomotive boiler. Show the positions of fusible plug, blow off cock, feed check valve and superheater. Mention the function of each. Describe the method of obtaining draught in this boiler.
6. Give an outline sketch showing the arrangement of water tubes and furnace of a Babcock and Wilcox boiler. Indicate on it the path of the flue gases and water circulation. Show the positions of fusible plug, blow off cock and superheater. Mention the function of each.
7. Explain why the superheater tubes are flooded with water at the starting of the boilers ?
8. Mention the chief advantages and disadvantages of fire tube boilers over water tube boilers.
9. Discuss the chief advantages of water tube boilers over fire tube boilers.
10. What are the considerations which would guide you in selecting the type of boiler to be adopted for a specific purpose ?
11. Distinguish between water-tube and fire-tube boilers and state under what circumstances each type would be desirable.

Chapter 6

Steam Turbine

Steam turbine is one of the most important prime mover for generating electricity. This falls under the category of power producing turbo-machines. In the turbine, the energy level of the working fluid goes on decreasing along the flow stream. Single unit of steam turbine can develop power ranging from 1 mW to 1000 mW. In general, 1 mW, 2.5 mW, 5 mW, 10 mW, 30 mW, 120 mW, 210 mW, 250 mW, 350 mW, 500 mW, 660 mW, 1000 mW are in common use. The thermal efficiency of modern steam power plant above 120 mW is as high as 38% to 40%.

The purpose of turbine technology is to extract the maximum quantity of energy from the working fluid, to convert it into useful work with maximum efficiency, by means of a plant having maximum reliability, minimum cost, minimum supervision and minimum starting time. This chapter deals with the types and working of various types of steam turbine. The construction details are given in chapter 15.

6.1. PRINCIPLE OF OPERATION OF STEAM TURBINE

The principle of operation of steam turbine is entirely different from the steam engine. In reciprocating steam engine, the pressure energy of steam is used to overcome external resistance and the dynamic action of steam is negligibly small. But the steam turbine depends completely upon the dynamic action of the steam. According to Newton's Second Law of Motion, the force is proportional to the rate of change of momentum (mass × velocity). If the rate of change of momentum is caused in the steam by allowing a high velocity jet of steam to pass over curved blade, the steam will impart a force to the blade. If the blade is free, it will move off (rotate) in the direction of force. In other words, the motive power in a steam turbine is obtained by the rate of change in moment of momentum of a high velocity jet of steam impinging on a curved blade which is free to rotate. The steam from the boiler is expanded in a passage or nozzle where due to fall in pressure of steam, thermal energy of steam is converted into kinetic energy of steam, resulting in the emission of a high velocity jet of steam which, Principle of working impinges on the moving vanes or blades of turbine (Fig. 6.1).

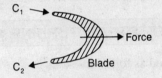

Fig. 6.1. Turbine Blade.

Attached on a rotor which is mounted on a shaft supported on bearings, and here steam undergoes a change in direction of motion due to curvature of blades which gives rise to a change in momentum and therefore a force. This constitutes the driving force of the turbine. This arrangement is shown. It should be realized that the blade obtains no motive force from the static pressure of the steam or from any impact of the jet, because the blade in designed such that the steam jet will glide on and off the blade without any tendency to strike it.

As shown in Fig. 6.2, when the blade is locked the jet enters and leaves with equal velocity, and thus develops maximum force if we neglect friction in the blades. Since the blade velocity is zero, no mechanical work is done. As the blade is allowed to speed up, the leaving velocity of jet from the blade reduces, which reduces the force. Due to blade velocity the work will be done and maximum work is done when the blade speed is just half of the steam speed. In this case, the steam velocity from the blade is near about zero *i.e.* it is trail of inert steam since all the kinetic energy of steam is converted into work. The force and work done become zero when the blade speed is equal to the steam speed. From the above discussion, it follows that a steam turbine should have a row of nozzles, a row of moving blades fixed to the rotor, and the casing (cylinder). A row of *nozzles and a raw of moving blades constitutes a stage of turbine.*

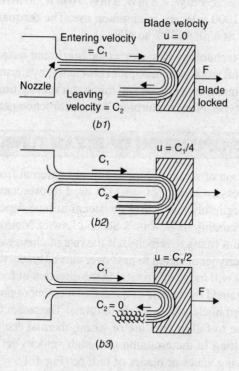

Fig. 6.2. Action of Jet on Blade.

6.2. CLASSIFICATION OF STEAM TURBINE

Steam turbine may be classified as follows: -

(A) On the Basis of Principle of Operation :

(*i*) Impulse turbine

(*a*) Simple, (*b*) Velocity stage, (*c*) Pressure stage, (*d*) combination of (*b*) and (*c*).

(*ii*) Impulse-reaction turbine

(*a*) 50% (Parson's) reaction, (*b*) Combination of impulse and reaction.

(*i*) **Impulse Turbine:** If the flow of steam through the nozzles and moving blades of a turbine takes place in such a manner that the steam is expanded only in nozzles and pressure at the outlet sides of the blades is equal to that at inlet side; such a turbine is termed as impulse turbine because it works on the principle of impulse. In other words, in impulse turbine, the drop in pressure of steam takes place only in nozzles and not in moving blades. This is obtained by making the blade passage of constant cross- section area

As a general statement it may be stated that energy transformation takes place only in nozzles and moving blades (rotor) only cause energy transfer. Since the rotor blade passages do not cause any acceleration of fluid, hence chances of flow separation are greater which results in lower stage efficiency.

(*ii*) **Impulse-Reaction Turbine:** In this turbine, the drop in pressure of steam takes place in *fixed* (*nozzles*) *as well as moving blades.* The pressure drop suffered by steam while passing through the moving blades causes a further generation of kinetic energy within the moving blades, giving rise to reaction and adds to the propelling force which is applied through the rotor to the turbine shaft. Since this turbine works on the principle of impulse and reaction both, so it is called impulse-reaction turbine. This is achieved by making the blade passage of varying cross-sectional area (*converging type*).

In general, it may be stated that energy transformation occurs in both fixed and moving blades. The rotor blades cause both energy transfer and transformation. Since there is an acceleration of flow in moving blade passage hence chances of separation of flow is less which results in higher stage efficiency.

(B) On the basis of "Direction of Flow" :

(*i*) Axial flow turbine, (*ii*) Radial flow turbine, (*iii*) Tangential flow turbine.

(*i*) **Axial Flow Turbine.** In axial flow turbine, the steam flows along the axis of the shaft. It is the most suitable turbine for large turbo-generators and that is why it is used in all modern steam power plants.

(*ii*) **Radial Flow Turbine.** In this turbine, the steam flows in the radial direction. It incorporates two shafts end to end, each driving a separate generator. A disc is fixed to each shaft. Rings of 50% reaction radial-flow bladings are fixed to each disk. The two sets of bladings rotate counter to each other. In this way, a relative speed of twice the running speed is achieved and every blade row is made to work. The final stages may be of axial flow design in order to achieve a larger area of flow. Since this type of turbine can be warmed and started quickly, so it is very suitable for use at times of peak load. Though this type of turbine is very successful in the smaller sizes but formidable design difficulties have hindered the development of large turbines of this type. In Sweden, however, composite radial/axial flow turbines have been built of outputs upto 275 MW. Sometimes, this type of turbine is also known as Liungstrom turbine after the name of its inventor B and F. Liungstrom of Sweden (Fig. 6.3).

(*iii*) **Tangential Flow Turbine.** In this type, the steam flows in the tangential direction. This turbine is very robust but not particularly efficient machine, sometimes used for driving power station auxiliaries. In this turbine, nozzle directs steam tangentially into buckets milled in the periphery of a single wheel, and on exit the steam turns

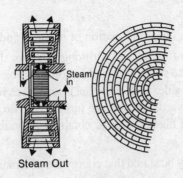

Fig. 6.3

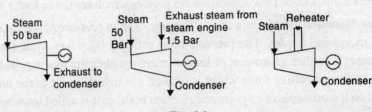

Fig. 6.4

through a reversing chamber, reentering bucket further round the periphery. This process is repeated several times, the steam flowing a helical path. Several nozzles with reversing chambers may be used around the wheel periphery.

(C) On the Basis of Means of Heat Supply:

(i) Single pressure turbine,

(ii) Mixed or dual pressure turbine

(iii) Reheated turbine.

(a) Single (b) Double

(i) **Single Pressure Turbine :** In this type of turbine, there is single source of steam supply.

(ii) **Mixed or Dual Pressure Turbine :** This type of turbines, use two sources of steam, at different pressures. The dual pressure turbine is found in nuclear power stations where it uses both sources continuously. The mixed pressure turbine is found in industrial plants (*e.g.*, rolling mill, colliery etc.) where there are two supplies of steam and use of one supply is more economical than the other; for example, the economical steam may be the exhaust steam from engine which can be utilised in the L. P. stages of steam turbine. Dual pressure system is also used in combined cycle.

(iii) **Reheated Turbine :** During its passage through the turbine steam may be taken out to be reheated in a reheater incorporated in the boiler and returned at higher tempera-ture to be expanded in (Fig. 6.6). This is done to avoid erosion and corrosion problems in the bladings and to improve the power output and efficiency. The reheating may be single or double or triple.

(D) On the Basis of Means of Heat Rejection :

(i) Pass-out or extraction turbine, (ii) Regenerative turbine, (iii) Condensing turbine, (iv) Non condensing turbine, (v) Back pressure or topping turbine.

(i) **Pass-out Turbine.** In this turbine, (Fig. 6.4), a considerable proportion of the steam is extracted from some suitable point in the turbine where the pressure is sufficient for use in process heating the remainder continuing through the turbine. The latter is controlled by separate valve-gear to meet the difference between the pass-out steam and electrical load requirements. This type of turbine is suitable where there is dual demand of steam-one for power and the other for industrial heating, for example sugar industries. Double pass-out turbines are sometimes used.

(*ii*) **Regenerative Turbine.** This turbine incorporates a number of extraction branches, through which small proportions of the steam are continuously extracted for the purpose of heating the boiler feed water in a feed heater in order to increase the thermal efficiency of the plant. Now a days, all steam power plants are equipped with reheating and regenerative arrangement.

(*iii*) **Condensing Turbine.** In this turbine, the exhaust steam is condensed in a condenser and the condensate is used as feed water in the boiler. By this way the condensing turbine allows the steam to expand to the lowest possible pressure before being condensed. All steam power plants use this type of turbine.

(*iv*) **Non-Condensing Turbine.** When the exhaust steam coming out from the turbine is not condensed but exhausted in the atmosphere is called non-condensing turbine. The exhaust steam is not recovered for feed water in the boiler.

(*v*) **Back Pressure or Topping Turbine.** This type of turbine rejects the steam after expansion to the lowest suitable possible pressure at which it is used for heating purpose. Thus back pressure turbine supplies power as well as heat energy.

The back pressure turbine generally used in sugar industries provides low pressure steam for heating apparatus, where as a topping turbine exhausts into a turbine designed for lower steam conditions.

(E) On the Basis of Number of Cylinder: Turbine may be classified as

(*i*) Single cylinder and (*ii*) Multi-cylinder.

(*i*) **Single Cylinder.** When all stages of turbine are housed in one casing, then it is called single cylinder. Such a single cylinder turbine uses one shaft.

(*ii*) **Multi-Cylinder.** In large output turbine, the number of the stages needed becomes so high that additional bearings are required to support the shaft. Under this circumstances, multi-cylinders are used.

(F) On the Basis of Arrangement of Cylinder Based on General Flow of Steam. (*i*) Single flow, (*ii*) Double flow, and (*iii*) Reversed flow

Single Flow. In a single flow turbines, the steam enters at one end, flows once [Fig. 6.5(*a*)] through

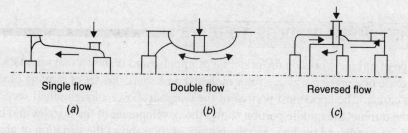

Fig. 6.5

the bladings in a direction approximately parallel to this axis, emerges at the other end. High pressure cylinder uses single flow. This is also common in small turbines.

Double Flow. In this type of turbines, the steam enters at the centre and divides, the two portions passing axially away from other through separate sets of blading on the same rotor Fig. 6.5(*b*). The low pressure cylinder normally uses double flow). This type of unit is completely balanced against the end thrust and gives large area of flow through two sets of bladings. This also helps in reducing the blade height as mass flow rate becomes half as compared to single flow for the same conditions.

Reversed Flow. Reversed flow arrangement is sometimes used in h.p, cylinder where higher temperature steam is used on the larger sets in order to minimise differential expansion *i.e.* unequal expansion of rotor and casing. The use of single, double and reversed flow is shown in the layout Fig. 6.5(*c*).

(G) On the Basis of Number of Shaft

(*i*) Tandem compound, (*ii*) Cross compound

(*i*) **Tandem Compound.** Most multi-cylinder turbines drive a single shaft and single generator Such turbines are termed as tandem compound turbines.

(*ii*) **Cross Compound.** In this type, two shafts are used driving separate generator. The may be one of turbine house arrangement, limited generator size, or a desire to run shafting at half speed. The latter choice is sometimes preferred so that for the same centrifugal stress, longer blades may be used, giving a larger leaving area, a smaller velocity and hence a small leaving loss.

(H) On the Basis of Rotational Speed

(*i*) constant speed turbines

(*ii*) Variable speed turbines

(*i*) **Constant Speed Turbines.** Requirements of rotational speed are extremely rigid in turbines which are directly connected to electric generators as these must be a-c unit except in the smallest sizes and must therefore run at speeds corresponding to the standard number of cycles per second and governed by the following equation :

$N = 120 \times$ Number of cycles per second $= 120$ f/p

Number of poles

The minimum number of poles, in a generator is two and correspondingly the maximum possible speed for 60 cycle is 3,600 rpm; for 50 c/s of frequency, the speeds would be 3,000, 1500 and 750 rpm for 2, 4 and 8 poles machines respectively.

(*ii*) **Variable Speed Turbines.** These turbines have geared units and may have practically any speed ratio between the turbine and the driven machine so that the turbine may be designed for its own most efficient speed. Such turbines are used to drive ships, compressors, blowers and variable frequency generators.

6.3. THE SIMPLE IMPULSE TURBINE

This type of turbine works on the principle of impulse and is shown diagrammatically. It mainly consists of a nozzle or a set of nozzles, a rotor mounted on a shaft, one set of moving blades attached to the rotor and a casing. The uppermost portion of the diagram shows a longitudinal section through the upper half of the turbine, the middle portion shows the development of the nozzles and blading *i.e.* the actual shape of the nozzle and blading, and the bottom portion shows the variation of absolute velocity and absolute pressure during flow of steam through passage of nozzles and blades. The example of this type of turbine is the de-Laval Turbine.

It is obvious from the figure that the complete expansion of steam from the steam chest pressure to the exhaust pressure or condenser pressure takes place only in one set of nozzles *i.e.* the pressure drop takes place only in nozzles. It is assumed that the pressure in the recess between nozzles and blade remains the same. The steam at condenser pressure or exhaust pressure enters the blade and comes out at the same pressure *i.e.* the pressure of steam in the blade passages remains approximately constant and equal to the condenser pressure. Generally, converging-diverging nozzles are used. Due to the relatively large ratio of expansion of steam in the nozzles, the steam leaves the nozzles at a very high velocity (supersonic), of about 1100 m/s. It is assumed that the velocity remains constant in the recess between the nozzles and the blades. The steam at such a high velocity enters the blades and reduces along the passage of blades and comes out with an appreciable amount of velocity (Fig. 6.6).

As it has been already shown, that for the good economy or maximum work, the blade speeded should be one half of the steam speed so blade velocity is of about 500 m/s which is very en high. This results in a very high rotational speed, reaching 30,000 r.p.m. Such high rotational speeds can only be utilised to drive generators or machines with large reduction gearing arrangements.

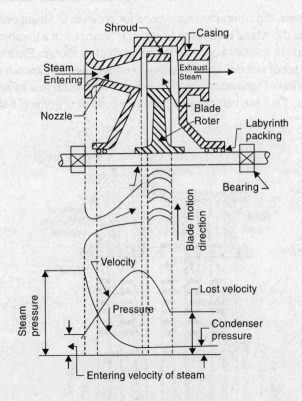

Fig. 6.6. Impulse Turbine.

In this turbine, the leaving velocity of steam is also quite appreciable resulting in an energy loss, called "carry over loss" or "leaving velocity loss". This leaving loss is so high that it may amount to about 11 percent of the initial kinetic energy. This type of turbine is generally employed where relatively small power is needed and where the rotor diameter is kept fairly small.

6.4. COMPOUNDING OF IMPULSE TURBINE

Compounding is a method for reducing the rotational speed of the impulse turbine to practical limits. As we have seen, if the high velocity of steam is allowed to flow through one row of moving blades, it produces a rotor speed of about 30,000 r.p.m. which is too high for practical use. Not only this, the leaving loss is also very high. It is therefore essential to incorporate some improvements in the simple impulse turbine for practical use and also to achieve high performance. This is possible by making use of more than one set of nozzles, blades, rotors, in a series, keyed to a common shaft, so that either the steam pressure or the jet velocity is absorbed by the turbine in stages. The leaving loss also will then be less. This process is called compounding of steam turbines. There are three main types

(a) Pressure-compounded impulse turbine.

(b) Velocity-compounded impulse turbine.

(c) Pressure and velocity compounded impulse turbine.

6.5. PRESSURE COMPOUNDED IMPULSE TURBINE

In this type of turbine, the compounding is done for pressure of steam only *i e.* to reduce the high rotational speed of turbine the whole expansion of steam is arranged in a number of steps by employing a number of simple turbine in a series keyed on the same shaft as shown. Each of these simple impulse turbine consisting of one set of nozzles and one row of moving blades is known as a stage of the turbine and thus this turbine consists of several stages. The exhaust from each row of moving blades enters the succeeding set of nozzles. Thus we can say that this arrangement is nothing but splitting up the whole pressure drop (Fig. 6.7).

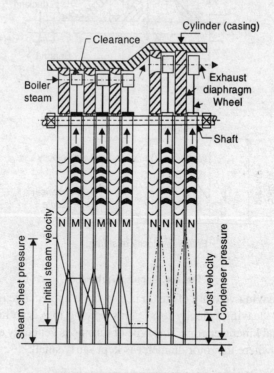

Fig. 6.7. Pressure Compounded Impulse Turbine.

from the steam chest pressure to the condenser pressure into a series of smaller pressure drop across several stages of impulse turbine and hence this turbine is culled, pressure-compound impulse turbine.

The pressure and velocity variation are also shown. The nozzles are fitted into a diaphragm which is locked in the casing. This diaphragm separates one wheel chamber from another. All rotors are

STEAM TURBINE

mounted on the same shaft and the blades are attached on the rotor. The rotor (*i.e.* disc) may be keyed to the shaft or it may be integral part of shaft.

The expansion of steam only takes place in the nozzles while pressure remains constant in the moving blades because each stage is a simple impulse turbine. So it is obvious from the pressure curve that the space between any two consecutive diaphragms is filled with steam at constant pressure and the pressure on either side of the diaphragm is different. Since the diaphragm is a stationary part, there must be clearance between the rotating shaft and the diaphragm. The steam tends to leak through this clearance for which devices like labyrinth packings, etc. are used.

Since the drop in pressure of steam per stage is reduced, so the steam velocity leaving the nozzles and entering the moving blades is reduced which reduces the blade velocity. Hence for good economy or maximum work shaft speed is significantly reduced so as be reduced by increasing the number of stages according to ones need. The leaving velocity of the last stage of the turbine is much less compared to the de Laval turbine and the leaving loss amounts to about 1 to 2 percent of the initial total available energy. This turbine was invented by the late prof L. Rateau and so it is also known as Rateau Turbine.

6.6. SIMPLE VELOCITY-COMPOUNDED IMPULSE TURBINE

In this type of turbine, the compounding is done for velocity of steam only *i.e.* drop in velocity is arranged in many small drops through many moving rows of blades instead of a single row of moving blades. It consists of a nozzle or a set of nozzles and rows of moving blades attached to the rotor or wheel and rows of fixed blades attached to casing as shown in Fig. 6.8.

The fixed blades are guide blades which guide the steam to succeeding rows of moving blades, suitably arranged between the moving blades and set in a reversed manner. In this turbine, three rows or rings of moving blades are fixed on a single wheel or rotor and this type of wheel is termed as the three row wheel. There are two blades or fixed blades placed between Lint first and the second and the second and third rows of moving blades respectively.

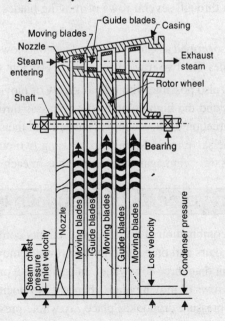

Fig. 6.8. Velocity Compounded Impulse Turbine.

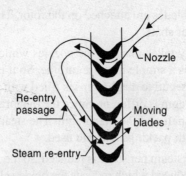

Fig. 6.9. Flow of Steam on Blades.

The whole expansion of steam from the steam chest pressure to the exhaust pressure takes place in the nozzles only. There is no drop in either in the moving blades or the fixed *i.e.* the pressure remains constant in the blades as in the simple impulse turbine. The steam velocity from the exit of the nozzle is very high as in the simple impulse turbine. Steam with this high velocity enters the first row of moving blades and on passing through these blades, the Velocity slightly reduces *i.e.* the steam gives up a part of its kinetic energy and reissues from this row of blades with a fairly high velocity. It then enters the first row of guide blades which directs the steam to the second row of moving blades. Actually, there is a slight drop in velocity in the fixed or guide blades due to friction. On passing through the second row of moving blades some drop in velocity again occurs *i.e.* steam gives up another portion of its kinetic energy to the rotor. After this, it is redirected again by the second row of guide lades to the third row of moving blades where again some drop in velocity occurs and finally the steam leaves the wheel with a certain velocity in a more or less axial direction. compared to the simple impulse turbine, the leaving velocity is small and it is about 2 percent of initial total available energy of steam.

So we can say that this arrangement is nothing but splitting up the velocity gained from the exit of the nozzles into many drops through several rows of moving blades and hence the name velocity-compounded

This type of turbine is also termed as Curtis turbine. Due to its low efficiency the three row wheel is used for driving small machines The two row wheel is more efficient than the three-row wheel.

velocity compounding is also possible with only one row of moving blades. The whole pressure drop takes place in the nozzles and the high velocity steam passes through the moving blades into a reversing chamber where the direction of the steam is changed and the same steam is arranged to pass through the moving blade of the same rotor. So instead of using two or three rows of moving blades, only one row is required to pass the steam again and again; thus in each pass velocity decreases.

6.7 PRESSURE AND VELOCITY COMPOUNDED IMPULSE TURBINE

This type of turbine is a combination of pressure and velocity compounding and is diagrammatically. There are two wheels or rotors and on each, only two rows of moving blades are attached cause two-row wheel are more efficient than three-row wheel. In each wheel or rotor, velocity drops *i.e.* drop in velocity is achieved by many rows of moving blades hence it is velocity compounded. There are two sets of nozzles in which whole pressure drop takes place *i.e.* whole pressure drop has been divided in small drops, hence it is pressure-compounded

In the first set of nozzles, there is some decrease in pressure which gives some kinetic energy to the steam and there is no drop in pressure in the two rows of moving blades of the first wheel and in the first row of fixed blades. Only, there is a velocity drop in moving blades though there is also a slight drop in velocity due to friction in the fixed blades. In second set of nozzles, the remaining pressure drop takes place but the velocity here increases and the drop in velocity takes place in the moving blades of the second wheel or rotor. Compared to the pressure-com-pounded impulse turbine this arrangement was more popular due to its simple construction. It is, however, very rarely used now due to its low efficiency.

6.8. IMPULSE-REACTION TURBINE

As the name implies this type of turbine utilizes the principle of im-pulse and reaction both. Such a type of turbine is diagrammatically shown. There are a number of rows of moving blades attached to the rotor and an equal number of fixed blades attached to the casing.

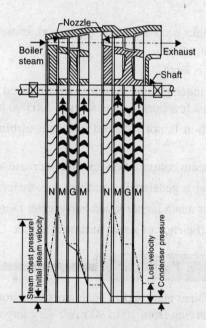

Fig. 6.10. Impulse Reaction Turbine.

In this type of turbine, the fixed blades which are set in a reversed manner compared to the moving blades, corresponds to nozzles mentioned in connection with the impulse turbine. Due to the row of fixed blades at the entrance, instead of the nozzles, steam is admitted for the whole circumference and hence there is all-round or complete admission. In passing through the first row of fixed blades, the steam undergoes a small drop in pressure and hence its velocity somewhat increases. After this it then enters the first row of moving blades and just as in the impulse turbine, it suffers a change in direction and therefore in momentum. This momentum gives rise to an impulse on the blades.

But in this type of turbine, the passage of the moving blades is so designed (converging) that there is a small drop in pressure of steam in the moving blades which results in a increase in kinetic energy of steam. This kinetic energy gives rise to reaction in the direction opposite to that of added velocity. Thus, the gross propelling force or driving force is the vector sum of impulse and reaction forces. Commonly, this type of turbine is called Reaction Turbine. It is obvious from the Fig. 6.10 that there is a gradual drop in pressure in both moving blades and fixed blades.

As the pressure falls, the specific volume increases and hence in practice, the height of blades is increased in steps *i.e.* say upto 4 stages it remains constant, then it increases and remains constant for the next two stages.

In this type of turbine, the steam velocities are comparatively moderate and its maximum value is about equal to blade velocity. In general practice, to reduce the number of stages, the steam velocity is arranged greater than the blade velocity. In this case the leaving loss is about 1 So 2 per cent of the total initial available energy. This type of turbine is used mostly in all power plants where it is great success. An example of this type of turbine is the Parsons-Reaction Turbine. The power plants 30 MW and above are all impulse-reaction type.

6.9 ADVANTAGES OF STEAM TURBINE OVER STEAM ENGINE

The various advantages of steam turbine are as follows :

(*i*) It requires less space.

(*ii*) Absence of various links such as piston, piston rod, cross head etc. make the mechanism simple. It is quiet and smooth in operation,

(*iii*) Its over-load capacity is large.

(*iv*) It can be designed for much greater capacities as compared to steam engine. Steam turbines can be built in sizes ranging from a few horse power to over 200,000 horse power in single units.

(*v*) The internal lubrication is not required in steam turbine. This reduces to the cost of lubrication.

(*vi*) In steam turbine the steam consumption does not increase with increase in years of service.

(*vii*) In steam turbine power is generated at uniform rate, therefore, flywheel is not needed.

(*viii*) It can be designed for much higher speed and greater range of speed.

(*ix*) The thermodynamic efficiency of steam turbine is higher.

6.10. STEAM TURBINE CAPACITY

The capacities of small turbines and coupled generators vary from 500 to 7500 kW whereas large turbo alternators have capacity varying from 10 to 90 mW. Very large size units have capacities up to 500 mW.

Generating units of 200 mW capacity are becoming quite common. The steam consumption by steam turbines depends upon steam pressure, and temperature at the inlet, exhaust pressure number of bleeding stages etc. The steam consumption of large steam turbines is about 3.5 to 5 kg per kWh.

Turbine kW = Generator kW / Generator efficiency

Generators of larger size should be used because of the following reasons:

(*i*) Higher efficiency.

(*ii*) Lower cost per unit capacity.

(*iii*) Lower space requirement per unit capacity. 3.45.1 Nominal rating.

It is the declared power capacity of turbine expected to be maximum load.

STEAM TURBINE

6.11 CAPABILITY

The capability of steam turbine is the maximum continuous out put for a clean turbine operating under specified throttle and exhaust conditions with full extraction at any openings if provided.

The difference between capability and rating is considered to be overload capacity. A common practice is to design a turbine for capability of 125% nominal rating and to provide a generator that will absorb rated power at 0.8 power factor. By raising power factor to unity the generator will absorb the full turbine capability.

6.12 STEAM TURBINE GOVERNING

Governing of steam turbine means to regulate the supply of steam to the turbine in order to maintain speed of rotation sensibly constant under varying load conditions. Some of the methods employed are as follows :

(*i*) Bypass governing.   (*ii*) Nozzle control governing.   (*iii*) Throttle governing.

In this system the steam enters the turbine chest (C) through a valve (V) controlled by governor. In case of loads of greater than economic load a bypass valve (Vi) opens and allows steam to pass from the first stage nozzle box into the steam belt (S).

In this method of governing the supply of steam of various nozzle groups N_1, N_2, and N_3 is regulated by means of valves V_1, V_2 and V_3 respectively.

In this method of governing the double beat valve is used to regulate the *flow* of steam into the turbine. When the load on the turbine decreases, its speed will try to increase. This will cause the *fly* bar to move outward which will in return operate the lever arm and thus the double beat valve will get moved to control the supply of steam to turbine. In this case the valve will get so adjusted that less amount of steam flows to turbine.

6.13 STEAM TURBINE PERFORMANCE

Turbine performance can be expressed by the following factors :

(*i*) The steam *flow* process through the unit-expansion line or condition curve.

(*ii*) The steam *flow* rate through the unit.

(*iii*) Thermal efficiency.

(*iv*) Losses such as exhaust, mechanical, generator, radiation etc.

Mechanical losses include bearing losses, oil pump losses and generator bearing losses. Generator losses include will electrical and mechanical losses. Exhaust losses include the kinetic energy of the steam as it leaves the last stage and the pressure drop from the exit of last stage to the condenser stage.

For successful operation of a steam turbine it is desirable to supply steam at constant pressure and temperature. Steam pressure can be easily regulated by means of safety valve fitted on the boiler. The steam temperature may try to fluctuate because of the following reasons :

(*i*) Variation in heat produced due to varying amounts of fuel burnt according to changing loads.

(*ii*) Fluctuation in quantity of excess air.

(*iii*) Variation in moisture content and temperature of air entering the furnace.

(iv) Variation in temperature of feed water.

(v) The varying condition of cleanliness of heat absorbing surface.

The efficiency of steam turbines can be increased:

(i) By using super heated steam.

(ii) Use of bled steam reduces the heat rejected to the condenser and this increases the turbine efficiency.

6.14 STEAM TURBINE TESTING

Steam turbine tests are made for the following:

(i) Power

(ii) Valve setting

(iii) Speed regulation

(iv) Over speed trip setting

(v) Running balance.

Steam condition is determined by pressure gauge, and thermometer where steam is super heated. The acceptance test as ordinarily performed is a check on (a) Output, (b) Steam rate or heat consumption, (c) Speed regulation, (d) Over speed trip setting.

Periodic checks for thermal efficiency and load carrying ability are made. Steam used should be clean. Unclean steam represented by dust carry over from super heater may cause a slow loss of load carrying ability.

Thermal efficiency of steam turbine depends on the following factors:

(i) Steam pressure and temperature at throttle valve of turbine.

(ii) Exhaust steam pressure and temperature.

(iii) Number of bleedings.

Lubricating oil should be changed or cleaned after 4 to 6 months.

6.15 CHOICE OF STEAM TURBINE

The choice of steam turbine depends on the following factors :

(i) Capacity of plant

(ii) Plant load factor and capacity factor

(iii) Thermal efficiency

(iv) Reliability

(v) Location of plant with reference to availability of water for condensate.

6.16 STEAM TURBINE GENERATORS

A generator converts the mechanical shaft energy it receive from the turbine into electrical energy. Steam turbine driven a.c. synchronous generators (alternators) are of two or four pole designs These are three phase measuring machines offering economic, advantages in generation and transmis

STEAM TURBINE

sion. Generator losses appearing as heat must be constantly removed to avoid damaging the windings. Large generators have cylindrical rotors with minimum of heat dissipation surface and so they have forced ventilation to remove the heat. Large generators generally use an enclosed system with air or hydrogen coolant. The gas picks up the heat from the generator any gives it up to the circulating water in the heat exchanger.

6.17 STEAM TURBINE SPECIFICATIONS

Steam turbine specifications consist of the following:

(*i*) Turbine rating. It includes :

(*a*) Turbine kilowatts

(*b*) Generator kilovolt amperes

(*c*) Generator Voltage

(*d*) Phases

(*e*) Frequency

(*f*) Power factor

(*g*) Excitor characteristics.

(*ii*) Steam conditions. It includes the following:

(*a*) Initial steam pressure, and Temperature

(*b*) Reheat pressure and temperature

(*c*) Exhaust pressure.

(*iii*) Steam extraction arrangement such as automatic or non-automatic extraction.

(*iv*) Accessories such as stop and throttle valve, tachometer etc.

(*v*) Governing arrangement.

SOLVED EXAMPLES

Example 1. *In an impulse steam turbine, steam is accelerated through nozzle from rest. It enters the nozzle at 9.8 bar dry and saturated. The height of the blade is 10 cm and the nozzle angle is 15°. Mean blade velocity is 144 m/s. The blade velocity ratio is 0.48 and blade velocity coefficient is 0.97. Find:*

(1) Isentropic heat drop.

(2) Energy lost in the nozzles and in moving blades due to friction.

(3) Energy lost due to finite velocity of steam leaving the stage.

(4) Mass flow rate.

(5) Power developed per stage.

(6) Diagram and stage efficiency. Take: Nozzle efficiency = 92%

Blade angles at inlet = Blade angles at out let Speed = 3000 rev/min

Solution. $\text{V.R.} = \dfrac{V_b}{V_1} = 0.48$

Now, $V_b = 144$ m/s; $V_1 = 144 = 300$ m/s 0.48

$V^2/2 \times 10^3$ = Isentropic heat drop × Nozzle efficiency

(1) Isentropic heat drop = $\dfrac{300^2}{(2 \times 10^3 \times 0.92)}$ = 48.9 kJ/kg

(2) Energy lost in nozzles = Isentropic heat drop × $(1 - \eta^n)$ = 48.9 × (1 – 0.92)
$\qquad\qquad\qquad\qquad\qquad = 3.91$ kJ/kg

Energy lost in moving blades due to friction = $\dfrac{Vr_1 - Vr_2^2}{(2 \times 10^3)}$ kJ/kg

Now $Vr_0 = 0.97 \times Vr_1$

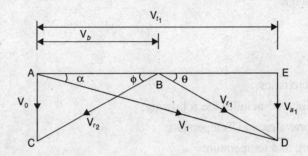

To draw velocity triangles, AB = V_b = 144 m/s, $\alpha = 15°$, $V_1 = 300$ m/s With this triangle ABD can be completed.

Measure V_{r_1}, and θ, $V_{r_1} = 168$ m/s, $\theta = 29°$

$V_{r_0} = 0.97 \times 168 = 163$ m/s, and $\theta = 29°$.

Velocity triangle ABC can be completed

Energy lost in moving blades due to friction = $\dfrac{168^2 - 163^2}{(2 \times 10^3)}$ = 0.83 kJ/kg 2×10^3

(3) Energy lost due to finite velocity of steam leaving the stage $\dfrac{V_0^2}{(2 \times 10^3)}$

From velocity triangle, $V_0 = 80$ m/s

Energy lost = $\dfrac{80^2}{(2 \times 10^3)}$

(4) $V_b = \pi D n/60$

$\qquad 144 = \pi \times D \times \dfrac{3000}{60}$

$\qquad D = 0.917$ m

STEAM TURBINE

Area of flow, $A = \pi D h = 3.14 * 0.917 * \dfrac{10}{100} = 0.288 \text{ m}^2$

Now from steam table, for steam dry and saturated at 9.8 bar,
$V_s = 0.98 \text{ m}^3/\text{kg}$
$V_{a1} = 77.65$ m/s from velocity triangles

Mass flow rate $= V_{a1} \times \dfrac{A}{V_s} = 77.65 * \dfrac{0.288}{0.198} = 112.95$ kg/s

(5) Power $= m \times V_b \times (V_{r1} - V_{r0})$ kW
from velocity triangle $V_{r1} - V_{r0} = 288$ m/s

power $= 112.95 \times 144 \times \dfrac{288}{1000} = 4682$ kW

(6) Diagram efficiency $= 2 \times V_b \times \dfrac{(V_{r1} - V_{r0})}{V_1^2} = 92.16$

stage efficiency = diagram efficiency $\times \eta_n = 0.9216 \times 0.92$

Example 2. *An impulsive stage of a steam turbine is supplied with dry and saturated steam at 14.7 bar. The stage has a single row of moving blades running at 3600 rev/min. The mean diameter of the blade disc is 0.9 m. The nozzle angle is 15° and the axial component of the absolute velocity leaving the nozzle is 93.42 m/s. The height of the nozzles at their exit is 100 mm. The nozzle efficiency is 0.9 and the blade velocity co-efficiency is 0.966. The exit angle of the moving blades is 2° greater than at the inlet. Determine:*

(1) The blade inlet and outlet angles.

(2) The isentropic heat drop in the stage.

(3) The stage efficiency.

(4) The power developed by the stage.

Solution. Mean blade velocity, $V_b = \dfrac{\pi D N}{60} = 3.14 \times 0.9 \times \dfrac{3600}{60} = 169.65$ m/s

$$\alpha = 15°\ ;\ V_{a1} = 93.42 \text{ m/s}$$

Now $\quad V_{a1} = V_1 \sin \alpha\ ;\ V_1 = \dfrac{V_{a1}}{\sin \alpha} = 360.95$ m/s

With this, inlet velocity triangle can be completed

From there: $\theta = 29.5°\ V_{r1} = 202.5$ m/s

$\quad\quad \varphi = 29.5 + 2 = 31.5\ ;\quad V_{r0} = 0.966 \times V_{r1} = 195$ m/s

With this, the outlet velocity triangle can be completed

$$\theta = 29.5°,\ \varphi = 31.5°$$

Now for dry and saturated steam at 14.7 bar; $V_s = 0.1392 \text{ m}^3/\text{kg}$

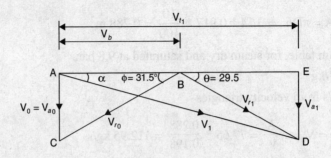

$$m = A \times \frac{V_{a1}}{V_s} = \pi Dh \times \frac{V_{a1}}{V_s} = 196.82 \text{ kg/s}$$

$$\text{power} = m \times V_b \times \frac{(V_{t1} - V_{t0})}{1000} \text{ kW}$$

from velocity triangle

$$(V_{t1} - V_{t0}) = 348.65 \text{ m/s}$$

$$\text{power} = 196.82 \times 169.65 \times \frac{348.65}{1000} = 11637.6 \text{ kW}$$

$$\text{Blade efficiency} = 2 \times v_b \times \frac{(V_{t1} - V_{t0})}{V_1^2}$$

$$\eta_b = 2 \times 169.64 \times \frac{348.65}{(360.95)^2}$$

$$= 90.79 \%$$

$$\eta_s = \eta_b \times \eta_n = 0.9079 \times 0.9 = 0.817$$

$$\text{heat drop for the stage} = \frac{15822.7}{0.817} = 14244.3 \text{ kJ/s}$$

$$= 72.37 \text{ kJ/kg}$$

Example 3. *In a simple steam Impulse turbine, steam leaves the nozzle with a velocity of 1000 m/s at an angle of 20° to the plane of rotation. The mean blade velocity is 60% of velocity of maximum efficiency. If diagram efficiency is 70% and axial thrust is 39.24 N/kg of steam/sec, estimate:*

(1) Blade angles.

(2) Blade velocity co-efficient.

(3) Heat lost in kJ in friction per kg.

Solution. for max. efficiency,

$$\text{V.R} = \frac{V_b}{V_1} = \cos \frac{\alpha}{2}$$

STEAM TURBINE

$$\frac{V_b}{V_1} = \cos \frac{20°}{2} = 0.47$$

For the present problem = $V_b = 0.6 \times 0.47 \times V_1$
$$= 282 \text{ m/s}$$

$\alpha = 20°$, $V_1 = 1000$ m/s

with this, the inlet velocity triangle can be completed. From here
$V_{r1} = 740$ m/s, $V_{a1} = 350$ m/s, $V_{t1} = 940$ m/s

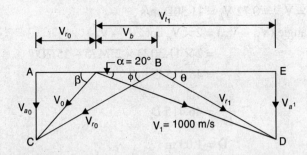

now axial thrust = $39.24 = V_{a1} - V_{a0}$
$V_{a0} = 350 - 39.24 = 310.76$ m/s

Now diagram efficiency = $0.70 = 2 \times V_b \times \frac{(V_{t1} - V_{t0})}{V_1^2}$

From here, $V_{t0} = 301$ m/s

Now the outlet velocity can be drawn

From here
$\theta = 280$, $\varphi = 28°$, $V_{r0} = 660$ m/s

Blade velocity co-efficient $\frac{V_{r0}}{V_{r1}} = 660/740 = 0.89$

Heat lost in kJ in friction per kg per s.

$$= \frac{V_{r1}^2 - V_{r0}^2}{2 \times 1000} = \frac{(740)^2 - (660)^2}{2000} = 55.94 \text{ kJ/kg/s}$$

Example 4. *A reaction steam turbine runs at 300 rev/min and its steam consumption is 16500 kg/hr. The pressure of steam at a certain pair is 1.765 bar (abs.) and its dryness fraction is 0.9. and the power developed by the pair is 3.31 kW. The discharge blade tip angel both for fixed and moving blade is 20° and the axial velocity of flow is 0.72 of the mean moving blade velocity. Find the drum diameter and blade height. Take the tip leakage as 8%, but neglect area blocked by blade thickness.*

Solution. $V_b = \frac{\pi DN}{60} = 15.7 \text{ D m/s}$

Steam flow rate through blades $= 0.92 \times \dfrac{16500}{3600} = 4.216$ kg/s

Power output $= 3.31 = m \times V_b \times \dfrac{(V_{t1} - V_{t0})}{1000}$

$(V_{t1} - V_{t0}) = 3.31 \times \dfrac{1000}{(4.216 \times 15.7 \, D)} = 50/D$ m/s

Flow velocity
$$V_{a1} = V_{a0} = 0.72 \, V_b = 11.30 D \text{ m/s}$$
From velocity triangle $(V_{t1} - V_{t0}) = 2 \times V_a \cot 20° - V_b$
$$= 2 \times 11.30 \, D \times 2.7475 - 15.7 D$$
$$= 46.415 \, D \text{ m/s}$$

$$\dfrac{50}{D} = 46.415 \, D$$

$$D = 1.03 \text{ m}$$

Flow velocity $\quad V_a = 11.30 \times 1.03 = 11.64$ m/s

Now V_s (at 1.765 bar and 0.9 dry, from steam table)
$$= 0.9975 \times 0.9 \text{ m}^3/\text{Kg}$$

$$A = \pi d h = m \times \dfrac{V_s}{V_a}$$

$$h = 4.216 \times 0.9957 \times \dfrac{0.9}{11.64} \times \pi \times 1.03$$

$$= 0.1 \text{ m}$$

Example 5. *At a particular ring of a reaction turbine the blade speed is 67 m/s and the flow of steam is 4.54 kg/s, dry saturated, at 1.373 bar. Both fixed and moving blades have inlet and exit angles of 35° and 20° respectively.*

Determine:

(a) Power developed by the pair of rings.

(b) The required blade height which is to be one tenth of the mean blade ring diameter.

(c) The heat drop required by the pair if the steam expands with an efficiency of 80%.

Solution. $V_b = 67$ m/s, $m = 4.54$ kg/s $\varphi = \alpha = 20°$, $\theta = \beta = 35°$

With this given data, the velocity triangles can be drawn From the velocity triangles,

$$V_{t1} - V_{t0} = 212 \text{ m/s}$$

STEAM TURBINE   215

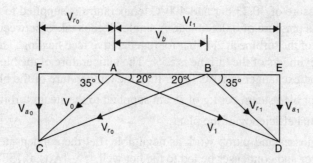

$$\text{Power} = m \times V_b \times \frac{(V_{t1} - V_{t0})}{1000} \text{ kW}$$

$$= 4.54 \times 67 \times \frac{212}{1000} = 64.47 \text{ kW}$$

at 1.373 bar, dry saturated
$$V_s = 1.259 \text{ m}^3/\text{Kg} \qquad \text{(from steam table)}$$

Now $\qquad m = \pi D \times D \times \dfrac{V_a}{V_s} \times 10$

From velocity triangle,
$$Va = 50 \text{ m/s}$$

$$D^2 = 4.54 \times 1.259 \times \frac{10}{p \times 50} = 0.3$$

$$D = 0.5477 \text{ m}$$
$$h = 5.477 \text{ cm}$$

Heat drop required $= \dfrac{74.47}{0.8} = 80.58$ kJ/kg.

EXERCISE

1. Steam is supplied to a turbine at a pressure of 58.42 bar abs and tetnperature of 440°C. It is expanded in a H.P. turbine to 6.865 bar abs., the internal efficiency of the turbine being 0.85. The steam is then reheated at constant pressure upto 300°C. Its is then expanded to 0.049 bar abs. In L.P. turbine having internal efficiency of 0.80. If the mechanical efficiency of the turbine is 98% and alternator efficiency is 96%, calculate the amount of steam generated by the boiler per kWh output. [**Ans.** 3.68 kg/hr]

2. A steam power plant working on regenerative heating cycle utilizes steam at 41.2 bar and 400°C and the condenser pressure is 0.944 bar vacuum. After expansion in the turbine to 4.90 bar, a part of steam is extracted from the turbine for heating feed water from condenser in the open heater. Draw the cycle on T-0 diagram and find thermal efficiency of the plant. Assume the heat drop to be isentropic and the atmospheric pressure may be taken as 1.013 bar.

[**Ans.** 38.89%]

3. Steam at pressure of 30.23 bar and 400°C temperature is supplied to a steam turbine and is exhausted at a pressure of 0.06865 bar. A single bleed is taken between the H.P. cylinder and L.P. cylinder of the turbine at 2.45 bar for regenerative feed heating. The isentropic efficiency for both the cylinders of the turbine is 85%. The temperature of the bleed condensate coming out of the heat exchanger is 10°C lower than the temperature of the bled steam. Determine:

   (a) Amount of bled steam per kg of steam supplied to the steam turbine.

   (b) The thermal efficiency of the plant.

   Consider no losses and pump work as negligible. Let the condensate coming out from the heat exchanger and condenser be led to the hot well.   [**Ans.** 0.158 kg/kg of steam, 32%]

4. In a condenser test, the following observations were made:

   Vacuum = 69 cm of Hg

   Barometer = 75 cm of Hg

   Mean temperature of condenser = 35°C

   Hot well temperature = 28°C

   Amount of cooling water = – 50,000 kg/hr

   Inlet temperature = 17°C

   Outlet temperature = 30°C

   Amount of condensate per hour = 1250 kg

   Find

   (a) the amount of air present per m^3 of condenser volume.

   (b) the state of steam entering the condenser.

   (c) the vacuum efficiency.

   R for air = 287 J/kgK.   [**Ans.** 0.0267 kg, 0.883, 97.48%]

5. In a power plant, steam at a pressure of 27.46 bar abs. and temperature 370°C is supplied by the main boilers. After expansion in the H.P. turbine to 5.$84 bar abs., the steam is removed and reheated to 370°C. Upon completing expansion in L.P. turbine, the steam is exhausted at a pressure of 0.0392 bar abs. Find the efficiency of the cycle with and without reheating.

   [**Ans.** 37.19%, 36.38%]

6. A fuel contains the following percentage of combustibles by mass C: 84%, H_2 : 4.1 %. If the air used from burning of coal in a boiler is 16.2 kg per kg of fuel, find the total heat carried away by dry flue gases if they escape at 300°C. The specific heats of CO_2, O_Z, N_Z are 0.892, 0.917 and 1.047 respectively. Find the minimum amount of air required from the complete combustion of 1 kg of this fuel and the excess O_2 supplied. [**Ans.** 4806.68 kJ, 11.16, 1.158]

7. A sample of coal has the percentage analysis by mass, C-85%, H-5%, incombustible-10%. In a combustion chamber, the coal is burnt with a quantity of air 50% in excess of that theoretically required for complete combustion. Obtain the volumetric composition of dry flue gases. When final temperature of flue gases is 307°C and boiler house temperature is 27°C, estimate the maximum quantity of heat available for steam raising per kg of coal. Take C_P for dry flue gases as 1.005 and assume that the total heat of vapour in flue gases as 2683.87 kJ/kg. The C.V. of coal is 32784.2 kJ/kg.

Chapter 7

Fuels and Combustion

7.1 INTRODUCTION

The components of fossil-fueled steam generators that dealt with the working fluid (water and steam) and with the air and flue gases. We deferred discussing the fuel aspects because they require independent treatment. There is a rather wide variety of fuels. Their preparation and feeding, often outside the steam generators, and their methods of firing deserve special attention.

The increasing worldwide demand for energy has focused attention on fuels, their availability and environmental effects. The fuels available to utility industry are largely nuclear and fossil, both essentially nonrenewable. Nuclear fuels originated with the universe, and it takes nature millions of years to manufacture fossil fuels.

Fossil fuels originate from the earth as a result of the slow decomposition and chemical conversion of organic material. They come in three basic forms: *solid* (coal). *liquid* (oil), and natural gas. Coal represents the largest fossil-fuel energy resource in the world. In the United States today (1983), it is responsible for about 50 percent of electric-power generation. Oil and natural gas are responsible for another 30 percent. The remaining percentage is mostly due to nuclear and hydraulic generation. Natural gas, however, is being phased out of the picture in the United States because it must be conserved for essential industrial and domestic uses.

New combustible-fuel options include the so-called synthetic fuels, or synfuels, which are liquids and gases derived largely from coal, oil shale, and tar sands. A tiny fraction of fuels used today are industrial by-products, industrial and domestic wastes. and biomass.

This chapter will cover the combustible fuels available to the utility industry, both natural (fossil) and synthetic, and their preparation and firing systems. Nuclear fuels and d renewable energy sources, and the environmental aspects of power generation in general will be covered later in this text.

7.2 COAL

Coal is a general term that encompasses a large number of solid organic minerals with widely differing compositions and properties, although all are essentially rich in amorous (without regular structure) elemental carbon. It is found in stratified deposits at different and often great depths, although sometimes near the surface. It is estimated that in the United States there are 270,000 million tons of recoverable reserves (those that can be mined economically within the foreseeable future) in 36 of the 50 states. this accounts for about 30 percent of the world's total.

There are many ways of classifying coal according to its chemical and physical properties The most accepted system is the one used by the American Society for Testing and Materials (ASTM), which classifies coals by grade or rank according to the degree of metamorphism (change in form and structure under the influences of pressure, and water), ranging from the lowest state, lignite, to the highest, anthrasite (ASTM D 388). These classifications are briefly described below in a de ..riding order.

Anthracite. This is the highest grade of coal. It cotains a high content, 86 to 98 mass percent of fixed carbon (the carbon content in the elemental state) on a dry, mineral matter-free basis and a low content of volatile matter, less than 2 to 14 mass percent chiefly methane, CH_n. Anthracite is a shiny black, dense, hard, brittle coal that — borders on graphite at the upper end of fixed carbon. It is slow-burning and has a heating value just below that of the highest for bituminous coal . Its use in steam generators is largely confined to burning on stokers, and rarely in pulverized form. In the United States it is mostly found in Pennsylvania.

The anthracite rank of coal is subdivided into three groups. In descending order of fixed-carbon percent, they are meta-anthracite, greater than 98 percent anthracite, 92 to 98 percent; and semi-anthracite, 86 to 92 percent.

Bituminous coal. The largest group, bituminous coal is a broad class of coals containing 46 to 86 mass percent of fixed carbon and 20 to 40 percent of volatile matter of more complex content than that found in anthracite. It derives its name from bitumen, an asphaltic residue obtained in the distillation of some fuels. Bituminous coals range in heating value from 11,000 to more than 14,000 Btu/lbm (about 25,600 to 32,600 kJ/kg). Bituminous coals usually burn easily, especially in pulverized form.

The bituminous rank is subdivided into five groups: low-volatile, medium-volatile, and high-volatile A, B, and C. The lower the volatility, the higher the heating value. The low-volatility group is grayish black and granular in structure, while the high volatility groups are homogeneous or laminar.

Subbituminous coal. This is a class of coal with generally lower heating values than bituminous coal, between 8300 to 11,500 Btu/lbm (about 19,300 to 26,750 kJ/kg). It is relatively high in inherent moisture content, as much as 15 to 30 percent, but often low in sulfur content. It is brownish black or black and mostly homogeneous in structure. Subbituminous coals are usually burned in pulverized form. The subbituminous rank is divided into three groups: A, B, and C.

Lignite. The lowest grade of coal, lignite derives its name from the Latin lignum, which means "wood." It is brown and laminar in structure, and remnants of wood fiber are often visible in it. It originates mostly from resin-rich plants and is therefore high in both inherent moisture, as high as 30 percent, and volatile matter. Its heating value ranges between less than 6300 to 8300 Btu/lb. (about 14,650 to 19,300 kJ/kg). Because of the high moisture content and low heating value, lignite it is not economical to transport over long distances and it is usually burned by utilities at the mine site. The lignite rank is subdivided into two groups: A and B.

Peat. Peat is not an ASTM rank of coal. It is, however, considered the first geological step in coal's formation. Peat is a heterogeneous material consisting of decomposed plant matter and inorganic minerals. It contains up to 90 percent moisture. Although not attractive as a utility fuel, it is abundant in many parts of the world. Several states in the United States have large deposits. Because of its abundance, it is used in a few countries (Ireland, Finland, the USSR) in some electric generating plants and in district heating.

7.3 COAL ANALYSIS

There are two types of coal analysis: proximate and ultimate, both done on a mass percent basis. Both of these methods may be based on: an as received basis, useful

for combustion calculations; a moisture free basis, which avoids variations of the moisture content even in the same shipment and certainly in the different stages of pulverization; and a dry mineral-matter free basis, which circumvents the problem of the ash content's not being the same as the mineral matter in the coal.

7.3.1 PROXIMATE ANALYSIS

This is the easier of two types of coal analysis and the one which supplies readily meaningful information for coal's use in steam generators. The basic method for proximate analysis is given by ANSI/ASTM Standards D 3172. It determines the mass percentages of fixed carbon, volatile matter, moisture, and ash. Sulfur is obtained m a separate determination.

Fixed carbon is the elemental carbon that exists in coal. In proximate analysis, its determination is approximated by assuming it to be the difference between the original sample and the sum of volatile matter, moisture, and ash.

The volatile matter is that portion of coal, other than water vapor, which is driven off when the sample is heated in the absence of oxygen in a standard test (up to 1750°F or 7 min). It consists of hydrocarbon and other gases that result from distillation and decomposition.

Moisture is determined by a standard procedure of drying in an oven. This does not account for all the water present, which includes combined water and water of hvdration. There are several other terms for moisture in coal. One, *inherent moisture*, that existing in the natural state of coal and considered to be part of the deposit, excluding surface water.

Ash is the inorganic salts contained in coal. It is determined in practice as the noncombustible residue after the combustion of dried coal in a standard test (at 1380°F).

Sulfur is determined separately in a standard test, given by ANSUASTM Standards D 2492. Being combustible, it contributes to the heating value of the coal. It forms oxides which combine with water to form acids. These cause corrosion problems in the back end of steam generators if the gases are cooled below the dew point, as well as environmental problems.

7.3.2 ULTIMATE ANALYSIS

A more scientific test than proximate analysis, ultimate analysis gives the mass percentages of the chemical elements that constitute the coal. These include carbon, hydrogen, nitrogen, oxygen, and sulfur. Ash is determined as a whole, sometimes in a separate analysis. Ultimate analysis is given by ASTM Standards D 3176.

7.3.3 HEATING VALUE

The heating value, Btu/lbm or·J/kg of fuel, may be determined on as-received, dry, or dry-and-ash-free basis. It is the heat transferred when the products of complete **American National Standards Institute/American Society for Testing and Materials.**

combustion of a sample of coal or other fuel are cooled to the initial temperature of air and fuel. It is determined in a standard test in a bomb calorimeter given by ASTM Standards D 2015. There are two determinations: the *higher (or gross) heating value* (HHV) assumes that the water vapor in the products condenses and thus includes the latent heat of vaporization of the water vapor in the products; the *lower heating value* (LHV) does not. The difference between the two is given by

$$LHV = HHV - m_w h_{fg} \quad \ldots(7.1a)$$

or
$$LHV = HHV - 9 m_{H_2} h_{fg} \quad \ldots(7.1b)$$

where

m_w = mass of water vapor in products of combustion per unit mass of fuel (due to the combustion of H_Z in the fuel *i.e.,* not including initial H_2O in fuel).

m_{H_2} = mass of original hydrogen per unit mass of fuel, known from ultimate analysis.

h_{fg} = latent heat of vaporization of water vapor at its partial pressure in the combustion products, Btu/lb. H_2O or J/kg H_2O.

The partial pressure of water vapor in the products of combustion is obtained by multiplying the mole faction of H_2O in the products, which is obtained from the combustion equation in the usual manner, by the total pressure of the products. The 9 in Eq. (7.1b) is the ratio of the molecular masses of H_2O and H_2 and represents the mass of H_2O vapor obtained from a unit mass of H_2.

Because gases are not usually cooled down below the dew point in steam generators (or engines), it does not seem fair to charge them with the higher heating value in calculating energy balances and efficiencies of cycles or engines. Some, however, argue that they should be charged with the total energy content of the fuel. A uniform standard had to be agreed upon, whereupon everybody uses the HHV in energy balances and efficiency calculations. (The LHV is the standard used in European practice, however.)

As indicated above, heating values are obtained by testing. However, a formula of the *Dulong* type (which does not include the effects of dissociation) is used to give approximate higher heating values of anthracite and bituminous coals in Btu/lbm.

$$HHV = 14{,}600C + 62{,}000(H - O/8) + 4050S \quad \ldots(7.2)$$

where C, H, O, and S are the mass fractions of carbon, hydrogen, oxygen, and sulfur, respectively, in the coal. For lower-rank fuels, the above formula usually underestimates the HHV.

Table 7.1 gives the proximate and ultimate analyses of some typical U.S. coals.

Table 7.1 : Proximate and ultimate analysis of some U.S. coals

Analysis mass percent	Anthractive	Bituminous medium velocity	Subiluminous	Lagnite
		Proximate		
Fixed carbon	83.8	70.0	45.9	30.8
Volatile matter	5.7	20.5	30.5	28.2
Moisture	2.5	3.3	19.6	34.8
Ash	8.0	6.2	4.0	6.2

(*Contd.*)

FUELS AND COMBUSTION

Analysis mass percent	Anthractive	Bituminous medium velocity	Subiluminous	Lagnite
		Ultimate		
C	83.9	80.7	58.8	42.4
H_2	2.9	4.5	3.8	2.8
S	0.7	1.8	0.3	0.7
O_2	0.7	2.4	12.2	12.4
N_2	1.3	1.1	1.3	0.7
H_2O	2.5	3.3	19.6	34.8
		HHV		
Btu/lbm	13.320	14.310	10.130	7.210

7.4 COAL FIRING

Since the old days of feeding coal into a furnace by hand, several major advances have been made that permit increasingly higher rates of combustion.

The earliest in the history of steam boilers were *mechanical stokers*, and several types are still being used for small and medium-sized boilers. All such stokers are designed to continuously feed coal into the furnace by moving it on a grate within the furnace and also to remove ash from the furnace.

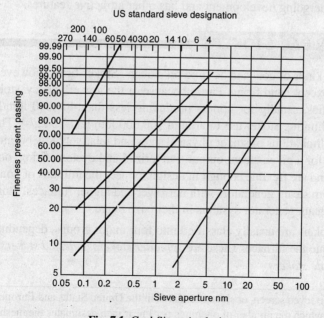

Fig. 7.1. Coal Sieve Analysis.

Fig. 7.1 Coal sieve analysis. (A) pulverized-coal sample; (B) coal range for cyclone firing; (C) coal as fired.

Pulverized-coal firing was introduced in the 1920s and represented a major increase in combustion rates over mechanical stokers. It is widely used today. To prepare the coal for use in pulverized firing, it is crushed and then ground to such a fine powder that approximately 70 percent of it will pass a 200 mesh sieve* (Fig. 7.1). It is suitable for a wide variety of coal, particularly the higher-grade ones. Advantages of pulverized coal firing are the ability to use any size coal; good variable-load response; a lower requirement for excess air for combustion, resulting in lower fan power consumption; !lower carbon loss; higher combustion temperatures and improved thermal efficiency; lower operation and maintenance costs; and the possibility of design for multiple-fuel combustion (oil, gas, and coal).

In the late 1930s *cyclone furnace firing* was introduced and became the third major advance in coal firing. It is now also widely used though for a lesser variety of uses than is pulverized coal. In addition to those advantages already mentioned for pulverized-coal firing, cyclone firing provides several other advantages. These are the obvious savings in pulverizing equipment because coal need only be crushed, reduction in furnace size, and reduction in fly ash content of the flue gases. Coal size for cyclone furnace firing is accomplished in a simple crusher and covers a wide band, with approximately 95 percent of it passing a 4-mesh sieve (Fig. 7.1).

Most recently, *fluidized-bed combustion* has been introduced. In this type of firing, crushed particles of coal are injected into the fluidized bed so that they spread across an air distribution grid. The combustion air, blown through the grid, has an upward velocity sufficient to cause the coal particles to become fluidized, *i.e.* held in suspension as they burn. Unburned carbon leaving the bed is collected in a cyclone separator and returned back to the bed for another go at combustion. The main advantage of fluidized-bed combustion is the ability to desulfurize the fuel during combustion in order to meet air quality standards for sulfur dioxide emissions. (Other methods are the use of low-sulfur coal, desulfurization of coal before it is burned, and removal of SO_2 from the flue gases by the use of scrubbers). Desulfurization is accomplished by the addition of limestone directly to the bed. Fluidized-bed combustion is still undergoing development and has other attractive features.

7.5 MECHANICAL STOKERS

Almost all kinds of coal can be fired on stokers. Stoker firing, however, is the least efficient of all types of firing except hand firing. Partly because of the low efficiency. stoker firing is limited to relatively low capacities, usually for boilers producing less than 400,000 lbm/h (50 kg/s) of steam, though designers are limiting stoker use to around 100,000 lbm/h (12.6 kg/s). These capacities are the result of the practical limitations of stoker physical sizes and relatively low burning rates which require a large furnace width for a given steam output. Pulverized and cyclone firing, on the other hand, have higher burning rates and are flexible enough in design to meet the millions of pounds per hour of steam requirements of modern steam generators with narrower and higher furnaces. Stokers, however, remain an important part of steam generator systems in their size range.

Mechanical stokers are usually classified into four major groups, depending upon the method of introducing the coal into the furnace. These are *spreader stokers underfed stokers, vibrating-grate stokers, and traveling-grate stokers.*

---

* There are some seven screen, or sieve, standards in the United States and Europe. The one used here : the U.S. Standard Sieve, in which the number of openings per linear inch designates the mesh. A 100 mesh screen has 100 openings to the inch, or 10,000 openings per square inch. The higher the **mesh, the finer the screen.** The diameter of the wire determines the opening size.

FUELS AND COMBUSTION

The *spreader stoker* is the most widely used for steam capacities of 75,000 to 400,000 lbm/h (9.5 to 50 kg/s). It can burn a wide variety of coals from high-rank bituminous to lignite and even some by-product waste fuels such as wood wastes, pulpwood, bark, and others and is responsive to rapid load changes. In the spreader stoker coal is fed from a hopper to a number of feeder-distributor units, each of which has a reciprocating feed plate that transports the coal from the hopper over an adjustable spill plate to an overthrow rotor equipped with curved blades. There are a number of such feeder-distributor mechanisms that inject the coal into the furnace in a wide uniform projectile over the stoker grate (Fig. 7.2). Air is primarily fed upward through the grate from an air plenum below it. This is called *undergrate air*. The finer coal particles, between 25 and 50 percent of the injected coal, are supported by the upward airflow and are burned while in suspension. The larger ones fall to the grate and burn in a relatively thin layer. Some air, called *overfire air*, is blown into the furnace just above the coal projectile. Forced-draft fans are used for both undergrate and overfire air. The unit has equipment for collecting and reinjecting dust and controls for coal low and airflow to suit load demand on the steam generator.

The problem with stationary spreader stokers was the removal of ash, which was first done manually and then by shutting off individual sections of grates and their air supply for ash removal without affecting other sections of the stoker. The spreader stoker became widely accepted only after the introduction of the *continuous-ash-discharge traveling-grate* stoker in the late 1930s. Traveling-grate stokers, as a class, also include the so-called *chain-grate* stoker. They have grates, links, or keys joined n an endless belt that is driven by a motorized sprocket drive at one end and over an idle shaft sprocket mechanism at the other. Coal may be injected in the above manner or fed directly from a hopper onto the moving grate through an adjustable gate that regulates the thickness of the coal layer. Ash is discharged into an ash pit at either end depending upon the direction of motion of the traveling grate.

Continuous-cleaning grates that use reciprocating or vibrating designs have also been developed, as have underfeed stokers that are suitable for burning special types of coals. The continuous-ash-removal traveling-grate stoker, however, has high burning rates and remains the preferred type of stoker.

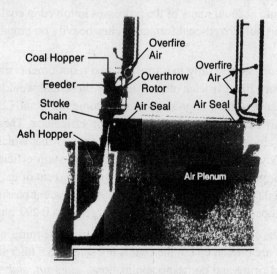

Fig. 7.2. Feeder Distributor Mechanism.

Ignition of the fresh coal in stokers, as well as its combustible volatile matter, driven off by distillation, is started by radiation heat transfer from the burning gases above. The fuel bed continues to burn and grows thinner as the stoker travels to the far end over the bend, where ash is discharged to the ash pit. Arches are sometimes built into the furnace to improve combustion by reflecting heat onto the coal bed.

7.6 PULVERIZED-COAL FIRING

The commercial development of methods for firing coal in pulverized form is a landmark in the history of steam generation. It made possible the construction of large, efficient, and reliable steam generators and power plants. The concept of firing "powdered" coal, as it was called in earlier times, dates back to Carnot, whose idea envisaged its use for the Carnot cycle; to Diesel, who used it in his first experiments on the engine that now bears his name; to Thomas Edison, who improved its firing in cement kilns, thus improving their efficiency and production; and to many others. It was not, however, until the pioneering efforts of John Anderson and his associates and the forerunner of the present Wisconsin Electric Power Company that pulverized coal was used successfully in electric generating power plants at their Oneida Street and Lakeside Stations, Milwaukee, Wisconsin.

The impetus for the early work on coal pulverization stemmed from the belief that, if coal were made fine enough, it would burn as easily and efficiently as a gas. Further inducements came from an increase in oil prices and the wide availability of coal, which makes the present situation sound rather like history repeating itself. Much theoretical work on the mechanism of pulverized-coal combustion began in the early 1920s. The mechanism of crushing and pulverizing has not been well under-stood theoretically and remains a matter of controversy even today. Probably the most accepted law is one published in 1867 in Germany, called *Rittinger's law*, that states that the work needed to reduce a material of a given size to a smaller size is proportional to the surface area of the reduced size. This, and other laws, however, do not take into account many of the processes involved in coal pulverization, and much of the progress in developing pulverized-coal furnaces relies heavily on empirical correlations and designs.

To burn pulverized coal successfully in a furnace, two requirements must be met: (1) the existence of large quantities of very fine particles of coal, usually those that would pass a 200-mesh screen, to ensure ready ignition because of their large surface-to-volume ratios and (2) the existence of a minimum quantity of coarser particles to ensure high combustion efficiency. These larger coarse particles should contain a very small amount larger than a given size, usually that which would be retained on a 50 mesh screen, because they cause slagging and loss of combustion efficiency. Line A in Fig. 7.1 represents a typical range for pulverized coal. It shows about 80 percent of the coal passing a 200 mesh screen that corresponds to a 0.074 mm opening and about 99.99 percent passing a 50 mesh screen that corresponds to a 0.297 mm opening, *i.e.* only 0.1 percent larger than 0.297 mm.

The size of bituminous coal that is shipped as it comes from the mine, called *run of mine coal*, is about 8 in. Oversized lumps are broken up but the coal is not screened. Other sizes are given names like *lump*, which is used in hand firing and domestic applications, *egg*, *nut*, *stoker*, and *slack*. [Anthracite coal has similar designations, ranging from *broken* to *buckwheat* and *rice*, ASTM D 310].

Coal is usually delivered to a plant site already sized to meet the feed size required by the pulverizing mill or the cyclone furnace. If the coal is too large, however it must go through *crushers*, which are

FUELS AND COMBUSTION

part of the plant coal-handling system and are usually located in a crusher house at a convenient transfer point in the coal-conveyor system.

Crushers. Although there are several types of commercially available coal crushers, a few stand out for particular uses. To prepare coal for pulverization, the *ring crusher, or granulator* (Fig. 7.3) and the hammer mill (Fig. 7.4) are preferred. The coal is fed at the top and is crushed by the action of rings that pivot off center on a rotor or by swinging hammers attached to it. Adjustable screen bars determine the maximum size of the discharged coal. Wood and other foreign material is also crushed, but a trap is usually provided to collect tramp iron (metal and other hard-to-crush matter.) Ring crushers and hammer mills are used off or on plant site. They reduce run-of-mine coals down to sizes such as 3/4 × 0 in. Thus they discharge a large amount of fines suitable for further pulverization, but not for cyclone-furnace firing. For the latter, a crusher type called the reversible *hammer mill* is preferred.

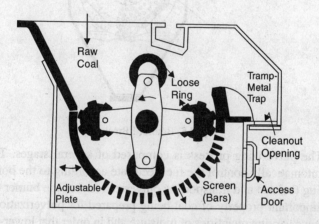

Fig. 7.3. Ring Crushers.

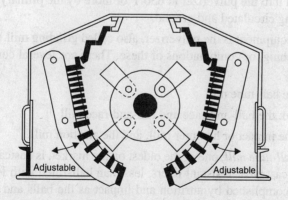

Fig. 7.3. Hammer Hill.

A third type, the *Bradford breaker* (Fig. 7.5), is used for large-capacity work. It is composed of a large cylinder consisting of perforated steel or screen plates to which lifting shelves are attached on the inside. The cylinder rotating slowly at about $2f$ r/min receives feed at one end. The shelves lift the coal, and the breaking action io accomplished by the repeated dropping of the coal until its size permits it to be discharged through the perforations, whose size determines the size of the discharge coal. The quantity of fines is limited because the crushing force, due to gravity is not large. Bradford breakers easily reject foreign matter and produce relatively uniform, size coals. They are usually used at the mine but may also be used at the plant.

Other simple devices called roll *crushers*, which have single or double rolls or rotors equipped with teeth, have been used but have not proven very satisfactory. because of their inability to produce coal of uniform size.

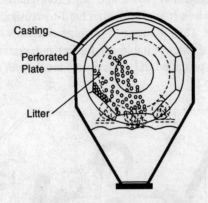

Fig. 7.5. Bradford Breaker.

Pulverizers. The pulverizing process is composed of several stages. The first is the *feeding* system, which must automatically control the fuel-feed rate according to the boiler demand and the air rates required for drying (below) and transporting pulverized fuel to the burner (primary air). The next stage is *drying*. One important property of coal being prepared for pulverization! is that it be dry and dusty. Because coals have varying quantities of moisture and in order that lower-rank coals can be used, *dryers* are an integral part of pulverizing equipment. Part of the air from the steam-generator air preheater, the primary air, is forced into the pulverizer at 650°F or more by the primary air fan. There it is mixed with the coal as it is being circulated and ground.

The heart of the equipment is the pulverizer, also called grinding mill. Grinding is accomplished by impact, attrition, crushing, or combinations of these. There are several commonly used pulverizers, classified by speed:

(1) low-speed: the ball-tube mill

(2) medium-speed: the ball-and-race and roll-and-race mill

(3) high-speed: the impact or hammer mill, and the attrition mill.

The low-speed *ball-tube mill*, one of the oldest on the market, is basically a hollow cylinder with conical ends and heavy-cast wear-resistant liners, less than half-filled with forged steel balls of mixed size. Pulverization is accomplished by attrition and impact as the balls and coal ascend and fall with cylinder rotation. Primary air is circulated over the charge to carry the pulverized coal to classifiers (below). The ball tube mill is dependable and requires low maintenance, but it is larger and heavier n construction, consumes more power than others, and because of poor air circulation, works less efficiently with wet coals. It has now been replaced with more efficient types

The medium-speed *ball-and-race* and *roll-and race* pulverizers are the type in most use nowadays. They operate on the principles of crushing and attrition. Pulverzation takes place between two surfaces, one rolling on top of the other. The rolling elements may be balls or ring-shaped rolls that rolls that roll between two races, in the manner of a ball bearing. Fig. 7.6 shows an example of the former. The balls are between a top stationary race or ring and a rotating bottom ring, which is driven by the vertical, aft of the pulverizer. Primary air causes coal feed to circulate between the grinding elements, and when it becomes fine enough, it becomes suspended in the air and is : carried to the classifier. Grinding pressure is varied for the most efficient grinding of various coals by externally adjustable springs on top of the stationary ring. The ball- and race pulverizer has ball circle diameters varying between 17 and 76 in and capacities between 1J to 20 tons/h. The roll-and-race pulverizer is operated at lower speeds and larger sizes. A typical one has an 89-in ball circle diameter, a 12 ft diameter, and a 22.5 ft height overall, weighs 150 tons, and is driven by a 700 hp motor. 3 oth types are suitable for direct-firing systems (see below).

High-speed pulverizers use *hammer beaters* that revolve in a chamber equipped with high-wear-resistant liners. They are mostly used with low-rank coals with high-moisture content and use flue gas for drying. They are not widely used for pulverized coal systems.

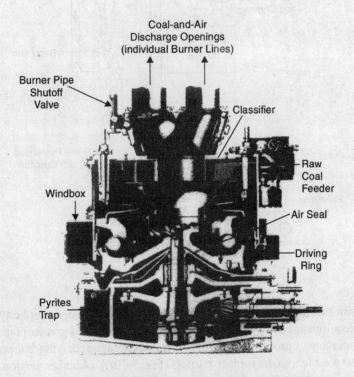

Fig. 7.6

The classifier referred to above is located at the pulverizer exit. It is usually a cyclone with adjustable inlet vanes. The classifier separates oversized coal and returns it to the grinders to maintain the proper fineness for the particular application and coal used. Adjustment is obtained by varying the gas-suspension velocity in the classifier by adjusting the inlet vanes.

The Pulverized-Coal System. A total pulverized-coal system comprises pulverizing, delivery, and burning equipment. It must be capable of both continuous operation and rapid change as required by load demands. There are two main systems: the bin or storage system and the direct-firing system.

The *bin* system is essentially a batch system by which the pulverized coal is prepared away from the furnace and the resulting pulverized-coal-primary-air mixture goes to a cyclone separator and fabric bag filter that separate and exhaust the moisture laden air to the atmosphere and discharge the pulverized coal to storage bins (Fig. 7.7). From there, the coal is pneumatically conveyed through pipelines to utilization bins near the furnace for use as required. The bin system was widely used before pulverizing equipment became reliable enough for continuous steady operation. Because of the many stages of drying, storing, transporting, etc., the bin system is subject to fire hazards. Nevertheless, it is still in use in many older plants. It has, however, given way to the direct-firing system, which is used exclusively in modern plants.

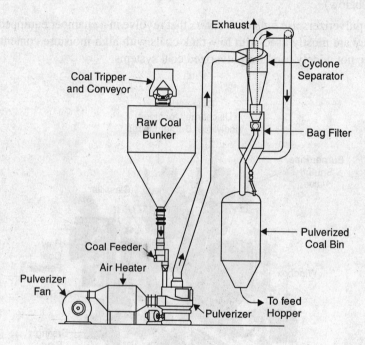

Fig. 7.7

Compared with the bin system the *direct-firing system* has greater simplicity and hence greater safety, lower space requirements, lower capital and operating costs, and greater plant cleanliness. As its name implies, it continuously processes the coal from the storage receiving bunker through a feeder, pulverizer, and primary-air fan, to the furnace burners (Fig. 7.7(a)). (Another version of this system, less used, places the fan on the outlet side of the pulverizer. Fuel flow is suited to load demand by a combination of controls on the feeder and on the primary-air fan in order to give air-fuel ratios suitable for the various steam-generator loads. The control operating range on any one direct firing pulverizer system is only about 3 to 1. Large steam generators are provided with more than one pulverizer system, each feeding a number of burners, so that a wide control range is possible by varying the number of pulverizers and the load on each

Burners A pulverized-coal *burner* is not too dissimilar to an oil burner. The latter must atomize the liquid fuel to give a large surface-to-volume ratio of fuel for proper interaction with the combustion air. A pulverized-coal burner already receives dried pulverized coal in suspension in the primary air and mixes it with the main combustion air from the steam-generator air preheater. The surface-to-volume ratio

FUELS AND COMBUSTION

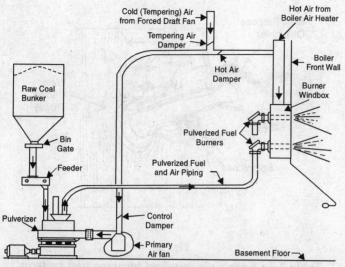

Fig. 7.7(a)

of pulverized coal or fineness requirements vary, though not too greatly, from coal to coal (the higher the fixed carbon, the finer the coal). For example, pulverized coal with 80 percent passing a 200 mesh screen and 99.5 percent passing a 50-mesh screen possesses a surface area of approximately 1500 cm^2/g with more than 97 percent of that surface area passing the 200-mesh screen.

The fuel burners may be arranged in one of two configurations. In the first, individual burners, usually arranged horizontally from one or opposite walls, are independent of each other and provide separate flame envelopes. In the second, the burners are arranged so that the fuel and air injected by them interact and produce a single flame envelope. In this configuration the burners are such that fuel and air are injected from the four corners of the furnace along lines that are tangents to an imaginary horizontal circle within the furnace, thus causing a rotative motion and intensive mixing and a flame envelope that fills the furnace area. Vertical firing is also used but is more complex and used only for hard-to-ignite fuels.

The burners themselves can be used to burn pulverized coal only (Fig. 7.8) or all three primary fuels, *i.e.* pulverized coal, oil, or gas (Fig. 7.9). In Fig. 7.8, the coal impeller promotes the mixing of fuel with the primary air and the tangential door built into the wind box provide turbulence of the main combustion, or secondary, air to help mix it with the fuel-primary-air mixture leaving the impeller.

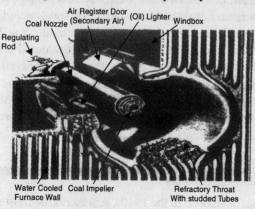

Fig. 7.8

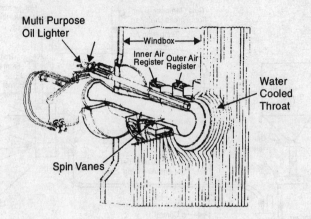

Fig. 7.9

Table 7.2 : Excess air required by some fuel systems

Fuel	System	Excess air, %
Coal	Pulverized, completely water-cooled furnace	15—20
	Pulverized, partially water-cooled furnace	15—40
	Spreader stoker	30—60
	Chain grate and travelling stoker	15—50
	Crushed, cyclone furnace	10—15
Fuel oil	Oil burners	5—10
	Multifuel burners	10—20
Gas :	Gas burners	5—10
	Multifuel burners	7—12

The total air-fuel ratio is greater than stoichrometric (chemically correct) but just enough to ensure complete combustion without wasting energy by adding too much sensible heat to the air. Table 7.2 gives the range of excess air, percent of theoretical, necessary for good combustion of some fuels. Initial ignition of the burners is accomplished in a variety of ways including a light-fuel oil jet, itself spark-ignited. This igniter is usually energized long enough to ensure a self-sustaining flame. The control equipment ranges from manual to a remotely operated programmed sequence. The igniters may be kept only for seconds in the case of fuel oil or gas. In the case of pulverized coal, however, they are usually kept much longer, sometimes for hours, until the combustion-zone temperature is high enough to ensure a self-sustaining flame. It may also be necessary to activate the igniter at very light loads especially for coals of low volatility. The impeller is the part of the burner that is subject to severe maintenance problems and is usually replaced once a year or so.

7.7 CYCLONE FURNACES

Cyclone-furnace firing, developed in the 1940s, represents the most significant step in coal firing since the introduction of pulverized-coal firing in the 1920s. It is now widely used to burn poorer grade of coal that contain a high ash content with a minimum of 6 percent to as high as 25 percent, and a high volatile matter, more than 15 percent, to obtain the necessary high rates of combustion. A wide range of moisture is allowable with pre-drying. One limitation is that ash should not contain a high sulfur content or a high Fe_2O_3; $(CaO + MgO)$ ratio. Such a coal has a tendency to form high ash-fusion temperature materials such as iron and iron sulfide in the slag, which negates the main advantage of cyclone firing.

The main advantage is the removal of much of the ash, about 60 percent, ao molten slag that is collected on the cyclone walls by centrifugal action and drained off the bottom to a slag-disintegrating tank below. Thus only 40 percent ash leave, with the flue gases, compared with about 80 percent for pulverized-coal firing. this materially reduces erosion and fouling of steam-generator surfaces as well as the size of dust-removal precipitators or bag houses at steam-generator exit. Other advantages are that only crushed coal is used and no pulverization equipment is needed and that the boiler size is reduced. Cyclone-furnace firing uses a range of coal sizes averaging 95 percent passing a 4-mesh screen.

The disadvantages are higher forced-draft fan pressures and therefore higher power requirements, the inability to use the coals mentioned above, and the formation of relatively more oxides of nitrogen, NO_2 which are air pollutants, in the combustion process.

The cyclone is essentially a water-cooled horizontal cylinder (Fig. 7.10) located outside the main boiler furnace, in which the crushed coal is fed and fired with very high rates of heat release. Combustion of the coal is completed before the resulting hot gases enter the boiler furnace. The crushed coal is fed into the cyclone burner at left along with primary air, which is about 20 percent of combustion or secondary air. The primary air enters the burner tangentially, thus imparting a centrifugal motion to the coal. The secondary air is also admitted tangentially at the top of the cyclone at high speed, imparting further centrifugal motion. A small quantity of air, called tertiary air, is admitted at the center.

The whirling motion of air and coal results in large heat-release-rate volumetric densities, between 450,000 and 800,000 Btu/(h.ft) (about 4700 to 8300 kW/m^3), and high combustion temperatures, more than 3000°F (1650°C). These high temperatures melt the ash into a liquid slag that covers the surface of the cyclone and eventually drains through the slag-tap opening to a slag tank at the bottom of the boiler

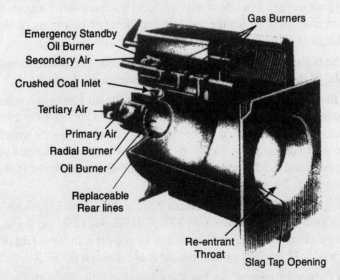

Fig. 7.10

furnace, where it is solidified and broken for removal. The slag layer that forms on the walls of the cyclone provides insulation against too much heat loss through the walls and contributes to the efficiency of cyclone firing. The high temperatures also explain the large production of NO, in the gaseous combustion products. These gases leave the cyclone through the throat at right and enter the main boiler furnace. Thus combustion takes place in the relatively small cyclone, and the main boiler furnace has the sole function of heat transfer from the gases to the water-tube walls. Cyclone furnaces are also suitable for fuel-oil and gaseous-fuel firing.

Initial ignition is done by small retractable oil or gas burners in the secondary air ports.

Like pulverized-coal systems, cyclone firing systems can be of the bin, or storage. or direct-firing types, though the bin type is more widely used, especially for most bituminous coals, than in the case of pulverized coal. The cyclone system uses either one-wall, or opposed-wall, firing, the latter being preferred for large steam generators The size and number of cyclones per boiler depend upon the boiler size and the desired load response because the usual load range for good performance of any one cyclone is from 50 to 100 percent of its rated capacity. Cyclones vary in size from 6 to 10 fv in diameter with heat inputs between 160 to 425 million Btu/h (about 47,000 to 125,Oa_ kW), respectively.

The cyclone component requiring the most maintenance is the burner, which is subjected to erosion by the high velocity of the coal. Erosion is minimized by the us; of tungsten carbide and other erosion-resistant materials for the burner liners, which are usually replaced once a year or so.

EXERCISES

1. A sample of coal has the following molal analysis C 67.35%, H_2 26.26%, O_2 2.28%, N_2 0.57%, S 1.37%, H_2O 2.17%. Write the complete combustion equation in stoichiometric air and calculate the coal ultimate analysis, mass percent.

2. Write the complete combustion equation for the anthracite coal, assuming stoichiometric air and find the dew point, degrees centigrade, of the combustion products if the total pressure is 1 bar.

3. H_2 burns in pure oxygen in a chemically correct (stoichiometric) mixture. Write the combustion equation and calculate (a) the mass of products per unit mass of H_2, and (b) the lower heating value of H_2 if its higher heating value is 61,100 Btu/lbm.

4. Calculate the higher and lower heating values, in Btus per pound mass, using the Dulong-type formula, of the anthracite coal, if the total pressure is 1 atm.

5. A gaseous fuel that is derived from coal has the following ultimate volumetric analysis: H_2 47.9%, methane (CH_4) 33.9%, ethylene (C_2H_4) 5.2%, CO 6.1%, CO_2 2.6%, N_2 3.7%, and O_2 0.6% It burns in 110 percent of theoretical air. Calculate (a) the volume flow rate of air required per unit volume flow rate of the gas when both are measured at the same pressure and temperature, and (b) the dew point of the combustion products, in degrees fahrenheit, if the total pressure is 2 atm.

6. 10,000 U.S. gal of a fuel oil are burned per hour in 20 percent excess air. The fuel oil has the following ultimate analysis by mass: C 87%, S 0.9%, H_2 12%, ash 0.1%. Write the combustion equation and find the volume flow rate of air required, in cubic feet per minute, if the fuel has a density of 7.73 Ibm u.s gal and the air is at t atm and 60°F.

7. A southern California natural gas has the following ultimate analysis by mass: H_2 23.3%. CH_4 72% N_2 0.76%, and O_2 1.22%. The flue gases have the following volumetric analysis: H_2O 15.583% SSWc. CO_2 8.387%, O_2 3.225%, N_2 72.805%. Calculate (a) the percent

theoretical air used in combustion and (b) the dew point, in degrees centigrade, if the flue gases are at 2 bars.

8. A fuel oil composed only of carbon, hydrogen, and sulfur is used in a steam generator. The volumetric flue gas analysis on a dry basis is: CO_2 11.7%, CO 0.440%, O_2 4.002%, SO_2 0.176, and N_2 83-682 Find (a) the fuel mass composition, (b) the air-fuel ratio by mass, (c) the excess air used, in percent, and (d) the dew point, in degrees centigrade, of the flue gases if their pressure is 2 bar.

9. A fuel oil burned in a steam generator has a composition which may be represented by $C_{14}H_{30}$. A dry basis flue-gas analysis shows the following volumetric composition: CO_2 11.226%. O_2 4.145% CO 0.863% N_2 83.766%. Write the complete combustion equation for 1 mol of fuel and calculate (a) the air-to-fuel ratio by mass, (b) the excess air, in percent, and (c) the mass of water vapor in the flue gases per unit mass of fuel.

10. A crushed bituminous coal to be used in a fluidized-bed combustion chamber caries in size between 1/4 and 3/4 in and has a density of 80 Ibm/ft^3. The coefficient of drag when fiuidized is 0.60. Calculate (a) the minimum gas velocity that fluidizes all the coal if the gas is at 1600°F and 9-atm pressure, and (b) the pressure drop in the bed, psi. Assume that the coal in the collapsed state has a height of 2 ft and a porosity of 0.25 and that the gas density can be approximated by that of pure air.

11. 10,000 tons of coal are burned in a powerplant per day. The coal has an as-received ultimate analysis of C 75%, H_2 5%, O_2 6.7%, H_2O 2.5%, S 2.3 %, N_2 1.5%, ash 7.0%. It burns in excess air in a fluidized bed combustor. Calculate (a) the mass of calcium carbonate to be added, in tons per day, and (b) the mass of calcium sulfate to be disposed of, in tons per day. (The molecular mass of calcium = 40.)

12. Write the chemical formula and sketch the molecule for the following hydrocarbons: (a) ethane, (b) ethene or ethylene, (c) decane, (d) iso-decane (2,2,3,3 tetramethyl hexane), (e) pentatriacontane (do not sketch), (f) isobutene (2-methyl propene), (g) I,5-heptadiene (the numbers indicate the positions of the carbon atoms that precede double bonds), (h) cyclohexane, (i) naphthalene, (j) 1-methyl napthalene (a methyl radical CH_3 attached to a carbon atom instead of a hydrogen atom), (k) tetracontane (do not sketch), and (l) dotriacontahectane (do not sketch).

13. A coal-oil mixture (COM) is composed of the bituminous medium volatility coal listed and a distillate oil no. 2 that has an ultimate analysis on a mass basis of C 87.2%, H_2 12.5%, S 0.3%, N_2.

Chapter 8

Diesel Power Plant

8.1 INTRODUCTION

The oil engines and gas engines are called Internal Combustion Engines. In IC engines fuels burn inside the engine and the products of combustion form the working fluid that generates mechanical power. Whereas, in Gas Turbines the combustion occurs in another chamber and hot working fluid containing thermal energy is admitted in turbine.

Reciprocating oil engines and gas engines are of the same family and have a strong resemblance in principle of operation and construction.

The engines convert chemical energy in fuel in to mechanical energy.

A typical oil engine has:

1. Cylinder in which fuel and air are admitted and combustion occurs.
2. Piston, which receives high pressure of expanding hot products of combustion and the piston, is forced to linear motion.
3. Connecting rod, crankshaft linkage to convert reciprocating motion into rotary motion of shaft.
4. Connected Load, mechanical drive or electrical generator.
5. Suitable valves (ports) for control of flow of fuel, air, exhaust gases, fuel injection, and ignition systems.
6. Lubricating system, cooling system

In an engine-generator set, the generator shaft is coupled to the Engine shaft.

The main differences between the gasoline engine and the diesel engine are:

- A gasoline engine intakes a mixture of gas and air, compresses it and ignites the mixture with a spark. A diesel engine takes in just air, compresses it and then injects fuel into the compressed air. The heat of the compressed air lights the fuel spontaneously.
- A gasoline engine compresses at a ratio of 8:1 to 12:1, while a diesel engine compresses at a ratio of 14:1 to as high as 25:1. The higher compression ratio of the diesel engine leads to better efficiency.
- Gasoline engines generally use either carburetion, in which the air and fuel is mixed long before the air enters the cylinder, or port fuel injection, in which the fuel is injected just prior to the intake stroke (outside the cylinder). Diesel engines use direct fuel injection to the diesel fuel is injected directly into the cylinder.

DIESEL POWER PLANT

The diesel engine has no spark plug, that it intakes air and compresses it, and that it then injects the fuel directly into the combustion chamber (direct injection). It is the heat of the compressed air that lights the fuel in a diesel engine.

The injector on a diesel engine is its most complex component and has been the subject of a great deal of experimentation in any particular engine it may be located in a variety of places. The injector has to be able to withstand the temperature and pressure inside the cylinder and still deliver the fuel in a fine mist. Getting the mist circulated in the cylinder so that it is evenly distributed is also a problem, so some diesel engines employ special induction valves, pre-combustion chambers or other devices to swirl the air in the combustion chamber or otherwise improve the ignition and combustion process.

One big difference between a diesel engine and a gas engine is in the injection process. Most car engines use port injection or a carburetor rather than direct injection. In a car engine, therefore, all of the fuel is loaded into the cylinder during the intake stroke and then compressed. The compression of the fuel/air mixture limits the compression ratio of the engine, if it compresses the air too much, the fuel/air mixture spontaneously ignites and causes knocking. A diesel compresses only air, so the compression ratio can be much higher. The higher the compression ratio, the more power is generated.

Some diesel engines contain a glow plug of some sort. When a diesel engine is cold, the compression process may not raise the air to a high enough temperature to ignite the fuel. The glow plug is an electrically heated wire (think of the hot wires you see in a toaster) that helps ignite the fuel when the engine is cold so that the engine can start.

Smaller engines and engines that do not have such advanced computer controls, use glow plugs to solve the cold-starting problem.

We recommend diesels due to their:

(a) Longevity-think of an 18 wheeler capable of 1,000,000 miles of operation before major service)

(b) Lower fuel costs (lower fuel consumption per kilowatt (kW) produced)

(c) Lower maintenance costs-no spark system, more rugged and more reliable engine,

Today's modern diesels are quiet and normally require less maintenance than comparably sized gas (natural gas or propane) units. Fuel costs per kW produced with diesels is normally thirty to fifty percent less than gas units.

1800 rpm water-cooled diesel units operate on average 12–30,000 hours before major maintenance is required. 1800 rpm water-cooled gas units normally operate 6–10,000 hours because they are built on a lighter duty gasoline engine block.

3600 rpm air-cooled gas units are normally replaced not overhauled at 500 to 1500 hours.

Because the gas units burn hotter (higher btu of the fuel) you will see significantly shorter lives than the diesel units.

Diesel engine power plants are installed where

1. Supply of coal and water is not available in desired quantity.

2. Where power is to be generated in small quantity for emergency services.

3. Standby sets are required for continuity of supply such as in hospital, telephone exchange.

It is an excellent prime mover for electric generator capacities of from 100 hp to 5000 hp. The Diesel units used for electric generation are more reliable and long - lived piece of equipment compared with other types of plants.

8.2 OPERATING PRINCIPLE

All the gas engines and oil engines operate in the same general way. The working fluid undergoes repeated cycles. A thermodynamic cycle is composed of a series of sequential events in a closed loop on P-V or T-S diagram. A typical cycle has following distinct operations

1. Cylinder is charged
2. Cylinder contents are compressed
3. Combustion (Burning) of charge, creation of high pressure pushing the piston and expansion of products of combustion.
4. Exhaust of spent products of combustion to atmosphere.

The route taken for these steps is illustrated conveniently on P-V diagram and T-S diagram for the cycle.

Various types of Gas Engines and Oil Engines have been developed and are classified on the basis of their operating cycles. Cycles are generally named after their Inventors e.g. Carnot Cycle; Diesel Cycle; Otto Cycle; Sterling Cycle; Bryton Cycle; Dual Cycle, etc.

New cycles are being developed for fuel saving and reduction of pollution.

Two principal categories of IC Engines are:

—Four Stroke Engines
—Two Stroke Engines

In a Four Stroke Engine Cycle, the piston strokes are used to obtain the four steps (intake, compression, expansion, exhaust) and one power stroke in two full revolutions of crankshaft. In a Two Stroke Engine Cycle, one power stroke is obtained during each full revolution of the crankshaft.

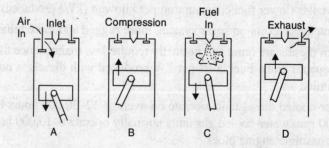

A. Air Charging   B. Compression   C. Ignition   D. Exhaust

Fig. 8.1. Diagrams Illustrating Operation of an I.C. Engine.

This is achieved by using air pressure slightly above atmospheric to blow out exhaust gases out of the cylinder and fill the fresh charge (scavenging). The methods of scavenging include: Crankcase scavenging; blower scavenging. Other methods include Super Charging; Turbo Charging.

8.3 BASIC TYPES OF IC ENGINES

Although alike in main mechanical aspects, the oil engines differ from gas engines in fuels and fuel handling *i.e.*, when fuel and air are injected how much charge is compressed and how ignited. Many variants exist.

8.3.1 TWO-STROKE, SPARK IGNITION GAS ENGINES/PETROL ENGINES

The well-known automobile engine fueled with petrol (also called Gas) and Natural Gas Engine, Bio-gas Engine is of this category. The low compression gas engine (petrol engine/natural gas engine) mixes fuel and air, outside the cylinder, before compression. With the automobile engine, a carburator is used for mixing the fuel and air and the mixture is injected in the cylinder. In a Natural Gas Engine, a mixing valve is used for the same purpose instead of the carburator.

In the mixture, the gas fuel and air proportion is almost perfect to produce complete combustion without excess air. This mixture flows into the cylinder and is then compressed. Near the end of the compression stroke, an electric spark ignites the inflammable mixture, which burns rapidly. The pressure in the cylinder rises rapidly and acts on the piston area and the piston is forced to move down on its power stroke.

Since the compressed gas mixture rises in pressure during the compression stroke, the mixture may get pre-ignited before the sparking resulting in loss of power. Hence compression pressure must be limited in this type of engine. Compression Ratio is therefore an important parameter in establishing combustion without pre-ignition.

The compression ratio is the ratio of cylinder volumes at the start and at the end of compression stroke.

In general, higher the compression ratio, higher will be the maximum pressure reached during combustion and higher is the efficiency of the engine.

Although it is desirable to have a high compression ratio, the nature of fuel imposes limits in engines where a nearly perfect mixture is compressed.

With natural gas for example the compression ratio might be about 5:1 and compression pressure of about 8 bar, pre-ignition being the limiting factor.

8.3.2 DIESEL ENGINES/HEAVY OIL ENGINES

In contrast to the engines in which the fuel and air mixes before compression, in diesel engines: air is compressed as the compression stroke begins and the fuel enters the cylinder at the end of compression stroke. Heat of compression is used for ignition of fuel.

In a typical diesel engine, air is compressed to about 30 bars, which increases the temperature when finely atomised diesel fuel oil is sprayed into the heated air, it ignites and burns. High compression ratio is therefore essential for reliable combustion and high efficiency. Compression ratios above those needed to achieve ignition do not improve the efficiency.

The pressure ratio depends on engine speed, cylinder size and design factors. Typical compression pressures in diesel engines range from 30 bar to 42 bar. Small high-speed engines have higher compression pressures.

8.3.3 DUEL FUEL ENGINES

In a duel fuel engine, a small quantity of pilot oil is injected near the end of the compression stroke. It is ignited by the compression and the mixture burns like standard diesel fuel. The pilot oil burning provides enough heat to the mixture of gas/air. Precise control of pilot oil injection and a separate set of fuel pumps and nozzles are added. Means are provided to reduce air quantity at partial loads.

8.3.4 HIGH COMPRESSION GAS ENGINES

With operation solely on gas, without duel mixtures and pilot oil, the high Compression Gas Engines of today use slightly richer mixtures of fuel and air, with lower compression ratios than duel fuel engines. The compression ratios are higher than conventional gas engines and lower than duel fuel engines. There is no need of pilot oil.

8.4 ADVANTAGE OF DIESEL POWER PLANT

The advantages of diesel power plants are listed below.
1. Very simple design also simple installation.
2. Limited cooling water requirement.
3. Standby losses are less as compared to other Power plants.
4. Low fuel cost.
5. Quickly started and put on load.
6. Smaller storage is needed for the fuel.
7. Layout of power plant is quite simple.
8. There is no problem of ash handling.
9. Less supervision required.
10. For small capacity, diesel power plant is more efficient as compared to steam power plant.
11. They can respond to varying loads without any difficulty.

8.5 DISADVANTAGE OF DIESEL POWER PLANT

The disadvantages of diesel power plants are listed below.
1. High Maintenance and operating cost.
2. Fuel cost is more, since in India diesel is costly.
3. The plant cost per kW is comparatively more.
4. The life of diesel power plant is small due to high maintenance.
5. Noise is a serious problem in diesel power plant.
6. Diesel power plant cannot be constructed for large scale.

8.6 APPLICATION OF DIESEL POWER PLANT

Since there are many disadvantage of diesel power plant, although the plant find wide application in the following fields.
1. They are quite suitable for mobile power generation and are widely used in transportation systems consisting of railroads, ships, automobiles and aeroplanes.
2. They can be used for electrical power generation in capacities from 100 to 5000 H.P.
3. They can be used as standby power plants.
4. They can be used as peak load plants for some other types of power plants.

DIESEL POWER PLANT

5. Industrial concerns where power requirement are small say of the order of 500 kW, diesel power plants become more economical due to their higher overall efficiency.

8.7 GENERAL LAYOUT OF DIESEL POWER PLANT

General layout of diesel power plant is as shown in Fig. 8.2.

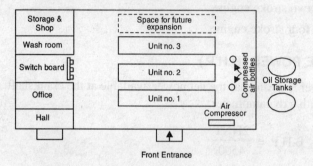

Fig. 8.2. General Layout of Diesel Power Plant.

Generally the units are placed in parallel lines as shown in fig. In any plant some space is always provided for further expansion. Also sufficient space should provide for maintenance of diesel engine. Proper ventilation is also provided in power plant. Storage of fuel for power plant is always provided outside the main building.

8.8 PERFORMANCE OF DIESEL ENGINE

The performance of the diesel engine means the power and efficiency. The engine develops as the various parameters of the engine, *e.g.* piston speed, air-fuel ratio, compression ratio, inlet air-pressure and temperature are varied.

The two usual conditions under which I.C. engines are operated are: (1) constant speed with variable load, and (2) variable speed with variable load. The first situation is found in a.c. generator drives and the second one in automobiles, railway engines and tractors etc. A series of tests are carried out on the engine to determine its performance characteristics, such as: indicated power (I.P.), Brake power (B.P.), Frictional Power (F.P.), Mechanical efficiency (η_m), thermal efficiency, fuel consumption and also specific fuel consumption etc. Below, we shall discussed how these quantities are measured:

8.8.1 INDICATED MEAN EFFECTIVE PRESSURE (IMEP)

In order to determine the power developed by the engine, the indicator diagram of engine should be available. From the area of indicator diagram it is possible to find an average gas pressure that while acting on piston throughout one stroke would account for the network done. This pressure is called indicated mean effective pressure (I.M.E.P.).

8.8.2 INDICATED HORSE POWER (IHP)

The indicated horse power (I.H.P.) of the engine can be calculated as follows:

$$\text{I.H.P.} = \frac{(P_m . L.A.N.n)}{(4500 \times k)}$$

where P_m = I.M.E.P. in kg/cm^2
L = Length of stroke in metres
A = Piston areas in cm^2
N = Speed in R.P.M.
n = Number of cylinders
k = 1 for two stroke engine
= 2 for four stroke engine.

8.8.3 BRAKE HORSE POWER (B.H.P.)

Brake horse power is defined as the net power available at the crankshaft. It is found by measuring the output torque with a dynamometer.

$$\text{B.H.P.} = \frac{2\Pi NT}{4500}$$

where T = Torque in kg.m.
N = Speed in R.P.NT.

8.8.4 FRICTIONAL HORSE POWER (F.H.P.)

The difference of I.H.P. and B.H.P. is called F.H.P. It is utilized in overcoming frictional resistance of rotating and sliding parts of the engine.

$$\text{F.H.P.} = \text{I.H.P.} - \text{B.H.P.}$$

8.8.5 INDICATED THERMAL EFFICIENCY (η_i)

It is defined as the ratio of indicated work to thermal input.

$$\eta_i = \frac{(\text{I.H.P.} \times 4500)}{(W \times C_v \times J)}$$

where W = Weight of fuel supplied in kg per minute.
C_V = Calorific value of fuel oil in kcal/kg.
J = Joules equivalent = 427.

8.8.6 BRAKE THERMAL EFFICIENCY (OVERALL EFFICIENCY)

It is defined as the ratio of brake output to thermal input.

$$\eta_b = \frac{(\text{B.H.P.} \times 4500)}{(W \times C_v \times J)}$$

8.8.7 MECHANICAL EFFICIENCY (η_m)

It is defined as the ratio of B.H.P. to I.H.P. Therefore,

$$\eta_m = \frac{\text{B.H.P.}}{\text{I.H.P.}}$$

DIESEL POWER PLANT 241

8.9 FUEL SYSTEM OF DIESEL POWER PLANT

The fuel is delivered to the plant by railroad tank car, by truck or by barge and tanker and stored in the bulk storage situated outdoors for the sake of safety. From this main fuel tank, the fuel oil is transferred to the daily consumption tank by a transfer pump through a filter. The capacity of the daily consumption should be atleast the 8-hour requirement of the plant. This tank is located either above the engine level so that the fuel flows by gravity to the injection pump or below the engine level and the fuel oil is delivered to the injection pump by a transfer pump driven from the engine shaft, Fig. 8.3. Fuel connection is normally used when tank-car siding or truck roadway is above tank level. If it is below tank level, then, an unloading pump is used to transfer fuel form tank car to the storage tank (dotted line).

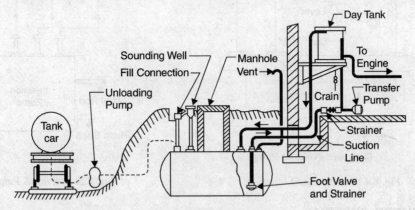

Fig. 8.3. Fuel System.

The five essential functions of a fuel injection system are:

1. To deliver oil from the storage to the fuel injector.
2. To raise the fuel pressure to the level required for atomization.
3. To measure and control the amount of fuel admitted in each cycle.
4. To control time of injection.
5. To spray fuel into the cylinder in atomized form for thorough mixing and burning.

The above functions can be achieved in a variety of ways. The following are the systems, which are usual on power station diesels:

1. Common Rail.
2. Individual Pump Injection.
3. Distributor.

1. COMMON RAIL INJECTION

A typical common rail injection system is shown in Fig. 8.4. It incorporates a pump with built in pressure regulation, which adjusts pumping rate to maintain the desired injection pressure. The function of the pressure relief and timing valves is to regulate the injection time and amount. Spring-loaded spray valve acts merely as a check. When injection valve lifts to admit high-pressure fuel to spray valve, its needle rises against the spring. When the pressure is vented to the atmosphere, the spring shuts the valve.

2. INDIVIDUAL PUMP INJECTION

In this system, each fuel nozzle is connected to a separate injection pump, Fig.8.5. The pump itself does the measuring of the fuel charge and control of the injection timing. The delivery valve in the nozzle is actuated by fuel-oil pressure.

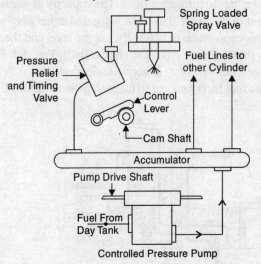
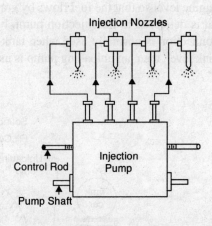

Fig. 8.4. Common Rail Injection. **Fig. 8.5.** Individual Pump Injection.

3. DISTRIBUTOR SYSTEM

This system is shown in Fig. 8.6. In this system, the fuel is metered at a central point *i.e.*, the pump that pressurizes, meters the fuel and times the injection. From here, the fuel is distributed to cylinders in correct firing order by cam operated poppet valves, which open to admit fuel to nozzles.

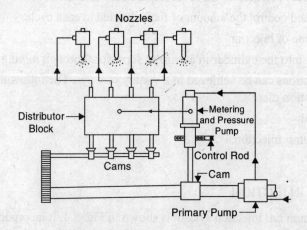

Fig. 8.6. Distribution System.

8.10 LUBRICATION SYSTEM OF DIESEL POWER PLANT

Since frictional forces causes wear and tear of rubbing parts of the engine and thereby the life of the engine is reduced. So the rubbing part requires that some substance should be introduced between the rubbing surfaces in order to decrease the frictional force between them. Such substance is called lubricant. The lubricant forms a thin film between the rubbing surfaces. And lubricant prevents metal-to-metal contact. So we can say "Lubrication is the admission of oil between two surface having relative motion".

The main function of lubricant is to,

1. To reduce friction and wear between the parts having relative motion by minimizing the force of friction and ensures smooth running of parts.
2. To seal a space adjoining the surfaces such as piston rings and cylinder liner.
3. To clean the surface by carrying away the carbon and metal particles caused by wear.
4. To absorb shock between bearings and other parts and consequently reduce noise.
5. To cool the surfaces by carrying away heat generated due to friction.
6. It helps the piston ring to seal the gases in the cylinder.
7. It removes the heat generated due to friction and keeps the parts cool.

The various parts of an engine requiring lubrication are;

1. Cylinder walls and pistons.
2. Main crankshaft bearings.
3. Piston rings and cylinder walls.
4. Big end bearing and crank pins.
5. Small end bearing and gudgeon pin bearings.
6. Main bearing cams and bearing valve tappet and guides
7. Timing gears etc.
8. Camshaft and cam shaft bearings.
9. Valve mechanism and rocker arms.

A good lubricant should possess the following properties:

1. It should not change its state with change in temperature.
2. It should maintain a continuous film between the rubbing surfaces.
3. It should have high specific heat so that it can remove maximum amount of heat.
4. It should be free from corrosive acids.
5. The lubricant should be purified before it enters the engine.
6. It should be free from dust, moisture, metallic chips, etc.
7. The lubricating oil consumed is nearly 1% of fuel consumption.
8. The lubricating oil gets heated because of friction of moving parts and should be cooled before recirculation.

The cooling water used in the engine may be used for cooling the lubricant. Nearly 2.5% of heat of fuel is dissipated as heat, which is removed by the lubricating oil.

The various lubricants used in engines are of three types:

1. Liquid Lubricants or Wet sump lubrication system.
2. Solid Lubricants or Dry sump lubrication system.
3. Semi-solid Lubricants or Mist lubrication system.

Liquid oils lubricants are most commonly used. Liquid lubricants are of two types:

(*a*) Mineral oils

(*b*) Fatty oils.

Graphite, white lead and mica are the solid lubricants.

Semi solid lubricants or greases as they are often called are made from mineral oils and fatty-oils.

8.10.1 LIQUID LUBRICANTS OR WET SUMP LUBRICATION SYSTEM

These systems employ a large capacity oil sump at the base of crank chamber, from which the oil is drawn by a low-pressure oil pump and delivered to various parts. Oil then gradually returns back to the sump after serving the purpose.

(*a*) **Splash system.** This system is used on some small four strokes, stationary engines. In this case the caps on the big ends bearings of connecting rods are provided with scoops which, when the connecting rod is in the lowest position, just dip into oil troughs and thus directs the oil through holes in the caps to the big end bearings. Due to splash of oil it reaches the lower portion of the cylinder walls, crankshaft and other parts requiring lubrication. Surplus oil eventually flows back to the oil sump. Oil level in the troughs is maintained by means of an oil pump which takes oil from sump, through a filter.

Splash system is suitable for low and medium speed engines having moderate bearing load pressures. For high performance engines, which normally operate at high bearing pressures and rubbing speeds this system does not serve the purpose.

(*b*) **Semi-pressure system.** This method is a combination of splash and pressure systems. It incorporates the advantages of both. In this case main supply of oil is located in the base of crank chamber. Oil is drawn from the lower portion of the sump through a filter and is delivered by means of

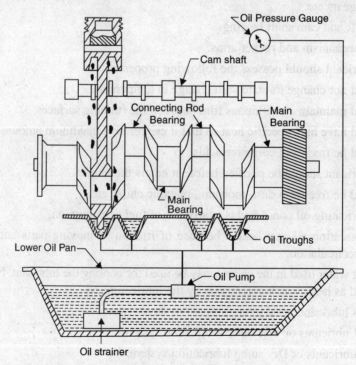

Fig. 8.7. Semi Pressure System.

a gear pump at pressure of about 1 bar to the main bearings. The big end bearings are lubricated by means of a spray through nozzles. Thus oil also lubricates the cams, crankshaft bearings, cylinder walls and timing gears. An oil pressure gauge is provided to indicate satisfactory oil supply.

The system is less costly to install as compared to pressure system. It enables higher bearing loads and engine speeds to be employed as compared to splash system.

(*c*) **Full pressure system.** In this system, oil from oil sump is pumped under pressure to the various parts requiring lubrication. Refer Fig. 8.8. The oil is drawn from the sump through filter and pumped by means of a gear pump. The pressure pump at pressure ranging delivers oil from 1.5 to 4 bar. The oil under pressure is supplied to main bearings of crankshaft and camshaft. Holes drilled through the main crankshafts bearing journals, communicate oil to the big end bearings and also small end bearings through holes drilled in connecting rods. A pressure gauge is provided to confirm the circulation of oil to the various parts. A pressure-regulating valve is also provided on the delivery side of this pump to prevent excessive pressure.

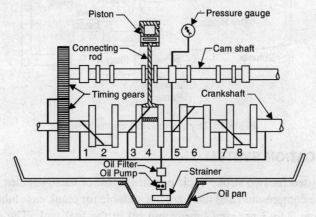

Fig. 8.8. Full Pressure System.

This system finds favour from most of the engine manufacturers as it allows high bearing pressure and rubbing speeds.

The general arrangement of wet sump lubrication system is shown in Fig. 8.9. In this case oil is always contained in the sump that is drawn by the pump through a strainer.

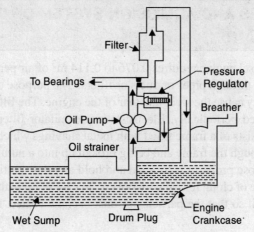

Fig 8.9. Wet sump lubrication system.

8.10.2 SOLID LUBRICANTS OR DRY SUMP LUBRICATION SYSTEM

Refer Fig. 8.10. In this system, the oil from the sump is carried to a separate storage tank outside the engine cylinder block. The oil from sump is pumped by means of a sump pump through filters to the storage tank. Oil from storage tank is pumped to the engine cylinder through oil cooler. Oil pressure may vary from 3 to 8 kgf/cm^2. Dry sump lubrication system is generally adopted for high capacity engines.

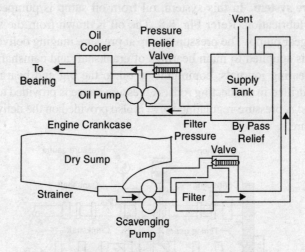

Fig. 8.10. Dry Sump Lubrications System.

8.10.3 MIST LUBRICATION SYSTEM

This system is used for two stroke cycle engines. Most of these engines are crank charged, *i.e.,* they employ crank case compression and thus, are not suitable for crank case lubrication. These engines are lubricated by adding 2 to 3 per cent lubricating oil in the fuel tank. The oil and fuel mixture is induced through the carburator. The gasoline is vaporized; and the oil in the form of mist, goes via crankcase into the cylinder. The oil that impinges on the crank case walls lubricates the main and connecting rod bearings, and rest of the oil that passes on the cylinder during charging and scavenging periods, lubricates the piston, piston rings and the cylinder.

8.11 AIR INTAKES AND ADMISSION SYSTEM OF DIESEL POWER PLANT

Generally a large diesel engine requires 0.076 to 0.114 m^3 of air per min per kw of power developed. The fresh air is drawn through pipes or ducts or filters. The purpose of the filter is to catch any air borne dirt as it otherwise may cause the wear and tear of the engine. The filters may be of dry or oil bath. The filters should be cleaned periodically. Electrostatic precipitator filters can also be used. Oil impingement type of filter consists of a frame filled with metal shavings which are coated with a special oil so that the air in passing through the frame and being broken up into a number of small filaments comes into contact with the oil whose property is to seize and hold any dust particles being carried by the air. The dry type of filter is made of cloth, felt, glass wool etc. In case of oil bath type of filter the air is swept over or through a pool of oil so that the particles of dust become coated. Lightweight steel pipe is the material for intake ducts.

Since the noise may be transmitted back to the outside air via the air intake. So, A silencer is needed in between the engine and the intake system.

There should be minimum pressure loss in the air intake system, otherwise specific fuel consumption will increase and the engine capacity is reduced.

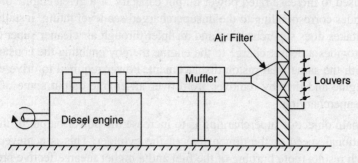

Fig. 8.11. Air Intake System.

The air intake system conveys fresh air through pipes or ducts to:
1. Air-intake manifold of four-stroke engine.
2. The scavenging pump inlet of a two-stroke engine.
3. The supercharger inlet of a supercharged engine.

The air intake may be located:
(*a*) Very Near the ground and outside the plant building.
(*b*) In the building roof.
(*c*) On the building roof.
(*d*) Inside the engine room.

Following precautions should be taken while constructing a suitable air intake system:
1. They do not locate the air-intakes inside the engine room.
2. Do not take air from a confined space as otherwise serious vibration problems can occur due to air pulsations.
3. Do not use air-intake line with too small a diameter or which is too long, otherwise engine starvation might occur.
4. Do not install air-intake filters in an inaccessible location.
5. Do not locate the air intake filters close to the roof of the engine room since serious vibrations of the roof may occur due to pulsating airflow through the filters.

8.12 SUPERCHARGING SYSTEM OF DIESEL POWER PLANT

The purpose of supercharging is to raise the volumetric efficiency above that value which can be obtained by normal aspiration.

Since the I.H.P. produced by an I.C. engine is directly proportional to the air consumed by the engine. And greater quantities of fuel to be added by increasing the air consumption permit and result in greater power produced by the engine. So, it is, therefore, desirable that the engine should take in the greatest possible mass of air. The supply of air is pumped into the cylinder at a pressure greater than the atmospheric pressure and is called supercharging. When greater quantity of air is supplied to an I.C engine it would be able to develop more power for the same size and conversely a small size engine fed with extra air would produce the same power as a larger engine supplied with its normal air feed. Supercharging is used to increase rated power output capacity of a given engine or to make the rating equal at high altitudes corresponding to the unsupercharged sea level rating. Installing a super charger between engine intakes does supercharging and air inlet through air cleaner super charger is merely a compressor that provides a denser charge to the engine thereby enabling the consumption of a greater mass of charge with the same total piston displacement. Power required to drive the super charger is taken from the engine and thereby removes from over all engine output some of the gain in power obtained through supercharging.

Since the main object of supercharging is to increase the power output of these engine without increasing its rotational speed or the dimensions of the cylinder. This is achieved by increasing the charge of air, which results more burning of the fuel and a higher mean effective pressure. So there are three possible methods that increase the air consumption of an engine,

1. To increasing the piston displacement, but this increases the size and weight of the engine, and introduces additional cooling problems.

2. Running the engine at higher speeds, which results in increased fluid and mechanical friction losses, and imposes greater inertia stresses on engine parts.

3. Increasing the density of the charge, such that a greater mass of charge is introduced into the same volume or same total piston displacement.

8.12.1 TYPES OF SUPERCHARGER

Supercharging is done by means of compressor; there are two types of compressors that may be used as super chargers. They are as follows:

1. Positive displacement type super chargers.
    (a) Piston Cylinder type
    (b) Roots blowers
    (c) Vane blower
2. Centrifugal type super chargers or turbo type.
3. Turbo type super chargers.

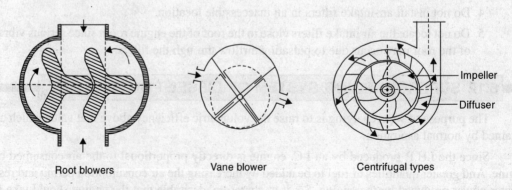

Root blowers   Vane blower   Centrifugal types

Fig. 8.12   Fig. 8.13

The blowers are usually driven from the engine cranks shaft by mean of Spur, helical or herringbone gears, silent chains or V-belts at a speed 2-3 times the engine speed. In the case of turbo-supercharger, the super charger is coupled to a gas turbine in which the exhaust of the engine is expanded.

The positive displacement types are used for low and medium speed engines with speeds not over 4000 rev/min. Positive displacement type used with many reciprocating engines in stationary plants, vehicles and marine installations.

The piston cylinder type is used on large and slow speed stationary engines, but their use is limited since they are bulkier, more expensive and less dependable than the rotary type blowers.

Centrifugal blowers are used both on low speed and high-speed engines. Centrifugal blowers driven by exhaust gas turbine are small and light and are used for stationary, locomotive, and marine and aircraft engine. The speed of the centrifugal type of blowers is high about 10000 to 15000 rev/min for low speed engines and 15000 to 30000 rev/min for high-speed engines such as aircraft engines.

Centrifugal type widely used as the supercharger for reciprocating engines, as well as compressor for gas turbines. It is almost exclusively used as the supercharger with reciprocating power plants for aircraft because it is relatively light and compact, and produces continuous flow rather than pulsating flow as in some positive displacement types.

8.12.2 ADVANTAGE OF SUPERCHARGING

Due to a number of advantages of supercharging the modern diesel engines used in diesel plants are generally supercharged. The various advantages of supercharging are as follows:

1. Power Increase. Mean effective pressure of the engine can be easily increased by 30 to 50% by supercharging which will result in the increase the power output.

2. Fuel Economy. Due to better combustion because of increased turbulence, better mixing of the fuel and air, and of an increased mechanical efficiency, the specific fuel consumption in most cases, though supercharging reduces not all.

3. Mechanical Efficiency. The mechanical efficiency referred to maximum load is increased since the increase of frictional losses with a supercharger driven directly from the engine is quite smaller as compared to the power gained by supercharging.

4. Fuel Knock. It is decreased due to increased compression pressure because increasing the inlet pressure decreases the ignition lag and this reduces the rate of pressure rise in the cylinder resulting in increasing smoothness of operation.

5. Volumetric Efficiency. Volumetric efficiency is increased since the clearance gases are compressed by the induced charge that is at a higher pressure than the exhaust pressure.

8.13 EXHAUST SYSTEM OF DIESEL POWER PLANT

The purpose of the exhaust system is to discharge the engine exhaust to the atmosphere outside the main building.

For designing of exhaust system of a big power plant, following points should be taken into consideration

1. Exhaust noise should be reduced to a tolerable degree.

2. To reduce the air pollution at breathing level, Exhaust should be exhausted well above the ground level

3. Pressure loss in the system should be reduced to minimum.

4. By use of flexible exhaust pipe, the vibrations of exhaust system must be isolated from the plant.

5. A provision should be made to extract the heat from exhaust if the heating is required for fuel oil heating or building heating or process heating.

In many cases, we have seen that the temperature of the exhaust gases under full load conditions may be of the order of 400°C. With the recovery of heat from hot jacket water and exhaust gases and its use either for heating oil or buildings in cold weather increase the thermal efficiency to 80%. Nearly 40% of the heat in the fuel can be recovered from the hot jacket water and exhaust gases. The heat from the exhaust can also be used for generating the steam at low pressure that can be used for process heating. Nearly 2 kg of steam at 8 kg/cm^2 can be generated per kW per hour, when the mass of exhaust gases can be taken as 10 kg/kW hr.

8.14 COOLING SYSTEM OF DIESEL POWER PLANT

During combustion process the peak gas temperature in the cylinder of an internal combustion engine is of the order of 2500 K. Maximum metal temperature for the inside of the combustion chamber space are limited to much lower values than the gas temperature by a large number of considerations and thus cooling for the cylinder head, cylinder and piston must therefore be provided. Necessity of engine cooling arises due to the following facts

1. During combustion period, the heat fluxes to the chamber walls can reach as high as 10 mW/m^2. The flux varies substantially with location. The regions of the chamber that are contacted by rapidly moving high temperature gases generally experience the highest fluxes. In region of high heat flux, thermal stresses must be kept below levels that would cause fatigue cracking. So temperatures must be less than about 400°C for cast iron and 300°C for aluminium alloy for water cooled engines. For air-cooled engines, these values are 270°C and 200°C respectively.

2. The gas side surface temperature of the cylinder wall is limited by the type of lubricating oil used and this temperature ranges from 160°C to 180°C. Beyond these temperature, the properties of lubricating oil deteriorates very rapidly and it might even evaporates and burn, damaging piston and cylinder surfaces. Piston seizure due to overheating resulting from the failure of lubrication is quite common.

3. The valves may be kept cool to avoid knock and pre-ignition problems which result from overheated exhaust valves (true for S.I. engines).

4. The volumetric and thermal efficiency and power output of the engines decrease with an increase in cylinder and head temperature.

Based on cooling medium two types of cooling systems are in general use. They are

(a) Air as direct cooling system.

(b) Liquid or indirect cooling system.

Air-cooling is used in small engines and portable engines by providing fins on the cylinder. Big diesel engines are always liquid (water/special liquid) cooled.

Liquid cooling system is further classified as

(1) Open cooling system

(2) Natural circulation (Thermo-system)

(3) Forced circulation system

(4) Evaporation cooling system.

8.14.1 OPEN COOLING SYSTEM

This system is applicable only where plenty of water is available. The water from the storage tank is directly supplied through an inlet valve to the engine cooling water jacket. The hot water coming out of the engine is not cooled for reuse but it is discharged.

8.14.2 NATURAL CIRCULATION SYSTEM

The system is closed one and designed so that the water may circulate naturally because of the difference in density of water at different temperatures. Fig. 8.14 shows a natural circulation cooling system. It consists of water jacket, radiator and a fan. When the water is heated, its density decreases and it tends to rise, while the colder molecules tend to sink. Circulation of water then is obtained as the water heated in the water jacket tends to rise and the water cooled in the radiator with the help of air passing over the radiator either by ram effect or by fan or jointly tends to sink. Arrows show the direction of natural circulation, which is slow.

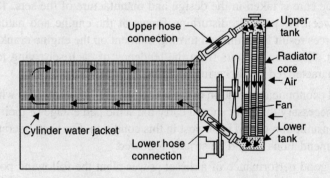

Fig. 8.14. Natural Circulation Cooling System.

8.14.3 FORCED CIRCULATION COOLING SYSTEM

Fig. 8.15 shows forced circulation cooling system that is closed one. The system consists of pump, water jacket in the cylinder, radiator, fan and a thermostat. The coolant (water or synthetic coolant) is circulated through the cylinder jacket with the help of a pump, which is usually a centrifugal type, and driven by the engine. The function of thermostat, which is fitted in the upper hose connection initially, prevents the circulation of water below a certain temperature (usually upto 85°C) through the radiation so that water gets heated up quickly.

Standby diesel power plants upto 200 kVA use this type of cooling. In the case of bigger plant, the hot water is cooled in a cooling tower and recirculated again. There is a need of small quantity of cooling make-up water.

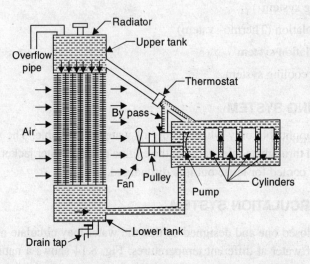

Fig. 8.15. Forced Circulation Cooling System.

8.15 DIESEL PLANT OPERATION

When diesel alternator sets are put in parallel, "hunting" or "phase swing may be produced due to resonance unless due care is taken in the design and manufacture of the sets. This condition occurs due to resonance between the periodic disturbing forces of the engine and natural frequency of the system. The engine forces result from uneven turning moment on the engine crank which are corrected by the flywheel effect. "Hunting" results from the tendency of each set trying to pull the other into synchronism and is characterized by flickering of lights.

To ensure most economical operation of diesel engines of different sizes when working together and sharing load it is necessary that they should carry the same percentage of their full load capacity at all times as the fuel consumption would be lowest in this condition. For best, operation performance the manufacturer's recommendations should be strictly followed.

In order to get good performance of a diesel power plant the following points should be taken care of:

1. It is necessary to maintain the cooling temperature within the prescribed range and use of very cold water should be avoided. The cooling water should be free from suspended impurities and suitably treated to be scale and corrosion free. If the ambient temperature approaches freezing point, the cooling water should be drained out of the engine when it is kept idle.

2. During operation the lubrication system should work effectively and requisite pressure and temperature maintained. The engine oil should be of the correct specifications and should be in a fit. Condition to lubricate the different parts. A watch may be kept on the consumption of lubricating oil as this gives an indication of the true internal condition of the engine.

DIESEL POWER PLANT

3. The engine should he periodically run even when not required to be used and should not be allowed to stand idle for more than 7 days.

4. Air litter, oil filters and fuel filters should be periodically serviced or replaced as recommended by the manufacturers or if found in an unsatisfactory condition upon inspection.

5. Periodical checking of engine compression and firing pressures and also exhaust temperatures should be made.

The engine exhaust usually provides a good indication of satisfactory performance of the engine. A black smoke in the exhaust is a. sign of inadequate combustion or engine over loading.

The loss of compression resulting from wearing old of moving parts lowers the compression ratio causing inadequate combustion. Taking indicator diagrams of the engine after reasonable intervals can check these defects.

8.16 EFFICIENCY OF DIESEL POWER PLANT

The efficiency of a diesel engine plant is enhanced by use of Turbo-compounded diesel engine and Heat Recovery Steam Generator (HRSG) and steam turbine generator.

A 24.8 mW Diesel Engine Generator Plant installed in Macau has two slow speed diesel engine generator units (1985). The overall efficiency of nearly 50% has been demonstrated. The high efficiency has been achieved by use of Turbine Generator, and Steam Turbine operated by exhaust gases of diesel engine (Fig. 8.16).

This is claimed to be the first Diesel Power Plant in the world in which exhaust heat is recovered and used for steam power generation. Thermal efficiency has been enhanced.

Fig. 8.17 shows the Heat Balance Diagram for the plant.

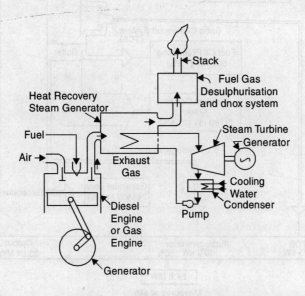

Fig. 8.16. Combined Cycle: Diesel Engine, HRSG and Steam Turbine.

Table 8.1. Technical Data about Diesel Engine-Steam Turbine Cogeneration Power Plant

No. of diesel engine generator units	2 × 24.4 mW
Rating of each unit at speed	24.8 mW at 100 rpm
No. of cylinders per engine	9
Bore/stroke	$\dfrac{800}{2300}$ mm
Diesel engine mitsui/MAN B and W	2 Stroke Low Speed
Generators 2 Nos. 11 kV, 50 Hz, 100 RPM	29.8 kVA each
Steam Turbine	0.67 mW
Gas Turbo Generator System	0.5 mW
Location	Macau, Canton, China

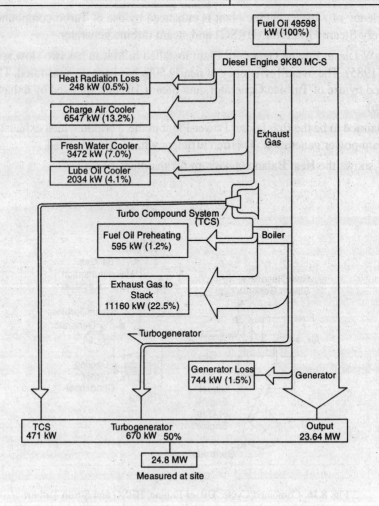

Fig. 8.17. Heat Balance of a Diesel Engine Power Plant.

DIESEL POWER PLANT

8.17 HEAT BALANCE SHEET

It is a useful method to watch the performance of the diesel power plant. Among all the heat supplied to an engine only part of it is converted into useful work, the remaining goes as waste. The distribution of the heat imparted to an engine is called as its heat balance. The heat balance of an engine depends on a number of factors among which load is primary importance. The heat balance of an internal combustion engine shows that the cooling water and exhaust gases carry away about 60-70% of heat produced during combustion of fuel. Heat balance sheet is a useful method to watch the performance of the plant. In order to draw the heat balance sheet of Diesel engine, the engine is run at constant load and constant speed and the indicator diagram is drawn with the help of indicator. The following quantities are noted:

1. The quantity of fuel consumed during a given period.
2. Quantity of cooling water and its outlet and inlet temperatures.
3. Weight of exhaust gases.
4. Temperature of exhausts gases.
5. Temperature of flue gases supplied.

To calculate the heat in various items proceed as follows.

Let

W = Weight of fuel consumed per minute in kg.

G = Lower calorific value of fuel, kcal per kg.

Then heat in fuel supplied per minute = WC_V kcal.

The energy supplied to Diesel engine in the form of fuel input is usually broken into following items:

(A) Heat Energy Absorbed in I.H.P. The heat energy absorbed in indicated horsepower, I.H.P. is found by the following expression:

Heat in L.H.P. per minute

$$(I.H.P. \times 4500)/J \text{ kcal}$$

(B) Heat Rejected to Colling in Water.

Let

W_1 = Weight of cooling water supplied per minute (kg)

T_1 = Inlet temperature of cooling water in °C

T_2 = Output temperature of cooling water in °C

Then heat rejected to cooling water = $W_1(T_2 - T_1)$

(C) Heat Carried Away by Exhaust Gases

Let W_2 = Weight of exhaust gases leaving per minute in kg.

(sum of weight of air and fuel supplied)

T_3 = Temperature of flue gases supplied per minute °C.

T_4 = Temperature °C of exhaust gases.

K_P = Mean specific heat at constant pressure of exhaust gases

The heat carried away by exhaust gases

$$= W_2 \times K_P \times (T_4 - T_3) \text{ kg cal.}$$

(D) Heat Unaccounted for (Heat Lost Due to Friction, Radiation etc.). The heat balance sheet is drawn as follows:

Item	Head units kcal	Percent
Heat in fuel supplied		
(a) Heat absorbed by I.H.P.		
(b) Heat rejected to cooling water		
(c) Heat carried away by exhaust gases		
(d) Heat unaccounted for (by difference)		
Total		

A typical heat balance sheet at full load for Diesel cycle (compression ignition) is as follows:
(1) Useful work = 30%
(2) Heat rejected to cooling water = 30%
(3) Heat carried away by exhaust gases = 26%
(4) Heat unaccounted (Heat lost due to friction, radiation etc.) = 10%.

SOLVED EXAMPLES

Example 1. *A diesel engine has a brake thermal efficiency of 30%. If the calorific value of fuel used in 10000 kcal kg, calculate the brake specific fuel consumption.*

Solution. η_b = Brake thermal efficiency = 0.3

I.H.B. hr = 632.5 kcal

η_b = H.P. hr equivalent/$(w \times C.V.)$

where, w = Specific fuel consumption per hr.

C.V. = Calorific value of fuel = 10,000 kcal/kg.

$$0.3 = \frac{632.5}{(w \times 10,000)}$$

w = 0.21 kg/H.P. hr.

Example 2. *A four-stroke diesel engine has a piston diameter of 16.5 cm and a stroke of 27 cm. The compression ratio is 14.3, the cut-off 4.23% of the stroke and the mean effective pressure 4.12 bar. The engine speed is 264 rev/min and the fuel consumption is 1.076 kg of oil per hour, having a calorific value of 39150 kJ/kg. Calculate the relative efficiency of the engine.*

Solution. I.P. = $\dfrac{P_m \, LAn}{(60 \times 1000 \times 2)}$, for two stroke engine

Given that;

$P_m = 4.12$ bar
$L = 0.27$ m

$$A = \frac{\pi}{4} \times (0.165)^2 = 0.214 \text{ m}^4$$

$n = 264$

$$\text{I.P.} = \frac{(4.12 \times 10^5 \times 0.27 \times 0.0214 \times 264)}{(60 \times 1000 \times 2)} = 5.24 \text{ kW}$$

Now, indicated thermal efficiency

$$= \frac{\text{(Heat equivalent of I.P. per hour)}}{\text{(Heat in fuel per hour)}}$$

$$= \frac{(5.24 \times 3600)}{(1.076 \times 39150)} = 44.78\%$$

Now air standard efficiency $= 1 - \left(\frac{r^{1-k}}{k}\right)\left\{\frac{(\rho^k - 1)}{(\rho - 1)}\right\}$

If clearance volume is taken as unity, then,

$r = 14.3$,
$\rho = 1 + \{(4.23 \times 13.3)100\} = 1.56$
$k = 1.4$

$$\text{Efficiency} = 1 - \left\{\frac{(14.3)^{-0.4}}{1.4}\right\}\left\{\frac{(1.56^{1.4} - 1)}{(1.56 - 1)}\right\} = 62\%$$

Now relative efficiency

$$= \frac{\text{Indicated Thermal efficiency}}{\text{Air standard efficiency}}$$

$$= \frac{0.4478}{0.62} = 72.23\%.$$

Example 3. *A six-cylinder two-stroke cycle marine diesel engine with 100 mm bore and 120 mm stroke delivers 200 B.H.P. at 2000. R.P.M. and uses 100 kg of fuel per hour. If I.H.P. is 240, determine the following:*

(a) Torque,

(b) Mechanical efficiency,

(c) Indicated specific fuel consumption.

Solution. (a) BHP = $\dfrac{2\Pi NT}{4500}$

Where,   T = torque
         N = RPM

$$200 = \dfrac{(2\pi \times 2000 \times T)}{4500}$$

T = 71.7 kg.m.

(b) η_m = Mechanical efficiency = $\dfrac{BHP}{IHP} = \dfrac{200}{240} = 0.83$

(c) Indicated specific fuel consumption = W/I.H.P.

Where,   W = Amount of fuel used per hour

Indicated specific fuel consumption = $\dfrac{100}{240}$

= 2.41 kg/IHP hour.

Example 4. *A diesel engine develops 200 H.P. to over come friction and delivers 1000 BHP. Air consumption is 90 kg per minute. The air fuel ratio is.15 to 1. Find the following:*

(a) IHP, (b) Mechanical efficiency, (c) Specific fuel consumption.

Solution. (a) BHP = 1000

FHP = 200

IHP = BHP + FHP = 1000 + 200 = 1200

(b) η_m = Mechanical efficiency = $\dfrac{BHP}{IHP} = \dfrac{1000}{1200}$

= 0.83 = 83%.

(c)   K = Air fuel ratio = 15

W = Air consumed per hour

= 90 × 60 = 5400 kg per hour

S = Amount of fuel consumed = $\dfrac{W}{K} = \dfrac{5400}{15}$

= 360 kg per hour.

Specific fuel consumption = $\dfrac{S}{IHP} = \dfrac{360}{1200}$

= 0.3 kg/IHP hr.

Example 5. *The brake thermal efficiency of a diesel engine is 30 percent. If the air to fuel ratio by weight is 20 and the calorific value of the fuel used is 41800 kJ/kg, what brake mean effective pressure may be expected at S.T.P. conditions ?*

DIESEL POWER PLANT

Solution. Brake thermal efficiency, $\eta_b = 30\%$

Air-fuel ratio by weight = 20

Calorific value of fuel used, $C = 41800$ kJ/kg

Brake mean effective pressure, $p_{mb} = ?$

Brake thermal efficiency = work produced/heat supplied

$0.3 =$ work produced/41800

Work produced per kg of fuel = $0.3 \times 41800 = 12540$ kJ

Mass of air used per kg of fuel = 20 kg

S.T.P. conditions refer to 1.0132 bar and 15°C

$$\text{Volume of air used} = \frac{mRT}{P} = \frac{(20 \times 287) \times (273+15)}{1.0132 \times 10^5}$$

$$= 16.31 \text{ m}^3$$

Brake mean effective pressure, $P_{mb} = \dfrac{\text{work done}}{\text{cylinder volume}}$

$$= \frac{(12540 \times 1000)}{16.31 \times 10^5} = 7.69 \text{ bar.}$$

Example 6. *A 2-cylinder C.I. engine with a compression ratio 13:1 and cylinder dimensions of 200mm × 250mm works on two stroke cycle and consumes 14kg/h of fuel while running at 300 r.p.m. The relative and mechanical efficiencies of engine are 65% and 76% respectively. The fuel injection is effected upto 5% of stroke. If the calorific value of the fuel used is given as 41800 kJ/kg, calculate the mean effective pressure developed.*

Solution. Refer Fig. 8.18.

Diameter of cylinder, $D = 200$ mm = 0.2 m

Stroke length, $L = 250 = 0.25$ m

Number of cylinders, $n = 2$

Compression ratio, $r = 14$

Fuel consumption = 14 kg/h

Engine speed, $N = 300$ r.p.m.

Relative efficiency, $\eta_{\text{relative}} = 65\%$

Mechanical efficiency, $\eta_{\text{mech}} = 76\%$

Cut-off = 5%n of stroke

Calorific value of fuel, $C = 41800$ kJ/kg

$k = 1$ for two-stroke cycle engine V_3

Cut-off ratio, $p = \dfrac{V_3}{V_2}$

Also, $V_3 - V_2 = 0.05 V_s = 0.05(V_1 - V_2)$

or, $V_3 - V_2 = 0.05(13V_2 - V_2)$, $\dfrac{V_1}{V_2} = 13$

or, $V_3 - V_2 = 0.06 V_2$

$$\dfrac{V_3}{V_2} = 1.6$$

$$\eta_{\text{air standard}} = 1 - \left\{\dfrac{1}{\gamma(r)^{\gamma-1}}\right\}\left[\dfrac{(\rho^{\gamma-1})}{(\rho-1)}\right]$$

$$= 1 - \left\{\dfrac{1}{1.4(14)^{1.4-1}}\right\}\left[\dfrac{(1.6^{1.4}-1)}{(1.6-1)}\right] = 0.615 = 61.5\%$$

$$\eta_{\text{relative}} = \dfrac{\eta_{\text{thermal}}}{\eta_{\text{air standard}}}$$

$$0.65 = \dfrac{\eta_{\text{thermal}}}{0.615}$$

$$\eta_{\text{thermal}} = 0.4$$

But, $\eta_{\text{thermal}} = \dfrac{\text{I.P.}}{(mf \times C)}$

$$0.4 = \dfrac{\text{I.P.}}{\left[\left(\dfrac{14}{3600}\right)41800\right]}$$

I.P. = 65 KW

$$\eta_{\text{mech}} = \dfrac{\text{B.P.}}{\text{I.P.}}$$

$$0.76 = \dfrac{\text{B.P.}}{65}$$

B.P. = 49.4 kW

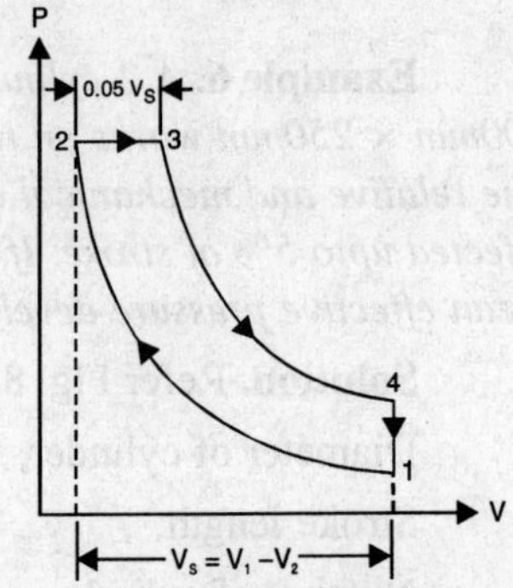

Fig. 8.18

Mean effective pressure can be calculated based on I.P. or B.P. of the engine

$$\text{I.P.} = \dfrac{(n.p_{\text{mi}}.\text{LAN}k.10)}{6}, \ p_{mi} = \text{indicated mean effective pressure}$$

$$65 = \dfrac{2 \times p_{mi} \times 0.25 \times \dfrac{\pi}{4}(0.2)^2 \times 300 \times 1 \times 10}{6}$$

$p_{mi} = 8.27$ bar

and brake mean effective pressure $(p_{mb}) = 0.76 \times 8.27 = 6.28$ bar.

DIESEL POWER PLANT

Example 7. *From the data given below, calculate indicated power, brake power and drawn heat balance sheet for a two stroke diesel engine run for 20 minutes at full load:*

r.p.m.	= 350
m.e.p.	= 3.1 bar
Net brake load	= 640N
Fuel consumption	= 1.52 kg
Cooling water	= 162 kg
Water inlet temperature	= 30°C
Water outlet temperature	= 55°C
Air used/hg of fuel	= 32 kg
Room temperature	= 25°C
Exhaust temperature	= 305°C
Cylinder bore	= 200 mm
Cylinder stroke	= 280 mm
Brake diameter	= 1 metre
Calorific value of fuel	= 43900 kJ/kg
Steam formed per kg of fuel in the exhaust	= 1.4 kg
Specific heat of steam in exhaust	= 2.09 kJ/kg K
Specific heat of dry exhaust gases	= 1.0 kJ/kg K

Solution. N = 350 r.p.m., p_{mi} = 3.1 bar, (W – S) = 640N, m_f = 1.52 kg, m_w = 162 kg, t_{w1} = 30°C, t_{w2} = 55°C, m_a = 32 kg/kg of fuel, t_r = 25°C, t_g = 305°C, D = 0.2 m, L = 0.28 m, D_b = 1 m, C = 43900 kJ/kg, c_{ps} = 2.09, c_{Pg} = 1.0,

k = 1 for two stroke cycle engine.

(1) Indicated power, I.P. = ?

$$\text{I.P.} = np_{mi}LANk \times \frac{10}{6}$$

$$= \frac{1 \times 3.1 \times 0.28 \times \left(\frac{\pi}{4}\right) \times 0.2^2 \times 350 \times 1 \times 10}{6} = \textbf{15.9 kW.}$$

(2) Brake power, B.P. = ?

$$\text{B.P.} = \frac{[(W-S)\,\Pi D_b N]}{(60 \times 1000)} = \frac{(640 \times \Pi \times 1 \times 350)}{60 \times 1000} = \textbf{11.73 kW}$$

Heat supplied in 20 minutes = 1.52 × 43900 = 66728 kJ

(*i*) Heat equivalent of I.P. in 20 minutes
= I.P. × 60 × 20 = 15.9 × 60 × 20 = 19080 kJ

(*ii*) Heat carried away by cooling water
= $m_w \times C_{pw} \times (t_{w2} - t_{w1})$ = 162 × 4.18 × (55 – 30) = 16929 kJ

Total mass of air = 32 × 1.52 = 48.64 kg

Total mass of exhaust gases = mass of fuel + mass of air

=1.52 + 48.64 = 50.16 kg

Mass of steam formed = 1.4 × 1.52 = 2.13 kg

Mass of dry exhaust gases = 50.16 – 2.13 = 48.03 kg

(iii) Heat carried away by dry exhaust gases

$$= m_g \times C_{pg} \times (t_g - t_r)$$
$$= 48.03 \times 1.0 \times (305 - 25) = 13448 \text{ kJ}$$

(iv) Heat carried away by steam

$$= 2.13 \,[h_f + h_{fg} + c_{ps}\,(t_{sup} - t_s)]$$

At 1.013 bar pressure (atmospheric assumed):

h_f = 417.5 kJ/kg

h_{fg} = 2257.9 kJ/kg

$$= 2.13\,[417.5 + 2257.9 + 2.09\,(305 - 99.6)] = 6613 \text{ kJ/kg}$$

[Neglecting sensible heat of water at room temperature]

Heat balance sheet (20 minute basis)

Item Heat supplied by fuel	kJ 66728	Percent 100
(i) Heat equivalent of I.P.	19080	28.60
(ii) Heat carried away by cooling water	16929	25.40
(iii) Heat carried away by dry exhaust gases	13448	20.10
(iv) Heat carried away steam in exhaust gases	6613	9.90
(v) Heat unaccounted for (by difference)	10658	16.00
Total	66728	100.00

Example 8. *The average indicated power developed in a C.I. engine is 13 kW/m³ of free air induced per minute. The engine is a three-liters four-stroke engine running at 3500 r.p.m., and has a volumetric efficiency of 81%, referred to free air conditions of 1.013 bar and 15°C. It is proposed to fit a blower, driven mechanically from the engine. The blower has an isentropic efficiency of 72% and works through a pressure ratio of 1.72. Assume that at the end of induction the cylinders contain a volume of charge equal to the swept volume, at the pressure and temperature of the delivery from the blower. Calculate the increase in brake power to be expected from the engine.*

Take all mechanical efficiencies as 78%.

Solution. Capacity of the engine = 3 liters = 0.003 m³

$$\text{Swept volume} = \left(\frac{3500}{2}\right) \times 0.003 = 5.25 \text{ m}^3/\text{min}.$$

Unsupercharged induced volume = $5.25 \times \eta_{vol.}$
$= 5.25 \times 0.81 = 4.25 \text{ m}^3$

Blower delivery pressure = $1.72 \times 1.013 = 1.74$ bar

Temperature after isentropic compression

$$= 288 \times (1.72)^{(1.4-1)/1.4} = 336.3 \text{ K} \qquad \left[\text{Since } \frac{T_2}{T_1} = \left(\frac{p_2}{p_1}\right)^{(\gamma-1)/\gamma}\right]$$

Blower delivery temperature = $288 + \dfrac{(336.3 - 288)}{0.72} = 355$ K $\qquad \left[\text{Since } \eta_{isen} = \dfrac{(T_2 - T_1)}{(T_2' - T_1)}\right]$

The blower delivery is 5.25 m³/min at 1.74 bar and 355 K.

Equivalent volume at 1.013 bar and 15°C

$$= \frac{(5.25 \times 1.74 \times 288)}{(1.013 \times 355)} = 7.31 \text{ m}^3/\text{min}.$$

Increase in induced volume = $7.31 - 4.25 = 3.06$ m³/min.

Increase in indicated power from air induced
$= 13 \times 3.06 = 39.78$ kW

Increase in I.P. due to the increased induction pressure

$$= \frac{[(1.74 - 1.013) \times 10^5 \times 5.25]}{(10^3 \times 60)} = 6.36 \text{ kW}$$

Total increase in I.P. = $39.78 + 6.36 = 46.14$ kW

Increase in engine B.P. = $\eta_{mech} \times 46.14 = 0.78 \times 46.14 = 35.98$ kW

From this must be deducted the power required to drive the blower Mass of air delivered by blower

$$= \frac{(1.74 \times 10^5 \times 5.25)}{(60 \times 287 \times 355)} = 0.149 \text{ kg/s}$$

Work input to blower = $mc_P(355 - 288) = 0.149 \times 1.005 \times 67$

Power required = $\dfrac{(0.149 \times 1.005 \times 67)}{0.78} = 12.86$ kW

Net increase in B.P. = $35.98 - 12.86 = 23.12$ kW.

THEORETICAL QUESTIONS

1. Draw the layout of diesel power plant.
2. Write a short notes on super charging.
3. What are the basic types of I.C. Engine ?
4. Discuss the advantage and disadvantage of a diesel engine.
5. State the applications of a diesel power plant?
6. Write a note on fuel system of diesel power plant.
7. Write a note on lubrication system of diesel power plant.
8. How air intake and admission system of diesel power plant works ?
9. What are the advantages of supercharger?
10. Write a note on exhaust system of diesel power plant.
11. Write a note on cooling system of diesel power plant.
12. Write a note on heat balance sheet.
13. Name and explain various types of fuel injection systems.

EXERCISES

1. A quality governed four-stroke, single cylinder gas engine has a bore of 146 mm and a stroke of 280 mm. At 475 r.p.m. and full load the net load on the friction brake is 433 N, and the torque arm is 0.45 m. The indicator diagram gives a net area of 578 mm^2 and a length of 70 mm with a spring rating of 0.815 bar/mm.

   Calculate:

   (*i*) The indicated power

   (*ii*) Brake power

   (*iii*) Mechanical efficiency. [**Ans.** (*i*) 12.5 kW (*ii*) 9.69 kW (*iii*) 77.596]

2. A single cylinder four-stroke gas engine has a bore of 178 mm and a stroke of 330 mm and is governed by hit and miss principle. When running at 400 r.p.m. at full load, indicator cards are taken which give a working loop mean effective pressure of 6.2 bar, and a pumping loop mean effective pressure of 0.35 bar. Diagrams from the dead cycle give a mean effective pressure of 0.62 bar. The engine was run light at the same speed (*i.e.*, with no load), and a mechanical counter recorded 47 firing strokes per minute.

   Calculate:

   (*i*) Full load brake power

   (*ii*) Mechanical efficiency of the engine. [**Ans.** (*i*) 13.54 kW; (*ii*) 84.7°10]

3. The efficiency ratio for the Otto engine is 0.60. The engine has four cylinder 7.5 cm by 11.4 cm stroke, having a compression ratio of 5, and consuming 7 kg of petrol per hour when running at 2000 rev/min. Taking calorific value of petrol as 44193.785 kJ/kg. Estimate the effective pressure. [**Ans.** 7.11 bar]

DIESEL POWER PLANT

4. In an engine working on the Otto cycle, the measured suction temperature was 100°C and the temperature at the end of compression was 300°C. 'Faking k for compression as 1.41, find the ideal efficiency and the compression ratio. **[Ans. 35%, 2.85]**

5. A four stroke gas engine develops 4.2 kW at 180 r.p.m. and at full load. Assuming the following data, calculate the relative efficiency based on indicated power and air-fuel ratio used. Volumetric efficiency = 87%, mechanical efficiency = 74%, clearance volume = 2100 cm^3, swept volume = 9000 cm" fuel consumption = 5 m^3/h, calorific value of fuel = 16750 kJ/m^3
[Ans. 50.2%, 7.456: 1]

6. A four-stroke cycle gas engine has a bore of 15.24 cm and a stroke of 22.86 cm. The compression ratio is 4 and the m.e.p. 3.43 bar. If the engine speed is 300 rev/min. and the thermal efficiency is 30%. Calculate the fuel consumption in m^3/kW hr. and the efficiency relative to the air-standard cycle. Calorific value of gas is 18297.2 kJ/m^3. **[Ans. 0.49 m^3, 69.7%]**

7. A two-cylinder, Single acting Diesel engine with a compression ratio of 14 and a cut-off ratio 1.8, works in the four-stroke cycle and uses 13.8 kg of oil per hour when running at 200 rev/min. If the relative efficiency is 0.6 (k = 1.4), find the I.P. and the m.e.p. of the engine. Cylinder diameter 30.48 cm, stroke 45.72 cm and the calorific value of oil 41870 kJ/kg.
[Ans. 78.3, 5.23 bar]

8. The following observations were recorded during a trial of a four-stroke engine with rope brake dynamometer:

Engine speed = 650 r.p.m.,

Diameter of brake drum = 600 mm,

Diameter of rope = 50 mm,

Dead load on the brake drum = 32 kg,

Spring balance reading = 4.75 kg.

Calculate the brake power. **[Ans. 5.9 kW]**

9. The following data refer to a four-stroke petrol engine:

Engine speed = 2000 r.p.m.

Ideal thermal efficiency = 35%,

Relative efficiency = 80%,

Mechanical efficiency = 85%,

Volumetric efficiency = 70%.

If the engine develops 29.42 kW brake power. Calculate the cylinder swept volume.
[Ans. 0.00185 m^3]

10. Derive an expression in terms of volume ratio for the ideal efficiency of the Diesel engine cycle, assuming constant specific heats. Calculate this ideal efficiency for an engine with a compression ratio of 15 and cutting off fuel at 5% of the stroke. **[Ans. 62%]**

11. During a 60 minutes trial of a single cylinder four stroke engine the following observations were recorded:

Bore = 0.3 m

Stroke = 0.45 m

Fuel consumption = 11.4 kg

Calorific value of fuel = 42000 kJ/kg

Brake mean effective pressure = 6.0 bar

Net load on brakes = 1500 N

r.p.m. = 300, brake drum diameter = 1.8 m

Brake rope diameter = 20 mm

Quantity of jacket cooling water = 600 kg

Temperature rise of jacket water = 55°C

Quantity of air as measured = 250 kg

Exhaust gas temperature = 420°C

C_P for exhaust gases = 1 kJ/kg K

Ambient temperature = 20°C.

Calculate: (i) Indicated power; (ii) Brake power; (iii) Mechanical efficiency; (iv) Indicated thermal efficiency.

Draw up a heat balance sheet on minute basis.

[**Ans.** (i) 47.7 kW, (ii) 42.9 kW, (iii) 89.9%, (iv) 35.86°b]

Chapter 9

Gas Turbine Powe Plant

9.1 INTRODUCTION

The gas turbine obtains its power by utilizing the energy of burnt gases and air, which is at high temperature and pressure by expanding through the several ring of fixed and moving blades. It thus resembles a steam turbine. To get a high pressure (of the order of 4 to 10 bar) of working fluid, which is essential for expansion a compressor, is required.

The quantity of the working fluid and speed required are more, so, generally, a centrifugal or an axial compressor is employed. The turbine drives the compressor and so it is coupled to the turbine shaft. If after compression the working fluid were to be expanded in a turbine, then assuming that there were no losses in either component the power developed by the turbine would be just equal to that absorbed by the compressor and the work done would be zero. But increasing the volume of the working fluid at constant pressure, or alternatively increasing the pressure at constant volume can increase the power developed by the turbine. Adding heat so that the temperature of the working fluid is increased after the compression may do either of these. To get a higher temperature of the working fluid a combustion chamber is required where combustion of air and fuel takes place giving temperature rise to the working fluid.

Thus, a simple gas turbine cycle consists of

(1) a compressor,
(2) a combustion chamber and
(3) a turbine.

Since the compressor is coupled with the turbine shaft, it absorbs some of the power produced by the turbine and hence lowers the efficiency. The network is therefore the difference between the turbine work and work required by the compressor to drive it.

Gas turbines have been constructed to work on the following: oil, natural gas, coal gas, producer gas, blast furnace and pulverized coal.

9.2 CLASSIFICATION OF GAS TURBINE POWER PLANT

The gas turbine power plants which are used in electric power industry are classified into two groups as per the cycle of operation.

(*a*) Open cycle gas turbine.
(*b*) Closed cycle gas turbine.

9.2.1 OPEN CYCLE GAS TURBINE POWER PLANT

A simple open cycle gas turbine consists of a compressor, combustion chamber and a turbine as shown in Fig. 9.1. The compressor takes in ambient air and raises its pressure. Heat is added to the air in combustion chamber by burning the fuel and raises its temperature.

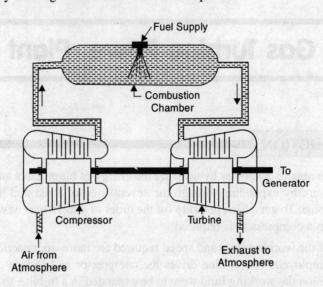

Fig. 9.1. Open cycle gas turbine.

The heated gases coming out of combustion chamber are then passed to the turbine where it expands doing mechanical work. Part of the power developed by the turbine is utilized in driving the compressor and other accessories and remaining is used for power generation. Since ambient air enters into the compressor and gases coming out of turbine are exhausted into the atmosphere, the working medium must be replaced continuously. This type of cycle is known as open cycle gas turbine plant and is mainly used in majority of gas turbine power plants as it has many inherent advantages.

(A) Advantages

1. **Warm-up time.** Once the turbine is brought up to the rated speed by the starting motor and the fuel is ignited, the gas turbine will be accelerated from cold start to full load without warm-up time.

2. **Low weight and size.** The weight in kg per kW developed is less.

3. **Fuels.** Almost any hydrocarbon fuel from high-octane gasoline to heavy diesel oils can be used in the combustion chamber.

4. Open cycle plants occupy comparatively little space.

5. The stipulation of a quick start and take-up of load frequently are the points in favour of open cycle plant when the plant is used as peak load plant.

6. Component or auxiliary refinements can usually be varied to improve the thermal efficiency and give the most economical overall cost for the plant load factors and other operating conditions envisaged.

7. Open-cycle gas turbine power plant, except those having an intercooler, does not require cooling water. Therefore, the plant is independent of cooling medium and becomes self-contained.

GAS TURBINE POWER PLANT

(B) Disadvantages

1. The part load efficiency of the open cycle plant decreases rapidly as the considerable percentage of power developed by the turbine is used to drive the compressor.

2. The system is sensitive to the component efficiency; particularly that of compressor. The open cycle plant is sensitive to changes in the atmospheric air temperature, pressure and humidity.

3. The open-cycle gas turbine plant has high air rate compared to the other cycles, therefore, it results in increased loss of heat in the exhaust gases and large diameter ductwork is necessary.

4. It is essential that the dust should be prevented from entering into the compressor in order to minimise erosion and depositions on the blades and passages of the compressor and turbine and so impairing their profile and efficiency. The deposition of the carbon and ash on the turbine blades is not at all desirable as it also reduces the efficiency of the turbine.

9.2.2 CLOSED CYCLE GAS TURBINE POWER PLANT

Closed cycle gas turbine plant was originated and developed in Switzerland. In the year 1935, J. Ackeret and C. Keller first proposed this type of machine and first plant was completed in Zurich in 1944.

It used air as working medium and had a useful output of 2 mW. Since then, a number of closed cycle gas turbine plants have been built all over the world and largest of 17 mW capacity is at Gelsenkirchen, Germany and has been successfully operating since 1967. In closed cycle gas turbine plant, the working fluid (air or any other suitable gas) coming out from compressor is heated in a heater by an external source at constant pressure. The high temperature and high-pressure air coming out from the external heater is passed through the gas turbine. The fluid coming out from the turbine is cooled to its original temperature in the cooler using external cooling source before passing to the compressor. The working fluid is continuously used in the system without its change of phase and the required heat is given to the working fluid in the heat exchanger.

The arrangement of the components of the closed cycle gas turbine plant is shown in Fig. 9.2.

(A) Advantages

1. The inherent disadvantage of open cycle gas turbine is the atmospheric backpressure at the turbine exhaust. With closed cycle gas turbine plants, the backpressure can be increased. Due to the control on backpressure, unit rating can be increased about in proportion to the backpressure. Therefore the machine can be smaller and cheaper than the machine used to develop the same power using open cycle plant.

2. The closed cycle avoids erosion of the turbine blades due to the contaminated gases and fouling of compressor blades due to dust. Therefore, it is practically free from deterioration of efficiency in service. The absence of corrosion and abrasion of the interiors of the compressor and turbine extends the life of the plant and maintains the efficiency of the plant constant throughout its life as they are kept free from the products of combustion.

3. The need for filtration of the incoming air which is a severe problem in open cycle plant is completely eliminated.

4. Load variation is usually obtained by varying the absolute pressure and mass flow of the circulating medium, while the pressure ratio, the temperatures and the air velocities remain almost constant. This result in velocity ratio in the compressor and turbine independent of the load and full load thermal efficiency maintained over the full range of operating loads.

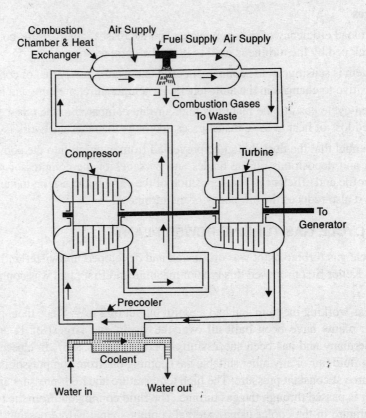

Fig. 9.2. Closed Cycle Gas Turbine Plant.

5. The density of the working medium can be maintained high by increasing internal pressure range, therefore, the compressor and turbine are smaller for their rated output. The high density of the working fluid further increases the heat transfer properties in the heat exchanger.

6. As indirect heating is used in closed cycle plant, the inferior oil or solid fuel can be used in the furnace and these fuels can be used more economically because these are available in abundance.

7. Finally the closed cycle opens the new field for the use of working medium (other than air as argon, CO_2, helium) having more desirable properties. The ratio γ of the working fluid plays an important role in determining the performance of the gas turbine plant. An increase in γ from 1.4 to 1.67 (for argon) can bring about a large increase in output per kg of fluid circulated and thermal efficiency of the plant.

The theoretical thermal efficiencies of the monoatomic gases will be highest for the closed cycle type gas turbine. Further, by using the relatively dense inert gases, such as argon, krypton and xenon, the advantage of smaller isentropic heat fall and smaller cross-sectional flow areas would be realised:

Whether CO_2 or Helium should be adopted as working medium is matter of controversy at present. Blade material poses a problem to use helium as working fluid. In case of CO_2, a new kind of compressor must be designed to compress the fluid. The main advantage of CO_2 is that it offers 40% efficiency at 700°C whereas helium would need 850°C or more to achieve the same efficiency. A helium turbine would also need to run faster imposing larger stresses on the rotor.

8. The maintenance cost is low and reliability is high due to longer useful life.

9. The thermal efficiency increases as the pressure ratio (R_p) decreases. Therefore, appreciable higher thermal efficiencies are obtainable with closed cycle for the same maximum and minimum temperature limits as with the open cycle plant.

10. Starting of plane is simplified by reducing the pressure to atmospheric or even below atmosphere so that the power required for starting purposes is reduced considerably.

(B) Disadvantages

1. The system is dependent on external means as considerable quantity of cooling water is required in the pre-cooler.

2. Higher internal pressures involve complicated design of all components and high quality material is required which increases the cost of the plant.

3. The response to the load variations is poor compared to the open-cycle plant.

4. It requires very big heat-exchangers as the heating of workings fluid is done indirectly. The space required for the heat exchanger is considerably large. The full heat of the fuel is also not used in this plant.

The closed cycle is only preferable over open cycle where the inferior type of fuel or solid fuel is to be used and ample cooling water is available at the proposed site of the plant.

However, closed cycle gas turbine plants have not as yet been used for electricity production. This is mainly a consequence of the limitations imposed by the unit size of heat exchanger. The use of a large number of parallel heat exchangers would practically eliminate the economic advantage resulting from increased plant size.

The inherent disadvantage of open cycle is the atmospheric backpressure, which limits the unit rating. This disadvantage can be eliminated in the closed cycle plant by increasing the backpressure of the cycle. With conventional closed cycle gas turbine plants, advantage can be taken of this only to a limited extent as the air heater limits the unit rating. This disadvantage does not apply to closed cycle plant with a nuclear reactor as heat source. Manufacturers of closed cycle gas turbine plant believe that with these sets, unit-rating up to 500 mW may be possible.

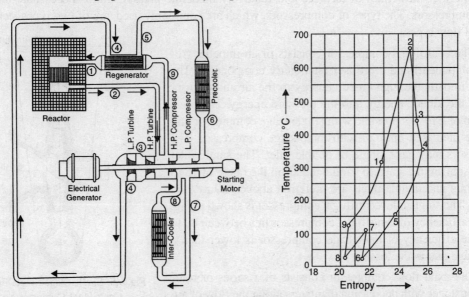

Fig. 9.3. Direct Turbine for Helium Cooled Fast Reactor **Fig. 9.4.** The Processes are Presented on T-s Diagram.

With the use of nuclear reactor as heating source for gas, the heat exchangers can be eliminated from the closed cycle plant and the above-mentioned limitation (number of heat exchangers) does not exist. The power density in the core of a helium cooled fast reactor is a few thousand times higher than in conventional gas heat exchanger. Thus, units of several thousands of megawatts designed for high gas pressures can be housed in a single pre-stressed concrete vessel.

A typical closed cycle gas turbine plant using helium as working medium and helium cooled fast breeder reactor is shown in Fig. 9.3 and corresponding T-s diagram is shown in Fig. 9.4.

A closed cycle gas turbine plant using helium as working medium is much smaller than of a conventional air-turbine plant of the same output. This is due to the better thermodynamic properties of helium relative to air and much higher pressures can be used in helium cooled fast reactor system. A helium-turbine used in closed cycle plant of 335 mW capacity at Switzerland is of 3.7 meter diameter and 14 meters long. The corresponding dimensions of the 17 mW air turbines at Gelsenkirohen plant are 2.6 meters in diameter and 9 meters long.

It is expected that in future, the combination of fast breeder reactors and gas turbines represent a very promising solution for future power generation. This is because of high breeding characteristics of the helium cooled fast reactors, which ensure continuity of low fuel cost while the use of closed cycle gas turbine plant is expected to reduce the capital investment of the plant.

Cost is also roughly proportional to weight. One can expect much cheaper turbo machinery than steam plant.

9.3 ELEMENTS OF GAS TURBINE POWER PLANT

It is always necessary for the engineers and designers to know about the construction and operation of the components of gas turbine plants.

9.3.1. COMPRESSORS

The high flow rates of turbines and relatively moderate pressure ratios necessitate the use of rotary compressors. The types of compressors, which are commonly used, are of two types, centrifugal and axial flow types.

The centrifugal compressor consists of an impeller (rotating component) and a diffuser (stationary component). The impeller imparts the high kinetic energy to the air and diffuser converts the kinetic energy into the pressure energy. The pressure ratio of 2 to 3 is possible with single stage compressor and it can be increased upto 20 with three-stage compressor. The compressors may have single or double inlet. The single inlet compressors are designed to handle the air in the range of 15 to 300 m^3/min and double inlets are preferred above 300 m^3/min capacity. The single inlet centrifugal compressor is shown in Fig. 9.5. The efficiency of centrifugal compressor lies between 80 to 90%. The efficiency of multistage compressor is lower than a single stage due to the losses.

The axial flow compressor consists of a series of rotor and stator stages with decreasing diameters along the flow of air.

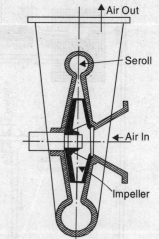

Fig. 9.5. Single Stage Single Entry Centrifugal Compressor.

The blades are fixed on the rotor and rotors are fixed on the shaft. The stator blades are fixed on the stator casing. The stator blades guide the air flow to the next rotor stage coming from the previous rotor stage. The air flows along the axis of the rotor. The kinetic energy is given to the air as it passes through the rotor and part of it is converted into pressure. The axial flow compressor is shown in Fig. 9.6. The number of stages required for pressure ratio of 5 is as large as sixteen or more.

A satisfactory air filter is absolutely necessary for cleaning the air before it enters the compressor because it is essential to maintain the designed profile of the aerofoil blades. The deposition of dust particles on the blade surfaces reduces the efficiency rapidly.

The advantages of axial flow compressor over centrifugal compressor are high isentropic efficiency (90-95%), high flow rate and small weight for the same flow quantity. The axial flow compressors are very sensitive to the changes in airflow and speed, which result in rapid drop in efficiency.

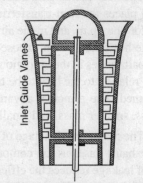

Fig. 9.6. Axial Flow Air Compressor.

In both types of compressors, it has been found that lowering of the inlet air temperature by 15 to 20°C gives almost 25% greater output with an increase of 5% efficiency.

9.3.2. INTERCOOLERS AND HEAT EXCHANGERS

The intercooler is generally used in gas turbine plant when the pressure ratio used is sufficiently large and the compression is completed with two or more stages. The cooling of compressed air is generally done with the use of cooling water. A cross-flow type intercooler is generally preferred for effective heat transfer.

The regenerators, which are commonly used in gas turbine plant, are of two types, recuperator and regenerator.

In a recuperative type of heat exchanger, the air and hot gases are made to flow in counter direction as the effect of counterflow gives high average temperature difference causing the higher heat flow.

A number of baffles in the path of airflow are used to make the air to flow in contact for longer time with heat transfer surface.

The regenerator type heat exchanger consists of a heat-conducting member that is exposed alternately to the hot exhaust gases and the cooler compressed air. It absorbs the heat from hot gases and gives it up when exposed to the air. The heat, capacity member is made of a metallic mesh or matrix, which is rotated slowly (40-60 r.p.m.) and continuously exposed to hot and cold air.

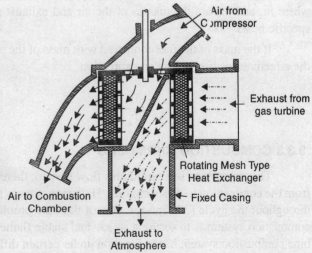

Fig. 9.7. Ritz Regenerative Heat Exchanger.

Prof. Ritz suggested the first application of regenerative heat exchanger to gas turbine plants of Germany and the heat exchanger was titled against his name. The arrangement of Ritz heat exchanger is shown in Fig. 9.7.

The heat-exchanging element A is slowly rotated by a drive from the gas turbine via shaft S. The rotation places the heat-transferring element A in the exhaust gas passage for one half of the time required for one r.p.m. and in the air supply passage for the remaining half. The heat element absorbs heat from the hot gases, when exposed to hot gases and gives out the same heat to the cold air when the heated part moves in the air region. By suitable design of the speed of rotation of transfer element and its mass in relation to the heat to be transferred, it is possible to secure a high effectiveness, values of 90% are claimed. The principal advantages claimed of this heat exchanger over the recuperative type are lightness, smaller mass, and small size for given effectiveness and low-pressure drop.

The major disadvantage of this heat exchanger is, there will be always a tendency for air leakage to the exhaust gases as the compressed air is at a much higher pressure than exhaust gases. This tendency of leakage reduces the efficiency gain due to heat exchanger. Therefore, the major problem in the design of this type of heat exchanger is to prevent or minimize the air loss due to leakage.

Recently very special seals are provided to prevent the air leakage. This seal stands at very high temperature and pressure and allows the freedom of movement.

The performance of the heat exchanger is determined by a factor known as effectiveness. The effectiveness of the heat exchanger is defined as

$$\varepsilon = \frac{\text{actual heat transfer to the air}}{\text{maximum heat transfer theoretically possible}}$$

The effectiveness is given by

$$\varepsilon = \frac{C_{pa} \, m_a \, (T_5 - T_2)}{C_{pg} \, m_g \, (T_4 - T_2)}$$

where m_a and m_g are the masses of the air and exhaust gases and C_{Pa} and C_{Pg} are the corresponding specific heats.

If the mass of the fuel compared with mass of the air, is neglected and $C_{Pa} = C_{Pg}$ is assumed, then the effectiveness is given by an expression

$$\varepsilon = \frac{T_5' - T_2}{T_4 - T_2}$$

9.3.3 COMBUSTION CHAMBERS

The gas turbine is a continuous flow system; therefore, the combustion in the gas turbine differs from the combustion in diesel engines. High rate of mass flow results in high velocities at various points throughout the cycle (300 m/sec). One of the vital problems associated with the design of gas turbine combustion system is to secure a steady and stable flame inside the combustion chamber. The gas turbine combustion system has to function under certain different operating conditions which are not usually met with the combustion systems of diesel engines. A few of them are listed below:

1. Combustion in the gas turbine takes place in a continuous flow system and, therefore, the advantage of high pressure and restricted volume available in diesel engine is lost. The chemical reaction takes place relatively slowly thus requiring large residence time in the combustion chamber in order to achieve complete combustion.

2. The gas turbine requires about 100:1 air-fuel ratio by weight for the reasons mentioned earlier. But the air-fuel ratio required for the combustion in diesel engine is approximately 15:1. Therefore, it is impossible to ignite and maintain a continuous combustion with such weak mixture. It is necessary to provide rich mixture fm ignition and continuous combustion, and therefore, it is necessary to allow required air in the combustion zone and the remaining air must be added after complete combustion to reduce the gas temperature before passing into the turbine.

3. A pilot or recirculated zone should be created in the main flow to establish a stable flame that helps to ignite the combustible mixture continuously.

4. A stable continuous flame can be maintained inside the combustion chamber when the stream velocity and fuel burning velocity are equal. Unfortunately most of the fuels have low burning velocities of the order of a few meters per second, therefore, flame stabilization is not possible unless some technique is employed to anchor the flame in the combustion chamber.

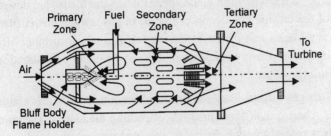

Fig. 9.8. Combustion Chamber with Upstream Injection with Bluff-body Flame Holder.

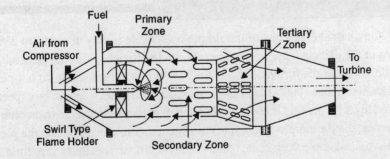

Fig. 9.9. Combustion Chamber with Downstream Injection and Swirl Holder.

The common methods of flame stabilization used in practice are bluff body method and swirl flow method. Two types of combustion chambers using bluff body and swirl for flame stabilization are shown in Fig. 9.8 and Fig. 9.9. The major difference between two is the use of different methods to create pilot zone for flame stabilization.

Nearly 15 to 20% of the total air is passed around the jet of fuel providing rich mixture in the primary zone. This mixture burns continuously in the primary (pilot) zone and produces high temperature gases. About 30% of the total air is supplied in the secondary zone through the annuals around the flame tube to complete the combustion. The secondary air must be admitted at right points in the combustion chamber otherwise the cold injected air may chill the flame locally thereby reducing the rate of reaction. The secondary air helps to complete the combustion as well as helps to cool the flame tube. The remaining 50% air is mixed with burnt gases in the "tertiary zone" to cool the gases down to the temperature suited to the turbine blade materials.

By inserting a bluff body in mainstream, a low-pressure zone is created downstream side that causes the reversal of flow along the axis of the combustion chamber to stabilize the flame.

In case of swirl stabilization, the primary air is passed through the swirler, which produces a vortex motion creating a low-pressure zone along the axis of the chamber to cause the reversal of flow. Sufficient turbulence must be created in all three zones of combustion and uniform mixing of hot and cold bases to give uniform temperature gas stream at the outlet of the combustion chamber.

9.3.4. GAS TURBINES

The common types of turbines, which are in use, are axial flow type. The basic requirements of the turbines are lightweight, high efficiency; reliability in operation and long working life. Large work output can be obtained per stage with high blade speeds when the blades are designed to sustain higher stresses. More stages of the turbine are always preferred in gas turbine power plant because it helps to reduce the stresses in the blades and increases the overall life of the turbine. More stages are further preferred with stationary power plants because weight is not the major consideration in the design which is essential in aircraft turbine-plant.

The cooling of the gas turbine blades is essential for long life as it is continuously subjected to high temperature gases. There are different methods of cooling the blades. The common method used is the air-cooling. The air is passed through the holes provided through the blade.

9.4 REGENERATION AND REHEATING

Generally, the thermal efficiency of the simple open cycle is only about 16 to 23% as lot of heat energy goes waste in the exhaust gases. Moreover the cycle efficiency directly depends upon the temperature of the inlet gases to the turbine. And as the metallurgical limitations do not permit the use of temperatures higher than about 1000°C, a sizeable increase in efficiency cannot be expected through the increased temperature of the gases. Of course, this efficiency handicap can be overcome by incorporating thermal refinements in the simple open cycle e.g. regeneration, reheating. But the plant will become complex in contrast to the simple open cycle plant which is compact, occupies very little space, does not need any water and can be quickly run up from cold. The thermal refinements can raise the plant efficiency to over 30% and thereby obliterate the advantage of fuel efficiency possessed by diesel or condensing steam power plants. These refinements are discussed below:

9.4.1 REGENERATION

In regeneration, the heat energy from the exhaust gases is transferred to the compressed air before it enters the combustion chamber. Therefore, by this process there will be a saving in fuel used in the combustion chamber if the same final temperature of the combustion gases is to be attained and also there will be a reduction of waste heat. Fig. 9.10. shows a regenerative cycle.

GAS TURBINE POWER PLANT

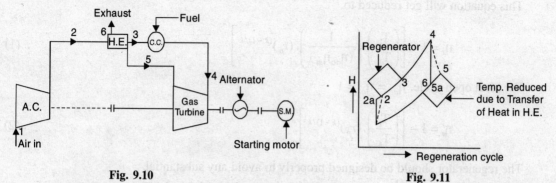

Fig. 9.10   Fig. 9.11

For regeneration to take place T_5 should be greater than T_2.

In the heat exchanger, the temperature of air is increased from T_2 to T_3, and the temperature of the exhaust gases is reduced from T_5 to T_6. If the regeneration is perfect, the air would be heated to the temperature of the exhaust gases entering the H.E. the effectiveness of the regeneration is defined as:

ε = effectiveness

$$= \frac{\text{Rise in air temperature}}{\text{Max. possible rise}} = \frac{T_3 - T_2}{T_5 - T_2}$$

For ideal regeneration,

$$T_3 = T_5 \text{ and } T_6 = T_2$$

The common values of effectiveness would be from 70 to 85%. The heating surface of the generator, as well as the dimensions and price of the gas turbine increases with the regeneration fraction. But to justify the regeneration economically, the effectiveness should atleast be 50%. The regenerative cycle has higher efficiency than the simple cycle only at low-pressure ratios. If the pressure ratio is raised above a certain limit, then the regenerator will cool the compressed air entering the combustion chamber instead of heating it and the efficiency of the regenerative cycle drops. This is clear from Fig. 9.12.

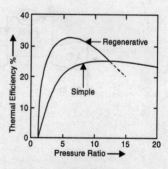

Fig. 9.12

It is clear from Fig. 9.11, that the compressor turbine works are not affected by regeneration. However, the heat to be supplied in the combustion chamber is reduced and also it is added at higher temperature as compared to the cycle without regeneration. Thus, the thermal efficiency of the cycle increases. It will be equal to,

$$\eta_t = \frac{C_p(T_4 - T_5) - C_p(T_2 - T_1)}{C_p(T_4 - T_3)}$$

For ideal regeneration, $T_3 = T_5$

$$\eta_t = 1 - \left[\frac{(T_2 - T_1)}{(T_4 - T_5)}\right]$$

This equation will get reduced to,

$$\eta_t = 1 - \left[\left(\frac{T_1}{T_4}\right)\cdot\left(\frac{1}{\eta_{ac}\eta_{at}}\right)\cdot(r_p)^{(k-1)/k}\right] \qquad \ldots(1)$$

For ideal open cycle, $\eta_{ac} = \eta_{at} = 1$

$$\eta_t = 1 - \left[\left(\frac{T_1}{T_4}\right)\cdot(r_p)^{(k-1)/k}\right] \qquad \ldots(2)$$

The regenerator should be designed properly to avoid any substantial.

Pressure loss in it, which might cancel out any gain in thermal efficiency. Because of some pressure loss in the regenerator, the turbine output and the net output will be slightly less than for the simple cycle.

9.4.2 REHEATING

In reheat cycle, the combustion gases are not expanded in one turbine only but in two turbines. The exhaust of the high-pressure turbine is reheated in a reheater and then expanded in a low-pressure turbine. By reheating, the power output of the turbine is increased but the cost of additional fuel may be heavy unless a heat exchanger is also used. A reheat cycle is shown in Fig. 9.13. Considering the adiabatic expansions, the total work done in the two turbines will be equal to: $(I_3 - I_{4a}) + (I_5 - I_{6a})$.

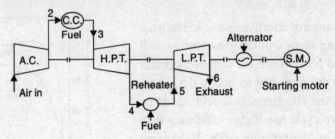

Fig. 9.13

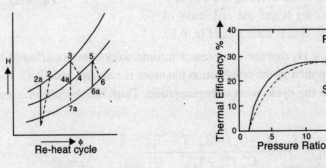

Fig. 9.14   Fig. 9.15

If the combustion gases were expanded in one turbine only down to point 7a for the same pressure ratio, then the work output would have been: $(I_3 - I_{7a})$.

Now the constant pressure lines on the H-Φ chart diverge away from the origin and converge towards the origin. Therefore the line 5–6a will be greater than 4a–7a. Hence reheating increases the power output. By reheating, the average temperature of heat addition is raised resulting in higher output and efficiency of the cycle. If reheat cycle is to be adopted then the pressure ratio must be high as at low pressure ratios, the thermal efficiency is lowered by reheating Fig. 9.14. Reheating reduces the airflow through the cycle resulting in decreased input to the compressor. For ideal reheating; the working fluid temperature after reheating is equal to the maximum permissible turbine inlet temperature. That is,

$$T_5 = T_3$$

The efficiency of the cycle will be given as,

$$\eta_t = \frac{(T_3 - T_4) + (T_5 - T_6) - (T_2 - T_1)}{(T_3 - T_2) + (T_5 - T_4)}$$

9.5 COGENERATION

Decentralized combined heat and power production-cogeneration is a very flexible and efficient way of utilizing fuels. Cogeneration based on biomass is environmentally friendly, and all kinds of biomass resources can be used.

The role combined heat and power production plays in Danish energy supply originates from the decision in 1978 to establish a national natural gas grid. At present the natural gas system is one factor blocking the utilization of biomass and natural gas in decentralized cogeneration plants, because a great part of the heat market is lost for decentralized cogeneration due to the individual gas supply.

In June 1986 it was decided that 450 mW decentralized heat and power plants should be established. These are very efficient and environmentally compatible, if they are based on natural gas or biomass. The interest in biomass as basis for combined heat and power production is caused partly by environmental considerations, and partly by the desire in agriculture and forestry to get rid of an increasing surplus of residue products, typically straw and wood chips.

But exceeding the problem with an insufficient heat market, the energy policy has caused that until now there has been no sufficiently purposeful and ambitious aiming at the cogeneration technologies, that first of all shall lead to an increased use of biomass in heat and power supply.

9.5.1 COGENERATION — WHY

There is a large political interest in changing the local heat supply to combined heat and power supply—this means cogeneration of heat and power.

It is a fundamental physical condition that not all-latent energy of a fuel can be converted into ractive power, *e.g.* to run a car. The main part of the energy is necessarily transformed to waste heat, which in the car example disappears by motor cooling and with the exhaust.

Cogeneration plants can be used in all situations where a given heat demands exists. This includes all together an extremely large number of district heating plants, institutions, co-operative building societies, industries, etc.

For the cogeneration technologies, the primary interest is due to, that a very large percentage of the fuel's energy content is utilized, typically 85-95%. This must be compared to the relatively low energy efficiency of centralized thermal power plants, the annual mean efficiency is about 55% in the ELSAM area (Jutland, Funen).

Another important reason for the interest in decentralized cogeneration is the possibility to utilize renewable bio fuels straw, wood, manure, etc. There are furthermore a few circumstances which are not that much noticed in the political debate.

First of all a large number of cogeneration plants increase the security of power supply. It is not usual that the large power units break down, but it happens. It is obvious that the consequences of missing a large unit are much more significant, than if it is one of the much smaller cogeneration plants.

Second there is a considerable energy loss from the power grid. In the ELSAM area it is good 7% in average. But this figure covers very large variations through the day, and furthermore depends very much on the voltage level. Thus the energy loss from the low-voltage grid is much larger than from the high-voltage grid. All in this entire means that *e.g.* on a winter day at 5 pm there is a large energy loss from the low-voltage grid.

Exactly because many of the cogeneration plants are coupled on the low-voltage grid, they also reduce the grid loss, which influence the overall energy efficiency.

9.5.2 COGENERATION TECHNOLOGIES

(*a*) **Gas Engines.** The most common type of combined heat and power production in Denmark is connected to gas-fired internal combustion engines, which is a well-known technology. They can be found on the market at sizes from 7 kW power to about 4 mW power, and the power efficiency is good 20% for the small engines and over 40% for the largest. As power production is viewed as the main purpose, it is important that the power efficiency is continuously increased.

The lower limit for a profitable cogeneration plant is a heat demand of 15,000 m natural gas per year and a power consumption of 50,000 kWh per year with the current engines at the market.

The gas engine fuel is mainly natural gas and it will remain like this for several years. A few plants are based on biogas, which will gain increased utilization, while various types of biogas plants are developed and established.

It is assumed that gas from thermal gasification of straw and wood will also spread as fuel for stationary cogeneration plants during the coming years.

There are some differences between cogeneration plants according to operation strategy. The larger plants, typically connected to a district heating plant or an industrial company, are mainly in operation during daytime at weekdays. It is because the payment for power is most favourable at that time, which again is due to that the capacity is paid for during the periods with high consumption. In these cases the cogeneration plant produces heat both for covering the actual consumption and for storage in large water storages. The storages are then emptied for heat at night and during the weekend. It is a political request that 90% of the annual heat consumption must be supplied from the engine; a gas boiler supplies the rest.

This operation strategy is only realistic when using natural gas as fuel, as there is enough at a certain time. Contrary to continuously gas production from a biogas or gasification plant. On the other hand, the demand for a variable power production will increase, when cogeneration plants with variable production are established.

The smaller plants are typically base load plants that operate day and night. They supply power to own installations and cover the power consumption. In this case the plant has 2 power meters; one that registers buy from the power utility when the consumption exceeds the production, and another that registers sale when own consumption is less than the actual production.

This type of cogeneration plants has severe environmental and resource advantages.

Natural gas is the least polluting of the fossil fuels. It is partly due to the relatively high hydrogen content that becomes water in the combustion. The CO_2 emission from natural gas is therefore smaller than from oil and coal.

The NO_x pollution from the engines are reduced according to authorities' demand by mounting a 3-way catalyzer or more often by using low-NO_x engines (lean burn). The smallest engines are excepted from these requirements.

According to resources, the advantage is as already mentioned a higher energy efficiency than at centralized thermal power plants.

(b) **Gas Turbines.** Some larger district heating plants have based their heat and power production on gas turbines. They can be regulated less than gas engines, and as they by mean of their size presuppose a large heat demand there will not be space for many new in the future. There are simply not that many cities with a sufficiently large heat demand. Apparently there is neither any product development-taking place to increase the power efficiency, as it is the case for gas engines.

Combined heat and power production based on steam

The Danish effort to increase the use of biomass mainly straw and wood as fuel in combined heat and power production increasingly draws the attention towards steam engines and steam turbines.

The steam engine is a well-known technology, but for different reasons it hasn't been developed for several years. One of the problems has been the contact between lubricating oil and steam. This problem has been solved with a new design of the steam generator, which is manufactured in Denmark and is just ready for the market.

The advantage of this cogeneration technology is that biomass can be combusted directly in the steam boiler and obtain the wanted steam pressure of 20-30 bars.

The disadvantage is that power efficiency will hardly exceed 15%. Therefore it is a question if the steam engine is able to compete with cogeneration based on gasified biomass in the longer term.

There seem to be better possibilities for steam turbines with a combination of direct stoking of biomass in the boiler, and superheating of the steam with natural gas. A Danish district heating plant is preparing a test plant based on this technology. Its advantage is significantly higher power efficiency than the steam engine.

(c) **The Stirling Engine.** The Stirling engine is a hot-air engine, named after the Scottish priest Stirling who invented it in 1817. Since then it has been designed and manufactured in a vast number of designs.

In spite of intensive and expensive research it is nearly without importance, as the research has been aimed at developing a car engine, which it is not suitable for.

On the other hand there are large perspectives in viewing it as a stationary combined heat and power plant. There is a growing understanding of this that has resulted in new research and production aimed at this. About 150 pieces have been made in batch production in India. This is a simple low-pressure design with a power efficiency of about 10%.

The Stirling engine has many advantages. In principle it is a very simple technology—also in the advanced version with helium instead of air and high mean pressure. Furthermore a big variety of fuels can be used, including concentrated solar heat and clean exhaust from *e.g.* a gas engine. With the materials used today, it demands about 700°C as optimum working temperature. And the hot air must be that clean, that coating does not occur at the heating surface. Finally, it is nearly noiseless and probably very stable in operation.

The description of the Stirling engine is to a high degree based on the research carried out at the Technical University of Denmark. A 10 kW power model with helium as medium and a mean pressure of 50 bar has been tested in summer 1992. The results are very promising and it is specially interesting that the power efficiency of this engine is about 30%.

The perspective of heat and power production based on the Stirling engine is that it can probably be produced in a range of 1 kW power to 150 kW power in the nearby future; in the longer term may be with an even higher output.

Such small cogeneration units can give the Stirling engine a tremendous distribution and have a revolutionary influence on our energy supply.

9.6 AUXILIARY SYSTEMS

Auxiliary systems are the backbone of the gas turbine plant. Without auxiliary system, the very existence of the gas turbine is impossible. It permits the safe working of the gas turbine. The auxiliary system includes starting, ignition, lubrication and fuel system and control.

9.6.1 STARTING SYSTEMS

Two separate systems-starting and ignition are required to ensure a gas turbine engine will start satisfactorily. During engine starting the two systems must operate simultaneously.

9.6.1.1 Types of Starter

The following are the various types of gas turbine starter.

(a) Electrical

(i) A.C. and (ii) D.C.

A.C. cranking motors are usually 3 phase induction types rated to operate on the available voltage and frequency.

D.C. starter motor takes the source of electrical energy from a bank of batteries of sufficient capacity to handle the starting load. Engaging or disengaging clutch is used.

(b) Pneumatic or Air Starter. Air starting is used mostly as it is light, simple and economical to operate. As air starter motor has a turbine rotor that transmits power through a reduction gear and clutch to the starter output shaft that is connected to the engine. The starter turbine is rotated by air pressure taken from an external ground supply, from an auxiliary power unit carried in the aircraft or from an engine that is running. An electrical control unit controls the air supply to the starter by opening the pressure-reducing valve. When an engine starter is selected and is automatically closed at a predetermined starter speed the clutch also automatically disengages as the engine accelerates. It is most suitable for natural gas pipeline gas turbine drive.

(c) Combustion Starter. It is in every respect a small gas turbine. It is a completely integrated system which incorporates a planetary reduction gear drive with over-running clutch. The unit is started with the electric starter. The starter turbine is directly geared to the gas turbine shaft through a reduction gear.

GAS TURBINE POWER PLANT

(*d*) **Hydraulic Starting Motor.** It consists of a hydraulic starter motor for main engine, an accumulator, a hydraulic pump motor for auxiliary power unit (A.P.U.).

Discharging the hydraulic accumulator to power a hydraulic pump motor starts APU. The hydraulic pump motor is driven with APU (Fig. 9.16) to start main engine and recharge accumulator. It is suitable for aircraft engine better.

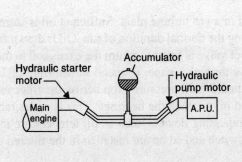

Fig. 9.16. Hydraulic Starter Motor.

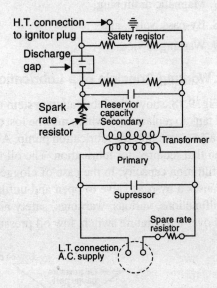

Fig. 9.17. Ignition System.

9.6.2 IGNITION SYSTEMS

Ignition system is utilized to initiate spark during the starting. Once it starts, the combustion is continuous and the working of ignition system is cut-off automatically. The following are the types of ignition system.

1. Capacitor discharge system.

    (*a*) High tension system and (*b*) Low tension system.

2. Induction system.

3. A. C. power circuits.

Here, A.C. power circuit ignition system is discussed. Fig. 9.17 shows the arrangement of A.C. power circuit. It receives an alternating current that is passed through a transformer and rectifier to charge a capacitor. When the voltage in the capacitor is equal to the breakdown value of a sealed discharge gap, the capacitor discharges the energy across the face of the ignition plug. Safety and discharge resistors are fitted in the circuit.

9.6.3 LUBRICATION SYSTEM

9.6.3.1 Elements of Lubrication System

The following are the elements of lubrication system of a gas turbine

1. Oil tank,
2. Oil pump,

3. Filter and strainer,
4. Relief valve,
5. Oil cooler,
6. Oil and pipe line,
7. Magnetic drain plug,
8. By-pass, valve, and
9. Warning devices.

9.6.3.2. Working Principle of a Lubrication System

Fig. 9.18. shows the lubrication system used in a gas turbine plant. Sufficient oil is stored in an external tank to replace that, which may be lost during the normal duration of run. Oil is drawn from the tank by a gear or rotor type lubrication pump. A relief valve is fitted to return the excess oil to the pump inlet than that required for lubrication. The oil flows through a paper or metal screen filter of 10 to 40 micron filtration capacity. In the case of clogged filter, the oil pressure drop across the filter increases which causes a by-pass valve to open and unfiltered oil to reach the bearings. This is undesirable. For such malfunctions, various; warnings, safety and indicating devices such as oil temperature indicator bulb, oil over temperature switch, low oil pressure switch and so on are installed in the filtered oil line.

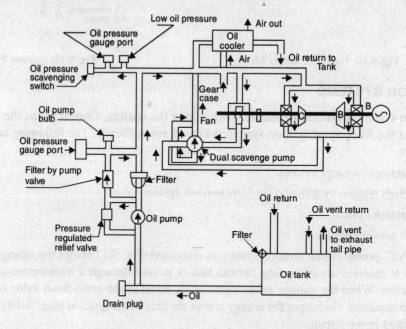

Fig. 9.18. Lubrication System for Gas Turbine.

The filtered oil flows to the bearings and gear case of turbine and after lubricating and cooling the oil washed components, it is returned to the oil tank through a cooler by a scavenge pump. Scavenge pumps have a greater displacement than the quantity of oil flowing into the cavity from which they are pumping. This introduces a negative pressure, which draws air into the sump through any leakage paths, precluding leakage of oil. The air-oil mixture returned through the cooler to the tank must be separated and for this it is recirculated in a separator and the air is permitted to escape. The separator vent usually

GAS TURBINE POWER PLANT

connects to the tail pipe so that the hot gases flowing out of exhaust will burn off any remaining oil vapor. This vent must be checked for obstruction regularly; otherwise the air may build high pressure in the tank and result in its failure. The separator is nothing but a simple baffle which causes the air-oil mixture to follow a devious path, the oil will tire to turning the corners and will find its way out through the vent into the tank, and the air will find its way out through the vent into the exhaust system. A drain plug of magnetic type is incorporated in the oil line before oil pump, which can collect any metal particles circulating in the oil system. A through study is made of debris collected by the plug to decide whether it is metal chips or else.

9.6.4 FUEL SYSTEM AND CONTROLS

9.6.4.1 Fuel System

Fig. 9.19 shows a fuel system used in gas turbine. Basically there are two sub-system-low pressure and high pressure. The low-pressure system consists of a fuel tank, boost pump, strainer, dual fine oil filters, Δp indicator and ordinary valves. There is a two stages filter process on the low-pressure side of the system. The purpose of 7.5 micron strainer is to collect the large particles impurities. The boost pump (centrifugal pump) is used to boost the pressure so that it may overcome the pressure drop in the two stage of filtration. In the fine filter, all small solid particles are collected as well as the water that may have passed through water separator located in the users fuel supply system. The dual filter arrangement allows for switching to an alternate element without having shutting down for the change. It is the Δp indicator, which warns the operator of the filter condition by indicating the red light for switch over. The red light glows as soon as the oil pressure drop through the working fine filter crosses over a pre-determined value.

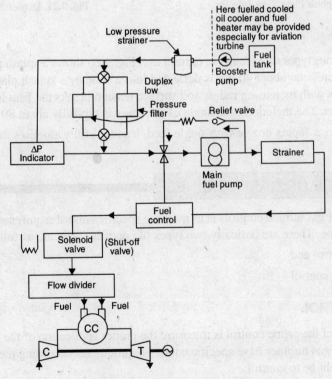

Fig. 9.19. Fuel System for Gas Turbine.

The high-pressure system consists of main fuel pump, relief valve, strainer, fuel control, shut-off valve, flow distributor and nozzles. The main fuel pump generates the oil pressure sufficient to be injected in the combustion chamber. The relief valve installed around the main fuel pump is to protect the fuel system from excessively high pressure created by the shut-off valve or clogging of the passage in the fuel system by transferring the excess fuel to the pump inlet. The function of shut-off valve is very simply to interrupt the supply of the fuel to the combustion chamber and affect a shut down of the turbine automatically or manually. It is also used for interrupting the flow of other fluids such as the gas supply pressure to the pneumatic starter, the flow to the pneumatically driven pumps and the supply to auxiliary fuel lines such as the starting fuel injector. Generally, the shut-off valve is of solenoid type. The gas turbine requires that the fuel should be well distributed in the various fuel injectors. Flow divider accomplishes the work. It is a pressure sensitive valve which positions the valve opening determined by the pressure at inlet to the flow divider.

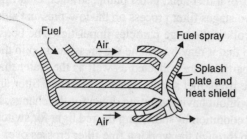

Fig. 9.20. Splash Plate Type Injector.

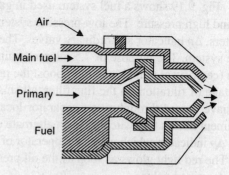

Fig. 9.21. Duplex Nozzle.

9.6.4.2 Injector

There are various types of injectors in current use. Fig. 9.20 shows a splash plate type injector. In this case, the fuel at pressure about 40 bar is deflected into a cone by a splash plate, without spin. The conical fuel film thins with increasing radius and surface tension breaks the film into droplets of about 50 to 100 μm diameter. The included cone angle of the spray is typically about 80° to 100°.

Fig. 9.21 shows a duplex nozzle with single feed. It successfully atomises the fuel varying over a flow range of 50 to 1.

9.7 CONTROL OF GAS TURBINES

The purpose of gas turbine controls is to meet the specific control requirements of users and safe operation of the turbine. There are basically two types of controls. They are as follows:

(A) Prime control and

(B) Protection control

9.7.1 PRIME CONTROL

The objective of the prime control is to ensure the proper application of the turbine power to the load. The users of the gas turbines have specific control requirements according the use of gas turbines. The requirements might be to control:

GAS TURBINE POWER PLANT

(1) The frequency of an a.c. generator,
(2) The speed of a boat or ship,
(3) The speed of an aircraft,
(4) The capacity or head of a pump or compressor,
(5) The road speed of a vehicle.

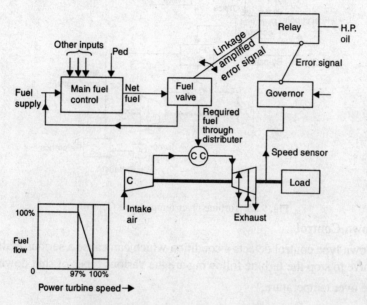

Fig. 9.22. Prime Control (Hydro-mechanical).

On a gas turbine a.c. generator set the prime objective of the control is to maintain the constant electrical frequency irrespective of the load. This is achieved by selecting the primary controller as a speed governor (a speed sensing device), which maintains the constant electrical loads. On a gas turbine driven pipeline compressor, the prime objective of the control is to maintain a constant pipeline pressure downstream of the driven compressor. In this case, the pipeline pressure is sensed and turbine power is varied to maintain a constant pipeline pressure for varying flow conditions. There is relatively constant relationship between turbine power and fuel flow, so in the prime control, the position of fuel valve is controlled.

In Hydro-mechanical Speed Governing System (Prime Control), the governing loop is similar to that discussed in the case of steam turbine. The system uses a centrifugal governor (mechanical) to sense the speed of the turbine and the reference is set for rated speed of the turbine (Fig. 9.22). The governor senses the speed through the accessory drive shaft and flyweights. Internally, the governor hydraulically amplifies the error signal and provides an actuator output position that can be linked to a fuel valve. The fuel valve either throttles or by passes the metered fuel from the main fuel control. The speed-sensing device may be electrical or hydraulic.

9.7.2 PROTECTIVE CONTROLS

The objective of the protective control is to ensure adequate protection for the turbine in preventing its operation under adverse conditions. Whenever, unsafe operating conditions are approached, the prime control is overtaken by the protective control to protect the turbine or driven equipment.

Basically, the protective control is of two types:
1. Shutdown control and
2. Modulating control

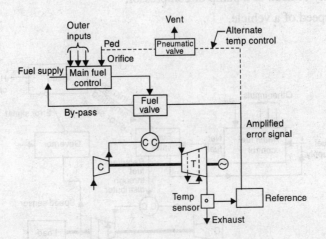

Fig. 9.23. Turbine Over temperature Control.

1. Shut Down Control

The shut down type control detects a condition which can cause a serious malfunction and actuate the shut-off valve to stop the turbine following are the various types of shut down controls.

(*a*) Turbine over temperature,

(*b*) Turbine over speed,

(*c*) Low lube oil pressure,

(*d*) High lube oil temperature, and

(*e*) excess vibration.

(*a*) **Turbine Over temperature.** Fig. 9.23 shows a turbine overtemperature control. Turbine temperature may be well sensed at the turbine inlet but the sensing device put at the turbine inlet goes wrong and therefore it is sensed at the turbine exhaust, which is also a indication of turbine inlet temperature. The temperature sensor may be a thermocouple, bimetal or mercury vapor. As soon as the turbine inlet/exhaust temperature increases a predetermined value, the relay system acts upon the shut-off valve and shut down the turbine by stopping the supply of fuel completely to the combustion chamber.

(*b*) **Turbine Overspeed.** The turbine overspeed control is similar to the prime speed control as shown in Fig. 9.22. In this case, the reference speed is the maximum allowable speed instead of rated speed. The speed sensor could be a centrifugal governor, a tachometer generator or magnetic pick up. If the speed increases a certain fixed value, the speed acts upon the relay system to shut down the fuel valve.

(*c*) **Low Lube Oil Pressure.** It is essential to protect the turbine from low lube oil pressure to ensure proper lubrication and cooling of the turbine bearings. This is accomplished by the use of a pressure switch in the lube oil supply line. The switch operates an alarm signal or shut down fuel valve if the oil pressure drops a safe valve.

GAS TURBINE POWER PLANT

(*d*) **High Lube Oil Temperature.** High lube oil temperature in the lubrication system is a dangerous signal as it is an indication of low lube oil supply or failure in bearings, gears, etc. This protection is generally accomplished through the use of a temperature-sensing device immersed in the lube oil. The sensing device may be a thermal switch, which triggers an alarm or shut down the turbine.

(*e*) **Excessive Vibration.** A slight increase in vibration is a cause of warning. Protection against vibration is accomplished by stalling one or moves vibration pick-ups. The output signal is fed to a monitoring device, which may shut down the turbine if the vibration increases a certain value.

(2) Modulating Controls

The purpose of modulating control is to sense an impending malfunction or a condition, which could adversely affect turbine life and make some modification to the operating condition of the turbine in order to alleviate the undesired conditions. An example of this control may be maximum turbine inlet temperature and maximum speed. The modulating control is more complex and more costly than the shutdown control. But it offers and advantage in allowing a turbine or turbine driven plant to continue operating when normally a shut down occurs. There are some conditions such as high vibrations and low lube oil pressure which cannot be taken care by corrective measures as in this case shut down of the turbine is essential.

Modulating controls for maximum turbine inlet temperature and maximum speed are slight modifications for over temperature and overspend controls.

9.8 GAS TURBINE EFFICIENCY

Gas turbines may operate either on a closed or on an open cycle. The majority of gas turbines currently in use operate on the open cycle in which the working fluid, after completing the cycle is exhausted to the atmosphere. The air fuel ratio used in these gas turbines is approximately 60:1.

The ideal cycle for gas turbine is Brayton Cycle or Joule Cycle. This cycle is of the closed type using a perfect gas with constant specific heats as a working fluid. This cycle is a constant pressure cycle and is shown in Fig. 9.24. On P-V diagram and in Fig. 9.25 on T-ϕ diagram. This cycle consists of the following processes:

The cold air at 3 is fed to the inlet of the compressor where it is compressed along 3-4 and then fed to the combustion chamber where it is heated at constant pressure along 4-1. The hot air enters the turbine at 1 and expands adiabatically along 1-2 and is then cooled at constant pressure along 2-3.

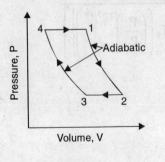

Fig. 9.24

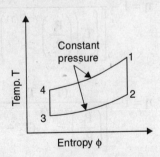

Fig. 9.25

Heat supplied to the system = $K_P(T_1 - T_4)$

Heat rejected from the system = $K_p(T_2 - T_3)$

where K_p = Specific heat at constant pressure,

Work done = Heat supplied – Heat rejected

$$= K_P(T_1 - T_4) - K_p(T_2 - T_3)$$

Thermal efficiency (η) of Brayton Cycle

$$\eta = \frac{\text{Work done}}{\text{Heat Supplied}} = \frac{[K_1\{(T_1 - T_4) - (T_2 - T_3)\}]}{[K_p(T_1 - T_4)]}$$

$$\eta = 1 - \frac{(T_2 - T_3)}{(T_1 - T_4)} \qquad \ldots(1)$$

For expansion 1-2

$$\frac{T_1}{T_2} = \left(\frac{P_1}{P_2}\right)^{(\gamma-1)/\gamma}$$

$$T_1 = T_2\left[\left(\frac{P_1}{P_2}\right)^{(\gamma-1)/\gamma}\right]$$

For compression 3-4

$$\frac{T_4}{T_3} = \left(\frac{P_4}{P_3}\right)^{(\gamma-1)/\gamma} = \left(\frac{P_1}{P_2}\right)^{(\gamma-1)/\gamma}$$

$$T_4 = T_3\left[\left(\frac{P_1}{P_2}\right)^{(\gamma-1)/\gamma}\right]$$

Substituting the values of T_1 and T_4 in equation (1), we get

$$\eta = 1 - \frac{(T_2 - T_3)}{\left[\left\{T_2\left(\left(\frac{P_1}{P_2}\right)^{(\gamma-1)/\gamma}\right)\right\} - \left\{T_3\left(\left(\frac{P_1}{P_2}\right)^{(\gamma-1)/\gamma}\right)\right\}\right]}$$

$$\eta = 1 - \frac{(T_2 - T_3)}{\left[\left(\frac{P_1}{P_2}\right)^{(\gamma-1)/\gamma}(T_2 - T_3)\right]}$$

GAS TURBINE POWER PLANT

9.8.1 EFFECT OF BLADE FRICTION

In a gas turbine there is always some loss of useful heat drop due to frictional resistance offered by the nozzles and blades of gas turbine thus resulting drop in velocity. The energy so lost in friction is converted into heat and, therefore, the gases get reheated to some extent. Therefore, the actual heat drop is less than the adiabatic heat drop as shown in Fig. 9.26, where 1-2' represents the adiabatic expansion and 1-2 represents the actual expansion.

Actual heat drop = $K_p(T_1 - T_2)$

Adiabatic heat drop = $K_p(T_1 - T_2')$

Adiabatic efficiency of turbine

$$= \frac{\text{Actual heat drop}}{\text{Adiabatic heat drop}} = \frac{[K_p(T_1 - T_2)]}{[K_p(T_1 - T_2)']} = \frac{(T_1 - T_2)}{(T_1 - T_2)'}$$

For adiabatic process 1 – 2'

$$\frac{T_2}{T_1} = \left(\frac{P_2}{P_1}\right)^{(\gamma-1)/\gamma}$$

In the compressor also reheating takes place, which causes actual heat increase to be more than adiabatic heat increase. The process 3-4 represents the actual compression while 3-4' represents adiabatic compression.

Adiabatic heat drop = $K_p(T'_4 - T_3)$

Actual heat drop = $K_p(T_4 - T_3)$

Adiabatic efficiency of compressor

$$= \frac{K_p(T'_p - T_3)}{K'_p(T_4 - T_3)} = \frac{T_4 - T_3}{T_4 - T_3}$$

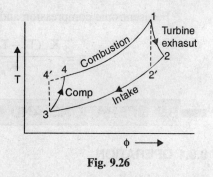

Fig. 9.26

9.8.2 IMPROVEMENT IN OPEN CYCLE

The open cycle for gas turbine is shown in Fig. 9.26. The fresh atmospheric is taken in at the point 3 and exhaust of the gases after expansion in turbine takes place at the point 2. An improvement in open cycle performance can he effected by the addition of a heat exchanger that raises the temperature of the compressed air entering the turbine by lowering exhaust gas temperature that is a waste otherwise. Less fuel is now required in the combustion chamber to attain a specified turbine inlet temperature. This is called a regenerative cycle (Fig. 9.27).

This regenerative cycle is shown on T-φ diagram in Fig. 9.28. Where φ = entropy.

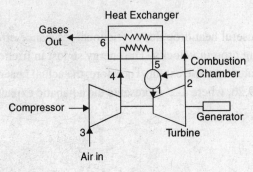
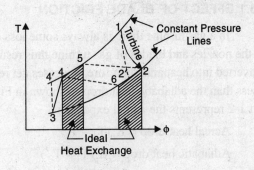

Fig. 9.27 Fig. 9.28

Heat supplied = $K_p(T_1 - T_3) = K_p(T_1 - T_2)$

Heat rejected = $K_p(T_5 - T_3) = K_p(T_4 - T_3)$

(η) Thermal efficiency of theoretical regenerative cycle

$$\eta = \frac{K_p(T_1 - T_2) - K_p(T_4 - T_3)}{K_p(T_1 - T_5)}$$

For isentropic compression and isentropic expansion thermal efficiency is given by

$$\eta = \frac{K_p(T_2 - T_2') - K_p(T_4 - T_3)}{K_p(T_1 - T_5)}$$

9.9 OPERATIONS AND MAINTENANCE PERFORMANCE

9.9.1 OPERATION

(*a*) **Starting.** Starting sequence of any gas turbine from rest to its rated speed requires a certain order of events to be accomplished either manually or automatically. The major steps in sequence are cranking, ignition, acceleration and governing.

The following is typical starting sequence of a gas turbine

1. Application of control power illuminates all the malfunctions lights.
2. Operate 'Reset switch' to reset malfunctions circuits: By doing so, malfunction lights go off and all control devices assume the condition for starting.
3. Operate "Start" switch to initiate starting sequence. By doing this, lube oil pump and cooling fan start. If there are separate switch for these, operate these.
4. When lube oil reaches a preset pressure, the starter is energized and cranking of the engine begins.
5. With the cranking of starting of starter, the engine and exhausts ducts are purged of any combustible gases that might be present.
6. During the cranking cycle, the fuel boost pump is used and operated to increase fuel pressure.

GAS TURBINE POWER PLANT

7. As soon as the fuel pressure has reached a prescribed minimum value, fuel and ignition switches are turned on provided a preset turbine speed has been reached.

8. The turbine accelerates due to combustion of fuel and assistance of cranking motor. At a preset value, say in the order of 70% of rated speed, the starter and ignition are cut-off automatically.

9. The turbine becomes self- sustaining and accelerates on its own to its governed speed till the governing system takes over the control.

(*b*) **Shut down.** To stop the gas turbine fuel supply should be turned off. This is accomplished by closing the fuel valve either manually or by de-energizing an electrically operated valve. In cases where sleeve bearings are used, circulation of lube oil to bearings after shutdown is necessary for cooling.

9.9.2 MAINTENANCE PERFORMANCE

The type of maintenance, which is done on the gas turbine, is the same as that of steam turbines. From the experiences of the most manufactures of gas turbine equipment forced outages are frequently caused-at least in part by inadequate maintenance. The basic purposes of a preventive maintenance programme are to reduce forced outages.

The following are the principal sub-systems of the gas turbine for which manufacturers present maintenance instructions:

1. Turbine gear
2. Starting
3. Clutches and coupling
4. Fuel system
5. Pneumatic system
6. Fire protection system
7. Control equipment
8. Generator-exciter
9. Electrical controls
10. Auxillary gear and main gear
11. Gas turbine
12. Lube oil system
13. Over speed protection
14. Temperature control and monitoring systems
15. Air conditioning system
16. Emergency power
17. Motors
18. Related station equipment.

Maintenance is carried out daily monthly, quarterly, semi-annually and annually.

So for the sequence of overhaul is concerned, it is similar to that of steam turbines with some exceptions.

9.10 TROUBLESHOOTING AND REMEDIES

Modern gas turbines are usually equipped with a very sophisticated protection system using microprocessor and computers, which gives a visual and audio alarm when any pre-established safe condition is violated. In all types of alarm, it is not necessary to shut the gas turbine down. If an alarm condition is of sufficient duration and magnitude the unit will trip and shut down automatically.

The following are the principal symptoms of gas turbine malfunctions and the most common causes of these malfunctions:

(a) Drop in compressor discharge pressure and subsequent drop in load:
1. Dirty intake screens
2. Dirty compressor blades
3. Loss of compressor blades
4. Damaged labyrinth seals

(b) Smoke or dark stack:
1. Burner nozzle-dirty or worn

Puffs of smoke indicate that carbon is building up around the fuel nozzle and then passing through the turbine creating rapid wear of blades, vanes and shrouds.

2. Uneven distribution of fuel to combustion chambers.
3. Combustion chambers damaged or out of position.

(c) Spread in turbine discharge temperatures:
1. Bad thermocouples
2. Uneven fuel to burners or dirty nozzles
3. Combustion chambers damaged or out of position
4. Unlit burners
5. Damaged burner nozzles

(d) High wheel space temperatures:
1. Cooling airlines plugged
2. Cooling air heat exchanger dirty, leaking water or loss of cooling water.
3. Bad thermocouples
4. Wheel space seals worn due to rubs by axial movement or rotor (worn thrust bearing), bowed shaft or casing out of round to open seal clearances
5. Cooling air supply not functioning properly.

(e) High turbine exhaust temperatures:
1. Loss of turbine blades or damaged inlet vanes
2. Bad thermocouples
3. Exhaust temperature controller out of adjustment
4. Increased blade tip clearances due to radial ribs.
5. Dirty air compressor.

(f) High Turbine Exhaust Pressure:
1. Turning vanes in turbine discharge duct damaged or missing.
2. Discharge silencer damaged.

GAS TURBINE POWER PLANT

(g) *Vibration:*
1. Indicating instrument out of adjustment
2. Loose shaft couplings
3. Bowed turbine shaft
4. Broken or missing turbine blades
5. Damaged bearings
6. Shaft mis-alignment

(h) *Loss of fuel pressure:*
1. Fuel control valve out of adjustment
2. Fuel strainers dirty
3. Fuel pump or compressor damaged

(i) *Light of failure:*
1. Faulty spark plug
2. Combustion chamber cross fire tubes out of place.
3. Electrical control out of adjustment
4. Fuel proportion out of adjustment
5. Fuel atomizing air out of proportion
6. Burner nozzles dirty or worn
7. Combustion chamber damaged

(j) *Machine 'Hunting':*
1. Worn governor and control parts
2. Fluctuating fuel controllers
3. Fluctuating exhaust temperature controllers
4. Hydraulic control valves leaking or strainers dirty.

(k) *Loss of oil pressure*
1. Filters
2. Pump failure
3. Leakage in pump.

9.11 COMBINED CYCLE POWER PLANTS

It has been found that a considerable amount of heat energy goes as a waste with the exhaust of the gas turbine. This energy must be utilized. The complete use of the energy available to a system is called the total energy approach. The objective of this approach is to use all of the heat energy in a power system at the different temperature levels at which it becomes available to produce work, or steam, or the heating of air or water, thereby rejecting a minimum of energy waste. The best approach is the use of combined cycles.

There may be various combinations of the combined cycles depending upon the place or country requirements. Even nuclear power plant may be used in the combined cycles.

Fig. 9.29 shows a combination of an open cycle gas turbine and steam turbine. The exhaust of gas turbine which has high oxygen content is used as the inlet gas to the steam generator where the combustion of additional fuel takes place. This combination allows nearer equality between the power outputs of the two units than is obtained with the simple recuperative heat exchanger. For a given total power output the energy input is reduced (*i.e.*, saving in fuel) and the installed cost of gas turbine per unit of power output is about one-fourth of that of steam turbine. In other words, the combination cycles exhibit higher efficiency. The greater disadvantages include the complexity of the plant, different fuel requirements and possible loss of flexibility and reliability. The most recent technology in the field of co-generation developed in USA utilizes the gaseous fuel in the combustion chambers produced by the gasification of low quality of coal. The system is efficient and the cost of power production per kW is less.

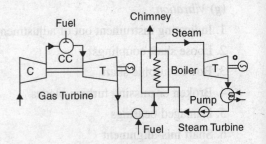

Fig. 9.29. Combined Cycle (Co-generation).

9.12 APPLICATIONS OF GAS TURBINE

1. Gas turbine plants are used as standby plants for the hydro-electric power plants.
2. Gas turbine power plants may be used as peak loads plant and standby plants for smaller power units.
3. Gas turbines are used in jet aircrafts and ships. Pulverised fuel fired plants are used in locomotive.

9.13 ADVANTAGES OF GAS TURBINE POWER PLANT

The economics of power generation by gas turbines is proving to be more attractive, due to low capital cost, and high reliability and flexibility in operation. Quick starting and capability of using wide variety of fuels from natural gas to residual oil or powdered coal are other outstanding features of gas turbine power plants. Major progress has been made in three directions namely increase in unit capacities of gas turbine units (50—100 mW), increase in their efficiency and drop in capital cost, (about Rs. 700 per kW installed). Primary application of gas turbine plant is to supply peak load. However gas turbine plants now-a-days are universally used as peak load, base lead as well as standby plants.

1. It is smaller in size and weight as compared to an equivalent steam power plant. For smaller capacities the size of the gas turbine power plant is appreciably greater than a high speed diesel engine plant but for larger capacities it is smaller in size than a comparable diesel engine plant. If size and weight are the main consideration such as in ships, aircraft engine and locomotives, gas turbines are more suitable.
2. The initial cost and operating cost of the plant is lower than an equivalent steam power plant. A thermal plant of 250 mW capacity cost about Rs. 250 crores. Presently whereas a gas turbine plant of that same-size cost nearly 70 crores.
3. The plant requires less water as compared to a condensing steam power plant.
4. The plant can be started quickly, and can be put on load in a very short time.

GAS TURBINE POWER PLANT

5. There are no standby losses in the gas turbine power plant whereas in steam power plant these losses occur because boiler is kept in operation even when the turbine is not supplying any load.
6. The maintenance of the plant is easier and maintenance cost is low.
7. The lubrication of the plant is easy. In this plant lubrication is needed mainly in compressor, turbine main bearing and bearings of auxiliary equipment.
8. The plant does not require heavy foundations and building.
9. There is great simplification of the plant over a steam plant due to the absence of boilers with their feed water evaporator and condensing system.

9.14 DISADVANTAGES

1. Major part of the work developed in the turbine is used to derive the compressor. Therefore, network output of the plant is low.
2. Since the temperature of the products of combustion becomes too high so service conditions become complicated even at moderate pressures.

SOLVED EXAMPLES

Example 1. *A gas turbine plant of 800 kW capacities takes the air at 1.01 bar and 15°C. The pressure ratio of the cycle is 6 and maximum temperature is limited to 700°C. A regenerator of 75% effectiveness is added in the plant to increase the overall efficiency of the plant. The pressure drop in the combustion chamber is 0.15 bars as well as in the regenerator is also 0.15 bars. Assuming the isentropic efficiency of the compressor 80% and of the turbine 85%, determine the plant thermal efficiency. Neglect the mass of the fuel.*

Solution. The arrangement of the components is shown in Fig. 9.30(*a*) and the processes are represented on T-s diagram as shown in Fig. 9.30(*b*).

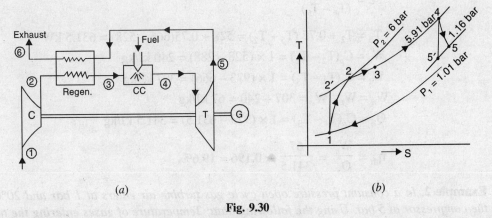

(a) (b)

Fig. 9.30

The given data is

$$T_1 = 15 + 273 = 288 \text{ K}$$
$$p_1 = 1.01 \text{ bar}$$

$$p_2 = 1.01 \times 6 = 6.06 \text{ bar}$$

$$R_p = \frac{P_2}{p_1} = 6$$

Pressure at point 4 = 6.06 − 0.15 = 5.91 bar

Applying isentropic law to the process 1 − 2

$$T_2' = T_1(R_P)^{(\gamma-1)/\gamma} = 288(6)^{0.286} = 480 \text{ K}$$

$$\eta_c = \frac{(T_{2'} - T_1)}{(T_2 - T_1)}$$

But

$$T_2 = T_1 + \eta_c(T_{2'} - T_1) = 288 + 0.8(480 - 288) = 528 \text{ K}$$
$$p_3 = 6.06 - 0.15 = 5.91 \text{ bar}$$

and

$$p_4 = 1.01 + 0.15 = 1.16 \text{ bar}$$

Applying isentropic law to the process 4 − 5'

$$T_{5'} = \frac{T_4}{\left[\left(\frac{P_3}{P_4}\right)^{(\gamma-1)/\gamma}\right]} = \frac{(700+273)}{\left[\left(\frac{5.91}{1.16}\right)^{0.286}\right]} = 612 \text{ K}$$

$$\eta_t = \frac{(T_4 - T_5)}{(T_4 - T_{5'})}$$

or,

$$T_5 = T_4 - \eta_t(T_4 - T_{5'})$$
$$= 973 - 0.85(973 - 612) = 666 \text{ K}$$

The effectiveness of the regenerator is given by

$$\varepsilon = \frac{(T_4 - T_5)}{(T_4 - T_5)}$$

$$T_3 = T_2 + 0.75(T_5 - T_2) = 528 + 0.75(666 - 528) = 631.5 \text{ kW}$$
$$W_c = C_p(T_2 - T_1) = 1 \times (528 - 288) = 240 \text{ kJ/kg}$$
$$W_t = C_p(T_4 - T_5) = 1 \times (973 - 666) = 307 \text{ kJ/kg}$$
$$W_n = W_t - W_c = 307 - 240 = 67 \text{ kJ/kg}$$
$$Q_S = C_p(T_4 - T_3) = 1 \times (973 - 631.5) = 341.5 \text{ kJ/kg}$$

$$\eta_{th} = \frac{W_n}{Q_s} = \frac{67}{341.5} = 0.196 = 19.6\%.$$

Example 2. *In a constant pressure open cycle gas turbine air enters at 1 bar and 20°C and leaves the compressor at 5 bar. Using the following data; Temperature of gases entering the turbine = 680°C, pressure loss in the combustion chamber = 0.1 bar, $\eta_{compressor}$ = 85%, $\eta_{turbine}$ = 80%, $\eta_{combustion}$ = 85%, γ = 1.4 and c_p = 1.024 kJ/kgK for air and gas, find:*

(1) The quantity of air circulation if the plant develops 1065 kW.

GAS TURBINE POWER PLANT

(2) Heat supplied per hg of air circulation.
(3) The thermal efficiency of the cycle. Mass of the fuel may be neglected.

Solution. $P_1 = 1$ bar
$P_2 = 5$ bar
$P_3 = 5 - 0.1 = 4.9$ bar
$P_4 = 1$ bar
$T_1 = 20 + 273 = 293$ K
$T_3 = 680 + 273 = 953$ K
$\eta_{compressor} = 85\%$
$\eta_{turbine} = 80\%$
$\eta_{combustion} = 85\%$

For air and gases: $c_{p'} = 1.024$ kJ/kgK
$y = 1.4$

Power developed by the plant,
$P = 1065$ kW

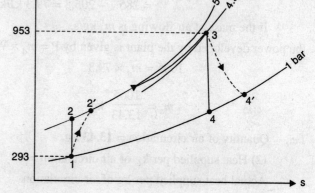

Fig. 9.31

(1) The quantity of air circulation, $m_a = ?$

For isentropic compression $1 - 2$,

$$\frac{T_2}{T_1} = \left(\frac{p_2}{p_1}\right)^{(\gamma-1)/\gamma} = \left(\frac{5}{1}\right)^{(1.4-1)/1.4} = 1.584$$

$T_2 = 293 \times 1.584 = 464$ K

Now,

$$\eta_{compressor} = \frac{(T_2 - T_1)}{(T_{2'} - T_1)} = 0.85$$

$$0.85 = \frac{(464 - 293)}{(T_{2'} - 293)}$$

$T_{2'} = 494$ K

For isentropic expansion process $3 - 4$,

$$\frac{T_4}{T_3} = \left(\left(\frac{P_4}{P_3}\right)^{(\gamma-1)/\gamma}\right) = \left(\left(\frac{1}{4.9}\right)^{(1.4-1)/1.4}\right) = 0.635$$

$T_4 = 953 \times 0.635 = 605$ K

Now, $\eta_{turbine} = \dfrac{(T_3 - T_{4'})}{(T_3 - T_4)} = 0.80$

$$\frac{0.85}{0.80} = \frac{(953 - T_{4'})}{(953 - 605)}$$

$$T_{4'} = 674.6 \text{ K}$$
$$W_{compressor} = C_p(T_{2'} - T_1) = 1.024(494 - 293) = 205.8 \text{ kJ/kg}$$
$$W_{turbine} = C_p(T_3 - T_{4'}) = 1.024(953 - 674.6) = 285.1 \text{ kJ/kg}$$
$$W_{net} = W_{turbine} - W_{compressor}$$
$$= 285.1 - 205.8 = 79.3 \text{ kJ/kg of air}$$

If the mass of air flowing is m_a kg/s,

the power developed by the plant is given by $P = m_a \times W_{net}$ kW

$$1065 = m_a \times 79.3$$

$$m_a = \frac{1065}{13.43} \text{ kg}$$

i.e., Quantity of air circulation = **13.43 kg.**

(2) Heat supplied per kg of air circulation = ?

Actual heat supplied per kg of air circulation

$$= \frac{c_p(T_3 - T_{2'})}{\eta_{combustion}} = \frac{1.024(953 - 494)}{0.85} = \mathbf{552.9 \text{ kJ/kg.}}$$

(3) Thermal efficiency of the cycle, $\eta_{thermal}$ = ?

$$\eta_{thermal} = \frac{\text{work output}}{\text{heat supplied}}$$

$$= \frac{79.3}{552.9} = 0.1434 \quad \text{or} \quad \mathbf{14.34\%.}$$

Example 3. *In an open cycle regenerative gas turbine plant, the air enters the compressor at 1 bar abs 32°C and leaves at 6.9 bar abs. The temperature at the end of combustion chamber is 816°C. The isentropic efficiencies of compressor and turbine are respectively 0.84 and 0.85. Combustion efficiency is 90% and the regenerator effectiveness is 60 percent, determine:*

(a) Thermal efficiency, (b) Air rate, (c) Work ratio.

Solution.
$$P_1 = 1.0 \text{ bar,}$$
$$T_1 = 273 + 32 = 305 \text{ K}$$
$$P_2 = P_{2a} = 6.9 \text{ bar}$$
$$T_4 = 816 + 273 = 1089 \text{ K}$$

$$\frac{T_{2a}}{T_1} = \left(\left(\frac{P_{2a}}{P_1}\right)^{(\gamma-1)/\gamma}\right)$$

$$= \left(\left(\frac{6.9}{1.0}\right)^{(1.4-1)/1.4}\right) = 1.736$$

$$T_{2a} = 1.736 \times 305 = 529.4 \text{ K}$$

GAS TURBINE POWER PLANT

Now, $\eta_{\text{compressor}} = \dfrac{(T_{2a} - T_1)}{(T_2 - T_1)} = 0.84$

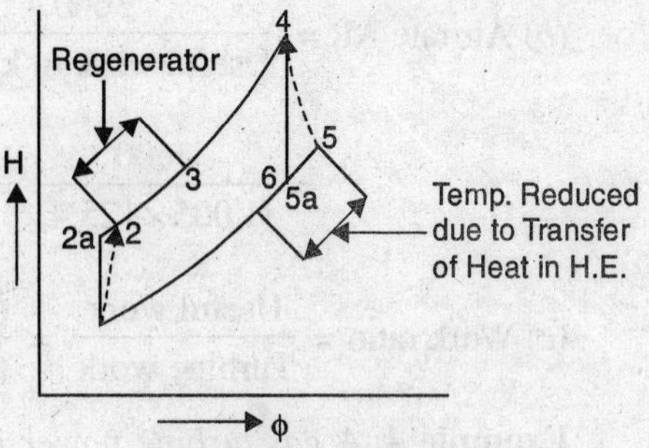

$0.84 = \dfrac{(529.4 - 305)}{(T_2 - 305)}$

$T_2 = 572.2 \text{ K}$

Again $\dfrac{T_4}{T_{5a}} = 1.736$

$T_{5a} = \dfrac{1089}{1.736} = 627.3 \text{ K}$

Now, $\eta_{\text{turbine}} = \dfrac{(T_4 - T_5)}{(T_4 - T_{5a})} = 0.85$

$T_4 - T_5 = 0.85(1089 - 627.3) = 392.4$

$T_5 = 1089 - 392.4 = 696.6 \text{ K}$

$0.84 = \dfrac{(529.4 - 305)}{(T_2 - 305)}$

$T_2 = 572.2 \text{ K}$

Again $\dfrac{T_4}{T_{5a}} = 1.736$

$T_{5a} = \dfrac{1089}{1.736} = 627.3 \text{ K}$

Now,

Regenerator efficiency $\eta_{rg} = \dfrac{(T_3 - T_2)}{(T_5 - T_2)}$

$T_3 - T_2 = 0.6 \times (696.6 - 572.2) = 74.65$

$T_3 = 572.2 + 74.65 = 646.85 \text{ K}$

(*a*) Thermal efficiency

$$\eta_t = \dfrac{\text{Useful workdone}}{\text{Heat supplied}} = \dfrac{[C_p(T_4 - T_5) - C_p(T_2 - T_1)]}{\left[\dfrac{C_p(T_4 - T_3)}{\eta_c}\right]}$$

$$\eta_t = \dfrac{(392.4 - 267.2)}{\left[\dfrac{(1089 - 646.85)}{0.90}\right]} = 25.48\%$$

Fig. 9.32 Regeneration cycle

(b) Air rate AR = $\dfrac{3600}{\text{Useful work in kW/kg}}$

$= \dfrac{3600}{(1.005 \times 125.4)} = 28.56$ kg/kW-hr

(c) Work ratio = $\dfrac{\text{Useful work}}{\text{Turbine work}} = \dfrac{(1.005 \times 125.2)}{(1.005 \times 392.4)} = \mathbf{0.32}.$

Example 4. *A gas turbine power plant is operated between 1 bar and 9 bar pressures and minimum and maximum cycle temperatures are 25°C and 1250°C. Compression is carried out in two stages with perfect intercooling. The gases coming out from HP. turbine are heated to 1250°C before entering into L.P. turbine. The expansions in both turbines are arranged in such a way that each stage develops same power. Assuming compressors and turbines isentropic efficiencies as 83%,*

(1) determine the cycle efficiency assuming ideal regenerator. Neglect the mass of fuel.

(2) Find the power developed by the cycle in kW if the airflow through the power plant is 16.5 kg/sec.

Solution. The arrangement of the components and the processes are shown in Fig. 9.33(*a* and *b*). The given data is

$T_1 = 25 + 273 = 298$ K $= T_3$ (as it is perfect intercooling),

$p_1 = 1$ bar and $p_3 = 9$ bar

$p_2 = \sqrt{p_1 p_3} = \sqrt{(1 \times 9)} = 3$ bar

$R_{P1} = R_{p2} = 3$

$\eta_{c1} = \eta_{c2} = \eta_{t1} = \eta_{t2} = 0.83,$

$T_6 = T_8 = 1250 + 273 = 1523$ K

$T_{10} = T_5$ (as perfect regenerator is given)

Applying isentropic law to the process $1 - 2'$

$T_{2'} = T_1 \left(\dfrac{P_2}{P_1}\right)^{(\gamma-1)/\gamma} = 298(3)^{0.286} = 408$ K

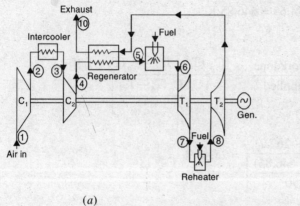
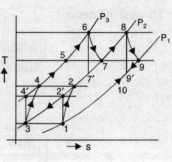

(a)   (b)

Fig. 9.33

$$\eta_{c1} = \frac{(T_{2'} - T_1)}{(T_2 - T_1)}$$

$$T_2 = \frac{T_1 + (T_{2'} - T_1)}{\eta_{c1}}$$

$$= \frac{298 + (408 - 298)}{0.83} = 430.5 \text{ K}$$

$$T_4 = T_2 = 430.5 \text{ K}$$

Applying isentropic law to the process $6 - 7'$

$$\frac{T_6}{T_{7'}} = \left(\frac{P_3}{P_2}\right)^{(\gamma-1)/\gamma} = (3)^{0.286} = 1.37 \text{ K}$$

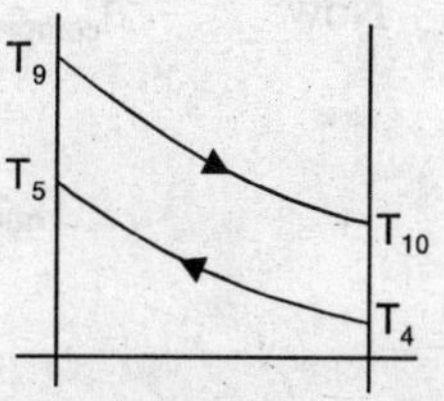

$$T_{7'} = \frac{1523}{1.37} = 1111 \text{ K}$$

$$\eta_{t1} = \frac{(T_6 - T_7)}{(T_6 - T_{7'})}$$

Fig. 9.34

$$T_7 = T_6 - \eta_{t1}(T_6 - T_{7'})$$
$$= 1523 - 0.83(1523 - 1111) = 1181 \text{ K}$$

$T_9 = T_7 = 1181$ K (as equal work is developed by each turbine)

$W_c = 2C_{Pa}(T_2 - T_1) = 2 \times 1(430.5 - 298) = 266$ kJ/kg

$W_t = 2C_{Pa}(T_6 - T_7) = 2 \times 1(1523 - 1181) = 687.5$ kJ/kg

$W_n = W_t - W_c = 687.5 - 266 = 421.5$ kJ/kg

When the ideal regeneration is given, then

$$\varepsilon = 1 \text{ therefore } T_5 = T_9 = 1181 \text{ K} = T_7$$

$$Q_S \text{ (heat supplied)} = 2C_{pa}(T_6 - T_5)$$
$$= 2 \times 1(1523 - 1181) = 684 \text{ kJ/kg}$$

(1) Thermal $\eta = \dfrac{W_n}{Q_s} = \dfrac{421.5}{684} = 0.615 = 61.5\%$

(2) Power developed by the plant $= W_n \times m = 421.5 \times 16.5 = 6954.75$ kW.

Example 5. *A gas-turbine power plant generates 25 MW of electric power. Air enters the compressor at 10°C and 0.981 bar and leaves at 4.2 bar and gas enters the turbine at 850°C. If the turbine and compressor efficiencies are each 80%, determine*

(1) The temperatures at each point in the cycle

(2) The specific work of the cycle

(3) The specific work of the turbine and the compressor

(4) The thermal efficiencies of the actual and ideal cycle

(5) The required airflow rate.

Solution. $T_1 = 273 + 20 = 293$ K

$T_3 = 273 + 850 = 1123$ K

$$T_{2a} = T_1 \left(\frac{P_2}{P_1}\right)^{(\gamma-1)/\gamma} = 293.(4.28)^{0.2857} = 443.9 \text{ K}$$

Similarly $T_{4a} = \dfrac{1123}{(4.28)^{0.2857}} = 741.25$ K

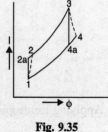

Fig. 9.35

Now $\eta_{compressor} = \dfrac{(T_{2a1} - T_1)}{(T_2 - T_1)}$

$\eta_{turbine} = \dfrac{(T_3 - T_4)}{(T_3 - T_{4a})}$

$$T_2 = \frac{T_1 + (T_{2a} - T_1)}{\eta_{compressor}} = \frac{293 + (443.9 - 293)}{0.8} = 481.6 \text{ K}$$

$T_4 = T_3 - \eta_{turbine}(T_3 - T_{4a})$

$= 1123 - 0.8(1123 - 741.25) = 817.6$ K

(2) and (3) specific work of compressor = $C_p (T_2 - T_1)$

$= 1.005(481.6 - 293) = 189.54$ kJ/kg

Specific work of turbine = $1.005 (T_3 - T_4)$

$= 1.005(1123 - 817.6) = 306.93$ kJ/kg

Net work = $306.93 - 189.54 = 117.4$ kJ/kg

(4) Thermal efficiency (η_t) of ideal cycle,

$$\eta_t = \frac{1-1}{\left(\dfrac{P_2}{P_1}\right)^{(\gamma-1)\gamma}} = 1 - 0.66 = 34\%$$

Thermal efficiency of actual cycle,

$$\eta_t = \frac{\text{(Heat supplied-Heat rejected)}}{\text{Heat supplied}}$$

$$= \frac{\{C_p(T_3 - T_2) - C_p(T_4 - T_1)\}}{\{C_p(T_3 - T_2)\}} = 1 - \frac{(T_4 - T_1)}{(T_3 - T_2)}$$

$$= 1 - \frac{(817.6 - 293)}{(1123 - 481)} = 1 - 0.818 = 18.20\%$$

(5) Air flow rate = $\dfrac{3600}{\text{net work output in kJ/kg}}$ kg/kW-hr.

$$= \left(\frac{3600}{117.4}\right) \times 25{,}000 \text{ kg/hr}$$

$$= \frac{(3600 \times 25000)}{(117.4 \times 3600)} \text{ kg/s} = 212.95 \text{ kg/s}$$

THEORETICAL QUESTIONS

1. Given the advantages and limitations of gas turbine power plant.
2. Given the application of gas turbine power plants.
3. Name the major components of a gas turbine plant.
4. Draw a simple line diagram for a simple open cycle gas turbine plant.
5. Derive an expression for the thermal efficiency.
6. Define Air-rate and work-ratio.
7. What is regeneration? Flow it improves the thermal efficiency of a simple open cycle gas turbine plant.
8. Define "effectiveness" of regeneration.
9. How "reheating" improves the thermal efficiency of a simple open cycle gas turbine plant ?
10. Discuss combined steam and gas turbine power plants.

EXERCISES

1. A simple, constant pressure gas turbine is designed for a pressure ratio of 5 to 1, and a turbine inlet temperature of 550°C. The adiabatic efficiency of compressing is 80% and that of expansion 85%, and there is a pressure loss of 0.0343bar through the combustion chamber. Calculate

   (a) the power per kg of air per sec.

   (b) the overall efficiency.

   Assuming the air to enter at 15°C and 1.01 bar. Take $k = 1.4$ and $C_n = 1.047$ for both air and combustion gases. Neglect the additional mass flow due to the fuel. [**Ans.** 65.47, 14.27%]

2. A gas turbine has a pressure ratio of 6/1 and a maximum cycle temperature of 600°C. The isentropic efficiencies of the compressor and turbine are 0.82 and 0.85 respectively. Calculate the power output in kilowatts of an electric generator geared to the turbine when the air enters the compressor at 15°C at the rate of 15 kg/s.

   Take: $c_p = 1.005$ kJ/kg K and $y = 1.4$ for the compression process, and take $c_p = 1.11$ kJ/kg K and $y = 1.333$ for the expansion process. [**Ans.** 920 kW]

3. In a gas turbine plant air at 10°C and 1.01 bar is compressed through a pressure ratio of 4:1. In a heat exchanger and combustion chamber the air is heated to 700°C while its pressure drops 0.14 bar. After expansion through the turbine the air passes through a heat exchanger, which cools the air through, 75% of maximum range possible, while the pressure drops 0.14

bar, and the air is finally exhausted to atmosphere. The isentropic efficiency of the compressor is 0.80 and that of turbine 0.85. Calculate the efficiency of the plant. [**Ans.** 22.76%]

4. In a gas turbine plant, air is compressed through a pressure ratio of 6:1 from 15°C. It is then heated to the maximum permissible temperature of 750°C and expanded in two stages each of expansion ratio $\sqrt{6}$, the air being reheated between the stages to 750°C. An heat exchanger allows the heating of the compressed gases through 75 percent of the maximum range possible. Calculate:

   (*i*) The cycle efficiency

   (*ii*) The work ratio

   (*iii*) The work per kg of air.

   The isentropic efficiencies of the compressor and turbine are 0.8 and 0.85 respectively.

   [**Ans.** (*i*) 32.75% (*ii*) 0.3852 (*iii*) 152 kJ/kg]

5. The gas turbine has an overall pressure ratio of 5:1 and a maximum cycle temperature of 550°C. The turbine drives the compressor and an electric generator, the mechanical efficiency of the drive being 97%. The ambient temperature is 20°C and the isentropic efficiencies of the compressor and turbine are 0.8 and 0.83 respectively. Calculate the power output in kilowatts for an air flow of 15 kg/s. Calculate also the thermal efficiency and the work ratio. Neglect changes are kinetic energy, and the loss of pressure in combustion chamber.

   [**Ans.** 655 kW; 12%; 0.168]

6. At the design speed the following data apply to a gas turbine set employing the heat exchanger: Isentropic efficiency of compressor = 75%, isentropic efficiency of the turbine = 85%, mechanical transmission efficiency = 99%, combustion efficiency = 98%, mass flow = 22.7 kg/s, pressure ratio = 6:1, heat exchanger effectiveness = 75%, maximum cycle temperature = 1000 K.

   The ambient air temperature and pressure are 15°C and 1.013 bar respectively. Calculate:

   (*i*) The net power output

   (*ii*) Specific fuel consumption

   (*iii*) Thermal efficiency of the cycle.

   Take the lower calorific value of fuel as 43125 kJ/kg and assume no pressure-loss in heat exchanger and combustion chamber. [**Ans.** (*i*) 2019 kW (*ii*) 0.4999 kg/kWh (*iii*) 16.7%]

Chapter 10

Nuclear Power Plant

10.1 INTRODUCTION

There is strategic as well as economic necessity for nuclear power in the United States and indeed most of the world. The strategic importance lies primarily in the fact that one large nuclear power plant saves more than 50,000 barrels of oil per day. At $30 to $40 per barrel (1982), such a power plant would pay for its capital cost in a few short years. For those countries that now rely on but do not have oil, or must reduce the importation of foreign oil, these strategic and economic advantages are obvious. For those countries that are oil exporters, nuclear power represents an insurance against the day when oil is depleted. A modest start now will assure that they would not be left behind when the time comes to have to use nuclear technology.

The unit costs per kilowatt-hour for nuclear energy are now comparable to or lower than the unit costs for coal in most parts of the world. Other advantages are the lack of environmental problems that are associated with coal or oil-fired power plants and the near absence of issues of mine safety, labor problems, and transportation bottle-necks. Natural gas is a good, relatively clean-burning fuel, but it has some availability problems in many countries and should, in any case, be conserved for small-scale industrial and domestic uses. Thus nuclear power is bound to become the social choice relative to other societal risks and overall health and safety risks.

Other sources include hydroelectric generation, which is nearly fully developed with only a few sites left around the world with significant hydroelectric potential. Solar power, although useful in outer space and domestic space and water heating in some parts of the world, is not and will not become an economic primary source of electric power.

Yet the nuclear industry is facing many difficulties, particularly in the United States, primarily as a result of the negative impact of the issues of nuclear safety waste disposal, weapons proliferation, and economics on the public and government The impact on the public is complicated by delays in licensing proceedings, court and ballot box challenges. These posed severe obstacles to electric utilities planning nuclear power plants, the result being scheduling problems, escalating and unpredictably costs, and economic risks even before a construction permit is issued. Utilities had a delay or cancel nuclear projects so that in the early 1980s there was a de fan moratorium on new nuclear plant commitments in the United States.

It is, however, the opinion of many, including this author, that despite these difficulties the future of large electric-energy generation includes nuclear energy as a primary, if not the main, source. The signs are already evident in many European and Asian countries such as France, the United Kingdom, Japan, and the U.S.S.R.

In a power plant technology course, it is therefore necessary to study nuclear energy: systems. We shall begin in this chapter by covering the energy-generation processes in nuclear reactors by starting with the structure of the atom and its nucleus and reactions that give rise to such energy generation. These include fission, fusion, aw different types of neutron-nucleus interactions and radioactivity.

10.2 GENERAL HISTORY AND TRENDS

10.2.1 MAJOR EVENTS

1945 : "Nuclear energy emerged from scientific obscurity and military secrecy."

1945-55 : "An enthusiastic vision developed of a future in which nuclear power would provide a virtually unlimited solution for the world's energy needs."

1955-73 : The pros and cons of nuclear energy were debated; however, the optimists prevailed and nuclear energy grew to become an important source of electricity.

Pros : Abundant, clean, and cheap energy. (We now know nuclear energy is not cheap.)

Cons : Large amounts of radioactivity are produced in the nuclear reactor, mishaps cannot be totally ruled out, and nuclear energy cannot be divorced from nuclear weapons. (Also, the long-term storage of nuclear wastes is now a very important issue.)

1955-65 : Many reactors designed, built, and put into operation.

1965-73 : Most of the US reactors were ordered during this period.

1973-85 : Many US reactors canceled during this period.

1970-90 : Most US reactors licensed to operate during this period.

1990-present : The number of nuclear reactors operating in the US and in the world leveled off, reaching a plateau. Few new reactors ordered and built.

Nuclear reactors started producing electricity in a significant way beginning about 1970 — just before the first international oil crisis in 1973. Thus, many countries saw nuclear energy as a means to reduce dependency on foreign oil. The US government saw nuclear energy as an important key to "energy independence."

However, the 1973 oil crisis lead to "side effects," which adversely affected nuclear energy:

Attention was focused worldwide on reducing energy consumption, including the consumption of electricity. (During the 1973-86 period, energy growth was erratic. Overall in the US, energy grew about as fast as the population, whereas electricity grew about as fast as the GNP, which means it grew faster than overall energy consumption, though not as fast as it had grown prior to 1973.

The oil crises reduced economic growth, thus, decreasing the demand for energy and electricity.

These effects reduced the demand for new nuclear plants. By 1973, the cost of nuclear energy was no longer regarded as "cheap," as had been touted in the early days of nuclear energy development, and safety concerns were starting to have an impact on the public view of nuclear energy. Also, nuclear energy was regarded as "establishments," and there were many protests against the establishment and its programs.

US nuclear energy capacity has been steady since the late 1980s. Currently, about 22% of US electricity is generated from nuclear energy (7.17 Quads). In 1994, there were 109 operating nuclear reactors in the US, with a total capacity of 99GWe. Currently, nuclear energy represents about 8% of the primary energy consumption in the US. However, coal is "king," generating about 55% of US electricity. Hydro generates about 10% of US electricity.

The US generates more electricity from nuclear energy than any other nation. However, France generates the greatest percentage of electricity from nuclear energy — about 75-80%. France is followed by Sweden. In 1994, Sweden generated about 50% of its electricity from nuclear energy, but now says it is getting out of nuclear energy electricity generation. The Swedish government claims this move will not increase its greenhouse gas emissions — a claim not believed in all circles.

Worldwide, for 1994, nuclear energy accounted for 6% of the primary energy consumption and 18% of the electricity generation. These numbers are just below the values for the US. 424 nuclear reactors operate worldwide, with a total capacity of 338GWe, spread over 30 countries.

In all but a few countries, nuclear energy growth was brought to a stop or at least to a crawl in the late 1980s and the 1990s. A summary of the reasons is:

- Reduction in oil and gas prices, especially since the late 1980s.
- Reduced growth in energy, compared to the pre-1973 period.
- Rising cost of nuclear energy.
- Increasing fears about nuclear energy.
- Campaigns against nuclear energy.

Public interest in nuclear energy began about 1944, grew strongly until about 1974, reached its peak then, and by 1994 dropped to a low level.

Is the age of nuclear energy over? Outside of a few countries, will more reactors be built? Has the verdict been given on nuclear energy?

10.2.2 WHAT MIGHT CHANGE THE CURRENT SITUATION?

Cost. Currently, nuclear energy is regarded as costly, and some costs are surely being passed on to future generations. The euphoric claims of the 1940s and 1950s regarding low cost nuclear energy have been discounted for at least two decades. The statement of the 1950s that nuclear energy would be "too cheap to meter" has haunted the industry. However, the text states that nuclear energy was cheaper than fossil energy for a period in the 1970s, and today is cheaper than fossil energy in some countries.

In the US, the long construction times, of about 10 years, have significantly driven up the cost. During construction period, capital is invested, interest payments occur, but no income from the sale of electricity occurs.

The development of factory-built, packaged, nuclear reactors, which could be purchased much as combined cycle combustion turbines are done today, would probably significantly reduce the cost. "From order to operation" within 2 or 3 years would be quite a change.

Standardization of nuclear reactor designs would likely significantly reduce the cost, and would likely increase safety.

Two things should be noted about US reactors. Many designs were developed and built. And most of the US reactors were ordered over a very short period of time, 1965 to 1973. Thus, during the 1970s and 1980s the opportunity to "get out the bugs," and for the better systems to evolve and win out didn't fully occur. With the benefit now of experience, with standardization, and with reduced order-to-start-up times, the cost of nuclear energy should come down.

Public Attitude. The public requires assurance that the industry truly has the issues of safety, fuel security, and waste disposal well under control. Perhaps the French experience will be convincing in this regard.

Greenhouse Effect. If the public comes to fear greenhouse warming, rather than simply having a concern about it, as currently the case, nuclear energy may be viewed more favorably. Coal is the "real" problem with respect to greenhouse gases. More electricity is produced by the burning of coal than by any other method. If the world continues to produce much of its electricity from coal, the evidence is fairly strong: CO_2 concentrations in the atmosphere will significantly increase, and greenhouse warming will occur (though the level of temperature increase is uncertain). Burning of all of the earth's fossil fuel resources would probably increase the atmospheric CO_2 concentration from the current level of 360 ppmv to about 1300 ppmv. 90% of this increase would be due to coal, since the oil and gas resources are small compared to the coal resources. The calculation assumes 4000 Gte (giga tonnes) of carbon in the earth's fossil fuel resources, an increase of 1ppmv CO_2 in the atmosphere for every 2.13 Gte of carbon burned, and a retention of 50% of the emitted CO_2 in the atmosphere. Since the start of the industrial age in the late 1700s, the CO_2 contention of the atmosphere has increased about 80 ppmv, and the mean temperature of the earth's atmosphere near the surface has increased about 1 degree F. If the temperature rise is assumed to be due to the CO_2 increase (which is debatable), a linear extrapolation implies a temperature increase of 12 degrees F for the 360 to 1300 ppmv CO_2 increase.

Demand for Eelectricity. Electricity is a desirable and convenient form of energy. Several factors could influence the demand for its generation, including its generation from nuclear power stations:

- Greater use of electricity, relative to heat, for manufacturing processes — a trend likely to continue and to drive up demand for electricity.
- Greater use of heat pumps for space heating. Significant growth here is problematic, since gas is cheap, and for many, heating with gas-fired furnaces is cheaper than converting to electric driven heat pumps.
- Electrification of transportation systems : Electric vehicles (EVs) and some types of hybrid electric vehicles (HEVs) depend on an external source of electricity. However, other types of HEVs and fuel cell powered vehicles generate electricity on board. It is too early to judge which system will evolve, or whether the internal combustion engine will retain predominance in a new form. Thus, a significant increase in electricity for the transportation sector is difficult to judge at present. See the front page of the Wall Street Journal for Monday, January 5, 1998 for an article on new power plants for automobiles.
- Combined cycle combustion turbines, fired on gas, are rapidly gaining popularity for generating electricity. Capital cost is relatively low, first law efficiency is high and will go higher (at least 60%), and order-to-start-up time is short. These systems may diminish the interest in new nuclear energy technology over the next one to two decades. Long term availability and price stability of the natural gas is the concern with respect to these systems. Also, they emit greenhouse gases, though the amount of CO_2 emitted per unit of electrical energy produced is less than one half that of a coal-fired electric power generating station.
- Renewable energy technology : What will be the growth of solar, wind, biomass, and other renewable energy technologies ? Will their cost competitiveness improve ? Are they as environmentally benign as thought ? Will they fill more than niche markets? Will technological breakthroughs occur ? Could they increase from 8% of US primary energy consumption (the current situation) to say the 20 to 30% level within 10 to 20 years ? If "yes," renewable energy may diminish the rejuvenation of the nuclear energy industry.

10.2.3 TECHNICAL HISTORY AND DEVELOPMENTS

Developments Prior to and During WW-2
- 1896: discovery of radioactivity.
- 1911: discovery of the nuclear atom.
- 1911: Rutherford noted the enormous amount of energy associated with nuclear reactions compared to chemical reactions.
- 1932: discovery of neutron.
- 1938: discovery of nuclear fission.
- 1939: researchers recognized that enough neutrons were released during fission reactions to sustain a chain reaction (in a pile of uranium and graphite). A chain reaction requires the release of two neutrons (or more) for every neutron used to cause the reaction.
- 1942 (Dec. 2): demonstration of the first operating nuclear reactor (200 Watts).
- 1943 (Nov.): 1 mW reactor put into operation at Oak Ridge, Tennessee.
- 1944 (Sept.): 200 mW reactor put into operation at Hanford, Washington—for the production of plutonium. This reactor was built in only 15 months.
- 1944 (Sept.): nuclear reactor for electricity generation proposed, using water for both cooling and neutron moderation. Essentially, this is the birth of nuclear energy for civilian use.

10.2.4 DEVELOPMENTS AFTER WW-2
- 1946: AEC (Atomic Energy Commission) established to oversee both military and civilian nuclear energy.
- 1953: Putman report/book, a thoughtful analysis of the case for nuclear energy for electricity production.
- 1953: US Navy began tests of the PWR (pressurized water reactor).
- 1957: 60 mW reactor at Shippingport, PA began to generate electricity for commercial use. The plant was built by the AEC, though Navy leadership played a predominant role.
- 1953-60: exploratory period: 14 reactors built, of many different designs, all but 3 under 100 mW size.
- 1960-65: only 5 reactors built.
- 1965-73: main period of ordering of nuclear reactors in the US. Size was much larger than before, many reactors of 600 to 1200 mW size.
- 1974: "honeymoon" over-nuclear energy no longer highly valued by the public.
- 1973-78: fall off in orders, with no US orders after 1978.
- 1974-85: cancellation of orders, over half of orders were canceled, or construction never brought to completion. Most reactors ordered prior to 1970 were built and brought on line. Many reactors ordered after 1970 never came on line they were canceled.
- 1970-90: most of US's reactors brought on line for commercial operation, indicating that most US reactors are 7 to 27 years old, or have 13 to 33 years of operation left, assuming a 40 year operating life.
- 1979: Three Mile Island accident. Reactor shut down.

- 1986: Chernobyl accident.
- Early 1990s: 7 nuclear reactors shut down, including 3 of early design and 4 of marginal performance. These shutdowns do not necessarily mean than a steady stream of reactors will be shut down before their nominal life of 40 years is reached.
- 1990s: Shoreham (Long Island) reactor shut down for good by public protest.

Capacity. Capacity factor (or capacity) = actual energy output integrated over a set period of time divided by the energy that would have occurred over the period of time if the reactor had been operated at rated power.

Routine maintenance and variations in demand limit maximum capacity to about 90%.

Long-term capacity over 80% is considered very good.

10.3 THE ATOMIC STRUCTURE

In 1803 John Dalton, attempting to explain the laws of chemical combination, propose his simple but incomplete atomic hypothesis. He postulated that all elements consists of indivisible minute particles of matter, atoms, that were different for different elements and preserved their identity in chemical reactions. In 1811 Amadeo Avo-gadro introduced the molecular theory based on the molecule, a particle of matter composed of a finite number of atoms. It is now known that the atoms are themselves composed of sub particles, common among atoms of all elements.

An atom consists of a relatively heavy, positively charged nucleus and a number of much lighter negatively charged electrons that exist in various orbits around the nucleus. The nucleus, in turn, consists of sub particles, called nucleons. Nucleons at primarily of two kinds: the neutrons, which are electrically neutral, and the proton: which are positively charged. The electric charge on the proton is equal in magnitude but opposite in sign to that on the electron. The atom as a whole is electrically neutral the number of protons equals the number of electrons in orbit. One atom may be transformed into another by losing or acquiring some of the above sub particles. Such reactions result in a change in mass Δm and therefore release (or absorb) large quantities of energy ΔE, according to Einstein's law

$$\Delta E = \frac{1}{g_c} \Delta m c^2 \qquad ...(10.1)$$

where c is the speed of light in vacuum and g_c is the familiar engineering conversion factor. Equation (10.1) applies to *all* processes, physical, chemical, or nuclear, in which energy is released or absorbed. Energy is, however, classified as *nuclear* if it is associated with changes in the atomic nucleus.

Figure 10.1 shows three atoms. Hydrogen has a nucleus composed of one proton, no neutrons, and one orbital electron. It is the only atom that has no neutrons. Deuterium has one proton and one neutron in its nucleus and one orbital electron. Helium contains two protons, two neutrons, and two electrons. The electrons exist in orbits, and each is quantitized as a lumped unit charge as shown. Most of the mass of the atom is in the nucleus. The masses of the three primary atomic sub particles are

Neutron mass m_n = 1.008665 amu

Proton mass m_P = 1.007277 amu

Electron mass m_e = 0.0005486 amu. The abbreviation amu, for *atomic mass unit*, is a unit of mass approximately equal to 1.66×10^{-27} kg, or 3.66×10^{-2} lb. These three particles are the primary building blocks of all atoms. Atoms differ in their mass because they contain varying numbers of them

NUCLEAR POWER PLANT

Atoms with nuclei that have the same number of protons have similar chemical and physical characteristics and differ mainly in their masses. They are called *isotopes*. For example, deuterium, frequently called *heavy hydrogen*, is an isotope of hydrogen. It exists as one part in about 6660 in naturally occurring hydrogen. When combined with oxygen, ordinary hydrogen and deuterium form *ordinary water* (or simply water) and *heavy water*, respectively.

The number of protons in the nucleus is called the *atomic number Z*. The total number of nucleons in the nucleus is called the mass *number A*.

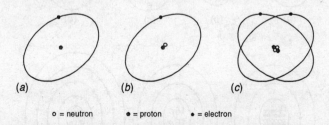

o = neutron • = proton • = electron

Fig. 10.1

As the mass of a neutron or a proton is nearly 1 amu, A is the integer nearest the mass of the nucleus which in turn is approximately equal to the atomic mass of the atom. Isotopes of the same element thus have the same atomic number but differ in mass number. Nucleus symbols are written conventionally as

$$_ZX^A$$

where X is the usual chemical symbol. Thus the hydrogen nucleus is $_1H^1$, deuterium is $_1H^2$ (and sometimes D), and ordinary helium is $_2He^1$. For particles containing no protons, the subscript indicates the magnitude and sign of the electric charge. The an electron is $-e^0$ (sometimes e or β) and a neutron is $_0n^1$. Symbols are also often written in the form He-4, helium-4, etc. Another system of notation, written as 7. will not be used in this text.

Fig. 10.2 shows, schematically, the structure of H^1, He^4 and some heavier atoms and the distribution of their electrons in various orbits. Two other particles of importance are the positron and the neutrino. The *positron* is a positively charged electron having the symbols $_{+1}e^0$, e^+ or β^+. The neutrino (little neutron) is a tiny electrically neutral particle that is difficult to observe experimentally. Initial evidence of its existence was based on theoretical considerations, nuclear reactions where a/3 particle of either kind is emitted or captured, the resulted energy (corresponding to the lost mass) was not all accounted for by the energy the emitted 13 particle and the recoiling nucleus. It was first suggested by Wolfgai Pauli in 1934 that the neutrino was simultaneously ejected in these reactions and the it carried the balance of the energy, often larger than that carried by the β particle itself. The importance of neutrinos is that they carry some 5 percent of the total energy produced in fission. This energy is completely react lost because neutrinos do not rea and are not stopped by any practical structural material. The neutrino is given the symbol u.

There are many other atomic sub particles. An example is the *mesons*, unstable positive, negative, or neutral particles that have masses intermediate between an electron and a proton. They are exchanged between nucleons and are thought to account for the forces between them. A discussion of these and other sub particles is, however beyond the scope of this book.

Electrons that orbit in the outermost shell of an atom are called *valence electron*. The outermost shell is called the *valence shell*. Thus, hydrogen has one valence electron and its K shell is the valence shell, etc. Chemical properties of an element are function of the number of valence electrons. The electrons play little or not part nuclear interactions.

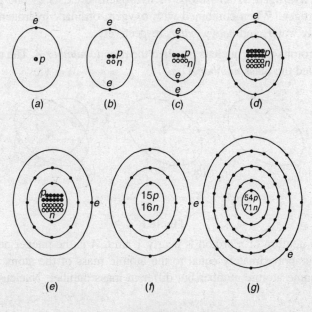

Fig. 10.2

10.4 SUMMARY OF NUCLEAR ENERGY CONCEPTS AND TERMS

10.4.1 SUMMARY OF FEATURES

1. Heat energy source is fission of radioactive material, (U-235)
2. Two typical plant designs:
   Pressurized water reactor (PWR) (U.S.)
   Boiling water reactor (BWR) (Russian)
3. Fuel pellets are in a large number of tubes (fuel rods)
4. Water circulates through core
5. Water converted to steam drives turbine
6. Turbine turns generator → electricity

10.4.2 FISSION

Unstable (radioactive) elements spontaneously split (radioactive decay), emitting high energy particles. Collision of particles with other atomic nuclei can trigger further nuclear decompositions. A small amount of mass is converted into a large amount of energy, when atomic nuclei are split.

Einstein equation: $E = mc^2$

Conversion of mass to energy. E = energy, m = mass converted, c = speed of light

10.4.3 CRITICAL MASS

There is a threshold mass of a radioactive isotope at which the flux density of radioactive particles will sustain a chain reaction. If this reaction is uncontrolled the result is an atomic bomb explosion. If the radiation fluxes are controlled and limited, we call it a nuclear reactor, which can be the basis of an electric power plant.

Types of Radiation	Atomic Weight	Charge
Alpha radiation (Helium nucleus)	4	+2
Beta radiation (Electron)	~0	−1
Neutron	1	0
Gamma ray	~0	0

10.4.4 ALPHA RADIATION

Alpha is quickly absorbed by matter because the particles have a large probability of collision with nuclei. Sources external to the human body cause radiation absorption within the thickness of the skin. Radiation from airborne particles in the lung are absorbed by surface membranes lining the lung. Alpha emitters ingested with food cause radiation absorption by the lining of the gut. The risk of genetic damage to adult organisms is very small because absorption takes place in surface cells.

10.4.5 BETA PARTICLES

Beta particles penetrate to the deepest parts of the body and can cause genetic damage and disrupt the function of cells anywhere in the body. Building walls and earthwork provide substantial shielding.

10.4.6 GAMMA PARTICLES

Gamma has the greatest penetration due to their small cross-section. Gamma particles can pass through ordinary materials. Effective shielding requires blankets of lead. Gamma radiation is a danger to all cells in the body.

10.4.7 URANIUM FISSION

$$_{92}U^{235} + {_0}n^1 \Rightarrow {_{92}}U^{236} \Rightarrow \text{Fission Products}$$
$$_{92}U^{238} + {_0}n^1 \Rightarrow {_{92}}U^{239} + \text{Gamma} \Rightarrow \text{Fission Products}$$
$$_{92}U^{239} \Rightarrow {_{93}}Np^{239} \Rightarrow {_{94}}Pu^{239}$$
$$\qquad\qquad\qquad \text{Neptunium} \qquad \text{Plutonium}$$

After many steps, (and a long time) the ultimate product is non-radioactive Lead atoms. The neutrons, whose absorbtion is indicated above, come from splitting of later fission products in reactions not shown here. Note that U-235 fission in the presence of U-238 causes the conversion of part of the U-238 into Plutonium-239 which can be concentrated to make an H-Bomb. Intermediate isotopes of health significance include Cesium-137, Iodine-131, Strontium-90 and many others.

10.4.8 HALF LIFE, T

Time for half the atomic nuclei to spontaneously split. The amount decays exponentially

$$N = N_o \exp(-t/T)$$

N = Amount of radioactive material,
N_o = Initial amount,
t = Elapsed time

10.5 ETHICAL PROBLEMS IN NUCLEAR POWER REGULATION

The Atomic Energy Commission (AEC), was formed to create a civilian nuclear energy industry, and had conflicting responsibilities:

- **Promoting Nuclear Power**
  —funded research in plant design
  —subsidized production of nuclear fuel
- **Regulating Plant Safety**
  —defined safety procedures, poor enforcement
  —inspecting, certifying plants
  —certifying operators, poor training
  As a result of these conflicting interests
- **No Long Term Waste Dispotal Plan was Completed**
  —wastes are still accumulating in temporary storage
  —radioactive waste? NIMBY
- **Future Termination/Cleanup Costs are not Factored into Current Electric Rates**
- **Power Companies are Largely Self-Regulated**
  —avoid reporting radiation release or do not monitor releases.
  —avoid safety regulations to save money.

Internal conflicts of the AEC were supposed to be resolved by splitting the promotional and regulatory duties between the new agencies:

Nuclear Regulatory Commission (NRC) – safety and standards

Dept. of Energy (DOE) – research, promotion, waste disposal, and fuel rod production.

10.6 CHEMICAL AND NUCLEAR EQUATIONS

Chemical reactions involve the combination or separation of whole atoms.

$$C + O_2 = CO_2$$

This reaction is accompanied by the release of about 4 electron volts (eV). An *electron volt* is a unit of energy in common use in nuclear engineering. 1 eV = 1.6021×10^{-19} joules (J) = 1.519×10^{-22} Btu = 4.44×10^{-26} kWh. 1 million electron volts (1 MeV) = 106 eV.

In chemical reactions, each atom participates as a whole and retains its identity. The molecules change. The only effect is a sharing or exchanging of valence electrons. The nuclei are unaffected. In

NUCLEAR POWER PLANT

chemical equations there are as many atoms of each participating element in the products (the right-hand side) as in the reactants (the left-hand side). Another example is one in which uranium dioxide (UO_2) is converted into uranium tetra fluoride (UF_4), called green salt, by heating it in an atmosphere of highly corrosive anhydrous (without water) hydrogen fluoride (HF), with water vapor (H_2O) appearing in the products

$$UO_2 + 4HF = 2H_2O + UF_4$$

Water vapor is driven off and UF_4 is used to prepare gaseous uranium hexafluoride (UF_6), which is used in the separation of the U^{235} and U^{238} isotopes of uranium by the gaseous diffusion method. (Fluorine has only one isotope, F^9, and thus combi-nations of molecules of uranium and fluorine have molecular masses depending only on the uranium isotope.)

Both chemical and nuclear reactions are either *exothermic* or *endothermic*, that is, they either release or absorb energy. Because energy and mass are convertible, Eq. (10.1), chemical reactions involving energy do undergo a mass decrease in exothermic reactions and a mass increase in endothermic ones. However, the quantities of energy associated with a chemical reaction are very small compared with those of a nuclear reaction, and the mass that is lost or gained is minutely small. This is why we assume a preservation of mass in chemical reactions, undoubtedly an incorrect assumption but one that is sufficiently accurate for usual engineering calculations.

In nuclear reactions, the reactant nuclei do not show up in the products, instead we may find either isotopes of the reactants or other nuclei. In balancing nuclear equations it is necessary to see that the same, or equivalent, nucleons show up in the products as entered the reaction. For example, if K, L, M, and N were chemical symbols, the corresponding nuclear equation might look like

$$_{Z1}K^{A1} + {}_{Z2}L^{A2} \longrightarrow {}_{Z3}M^{A3} + {}_{Z4}N^{A4}$$

To balance the following relationship must be satisfied.

$$Z_1 + Z_2 = Z_3 + Z_4$$
$$A_1 + A_2 = A_3 + A_4$$

Sometimes the symbols y or v are added to the products to indicate the emission of electromagnetic radiation or a neutrino, respectively. They have no effect on equation balance because both have zero Z and A, but they often carry large portions of the resulting energy.

Although the mass numbers are preserved in a nuclear reaction, the masses of the isotopes on both sides of the equation do not balance. Exothermic or endothermic energy is obtained when there is a reduction or an increase in mass from reactants to products, respectively.

10.7 NUCLEAR FUSION AND FISSION

Nuclear reactions of importance in energy production are fusion, fission, and radioactivity. Infusion, two or more light nuclei fuse to form a heavier nucleus. In fission, a heavy nucleus is split into two or more lighter nuclei. In both, there is a decrease in mass resulting in exothermic energy.

The same as in force = $\dfrac{1}{g}$, × mass × acceleration.

Table 10.1. Mass-energy Conversion factors

Mass	Energy				
	MeV	J	Bru	kWh	mW day
amu	931.478	1.4924×10^{-10}	1.4145×10^{-13}	4.1456×10^{-17}	9.9494×10^{-13}
kg	5.6094×10^{29}	8.9873×10^{16}	8.5184×10^{13}	2.4965×10^{10}	5.9916×10^{14}
lb_m	2.5444×10^{29}	4.0766×10^{16}	3.8639×10^{23}	1.1324×10^{10}	2.7177×10^{14}

10.7.1 FUSION

Energy is produced in the sun and stars by continuous fusion reactions in which four nuclei of hydrogen fuse in a series of reactions involving other particles that continually appear and disappear in the course of the reactions, such as He, nitrogen, carbon, and other nuclei, but culminating in one nucleus of helium and two positrons resulting in a decrease in mass of about 0.0276 amu, corresponding to 25.7 MeV.

$$4_1H^1 = {}_2He^4 + 2_{+1}e^0$$

The heat produced in these reactions maintains temperatures of the order of several million degrees in their cores and serves to trigger and sustain succeeding reactions. On earth, although fission preceded fusion in both weapons and power generation. the basic fusion reaction was discovered first, in the 1920s, during research on particle accelerators. Artificially produced fusion may be accomplished when two light atom fuse into a larger one as there is a much greater probability of two particles colliding than of four. The 4-hydrogen reaction requires, on an average, billions of years for completion, whereas the deuterium-deuterium reaction requires a fraction of a second. To cause fusion, it is necessary to accelerate the positively charged nuclei to high kinetic energies, in order to overcome electrical repulsive forces, by raising their temperature to hundreds of millions of degrees resulting in a plasma. The plasma must be prevented from contacting the walls of the container, and must be confined for a period of time (of the order of a second) at a minimum density. Fusion reactions are called *thermonuclear* because very high temperatures are required to trigger and sustain them. Table 10.2 lists the possible fusion reactions and the energies produced by them.

Table 10.2

Fusion reaction			Energy per reaction, MeV
Number	Reactants	Products	
1	D + D	T + p	4
2	D + D	$He^3 + n$	3.2
3	T + D	$He^4 + n$	17.6
4	He^3 + D	$He^4 + p$	18.3

n, p, D, and T are the symbols for the neutron, proton, deuterium and tritium respectively.

NUCLEAR POWER PLANT

Many problems have to be solved before an artificially made fusion reactor becomes a reality. The most important of these are the difficulty in generating and maintaining high temperatures and the instabilities in the medium (plasma), the conversion of fusion energy to electricity, and many other problems of an operational nature. Fusion power plants will not be covered in this text.

10.7.2 Fission

Unlike fusion, which involves nuclei of similar electric charge and therefore requires high kinetic energies, fission can be caused by the neutron, which, being electrically neutral, can strike and fission the positively charged nucleus at high, moderate, or low speeds without being repulsed. Fission can be caused by other particles, but neutrons are the only practical ones that result in a sustained reaction because two or three neutrons are usually released for each one absorbed in fission. These keep the reaction going. There are only a few fissionable isotopes U^{235}, Pu^{239} and U^{233} are fissionable by neutrons of all energies.

The immediate (prompt) products of a fission reaction, such as Xe^o and Sr^{y4} above, are called fission fragments. They, and their decay products, are called fission products. Fig. 10.4 shows fission product data for U^{235} by thermal and fast neutrons and for U^{233} and Pu^{239} by thermal neutrons 1841. The products are represented by their mass numbers.

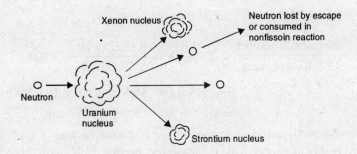

Fig. 10.3

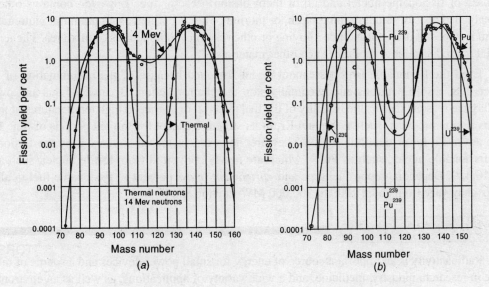

Fig. 10.4

10.8 ENERGY FROM FISSION AND FUEL BURN UP

There are many fission reactions that release different energy values. Another

$$_{92}U^{235} + {_0}n^1 \longrightarrow {_{56}}Ba^{137} + {_{36}}Kr^{97} + 2{_0}n^1 \qquad ...(1)$$

has the mass balance

$$235.0439 + 1.00867 \longrightarrow 136.9061 + 96.9212 + 2 \times 1.00867$$
$$236.0526 \longrightarrow 235.8446$$
$$\Delta m = 235.8446 - 236.0526 = -0.2080 \text{ amu} \qquad ...(2)$$

Thus $\qquad \Delta E = 931 \times -0.2080 = -193.6 \text{ MeV} = -3.1 \times 10^{-11} \text{ J} \qquad ...(3)$

On the average the fission of a U^{235} nucleus yields about 193 MeV. The same figure roughly applies to $U2^{33}$ and Pu^{239}. This amount of energy is prompt, *i.e.*, released at the time of fission. More energy, however, is produced because of (1), the slow decay of the fission fragments into fission products and (2) the nonfission capture of excess neutrons in reactions that produce energy, though much less than that of fission.

The *total energy,* produced *per* fission reaction, therefore, is greater than the prompt energy and is about 200 MeV, a useful number to remember.

The complete fission of 1 g of U^Z nuclei thus produces

$$\frac{\text{Avogadro's number}}{U^{235} \text{ isotope mass}} = 200 \text{ MeV} = \frac{0.60225 \times 10^{24}}{235.0439} \times 200$$

$$= 0.513 \times 10^{24} \text{ MeV} = 2.276 \times 10^{24} \text{ kWh}$$

$$= 8.190 \times 10^{10} \text{ J} = 0.948 \text{ MW-day}.$$

Another convenient figure to remember is that a reactor burning 1 g of fissionable material generates nearly 1 MW-day of energy. This relates to fuel burnup. Maximum theoretical burnup would therefore be about a million MW-day/ton (metric) of fuel. This figure applies if the fuel were entirely composed of fissionable nuclei and all of them fission. Reactor fuel, however, contains other non-fissionable isotopes of uranium, plutonium, or thorium. Fuel is defined as all uranium, plutonium, and thorium isotopes. It does not include alloying or other chemical compounds or mixtures. The term fuel material is used to refer to fuel plus such other materials.

Even the fissionable isotopes cannot be all fissioned because of the accumulation of fission products that absorb neutrons and eventually stop the chain reaction. Because of this-and owing to metallurgical reasons such as the inability of the fuel material to operate at high temperatures or to retain gaseous fission products [such as Xe and Kr, in its structure except for limited periods of time-burnup values are much lower than this figure. They are, however, increased somewhat by the fissioning of some fissionable nuclei, such as Pu^{23y}, which are newly converted from fertile nuclei, such as U^{238} (Sec. 10.4.7). Depending upon fuel type and *enrichment* (*mass* percent of fissionable fuel in all fuel), burnups may vary from about 1000 to 100,000 MW-day/ton and higher.

10.9 RADIOACTIVITY

Radioactivity is an important source of energy for small power devices and a source of radiation for use in research, industry, medicine, and a wide variety of applications, as well as an environmental concern.

Most of the naturally occurring isotopes are stable. Those that are not stable, *i.e.*, *radioactive*, are some isotopes of the heavy elements thallium (Z = 81), lead (Z = 82), and bismuth (Z = 83) and all the isotopes of the heavier elements beginning with polonium (Z = 84). A few lower-mass naturally occurring isotopes are radioactive, such as K^{40}, Rb^{87} and In^{115}. In addition, several thousand artificially produced isotopes of all masses are radioactive. Natural and artificial radioactive isotopes, also called *radioisotopes*, have similar disintegration rate mechanisms. Fig. 10.5 shows a Z-N chart of the known isotopes.

Radioactivity means that a radioactive isotope continuously undergoes spontaneous (*i.e.*, without outside help) disintegration, usually with the emission of one or more smaller particles from the *parent* nucleus, changing it into another, or daughter, nucleus. The parent nucleus is said to decay into the *daughter* nucleus. The *daughter* may or may not be stable, and several successive decays may occur until a stable isotope is formed. An example of radioactivity is

$$_{49}In^{115} = {}_{50}Sn^{115} + {}_{-1}e^0$$

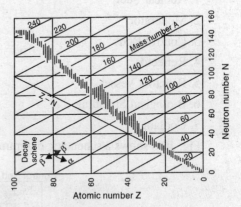

Fig. 10.5. Z-N Chart.

Radioactivity is *always* accompanied by a *decrease* in mass and is thus always exothermic. The energy liberated shows up as kinetic energy of the emitted particles and as γ radiation. The light particle is ejected at high speed, whereas the heavy one recoils at a much slower pace in an opposite direction.

Naturally occurring radio isotopes emit α, β, or γ particles or radiations. The artificial isotopes, in addition to the above, emit or undergo the following particles or reactions: positrons; orbital electron absorption, called K capture; and neutrons. In addition, neutrino emission accompanies β emission (of either sign).

Alpha decay. Alpha particles are helium nuclei, each consisting of two protons and two neutrons. They are commonly emitted by the heavier radioactive nuclei. An example is the decay of Pu^{239} into fissionable U^{235}

$$_{94}Pu^{239} = {}_{92}U^{235} + {}_{2}He^4$$

Beta decay. An example of β decay is

$$_{82}Pb^{214} = {}_{83}Bi^{214} + {}_{-1}e^0 + \nu$$

where ν, the symbol for the neutrino, is often dropped from the equation. The penetrating power of β particles is small compared with that of γ-rays but is larger than that of α particles. β- and α-particle decay are usually accompanied by the emission of γ radiation.

Gamma radiation. This is electromagnetic radiation of extremely short wavelength and very high frequency and therefore high energy. γ-rays and X-rays are physically similar but differ in their origin and energy: γ-rays from the nucleus, and X-rays from the atom because of orbital electrons changing orbits or energy levels. Gamma wave-lengths are, on an average, about one-tenth those of X-rays, although the energy ranges overlap somewhat. Gamma decay does not alter either the atomic or mass numbers.

10.10 NUCLEAR REACTOR

10.10.1 PARTS OF A NUCLEAR REACTOR

A nuclear reactor is an apparatus in which heat is produced due to nuclear fission chain reaction. Fig. 10.6 shows the various parts of reactor, which are as follows :

1. Nuclear Fuel
2. Moderator
3. Control Rods
4. Reflector
5. Reactors Vessel
6. Biological Shielding
7. Coolant.

Fig. 10.6 shows a schematic diagram of nuclear reactor.

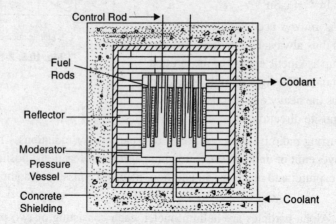

Fig. 10.6. Nuclear Reactor.

10.10.2 NUCLEAR FUEL

Fuel of a nuclear reactor should be fissionable material which can be defined as an element or isotope whose nuclei can be caused to undergo nuclear fission by nuclear bombardment and to produce a fission chain reaction. It can be one or all of the following

$$U^{233}, U^{235} \text{ and } Pu^{239}.$$

Natural uranium found in earth crust contains three isotopes namely U^{234}, U^{235} and U^{238} and their average percentage is as follows :

U^{238} — 99.3%

U^{235} — 0.7%

U^{234} — Trace

Out of these U^{235} is most unstable and is capable of sustaining chain reaction and has been given the name as primary fuel. U^{233} arid Pu^{239} are artificially produced from Th^{232} and U^{238} respectively and are called secondary fuel.

Pu^{239} and U^{233} so produced can be fissioned by thermal neutrons. Nuclear fuel should not be expensive to fabricate. It should be able to operate at high temperatures and should be resistant to radiation damage.

Uranium deposits are found in various countries such as Congo, Canada, U.S.A., U.S.S.R., Australia.

The fuel should be protected from corrosion and erosion of the coolant and for this it is encased in metal cladding generally stainless steel or aluminum. Adequate arrangements should be made for fuel supply, charging or discharging and storing of the fuel.

For economical operation of a nuclear power plant special attention should be paid to reprocess the spent: up (burnt) fuel elements and the unconsumed fuel. The spent up fuel elements are intensively radioactive and emits some neutron and gamma rays and should be handled carefully.

In order to prevent the contamination of the coolant by fission products, a protective coating or cladding must separate the fuel from the coolant stream. Fuel element cladding should possess the following properties :

1. It should be able to withstand high temperature within the reactor.
2. It should have high corrosion resistance.
3. It should have high thermal conductivity.
4. It should not have a tendency to absorb neutrons.
5. It should have sufficient strength to withstand the effect of radiations to which it is subjected.

Uranium oxide (UO_2) is another important fuel element. Uranium oxide has the following advantages over natural uranium:

1. It is more stable than natural uranium.
2. There is no problem or phase change in case of uranium oxide and therefore it can be used for higher temperatures.
3. It does not corrode as easily as natural uranium.
4. It is more compatible with most of the coolants and is not attacked by H_2, N_z.
5. There is greater dimensional stability during use.

Uranium oxide possesses following disadvantages :

1. It has low thermal conductivity.
2. It is more brittle than natural uranium and therefore it can break due to thermal stresses.
3. Its enrichment is essential.

Uranium oxide is a brittle ceramic produced as a powder and then sintered to form fuel pellets. Another fuel used in the nuclear reactor is uranium carbide (UC). It is a black ceramic used in the form of pellets.

Table indicates some of the physical properties of nuclear fuels.

Fuel	Thermal conductivity K-cal/m. hr°C	Specific heat kcal/kg °C	Density kg/m^3	Melting point (°C)
Natural uranium	26.3	0.037	19000	1130
Uranium oxide	1.8	0.078	11000	2750
Uranium carbide	20.6	—	13600	2350

10.10.3 MODERATOR

In the chain reaction the neutrons produced are fast moving neutrons. These fast moving neutrons are far less effective in causing the fission of U^{235} and try to escape from the reactor. To improve the utilization of these neutrons their speed is reduced. It is done by colliding them with the nuclei of other material which is lighter, does not capture the neutrons but scatters them. Each such collision causes loss of energy, and the speed of the fast moving neutrons is reduced. Such material is called Moderator. The slow neutrons (Thermal Neutrons) so produced are easily captured by the nuclear fuel and the chain reaction proceeds smoothly. Graphite, heavy water and beryllium are generally used as moderator.

Reactors using enriched uranium do not require moderator. But enriched uranium is costly due to processing needed.

A moderator should process the following properties :

1. It should have high thermal conductivity.
2. It should be available in large quantities in pure form.
3. It should have high melting point in case of solid moderators and low melting point in case of liquid moderators. Solid moderators should also possess good strength and machinability.
4. It should provide good resistance to corrosion.
5. It should be stable under heat and radiation.
6. It should be able to slow down neutrons.

10.10.4 MODERATING RATIO

To characterize a moderator it is best to use so called moderating ratio which is the ratio of moderating power to the macroscopic neuron capture coefficient. A high value of moderating ratio indicates that the given substance is more suitable for slowing down the neutrons in a reactor. Table 10.3 indicates the moderating ratio for some of the material used as moderator.

Table 10.3

Material	Moderating ratio
Beryllium	160
Carbon	170
Heavy Water	12,000
Ordinary Water	72

This shows that heavy water, carbon and, beryllium are the best moderators

Table 10.4

Moderator	Density (gm/cm^3)
H_2O	1
D_2O	11
C	1.65
Be	1.85

Table 10.5 shows some of the physical constants of heavy water and ordinary water

Table 10.5

Physical constant	D$_2$O	H$_2$O
Density at 293 K	1.1 gm/cm^3	0.9982 gm/cm^3
Freezing temperature	276.82	273
Boiling temperature	374.5	373 K
Dissociation Constant	0.3×10^{-14}	1×10^{-14}
Dielectric Constant at 293°K	80.5	82
Specific heat at 293°K	1.018	1

Control Rods. The Control and operation of a nuclear reactor is quite different from a fossil and fuelled (coal or oil fired) furnace. The furnace is fed continuously and the heat energy in the furnace is controlled by regulating the fuel feed, and the combustion air whereas a nuclear reactor contains as much fuel as is sufficient to operate a large power plant for some months. The consumption of this fuel and the power level of the reactor depends upon its neutron flux in the reactor core. The energy produced in the reactor due to fission of nuclear fuel during chain reaction is so much that if it is not controlled properly the entire core and surrounding structure may melt and radioactive fission products may come out of the reactor thus making it uninhabitable. This implies that we should have some means to control the power of reactor. This is done by means of control rods.

Control rods in the cylindrical or sheet form are made of boron or cadmium. These rods can be moved in and out of the holes in the reactor core assembly. Their insertion absorbs more neutrons and damps down the reaction and their withdrawal absorbs less neutrons. Thus power of reaction is controlled by shifting control rods which may be done manually or automatically.

Control rods should possess the following properties :
1. They should have adequate heat transfer properties.
2. They should be stable under heat and radiation.
3. They should be corrosion resistant.
4. They should be sufficient strong and should be able to shut down the reactor almost instantly under all conditions.
5. They should have sufficient cross-sectional area for the absorption.

10.10.5 REFLECTOR

The neutrons produced during the fission process will be partly absorbed by the fuel rods, moderator, coolant or structural material etc. Neutrons left unabsorbed will try to leave the reactor core ever to return to it and will be lost. Such losses should be minimized. It is done by surrounding the reactor core by a material called reflector which will send the neutrons back into the core. The returned neutrons can then cause more fission and improve the neutrons economy of the reactor. Generally the reflector is made up of graphite and beryllium.

10.10.6 REACTOR VESSEL

It is a. strong walled container housing the cure of the power reactor. It contains moderator, reflector, thermal shielding and control rods.

10.10.7 BIOLOGICAL SHIELDING

Shielding the radioactive zones in the reactor roan possible radiation hazard is essential to protect, the operating men from the harmful effects. During fission of nuclear fuel, alpha particles, beta particles, deadly gamma rays and neutrons are produced. Out oil these nc-1utroxrs and gamma rays are of main significance. A protection must be provided against them. Thick layers of lead or concrete are provided round the reactor for stopping the gamma rays. Thick layers of metals or plastics are sufficient to stop the alpha and beta particles.

10.10.8 COOLANT

Coolant. flows through and around the reactor core. It is used to transfer the large amount of heat produced in the reactor due to fission of the nuclear fuel during chain reaction. The coolant either transfers its heat to another medium or if the coolant used is water it takes up the heat and gets converted into steam in the reactor which is directly sent to the turbine.

Coolant used should be stable under thermal condition. It should have a low melting point and high boiling point. It should not corrode the material with which it comes in contact. The coolant should have high heat transfer coefficient. The radioactivity induced in coolant by the neutrons bombardment should be nil. The various fluids used as coolant are water (light water or heavy water), gas (Air, CO_2, Hydrogen, Helium) and liquid metals such as sodium or mixture of sodium and potassium and inorganic and organic fluids.

Power required to pump the coolant should be minimum. A coolant of greater density and higher specific heat demands less pumping power and water satisfies this condition to a great extent. Water is a good coolant as it is available in large qualities can be easily handled, provides some lubrication also and offers no unusual corrosion problems. But due to its low boiling point (212 F at atmospheric pressure) it is to be kept under high pressure to keep it in the liquid state to achieve a high that transfer efficiency. Water when used as coolant should be free from impurities otherwise the impurities may become radioactive and handling of water will be difficult.

10.10.9 COOLANT CYCLES

The coolant while circulating through the reactor passages take up heat produced due to chain reaction and transfer this heat to the feed water in three ways as follows :

(a) *Direct Cycle.* In this system coolant which is water leaves the reactor in the form of steam Boiling water reactor uses this system.

(b) *Single Circuit System.* In this system the coolant transfers the heat to the feed water in the steam generator. This system is used in pressurized reactor.

(c) *Double Circuit System.* In this system two coolant are used. Primary coolant after circulating through the reactor flows through the intermediate heat exchanger (IHX) and passes on its hest to the secondary coolant which transfers its heat in the feed water in the steam generator. This system is used in sodium graphite reactor and fast breeder reactor.

10.10.10 REACTOR CORE

Reactor core consists of fuel rods, moderator and space through which the coolant flows.

NUCLEAR POWER PLANT

10.11 CONSERVATION RATIO

It is defined as the ratio of number of secondary fuel atoms to the number of consumed primary fuel atoms. A reactor with a conversion ratio above unity is known as a breeder reactor. Breeder reactor produces more fissionable material than it consumes. If the fissionable material produced is equal to or less than the consumed, the reactor is called converter reactor.

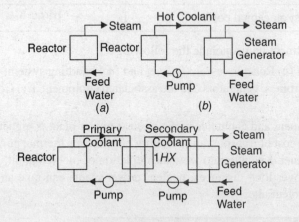

Fig. 10.7

10.12 NEUTRON FLUX

It is a measure of the intensity of neutron radiation and it is the number of neutrons passing through 1 cm^2 of a given target in one second. It is expressed as uv, where u is number of neutrons per cubic centimeter and u is velocity of neutrons in cm/sec.

10.13 CLASSIFICATION OF REACTORS

The nuclear reactors can be classified as follows :

1. **Neutron Energy.** Depending upon the energy of the neutrons at the time they are captured by the fuel to induce fissions, the reactors can be named as follows :

(a) *Fast Reactors.* In such reactors fission is brought about by fast (non moderated) neutrons.

(b) *Thermal Reactors or Slow Reactors.* In these reactors the fast moving neutrons are slowed down by passing them through the moderator. These slow moving neutrons are then captured by the fuel material to bring about the fission of fundamental research.

10.14 COST OF NUCLEAR POWER PLANT

Nuclear power plant is economical if used as base load power plant and run at higher load factors. The cost of nuclear power plant is more at low load factors. The overall running cost of a nuclear power plant of large capacity may be about 5 paisa per kWh but it may be as high 15 paisa per kWh if the plant is of smaller capacity. The capital cost of a nuclear power plant of larger capacity (say 250 mW) is nearly Rs. 2500 per kW installed. A typical sub-division of cost is as follows :

Item	Approximate Cost %
(a) Capital cost of land, building and equipment etc.	62%
(b) Fuel cost	22%
(c) Maintenance cost	6%
(d) Interest on capital cost	10%

The capital investment items include the following :

(i) Reactor Plant : (a) Reactor vessel, (b) Fuel and fuel handling system, (c) Shielding. (ii) Coolant system. (iii) Steam turbines, generators and the associated equipment. (iv) Cost of land and construction costs.

The initial investment and capital cost of a nuclear power plant is higher as compared to a thermal power plant. But the cost of transport and handling of coal for a thermal power plant is much higher than the cost of nuclear fuel. Keeping into view the depletion of fuel (coal, oil, gas) reserves and transportation of such fuels over long distances, nuclear power plants can take an important place in the development of power potentials.

10.15 NUCLEAR POWER STATION IN INDIA

The various nuclear power stations in India are as follows :

(i) **Tarapur Nuclear Power Station.** It is India's first nuclear power plant. It has been built at Tarapur 60 miles north of Bombay with American collaboration. It has two boiling water reactors each of 200 mW capacity and uses enriched uranium as its fuel. It supplies power to Gujarat and Maharashtra.

Tarapur power plant is moving towards the stage of using mixed oxide fuels as an alternative to uranium. This process involves recycling of the plutonium contained in the spent fuel. In the last couple of years it has become necessary to limit the output of reactors to save the fuel cycle in view of the uncertainty of enriched uranium supplies from the United States.

(ii) **Rana Pratap Sagar (Rajasthan) Nuclear Station.** It has been built at 42 miles south west of Kota in Rajasthan with Canadian collaboration. It has two reactors each of 200 mW capacity and uses natural uranium in the form of oxide as fuel and heavy water as moderator.

(iii) **Kalpakkam Nuclear Power Station.** It is the third nuclear power station in India and is being built at about 40 miles from Madras City. It will be wholly designed and constructed by Indian scientists and engineers. It has two fast reactors each of 235 mW capacity and will use natural uranium as its fuel.

The first unit of 235 mW capacity has started generating power from 1983 and the second 235 MW unit is commissioned in 1985. The pressurized heavy water reactors will use natural uranium available in plenty in India. The two turbines and steam generators at the Kalpakkam atomic power project are the largest capacity generating sets installed in our country. In this power station about 88% local machinery and equipment have been used.

(iv) **Narora Nuclear Power Station.** It is India's fourth nuclear power station and is being built at Narora in Bullandshahar District of Uttar Pradesh. This plant will initially have two units of 235 mW each and provision has been made to expand its capacity of 500 mW. It is expected to be completed by 1991.

This plant will have two reactors of the CANDUPHW (Canadian Deutrium-Uranium-Pressurised Heavy Water) system and will use natural uranium as its fuel. This plant will be wholly designed and constructed by the Indian scientists and engineers. The two units are expected to be completed by 1989 and 1990 respectively. This plant will use heavy water as moderator and coolant. This plant will provide electricity at 90 paise per unit. Compared to the previous designs of Rajasthan and Madras nuclear power plants the design of this plant incorporates several improvements. This is said to be a major effort towards evolving a standardized design of 235 mW reactors and a stepping stone towards the design of 500 mW reactors. When fully commissioned plant's both units will provide 50 mW to Delhi, 30 mW to Haryana, 15 mW to Himachal Pradesh, 35 mW to Jammu and Kashmir, 55 mW to Punjab, 45 mW to Rajasthan, 165 mW to Uttar Pradesh and 5 mW to Chandigarh. The distribution of remaining power will depend on the consumer's demands. In this plant one exclusion zone of 1.6 km radius has been provided where no public habitation is permitted. Moderate seismicity alluvial soil conditions in the region of Narora have been fully taken into account in the design of the structure systems and equipment in Narora power plant.

Narora stands as an example of a well coordinated work with important contributions from Bhabha Atomic Research Centre, Heavy Water Board, Nuclear Fuel Complex, Electronics Corporation of India Limited (ECIL) and other units of Department of Atomic Energy and several private and public sector industries Instrumentation and control systems are supplied by ECIL. Bharat Heavy Electrical Limited (BHEL) is actively associated with Nuclear Power Corporation of India. It has supplied steam generators, reactor headers and heat exchangers for Narora Atomic Power Plant (NAPP) 1 and 2 (2×235 MW).

NAPP is the forerunner of a whole new generation of nuclear power plants that will come into operation in the next decade. The design of this reactor incorporates several new safety features ushering in the state of the art in reactor technology. The design also incorporates two fast acting and independent reactor shut down systems conceptually different from those of RAPP and MAPP.

Some of the new systems introduced are as follows:

1. Emergency Core Cooling System (ECCS).
2. Double Containment System.
3. Primary Shut off rod System (PSS).
4. Secondary Shut off rod System (SSS).
5. Automatic Liquid Poison Addition System (ALPAS).
6. Post accident clean up system.

According to Department of Atomic Energy (DAE) the Narora Atomic Power Plant (NAPP) has the following features.

1. It does not pose safety and environmental problems for the people living in its vicinity. The safety measures are constantly reviewed to ensure that at all times radiation exposure is well within limits not only to the plant personnel but also to the public at large.
2. NAI'P design rneets all the requirement laid down in the revised safety standards. The design of power plant incorporates two independent fast acting shut down systems high pressure, intermediate pressure and low pressure emergency care cooling systems to meet short and long term requirements and double containment of the reactor building.

Narora Atomic Power Plant (NAPP) is pressurized heavy water reactor (PHWR) that has been provided with double containment. The inner containment is of pre-stressed concrete designed to withstand the full pressure of 1.25 kg/cm^2 that is likely to be experienced in the event of an accident. The outer containment is of reinforced cement concrete capable of withstanding the pressure of 0.07 kg/cm^2. The angular space between the two containments is normally maintained at a pressure below atmosphere to ensure that any activity that might leak past primary containment is vented out through the stock and not allowed to come out to the environment in the immediate vicinity of the reactor building. The primary and the secondary containments are provided with highly efficient filtration systems which filter out the active fission products before any venting is done.

The moment containment gets pressurized it gets totally sealed from the environment. Subsequently the pressure in the primary containment is brought down with the help of the following provisions.

1. Pressure suppression pool at the basement of the reactor building.
2. Special cooling fan units which are operated on electrical power obtainable from emergency diesel generators. The containment provisions are proof tested to establish that they are capable of withstanding the pressures that are expected in the case of an accident. Fig. 5.12 (a) shows primary and secondary containment arrangement.
3. The cooling water to all the heavy water heat exchangers is maintained in a closed loop so that failure in these do not lead to escape of radioactivity very little water from River (Ganga would be drawn for cooling purposes and most of water would be recycled.
4. The power plant has a waste management plant and waste burial facility within the plant area.
5. NAPP is the first pressurized heavy water reactor (PHWR) in the world to have been provided with double containment.
6. No radioactive effluent, treated or otherwise will be discharged into Ganga River. Therefore there will be no danger of pollution of the Ganga water.
7. An exclusion zone of 1.6 km radius around the plant has been provided where no habitation is permitted.
8. A comprehensive fire fighting system on par with any modern power station has been provided at NAPP.
9. NAPP has safe foundations. It is located on the banks of river Ganges an alluvial soil. The foundations of the plant reach upto a depth where high relative densities and bearing capacities are met. The foundations design can cater to all requirements envisaged during life of plant.) It is safe against earthquakes.
10. In the event of danger over heated care of the reactor would be diffused with in a few seconds by two features namely shut down through control rods followed by injection of boron rich water which will absorb the neutrons and stop their reaction in the core. This is in addition to other feature like double containment system provided in the reactor.

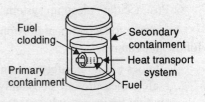

Fig. 10.8

Above features assure total radiation safety of the plant personnel, general public and the environment during the operation of power plant. With the completion of NAPP it would make a useful contribution to the North-grid thereby accelerating the pace of development in this region.

Narora Atomic Power Plant is the fourth atomic power plant to be commissioned in India. This power plant is meant to generate electricity and supply the same to the distribution system (grid) in Uttar Pradesh and other states in the northern region. It has two units each with a capacity of 235 mW of which about seven per cent will be used to run the in house equipment and the rest will be fed into the grid. The net output from the power plant will be about 435 mW. At this power plant all due precautions have been taken in the design, construction, commissioning and operation of the unit with safety as the over-riding consideration. Therefore there appears to be no danger to the public from the operation of this power plant.

(*v*) **Kakarpar Nuclear Power Plant.** This fifth nuclear power plant of India is to be located at Kakarpar near Surat in Gujarat. This power station will have four reactors each of 235 mW capacity.

The reactors proposed to be constructed at Kakarpar would be of the Candu type natural uranium fuelled and heavy water moderated reactors-incorporating the standardised basic design features of the Narora reactors suitably adapted to local conditions. The fuel for the power plant will be fabricated at the Nuclear Fuel complex, Hyderabad. The power plant is expected to be completed by 1991.

The Kakarpur unit has two fast shut down systems. The primary one works by cadmium shut off rods at 14 locations which drop down in case of heat build up and render the reactor sub-critical in two seconds. There are 12 liquid shut off rods as a back up, further backed by slow acting automatic liquid poison addition system which absorbs neutrons completely and stop the fissile reaction.

In case of sudden loss of coolant, heavy water inside the reactor, there is an emergency core cooling system which also stops the fissile reaction. Lastly, the pressure suppression system in which cool water under the reactor rises automatically to reduce pressure in case it increases and a double containment wall ensures that no radioactivity would be released at ground level even in case of an unlikely accident.

The Department of Atomic Energy (DAE) has also evolved emergency preparedness plans for meeting any accident even after all these safety measures. It ensures a high level of preparedness to face an accident including protecting the plant personnel and sur-rounding population. There is no human settlement for five km belt around a nuclear power installation as a mandatory provision.

(*vi*) **Kaiga Atomic Power Plant.** The sixth atomic power plant will be located at Kaiga in Karnataka. Kaiga is located away from human habitation and is a well suited site for an atomic power plant. It will have two units of 235 mW each. It is expected to be commissioned by 1995. This nuclear power plant will have CANDU type reactors. These reactors have modern systems to prevent accidents. The plant would have two solid containment walls-inner and outer to guard against any leakage. The inner containment wall could withstand a pressure of 1.7 kg/cm^2 and could prevent the plant from bursting. The outer containment walls of the reinforced cement concrete has been design to withstand pressure of 0.07 kg/cm^2. The annular space between the two containment walls would be maintained at a lower pressure below that of the atmosphere to ensure that no radioactivity leaked past the primary containments.

10.16 LIGHT WATER REACTORS (LWR) AND HEAVY WATER REACTORS (HWR)

Light water reactors use ordinary water (technically known as light water) as coolant and moderator. They are simpler and cheaper. But they require enriched uranium as their fuel. Natural uranium contains 0.6% of fissionable isotope U^{235} and 99.3% of fertile Lj^{23} and to use natural uranium in such reactors it is to be enriched to about 3%, U^{235} and for this uranium enrichment plant is needed which requires huge investment and high operational expenditure. Heavy water reactors use heavy water as their coolant and moderator. They have the advantage of using natural uranium as their fuel. Such reactors have some operation problem too. Heavy water preparation plants require sufficient investment and leakage of heavy water must be avoided as heavy water is very costly. Heavy water required in primary circuits must be 99% pure and this requires purification plants heavy water should not absorb moisture as by absorbing moisture it gets degraded. In order to have sufficient quantity of heavy water required for nuclear power plants, the work is fast progressing in our country on four heavy water plant. These plants are situated at Kotah (100 tonnes per year), Baroda (67.2 tonnes), Tuticorin (71.3 tonnes) and Talcher (67.2 tonnes per year). These plants will give our country an installed heavy water production capacity of about 300 tonnes per year.

10.16.1 Importance of Heavy Water

The nuclear power plants of **Kota in** Rajasthan, Kalpakkam in Tamil Nadu and Narora in U.P. use heavy water as coolant and moderator. All these projects have CANDU reactors using natural uranium as fuel and heavy water as moderator. After this enriched uranium natural water reactor at Tarapur, the CANDU reactors are the second generation of reactors in India's nuclear power programme. The CANDU reactor will produce plutonium which will be the core fuel for fast breeder reactor. In fact in breeder reactor heavy water is used as moderator.

A CANDU reactor of 200 mW capacity requires about 220 tonnes of heavy water in the initial stages and about 18 to 24 tonnes each year subsequently. Therefore, about one thousand tonnes of heavy water will be required to start the different nuclear power stations using heavy water. The total capacity of different heavy water plants will be about 300 tonnes per year if all the heavy water plant under construction start production. It is expected that heavy water from domestic production will be available from Madras and Narora atomic power plants. The management of the heavy water system is a highly complicated affair and requires utmost caution. Heavy water is present in ordinary water in the ratio 1 6000. One of the methods of obtaining heavy water is electrolysis of ordinary water.

ADVANTAGES OF NUCLEAR POWER PLANT

The various advantages of a nuclear power plant are as follows:

1. Space requirement of a nuclear power plant is less as compared to other conventional power plants are of equal size.
2. A nuclear power plant consumes very small quantity of fuel. Thus fuel transportation cost is less and large fuel storage facilities are not needed Further the nuclear power plants will conserve the fossil fuels (coal, oil, gas etc.) for other energy need.
3. There is increased reliability of operation.
4. Nuclear power plants are not effected by adverse weather conditions.
5. Nuclear power plants are well suited to meet large power demands. They give better performance at higher load factors (80 to 90%).
6. Materials expenditure on metal structures, piping, storage mechanisms are much lower for a nuclear power plant than a coal burning power plant.

    For example for a 100 mW nuclear power plant the weight of machines and mechanisms weight of metal structures, weight of pipes and fittings and weight of masonry and bricking up required are nearly 700 tonnes, 900 tonnes, 200 tonnes and 500 tonnes respectively wherea for a 100 mW coal burning power plant the corresponding value are 2700 tonnes, 1250 tonnes, 300 tonnes and 1500 tonnes respectively. Further area of construction site required aired for 100 mW nuclear power plant is 5 hectares whereas was for a 100 mW coal burning power plant the area of construction site is nearly 15 hectares.

7. It does not require large quantity of water.

DISADVANTAGES

1. Initial cost of nuclear power plant is higher as compared to hydro or steam power plant.
2. Nuclear power plants are not well suited for varying load conditions.
3. Radioactive wastes if not disposed carefully may have bad effect on the health of workers and other population.

   In a nuclear power plant the major problem faced is the disposal of highly radioactive waste in form of liquid, solid and gas without any injury to the atmosphere. The preservation of waste for a long time creates lot of difficulties and requires huge capital.
4. Maintenance cost of the plant is high.
5. It requires trained personnel to handle nuclear power plants.

10.17 SITE SELECTION

The various factors to be considered while selecting the site for nuclear plant are as follows :

1. **Availability of water.** At the power plant site an ample quantity of water should be available for condenser cooling and made up water required for steam generation. Therefore the site should be nearer to a river, reservoir or sea.
2. **Distance from load center.** The plant should be located near the load center. This will minimise the power losses in transmission lines.
3. **Distance from populated area.** The power plant should be located far away from populated area to avoid the radioactive hazard.
4. **Accessibility to site.** The power plant should have rail and road transportation facilities.
5. **Waste disposal.** The wastes of a nuclear power plant are radioactive and there should be sufficient space near the plant site for the disposal of wastes.

Safeguard against earthquakes. The site is classified into its respective seismic zone 1, 2, 3, 4, or 6. The zone 5 being the most seismic and unsuitable for nuclear power plants. About 300 km of radius area around the proposed site is studied for its past history of tremors, and earthquakes to assess the severest earthquake that could occur for which the foundation building and equipment supports are designed accordingly. This ensures that the plant will retain integrity of structure, piping and equipments should an earthquake occur. The site selected should also take into account the external natural events such as floods, including those by up-stream dam failures and tropical cyclones.

The most important consideration in selecting a site for a nuclear power plant is to ensure that the site-plant combination does not pose radio logical or any hazards to either the public, plant personnel on the environment during normal operation of plant or in the unlikely event of an accident.

The Atomic Energy Regulatory Board (AERB) has stipulated a code of practice on safety in Nuclear Power Plant site and several safety guide lines for implementation.

10.18 COMPARISON OF NUCLEAR POWER PLANT AND STEAM POWER PLANT

The cost of electricity generation is nearly equal in both these power plants. The other advantages and disadvantages are as follows :

(*i*) The number of workman required for the operation of nuclear power plant is much less than a steam power plant. This reduces the cost of operation.

(*ii*) The capital cost of nuclear power plant falls sharply if the size of plant is increased. The capital cost as structural materials, piping, storage mechanism etc. much less in nuclear power plant than similar expenditure of steam power plant. However, the expenditure of nuclear reactor and building complex is much higher.

(*iii*) The cost of power generation by nuclear power plant becomes competitive with cost of steam power plant above the unit size of about 500 mW.

10.19 MULTIPLICATION FACTOR

Multiplication factor is used to determine whether the chain reaction will continue at a steady rate, increase or decrease. It is given by the relation,

$$K = \frac{P}{(A+E)}$$

where K = Effective multiplication factor.

P = Rate of production of neutrons.

A = Combined rate of absorption of neutrons.

E = Rate of leakage of neutrons.

$K = 1$ indicates that the chain reaction will continue at steady rate (critical) $K > 1$ indicates that the chain reaction will be building up

(super critical) whereas $K < 1$ shows that reaction will be dying down (subcritical).

10.20 URANIUM ENRICHMENT

In some cases the reaction does not take place with natural uranium containing only 0.71% of U^{235}.

In such cases it becomes essential to use uranium containing higher content of U^{335}. This is called U^{235} concentration of uranium enrichment. The various methods of uranium enrichment are as

1. The gaseous diffusion method. This method is based on the principle that the diffusion or penetration molecular of a gas with a given molecular weight through a porous barrier is quicker than the molecules of a heavier gas. Non-saturated uranium hexa-flouride (UF^6) is used for gaseous diffusion. The diffusing molecules have small difference in mass. The molecular weight of U^{235} Fs = 235 + 6 × 19 = 349 and that U^{238} Fs = 352. The initial mixture is fed into the gap between the porous barrier. That part of the material which passes through the barrier is enriched product, enriched in U^{235} Fs molecules and the remainder is depleted product.

2. Thermal diffusion method. In this method (Fig. 10.9) a column consisting of two concentric pipes is used. Liquid UF^6 is filled in the space between the two pipes. Temperature of one of the pipes is kept high and that of other is kept low. Due to difference in temperature the circulation of the liquid starts, the liquid rising along the hot wall and falling along the cold wall. Thermal diffusion takes place in the column. The light U^{235} Fs molecules are concentrated at the hot wall and high concentration of U^{236} Fs is obtained in the upper part of the column.

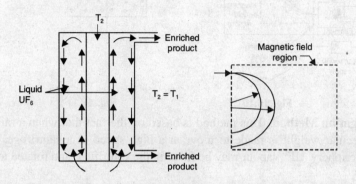

Fig. 10.9

3. Electromagnetic Method. This method is based on the fact that when ions moving at equal velocities along a straight line in the same direction are passed through a magnetic field, they are acted upon by forces perpendicular to the direction of ion movement and the field.

Let P = force acting on ion e = charge on ion

v = velocity of ion

H = magnetic field strength m = Ion mass

R = radius of ion path $P = euH$

As this force is centripetal

$$\therefore \quad P = \frac{mv^2}{R}$$

$$\therefore \quad \frac{mv^2}{R} = evH$$

$$\therefore \quad R = \frac{mv}{eH}$$

This shows that ions moving at equal velocities but different masses move along ng circumferences of different radii (Fig. 10.10). Fig. 10.11 shows an electromagnetic separation unit for uranium isotopes. A gaseous uranium compound is fed into the ion source, where neutral atoms are ionised with the help of ion bombardment. The ions produced come out in the form of narrow beam after passing through a number of silos. This beam enters the acceleration chamber. These ions then enter a separation chamber where a magnetic field is applied. Due to this magnetic field the ions of different masses move along different circumference.

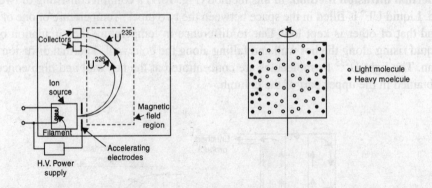

Fig. 10.10   Fig. 10.11

4. Centrifugation Method, This method is based on the fact that when a mixture of two gases with different molecular weight is made to move at a high speed in a centrifuge, the heavier gas is obtained near the periphery. UF^6 vapour may be filled in the centrifuge and rotated to separate uranium isotopes.

10.21 REACTOR POWER CONTROL

The power released in a nuclear reactor is proportional to the number of mole fissioned per unit time this number being in turn proportional to density of the neutron flux in the reactor. The power of a nuclear reactor can be controlled by shifting control rods which may be either actuated manually or automatically.

Power control of a nuclear reactor is simpler than that of conventional thermal power plant because power of a nuclear reactor is a function of only one variable whereas power of a thermal power plant depends on number of factors such as amount of fuel, its moisture content, air supply etc. This shows that power control of thermal plant requires measuring and regulating several quantities which is of course considerably more complicated.

10.22 NUCLEAR POWER PLANT ECONOMICS

Major factors governing the role of nuclear power are its economic development and availability of sufficient amount of nuclear fuel.

It is important to extract as much energy from a given amount of fuel as possible. The electrical energy extracted per unit of amount of fuel or expensive moderator might be called the "material efficiency". In a chain reactor the high material efficiency as well as high thermal efficiency leads to low over all energy cost.

Since the most attractive aspect of nuclear energy is the possibility of achieving fuel costs considerably below that for coal, all nuclear power system being considered for large scale power production involve breeding or regenerative systems. This program includes the development of the technology of low neutron absorbing structural materials such as zirconium, the use of special moderating materials such a D20 and the consideration of special problems associated with fast reactors. In so far as economic factors are concerned it is necessary to consider neutron economy in a general way such as that measured by the conversion ratio of the system. The conversion ratio is defined as the atoms of new fuel produced in fertile material per atom of fuel burnt. The conversion ratio varies with the reactor design. Its values for different reactors are indicated in table.

Type of reactor	Conversion ratio
BWR, PWR and SGR	1
Aqueous thorium breeder	1.2
Fast breeder reactor	1.6

10.23 SAFETY MEASURES FOR NUCLEAR POWER PLANTS

Nuclear power plants should be located far away from the populated area to avoid the radioactive hazard. A nuclear reactor produces α and (β particles, neutrons and γ-quanta which can disturb the normal functioning of living organisms. Nuclear power plants involve radiation leaks, health hazard to workers and community, and negative effect on surrounding forests.

At nuclear power plants there are three main sources of radioactive contamination of air.

(*i*) Fission of nuclei of nuclear fuels.

(*ii*) The second source is due to the effect of neutron fluxes on the heat carrier in the primary cooling system and on the ambient air.

(*iii*) Third source of air contamination is damage of shells of fuel elements.

This calls for special safety measures for a nuclear power plant. Some of the safety measures are as follows.

(*i*) Nuclear power plant should be located away from human habitation.

(*ii*) Quality of construction should be of required standards.

(*iii*) Waste water from nuclear power plant should be purified. The water purification plants must have a high efficiency of water purification and satisfy rigid requirements as regards the volume of radioactive wastes disposed to burial.

(*iv*) An atomic power plant should have an extensive ventilation system. The main purpose of this ventilation system is to maintain the concentration of all radioactive impurities in the air below the permissible concentrations.

(*v*) An exclusion zone of 1.6 km radius around the plant should be provided where no public habitation is permitted.

(*vi*) The safety system of the plant should be such as to enable safe shut down of the reactor whenever required. Engineered safety features are built into the station so that during normal operation as well as during a severe design basis accident the radiation dose at the exclusion zone boundary will be within permissible limits as per internationally accepted values. Adoption of a integral reactor vessel and end shield assemblies, two independent shut down systems, a high pressure emergency core cooling injection system and total double containment with suppression pool are some of the significant design improvements made in Narora Atomic Power Project (NAPP) design. With double containment NAPP will be able to withstand seismic shocks.

In our country right from the beginning of nuclear power programme envisaged by our great pioneer Homi Bhabha in peaceful uses of nuclear energy have adopted safety measures of using double containment and moderation by heavy water one of the safest moderators of the nuclear reactors.

(*vii*) Periodical checks be carried out to check that there is no increase in radioactivity than permissible in the environment.

(*viii*) Wastes from nuclear power plant should be carefully disposed off. There should be no danger of pollution of water of river or sea where the wastes are disposed.

In nuclear power plant design, construction, commissioning and operation are carried out as power international and national codes of protection with an overriding place given to regulatory processes and safety of plant operating personnel, public and environment.

10.24 SITE SELECTION AND COMMISSIONING PROCEDURE

In order to study prospective sites for a nuclear power plant the Department of Atomic Energy (DAE) of our country appoints a site selection committee with experts from the following:

1. Central Electricity Authority (CEA).
2. Atomic Minerals Division (AMD).
3. Health and safety group and the Reactor Safety Review group of the Bhabha Atomic Research Center (BARC).
4. Nuclear Power Corporation (NPC).

The committee carries out the study of sites proposed. The sites are then visited, assessed and ranked. The recommendations of the committee are then forwarded to DAE and the Atomic Energy Commission (AEC) for final selection.

The trend is to locate a number of units in a cluster at a selected site. The highest rated units in India are presently of 500 mW. The radiation dose at any site should not exceed 100 milligram per member of the public at 1.6 km boundary.

The commissioning process involves testing and making operational individually as well as in an integrated manner the various systems such as electrical service water, heavy water, reactor regulating and protection, steam turbine and generator. To meet the performance criteria including safe radiation levels in the plant area and radioactive effluents during operation the stage-wise clearance from Atomic Energy Regulatory Board (AERB) is mandatory before filling heavy water, loading fuel making the reactor critical, raising steam, synchronizing and reaching levels of 25%, 50%, 75% and 100% of full power. The commissioning period lasts for about two years.

10.25 MAJOR NUCLEAR POWER DISASTERS

Chernobyl — is near Kiev, Ukraine, in the former Soviet Union. Destroyed by steam and hydrogen explosions followed by fire, it caused many deaths on site, increased cancer rates in the thousands of square miles it contaminated.

Three Mile Island — Located 10 miles southeast of Harrisburg PA on the Susquehanna River. The accident, and radiation release, caused no immediate deaths. The cleanup cost more than $1.5 Billion.

The Three Mile Island accident occurred in 1979 and 1986, Chernobyl occurred essentially killing the expension of nuclear energy. No other nuclear power plants have been ordered in the US since the late 1970's.

T.M.I. Aaccount Chronology in Brief

1970's AEC LOFT (Loss of Fluid Test) research canceled as economy measure.

September 12, 1978 T.M.I. Unit #2 dedicated.

January 1979 TMI #2 began commercial operation.

March 26, 1979 Emergency core cooling pumps tested, with diverter valves switched to disconnect ECCS from reactor. Valves not switched back.

March 28, 1979, 4 a.m. Three Mile Island Incident began.

—Filter in inner loop switched offline to clean

—Pressure transient triggers shutdown sequence.

—Core overheats, pressure relief valve sticks open, in manual override

—Water in core begins leaking out open relief valve

—Emergency cooling pumps don't work !

—After more errors, 1/3 of core exposed, partial meltdown of fuel rods results.

—2nd day someone closes relief valve (unrecorded).. situation stabilizes

—hydrogen gas bubble forms.

—Governor/NRC, order partial evacuation

Cleanup/termination cost $1.5+ BILLION.

Cleanup after the Three Mile Island Accident. After the Accident it was necessary to dispose of the radioactive gases, water, and contaminated debris from radioactive plumbing etc. The water had to be filtered to separate and concentrate radioactive contaminants for disposal. After these were removed it was possible to begin dismantling the pressure vessel and extract the fuel rods. It was not until then that the inside of the core could be inspected. As the damaged reactor was brought under control, it was known from radiation monitoring that there was a significant amount of radioactive material in the bottom of the pressure vessel. In spite of this, the power company still maintained that the damage to the core had been minimal. When a robot with a video camera was lowered into the pressure vessel, four years after the accident, this is what it saw:

10.26 CHERNOBYL NUCLEAR POWER PLANT

Chernobyl is a town of 30,000 people, 70 miles north of Kiev, in the Ukraine. The V. I. Lenin nuclear power plant is located 10 miles from the town of Chernobyl. Adjacent to the plant is the town of Pripyat, which houses and services plant workers. The plant is on the Pripyat River, near its mouth into the Kiev reservoir.

The plant had 4 nuclear reactors, each with associated steam turbines and electric generators. Two additional units were under construction at the time of the accident, April 26, 1986. Each of these units was of the same Soviet design, designated RBMK-1000.

Chernobyl was the location of the world's worst nuclear power plant disaster. Massive amounts of radioactivity were released, a thousand square mile area will be uninhabitable for many decades.

10.26.1 REACTOR DESIGN : RBMK-1000

Boiling Water Reactor

Electric generating capacity 1000 Megawatt.

Thermal output of core about 2000 Megawatt.

1661 zirconium fuel rods, holding mix of U-238 and U-235; Plutonium-239 is a byproduct, which can be extracted by reprocessing the fuel rod material. Each fuel rod is enclosed in a heat transfer water channel.

211 Boron control rods with 8 fuel rods/control rod

Graphite core 1700 tons, made up of graphite bricks

10.26.2 CONTROL OF THE REACTOR

1. Graphite Core, moderates neutron flux from fuel rods
2. Boron control rods to reduce neutron flux for shutdown
3. Thermal transfer control — closed circuit water/steam loop, multiple water pumps Nitrogen/Helium gas within containment — low thermal conductivity and oxygen exclusion — pressure and gas mixture are controlled. emergency core cooling water system (ECCS)

10.26.3 CHERNOBYL REACTOR OPERATIONS

Computer for fine control, operator controls set points of feedback controllers Central power authority dictated operating levels in managing power grid Unnecessary shutdown meant 600,000 ruble revenue loss, firing of person responsible.

Plant engineers found the plant unstable at low power levels,

Local practice was to manually pull control rods if downward fluctuation threatened spontaneous shutdown. Response time to scram: 18 seconds (theoretically it was claimed to be 3 seconds).

Regulations against manual control routinely excepted.

10.26.4 ACCIDENT\SAFETY PLANS

Published odds million to 1 against an accident. Authoritarian control staff & engineers do not question safety. Accident planning was around a scenario of 1 or 2 fuel rod/water channels bursting. No plan included a graphite fire. Administration building had emergency bunker under it. Reactor building was a water tight containment building.

10.26.5 EVACUATION

Plant director had authority in principle to order evacuation of Pripyat. However a standing order made any nuclear accident a state secret.

10.27 SAFETY PROBLEMS IN CHERNOBYL REACTOR DESIGN

10.27.1 SYSTEM DYNAMICS

A problem with RMBK-1000 reactor design is that the time constants for changes in thermal output are short. Control depends on computer regulated feedback control systems. The human operator could not react fast enough to manually control it without the automatic controls.

NUCLEAR POWER PLANT

Neutron absorption and heat transfer coefficients are very different for water and steam, so neutron flux and thermal output changes rapidly as water in the tubes through the core makes a transition from hot water to steam.

10.27.2 ANOTHER SAFETY PROBLEM WITH THE DESIGN

The normal operating temperature of core tubes is greater than the ignition temperature of the graphite blocks of the core (carbon) in an O_2 atmosphere. Its normal environment is an atmosphere with no oxygen.

Heat exchange system :

One closed loop through reactor core and steam turbines

Secondary loop to condense steam to water after turbine

Construction problems :

Turbine building roof; specification said it should be fireproof. Materials for 1 km × 50 m fireproof roof was not available. Control cable conduits supposed to be fireproof. Material not available. Exception granted. Cement and tiles, etc. Quality control problems. Director had to prioritize uses, discard defective materials. Fittings often required remanufacture to meet specifications.

Hazard Potential of Water on Hot Graphite

Water Gas Reaction:

$$C + H_2O \Rightarrow CO + H_2$$

Often used as a H_2 generator in freshman chemistry labs, it has a similar hazard if not carefully controlled:

$$2\,CO + O_2 \Rightarrow 2\,CO_2$$
$$2\,H_2 + O_2 \Rightarrow 2\,H_2O$$

10.28 OTHER, EARLIER, SOVIET NUCLEAR ACCIDENTS

September 1982 — Chernobyl Unit 1, after 5 years service, was shut down for maintenance. Restarted with some valves closed. Result: no water flow in a few channels. Explosion in core, a few fuel rods melted. Some radioactivity escaped plant. No radiation survey was done outside plant. Streets of Pripyat were hosed down. No announcement to population. Emergency core cooling system saved plant. Chief Engineer, his deputy, and chief operator of the shift were all demoted and transferred.

1980 Kursh power station. RBMK-1000 plant had a power outage.

Reactor damaged because control rods and circulation driven by electric motors/pumps failed. Time delay to start diesel generators was 40 seconds During which, power surge damaged some fuel rods. Solution : design a special generator to tap turbine power as it spun down during shutdown, to power emergency equipment.

Oct. 1982. Armyansk nuclear power station. Explosion. Subsequent fire destroyed turbine building.

Fall 1983 Chernobyl Unit 4 startup. Certification team saw anomalous power surge when control rod insertion starts. Considered minor, had been seen in another reactor. No explanation, not documented.

June 1985 Balakovsky PWR power station, Valve burst, release of 300 degree C. steam, cooked 14 workmen. Safety regulations viewed as guidelines, chief engineer regularly made exceptions.

EXERCISES

1. What is a chain reaction ? How it is controlled ?
2. What is a nuclear reactor ? Describe the various parts of a nuclear reactor.
3. What are different types of reactors commonly used in nuclear power stations ? Describe the fast breeder reactor ?
4. Discuss the various factors to be considered while selecting the site for nuclear power station. Discuss its advantages and disadvantages.
5. Write short notes on the following:
    (a) Boiling water reactor (B.W.R.)
    (b) Pressurised water reactor (P.W.R.)
    (c) Multiplication factor.
    (d) Fertile and fissionable material.
6. What are the different components of a nuclear power plant ? Explain the working of a nuclear power plant. What are the different fuels used in such a power plant ?
7. What is a Homogeneous Reactor ? Describe a Homogeneous Aqueous Reactor (H.A.R.).
8. What is meant by uranium enrichment? Describe some methods of Uranium enrichment.
    Compare the economic (cost) of nuclear power plant with steam power plant.
    Explain the terms 'Breeding' and 'Burn up'.
9. State the properties of control rods.
10. Explain the properties of moderator used in a nuclear reactor. Explain the principle of operation of a sodium graphite reactor.
11. Discuss the factors which go in favour of nuclear power plant as compared to other types of power plants.

Chapter 11

Hydro-Electric Power Plants

11.1 INTRODUCTION

When rain water falls over the earth's surface, it possesses potential energy relative to sea or ocean towards which it flows. If at a certain point, the water falls through an appreciable vertical height, this energy can be converted into shaft work. As the water falls through a certain height, its potential energy is converted into kinetic energy and this kinetic energy is converted to the mechanical energy by allowing the water to flow through the hydraulic turbine runner. This mechanical energy is utilized to run an electric generator which is coupled to the turbine shaft. The power developed in this manner is given as:

$$\text{Power} = W.Q.H.\eta \text{ watts} \qquad \ldots(11.1)$$

where
W = Specific weight of water, N/m^3
Q = rate of water flow, m^3/sec.
H = Height of fall or head, m
η = efficiency of conversion of potential energy into mechanical energy.

The generation of electric energy from falling water is only a small process in the mighty heat power cycle known as "Hydrological cycle" or rain evaporation cycle". It is the process by which the moisture from the surface of water bodies covering the earth's surface is transferred to the land and back to the water bodies again. This cycle is shown in Fig. 11.1. The input to this cycle is the solar energy. Due to this, evaporation of water takes, lace from the water bodies. On cooling, these water vapours form clouds. Further cooling makes the clouds to fall down in the form of rain, snow, hail or sleet etc; known as precipitation. Precipitation includes all water that falls from the atmosphere to the earth's surface in any form. Major portion of this precipitation, about 2/3rd, which reaches the land surface is returned to the atmosphere by evaporation from water surfaces, soil and vegetation and through transpiration by plants. The remaining precipitation returns ultimately to the sea or ocean through surface or underground channels. This completes the cycle. The amount of rainfall which runs off the earth's land surface to form streams or 'rivers is useful for power generation. The precipitation that falls on hills and mountains in the form of snow melts during warmer weather as run-off and converges to form streams can also be used for power generation.

Hydro projects are developed for the following purposes:
1. To control the floods in the rivers.
2. Generation of power.
3. Storage of irrigation water.
4. Storage of the drinking water supply.

In India, the water resources development is concerned with the first three purposes. As reported by the Irrigation Commission (1972), the country average annual run-off is 178 million hectare meters. Of this, 29.2% is contributed by Ganga, 30.1% by Brahmputra and north eastern rivers, 11.8% by the west flowing rivers south of Tapti. The balance of 29% is contributed by Indus, the west and east flowing rivers of Central India and the east flowing rivers of South India. It is very apparent from the above analysis that the nation's water

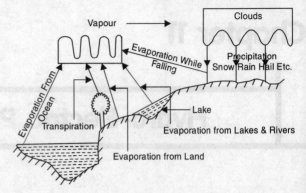

Fig. 11.1

resources are very much unevenly distributed. As the country is committed to socialistic development of economy, there is a great need to have as uniform as possible the distribution of water resources throughout the country. The two ways to achieve this goal are, inter basin transfer and joint use of surface and ground water. Under the first scheme Ganga-Cauvery link is under active proposal, by which 12715 cumecs of water will be diverted from Ganga by constructing a barrage near Patna in Bihar. Of this total quantity, 285 cumecs (10000 cusecs) would be supplied for 300 days out of a year, to the drought affected areas in South U.P and South Bihar which are in the Ganga basin itself. The remaining 1430 cumecs (50000 cusecs) of water for 150 days will be diverted from the basin to meet partially the water demand of lingering drought affected areas of Madhya Pradesh, Rajasthan, Gujarat, Maharashtra, Andhra Pradesh, Mysore and Tamil Nadu. On its route, the link would connect the proposed Bargi reservoir on Narmada, the proposed Champalli reservoir on Godavari, the under construction Srisailam reservoir on Krishna and would finally meet the river Cauvery at the existing Grand Anicut. Of course, there is a problem of high head pumping (380 meters) from Ganga which would have to be resolved.

In the remaining chapter, the problem of generation of electric power will be dealt with. There are two reasons for the extensive development of the water power. One is that more and more electric power is needed for industrial; agricultural, commercial and domestic purposes. The other is the high cost of coal and its dwindling reserves. A water power site is usually developed to supply electric power to a newly and a specially established industry or town or to provide additional power to an already existing or a proposed interconnected electric system. Before a water power site is considered for development, the following factors must be thoroughly analyzed:

1. The capital cost of the total plant.

2. The capital cost of erecting and maintaining the transmission lines and the annual power loss due to transformation and transmission of electric power since the water power plants are usually situated in hilly areas away from the load center.

3. The cost of electric generation compared with steam, oil or gas plants which can be conveniently set up near the load center.

Inspite of the above factors, the water power plants have the following advantages which make these suitable for large interconnected electric system:

1. The plant is highly reliable and its maintenance and operation charges are very low.

2. The plant can be run up and synchronized in a few minutes.

3. The load can be varied quickly and the rapidly changing load de-mands can be met without any difficulty.

4. The plant has no stand by losses.

5. No fuel charges.

6. The efficiency of the plant does not change with age.

7. The cost of generation of electricity varies little with the passage of time.

However, the hydro-electric power plants have the following disadvantages also:

1. The capital cost of the plant is very high.

2. The hydro-electric plant takes much longer in design and execution.

3. These plants are usually located in hilly areas far away from the load center.

4. Transformation and transmission costs are very high.

5. The output of a hydro-electric plant is never constant due to vagaries of monsoons and their dependence on the rate of water flow in a river.

11.2. RUN-OFF

Rain fall (used in a general sense) or "precipitation" may be defined as the total condensation of moisture that reaches the earth in any form. It includes all forms of rains, ice, snow, hail or sleet etc. "Evaporation" represents practically all of that portion of the rainfall that does not reach the point of ultimate use as stream flow. So, evaporation, includes all the rainfall that is returned to the atmosphere from land and water surfaces. Thus total evaporation is:

1. Evaporation from land and water surfaces.

2. Evaporation by transpiration which is the vaporization of water from the breathing pores of vegetable matter.

3. Atmospheric evaporation (evaporation while precipitation is falling).

Rain-fall is measured in terms of centimeters of water over a given area and over a given period (usually one year). The portion of the total precipitation that flows through the catchment area is known as "Run-off". The catchment area of a hydrosite is the total area behind the dam, draining water into the reservoir. Thus,

Run-off = Total precipitation − Total evaporation

Part of the precipitation is absorbed by the soil and seeps or percolates into ground and will ultimately reach the catchment area through the underground channels. Thus.

Total run-off = Direct run off over the land surface T Run-off through seepage.

The unit of run-off are m^3/s or day-second meter.

Day-second meter = Discharge collected in the catchment area at the rate of 1 in 3/S for one day
$$= 1 \times 24 \times 3600 = 86400 \ m^3/day.$$

The flow of run-off can also be expressed in cms. of water on the drainage area feeding the river site for a stated period, or km, cm of water per unit of time.

Factors Affecting Runoff

1. **Nature of Precipitation.** Short, hard showers may produce relatively little run-off. Rains lasting a longer time results in larger run-off. The soil tends to become saturated and the rate of seepage decreases. Also, the humid atmosphere lowers evaporation, resulting in increased run-off.

2. **Topography of Catchments Area.** Steep, impervious areas will produce large percentage of total run-off. The water will flow quickly and absorption and evaporation losses will be small.

3. **Geology of Area.** The run-off is very much affected by the types of surface soil and sub-soil, type of rocks etc. Rocky areas will give more run-off while pervious soil and sub-soil and soft and sandy area will give lesser run-off.

4. **Meteorology.** Evaporation varies with temperature, wind velocity and relative humidity. Run-off increases with low temperature, low wind velocity and high relative humidity and vice versa.

5. **Vegetation.** Evaporation and seepage are increased by cultivation. Cultivation opens and roughens the hard, smooth surface and promotes seepage. Thick vegetation like forests consumes a portion of the rain fall and also acts as obstruction for run-off.

6. **Size and Shape of Area.** Large areas will give more run-off. A wide area like a fan will give greater run-off, whereas, a narrow area like a leaf will give lesser run-off. In an area whose length is more than its width, the flow along its width will give more run-off than if the flow is along its length, since in the former case, seepage and evaporation will be less.

Measurement of Run-Off or Flow : The run-off or stream flow can be determined with the help of three methods:

1. **From Rain-Fall Records.** The run-off can be estimated from rain-fall records by multiplying the rain fall with "run-off coefficient" for the drainage area. The run-off coefficient takes into account the various losses and will depend upon the nature of the catchment area, as given below : in Table 11.1

Table 11.1

Drainage Area	Run-off-Coefficient
Commercial and industrial	0.90
Asphalt or concrete pavement	0.85
Forests	0.05 to 0.30
Parks, farmland and pastures	0.05 to 0.30

Then, Run-off = Rain fall × run-off co-efficient

This is not an accurate method of measuring run-off since the estimation of run-off co-efficient can not be very accurate.

2. **Empirical Formulas.** Empirical relations to determine the stream flow relate only to a particular site and can not be relied upon for general use.

3. **Actual Measurement.** Direct measurement by stream gauging at a given site for a long period is the only precise method of evaluation of stream flow. The flow is measured by selecting a channel of fixed cross-section and measuring the water velocity at regular intervals, at enough points in the cross-section for different water levels. The velocity of flow can

HYDRO-ELECTRIC POWER PLANTS

be measured with the help of current meter or float method. By integrating the velocities over the cross-section for each stage, the total flow for each stage can be calculated.

11.3. HYDROGRAPH AND FLOW DURATION CURVE

A hydrograph indicates the variation of discharge or flow with time. It is plotted with flows as ordinates and time intervals as abscissas. The flow is in m^3/sec and the time may be in hours, days, weeks or months.

A flow duration curve shows the relation between flows and lengths of time during which they are available. The flows are plotted as the ordinates and lengths of time as abscissas. The flow duration curve can be plotted from a hydrograph.

11.4. THE MASS CURVE

The use of the mass curve is to compute the capacity of the reservoir for a hydrosite. The mass curve indicates the total volume of run-off in second meter-months or other convenient units, during a given period. The mass curve is obtained by plotting cumulative volume of flow as ordinate and time (days, weeks by months) as abscissa. Fig. 11.2 shows a mass curve for a typical river for which flow data is given in Table 11.2.

The monthly flow is only the mean flow and is correct only at the beginning and end of the months. The variation of flow during each month is not considered. Cumulative daily flows, instead of monthly flows, will give a more accurate mass curve, but this involves an excessive amount of work. The slope of the curve at any point gives the flow rate in second-meter. Let us join two points X and Y on the curve. The slope of this line gives the average rate of flow during the period between X and Y. This will be = (Flow at Y-Flow at X)/Time Span

Table 11.2

Month	Mean monthly flow second-metre	Accumulative flow second metre-month	Month	Mean monthly flow second-metre	Accumulative flow second metre-month
April	7710	7710	October	3300	23280
May	3850	11560	November	2330	25610
June	860	12420	December	3880	29490
July	1040	13460	January	5200	34690
August	2500	15960	February	6000	40690
September	4020	19980	March	12,100	52790

Let the flow demand be, 3000 sec-meter. Then the line X-Y may be called as `demand line' or 'Use line'. If during a particular period, the slope of the mass Curve is greater than that of the demand line, it means more water is flowing into the reservoir than is being utilized, so the level of water in the reservoir will be increasing during that period and vice versa. Upto point X and beyond point Y the reservoir will be overflowing. being full at both X and Y.

The capacity of the reservoir is given by the maximum ordinate between the mass curve and the demand line. For the portion of mass curve between point X and Y, the storage capacity is about 4600 sec-meter-month. However, considering the entire mass curve, storage capacity will be about 15,400 sec-meter-months.

Note: The amount of water to be stored in a reservoir can also be determined with the help of hydrograph. A horizontal line giving the average flow demand is drawn. The portion of the curve above this line will indicate the amount of water that should be stored in the reservoir to be used during lean months. But, hydrograph for one year varies from year to year, the storage capacity should be determined with the help of the mass curve.

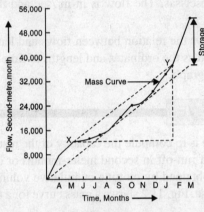

Fig. 11.2

11.5. SELECTION OF SITE FOR A HYDRO-ELECTRIC POWER PLANT

While selecting a suitable site, if a good system of natural storage lakes at high altitudes and with large catchment areas can be located, the plant will be comparatively economical. Anyhow the essential characteristics of a good site are: large catchment areas, high average rainfall and a favorable place for constructing the storage or reservoir. For this purpose, the geological, geographical and meteorological conditions of a site need careful investigation. The following factors should be given careful consideration while selecting a site for a hydro-electric power plant:

1. Water Available. To know the available energy from a given stream or river, the discharge flowing and its variation with time over a number of years must be known. Preferably, the estimates of the average quantity of water available should be prepared on the basis of actual measurements of stream or river flow. The recorded observation should be taken over a number of years to know within reasonable, limits the maximum and minimum variations from the average discharge. the river flow data should be based on daily, weekly, monthly and yearly flow ever a number of years. Then the curves or graphs can be plotted between tile river flow and time. These are known as hygrographs and flow duration curves.

The plant capacity and the estimated output as well as the need for storage will be governed by the average flow. The primary or dependable power which is available at all times when energy is needed will depend upon the minimum flow. Such conditions may also fix the capacity of the standby plant. The, maximum of flood flow governs the size of the headwords and dam to be built with adequate spillway.

2. Water-Storage. As already discussed, the output of a hydropower plant is not uniform due to wide variations of rain fall. To have a uniform power output, a water storage is needed so that excess flow at certain times may be stored to make it available at the times of low flow. To select the site of the dam ; careful study should be made of the geology and topography of the catchment area to see if the natural foundations could be found and put to the best use.

3. Head of Water. The level of water in the reservoir for a proposed plant should always be within limits throughout the year.

4. Distance from Load Center. Most of the time the electric power generated in a hydro-electric power plant has to be used some considerable distance from the site of plant. For this reason, to be economical on transmission of electric power, the routes and the distances should be carefully considered since the cost of erection of transmission lines and their maintenance will depend upon the route selected.

5. Access to Site. It is always a desirable factor to have a good access to the site of the plant. This factor is very important if the electric power generated is to be utilized at or near the plant site. The transport facilities must also be given due consideration.

11.6. ESSENTIAL FEATURES OF A WATER-POWER PLANT

A simplified flow sheet of a water power plant is shown in Fig. 11.3: The essential features of a water power plant are as below:

1. Catchment area.
2. Reservoir.
3. Dam and intake house.
4. Inlet water way.
5. Power house.
6. Tail race or outlet water way.

1. Catchment Area. The catchment area of a hydro plant is the whole area behind the dam, draining into a stream or river across which the dam has been built at a suitable place.

2. Reservoir. Whole of the water available from the catchment area is collected in a reservoir behind the dam. The purpose of the storing of water in the reservoir is to get a uniform power output throughout the year. A reservoir can be either natural or artificial. A natural reservoir is a lake in high mountains and an artificial reservoir is made by constructing a dam across the river.

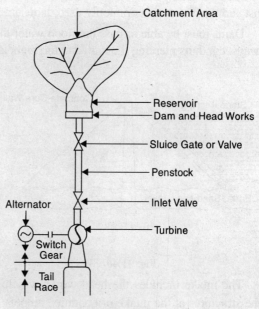

Fig. 11.3

3. Dam and Intake House. A dam is built across a river for two functions: to impound the river water for storage and to create the head of water. Dams may be classified according to their structural materials such as: Timber, steel, earth, rock filled and masonary. Timber and steel are used for dams of height 6 m to 12 m only. Earth dams are built for larger heights, upto about 100 m. To protect the dam from the wave erosion, a protecting coat of rock, concrete or planking must be laid at the water line. The other exposed surfaces should be covered with grass or vegetation to protect the dam from rainfall erosion. Beas dam at Pong is a 126.5 m high earth core-gravel shell dam in earth dams, the base is quite large as compared to the height. Such dams are quite suitable for a pervious foundation because the wide base makes a long seepage path. The earth dams have got the following advantages.

(*a*) Suitable for relatively pervious foundation.

(*b*) Usually less costlier than a masonary dam.

(*c*) If protected from erosion, this type of dam is the most permanent type of construction.

(*d*) It fits best in natural surroundings.

The following are the disadvantages of earth dams :

(*a*) Greater seepage loss than other dams.

(*b*) The earth dam is not suitable for a spillway, therefore, a supplementary spillway is required.

(*c*) Danger of possible destruction or serious damage from erosion by water either seeping through it or overflowing the dam.

The masonary dams are of three major classes: solid gravity dam, buttress dam and the arched dam.

The buttress or deck dam has an inclined upstream face, so that water pressure creates a large downward force which provides stability against overturning or sliding. An arch dam is preferable where a narrow canyon width is available. It can be anchored well and the water pressure against the arch will be carried by less concrete than with a straight gravity type. This dam has the inherent stability against sliding. The most commonly used dams are shown in Fig. 11.4.

Dams must be able to pass the flood water to avoid damage to them. This may be achieved by : spillways, conduits piercing the dam and the tunnels by passing the dam.

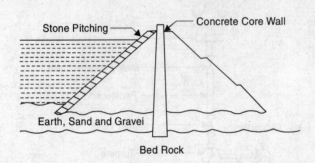

Fig. 11.4a

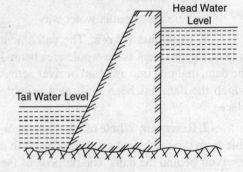

Fig. 11.4b

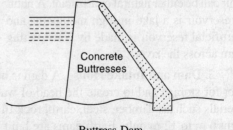

Buttress Dam
Fig. 11.4c

The intake includes the head works which are the structures at the intake of conduits, tunnels or flumes. These structures include booms, screens or trash racks, sluices for bypassing debris, and gates or valves for controlling the water flow. Booms prevent the ice and floating logs from going into the intake by diverting them to a bypass chute. Booms consist of logs tied end to end and form a floating chain. Screens or trash racks are fitted directly at the intake to prevent the debris from going into the intake. Debris cleaning devices should also be fitted on the trash racks. Gates and valves control the rate of water flow entering the intake.

The different types of gates are radial gates, sluice gates, wheeled gates, plain sliding gates, crest gates, rolling or drum gates etc. The various types of valves are rotary, spherical, butterfly or needle valves. A typical intake house is shown in Fig. 11.5. An air vent should be placed immediately below the gate and connected to the top of the penstock and taken to a level above the head water. When the head gates are closed and the water is drawn off through the turbines, air will enter into the penstock through the air vent and prevent the penstock vaccum which otherwise may cause collapsing of the pipe. A filler gate is also provided to balance the water pressure for opening the gate.

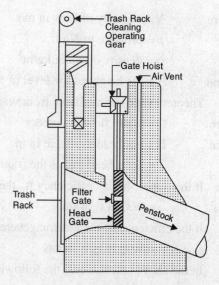

Fig. 11.5

4. Inlet Water Ways. Inlet water ways are the passages, through which the water is conveyed to the turbines from the dam. These may include tunnels, canals, flumes, forebays and penstocks and also surge tanks. A forebay is an enlarged passage for drawing the water from the reservoir or the river and giving it to the pipe lines or canals. Tunnels are of two types: pressure type and non-pressure type.

The pressure type enables the fall to be utilized for power production and these are usually lined with steel or concrete to prevent leakages and friction losses. The non-pressure type tunnel acts as a channel. The use of the surge tank is to avoid water hammer in the penstock. Water hammer is the sudden rise in pressure in the penstock due to the shutting off the water to the turbine. This sudden rise in pressure is rapidly destroyed by the rise of the water in the surge tank otherwise it may damage or burst the penstock.

5. Power House. The power house is a building in which the turbines, alternators and the auxiliary plant are housed.

6. Tail Race or Outlet Water Way. Tail race is a passage for discharging the water leaving the turbines, into the river and in certain cases, the water from the tail race can be pumped back into the original reservoir.

11.7. CALCULATIONS OF WATER POWER PLANTS

These calculations are concerned with the river or stream flow and the available head through which the water falls to generate the electric power. Water in motion possesses three forms of energy ; kinetic energy due to its velocity, pressure energy due to its pressure and potential energy due to its height.

$$\text{Kinetic energy} = \frac{V^2}{2}, \text{Nm per kg of water.}$$

$$\text{Pressure energy} = \frac{p}{\rho}, \text{Nm per kg of water.}$$

$$\text{Potential energy} = gH, \text{Nm per kg of water.}$$

where   V = velocity of flow in m/s,
          p = pressure in N/m^2,
          ρ = density of water kg/m^3
and   H = the height of the level of water above some datum level.

Theoretical power available from water = WQH, watts

where   Q = water flow in cumecs
and   H = net head available in m.
         = Total head minus the frictional losses.

If the turbine has an efficiency η_t, then the B.P. at turbine shaft
       = W.Q.H.η_t watts

It the efficiency of the electric generator is ηg, then the effective power at switch board
       = W.Q.H.$\eta_t.\eta_g$ watts ...(11.2)

In the above calculations, the following relations can be used to calculate the discharge.

1 cusecs = 1.3 sq.mile, ft. per year.

i.e. one foot of water over an area of 1.12 sq mile will give a discharge of one cusecs throughout the year, assuming the run off as 100%.

11.8. CLASSIFICATION OF HYDRO-PLANT

In hydro-plants, water is collected behind the dam. This reservoir of water may be classified as either storage or pondage according to the amount of water flow regulation they can exert. The function of the storage is to impound excess river flow during the rainy season to supplement the low rates of flow during dry seasons. They can meet the demand of load fluctuations for six months or even for a year. Pondage involves in storing water during low loads so that this water can be utilized for carrying the peak loads during the week. They can meet the hourly or weekly fluctuations of load demand. With poundage, the water level always fluctuates during operations It rises at the time of storing water, falls at the time 'off drawing water, remains constant when the load is constant.

The hydro-power plants can be classified as below:

1. Storage plant
   (*a*) High head plants
   (*b*) Low head plants
   (*c*) Medium head plants.
2. Run-of-river power plants
   (*a*) With pondage
   (*b*) Without pondage.
3. Pumped storage power Plants.

11.8.1 STORAGE PLANTS

These plants are usually base load plants. The hydro-plants cannot be classified directly on the basis of head alone as there is no clear line of demarcation between a high head and a medium head or between medium head and low head. The power plant can be classified on the basis of head roughly in the following manner:

(a) **High head plants.** About 100 m and above.

(b) **Medium head plants.** about 30 to 500 m.

(c) **Low head plants.** Upto about 50 m.

High Head Plants. Fig. 11.6 shows the elevation of a high head plant. The water is taken from the reservoir through tunnels which distribute the water to penstock through which the water is conveyed to the turbines. Alternately, the water from the reservoir can be taken to a smaller storage known as a forebay, by mans of tunnels. From the forebay, the water is then distributed to the penstocks. The function of the forebay is to distribute the water to penstocks leading to turbines. The inflow to the forebay is so regulated that the level in the forebay remains nearly constant. The turbines will thus be fed with under a constant static head. Thus, the forebays help to regulate the demand for water according to the load on the turbines. Trash racks are fitted at the inlets of the tunnels to prevent the foreign matter from going into the tunnels. In places; where it in not possible to construct forebays, vertical constructions known as `surge tanks' are built. The surge tanks are provided before the valve house and after the tunnel from the head works. The function of the surge tank is to prevent a sudden pressure rise in the penstock when the load on the turbines decreases and the inlet valves to the turbines are suddenly closed. In the valve house, the butterfly valves or the sluice type valves control the water flow in the penstocks and these valves are electrically driven. Gate valves are also there in the power house to control the water flow through the turbines. after flowing through tile turbines . The water is discharged to the tail race.

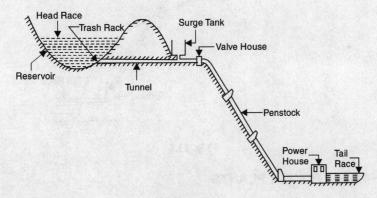

Fig. 11.6

Low Head Power Plants. These power plants are also known as Canal power plants. Such a plant is shown in Fig. 11.7.

A dam is built on the river and the water is diverted into a canal which conveys the water into a forebay from where the water is allowed to flow through turbines. After this, the water is again discharged into the river through a tail race. At the mouth of the canal, head gates are fitted to control the flow in the canal. Before the water enters the turbines from the forebay. It is made to flow through screens or trash-racks so that no suspended matter goes into the turbines. If there is any excess water due to increased flow in the river or due to decrease of load on the plant, it will flow over the top of the dam or a *waste weir can be constructed* along the forebay so that the excess water flows over it into the river. For periodic cleaning and repair of the canal and the forebay, a drain gate is provide on the side of the waste weir. The head gate is closed and the *drain gate* is opened so that whole of the water is drawn from the forebay and the canal for their cleaning and repair.

Medium Head Plants. If the head of water available is more than 50 m., then the water from the forebay is conveyed to the turbines through pen-stocks. Such a plant will then be named as a medium head plant. In these plants, the river water is usually tapped off to a forebay on one bank of the river as in the case of a low head plant. From the forebay, the water is then led to the turbines through penstocks. Such a layout is shown in Fig. 11.8.

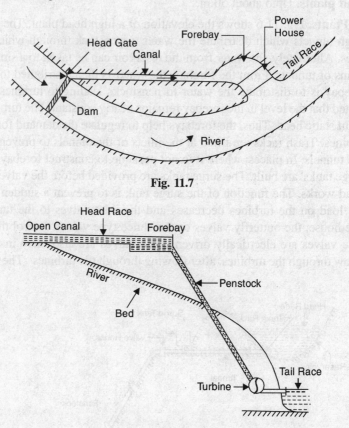

Fig. 11.7

Fig. 11.8

11.8.2. RUN-OF-RIVER POWER PLANTS

These plants can be classified as either without pondage or with pondage. A run-of-river plant without pondage has no control over river flow and uses the water as it comes. These plants usually supply peak load. During floods, the tail water level may become excessive rendering the plant inoperative. A run-of-river plant with pondage may supply base load or peak load power. At times of high water flow it may be base loaded and during dry seasons it may be peak loaded.

11.8.3. PUMPED STORAGE POWER PLANTS

These plants supply the peak load for the base load power plants and pump all or a portion of their own water supply. The usual construction would be a tail water pond and a head water pond connected through a penstock. The generating pumping plant is at the lower end. During off peak hours, some of the surplus electric energy being generated by the base load plant, is utilized to pump the water from tail water pond into the head water pond and this energy will be stored there. During times of peak load, this energy will be released by allowing the water to flow from the head water pond through the water turbine of the pumped storage plant. These plants can be used with hydro, steam and i.e. engine plants. This plant is nothing but a hydraulic accumulator system and is shown in Fig. 11.9. These plants can have either vertical shaft arrangement or horizontal shaft arrangement. In the older plants, there were separate motor driven pumps and turbine driven generators. The improvement was the pump and turbine on the same shaft with the electrical element acting as either generator or motor. The latest design is to use a Francis turbine which is just the reverse of centrifugal pump. When the water flows through it from the head water pond it will act as a turbine and rotate the generator. When rotated in the reverse direction by means of an electric motor, it will act as a pump to shunt the water from the tail water pond to the head water pond.

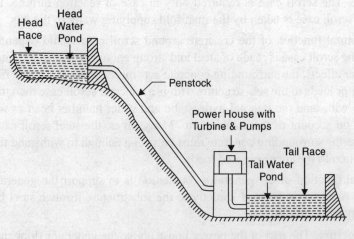

Fig. 11.9

The efficiency of such a plant is never 100 per cent. Some water may evaporate from the head water pond resulting in the reduction in the stored energy or there might be run off through the soil.

11.9 POWER HOUSE AND TURBINE SETTING

According to the location of the hydel power station, the power houses are classified as surface power house or underground power house. As the name implies, the underground power house is one which is built underground. A cavity is excavated inside earth surface where the sound rock is available to house the power station. A surface power house is one which is founded on earth's surface and its superstructure rests on the foundation.

The surface power house has been broadly divided into three subdivisions which is separated from the intake as mentioned below :

(*a*) Substructure ;   (*b*) Intermediate structure ;   (*c*) Super-structure.

(*a*) **Substructure.** The substructure of a power-house is defined as that part which extends from the top of the draft tube to the soil or rock. Its purpose is to house the passage for the water coming out of the turbine. In case of reaction turbines, the hydraulic function of the sub-structure is to provide a diverging passage (known as draft tube) where the velocity of the exit water is gradually reduced in order to reduce the loss in pushing out the water. In case of impulse turbine, such a draft tube is not required and only an exit gallery would serve the purpose.

The structural function of substructure is dual. The first function is to safely carry the superimposed loads of machines and other structures over the cavities. The second function is to act as transition foundation member which distributes heavy machine loads on the soil such that the obtainable ground pressures are within safe limits.

(*b*) **Intermediate structure.** The intermediate structure of a power house may be defined as that part of the power house which extends from the top of the draft tube to top of the generator foundation. This structure contains two important elements of the power house, one is the scroll case which feeds water to the turbine. The generator foundation rests on the scroll-case which is embedded in the concrete. The other galleries, adits and chambers also rest on the same foundation. Scroll or spiral case is a part of the turbine and it distributes water coming from penstock uniformly and smoothly through guide vanes to the turbine. The scroll case is required only in case of reaction turbine. In case of impulse turbine the place of scroll case is taken by the manifold supplying water to the jets.

(*c*) The structural function of the concrete around scroll case would depend upon the type of scroll case used. If the scroll case is made of steel and strong enough to withstand internal loads including the water hammer effects, the surrounding concrete acts more or less as a space fill and a medium to distribute the generator loads to the sub-structure. If it is a concrete scroll case then this concrete should be strong enough to withstand the internal hydrostatic and water hammer head as well as the external superimposed loads on account of the machine etc. Many times, the steel scroll case is used as water linear and in this case the surrounding concrete must be strong enough to withstand the internal hydraulic pressures in addition to the superimposed loads.

The structural function of the generator foundation is to support the generator. Arrangements may be made either to transmit the load 'directly to the substructure through steel barrel or through a column beam or slab arrangement.

(*c*) **Superstructure.** The part of the power house above the generator floor right upto the roof is known as superstructure. This part provides walls and roofs to power station and also provides an overhead travelling crane for handling heavy machine parts.

The arrangement of the power house is shown in Fig. 11.10.

Arrangement of Reaction and Impulse Turbines. Factors affecting the choice between horizontal and vertical setting of machines are : relative cost of plant, foundations, building space and layout of the plant in general.

Vertical machines offer many advantages over horizontal especially when there are great variations in tail-race level. Horizontal machines turbine-house should be above the tail-race level or the lower part of the house must be made watertight. In vertical machines, the weight of rotating parts acts in the same direction as axial hydraulic thrust. This requires a thrust bearing capable of carrying considerable heavy load. The efficiency of the vertical arrangement is 1 to 2% higher than for a similar horizontal arrangement. This is due to the absence of a suction bend near the runner. As the alternator being mounted above the turbine, it is completely free from flooding.

With the horizontal machines, there may be two turbines driving one generator and turbines would operate at a higher speed bringing about a smaller and lighter generator. The horizontal machines would occupy a greater length than the vertical but the foundations need not be so deep as required for vertical machines. The horizontal shaft machines require higher settings to reduce or to eliminate the cost of sealing the generator, the auxiliary electrical equipment and cable ducts against water.

HYDRO-ELECTRIC POWER PLANTS

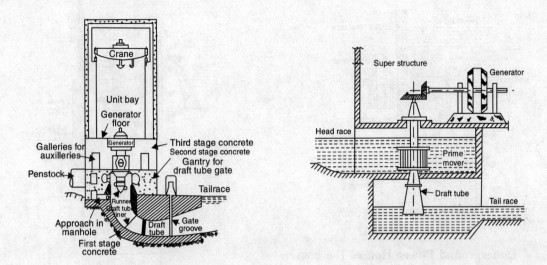

Fig. 11.10   Fig. 11.11

In actual cases, the arrangement of the machine (vertical or horizontal) is so chosen which will give the lowest cost of the station. The majority of impulse turbines are of the horizontal shaft types. The horizontal arrangement is simpler than vertical from constructional and maintenance point of view.

The overall height and width of the station will be relatively greater in case of vertical arrangement. The floor space occupied by horizontal shaft units is in general greater than that required for vertical shaft machines. Horizontal shaft arrangement is adopted in most cases, for Pelton wheels, mainly because this type of setting lends itself readily to the use of multiple runner units and secondly, because the resulting hydraulic conditions are not favourable with vertical machines.

There are mainly two principal types of setting as :

(1) open flume and
(2) cased turbines.

The open flume setting as shown in Fig. 11.12 and Fig. 11.13 (Rewalls power plant on black river at Watertown in U.S.A.) are chiefly used for low heads with concentrated falls or with a short canal. Open penstock setting is one where the entry to the runner has no casing but is placed in an open forebay. The runner should be placed at a convenient depth below the water surface such that eddies and suction of air through vertices will not take place. The turbine is completely submerged which results in a simple and comparatively cheap plant. The disadvantage of this arrangement is that the pit must be drained to enable inspection and maintenance to be carried out on the turbine and guide vane mechanism. The turbine should have an adequate water head above it, otherwise a sudden increase in load may draw the water to a dangerous level and allow air to enter. Such condition would break the vacuum in the draft tube and stop the turbine.

The cased turbines are further divided as concrete casing or steel plate casing as mentioned earlier. The width of the concrete flume should be kept as small as possible as design permits because the concrete approach flume often fixes the machine spacing. The concrete scrolls are limited to low head installations upto 20 metre heights. The complicated form work and reinforcement required for a concrete flume makes it expensive so that other methods of construction have to be used.

Steel plate scrolls are used for heads ranging from 10 m to 120 m. The arrangement of steel scroll is shown in Fig. 11.13.

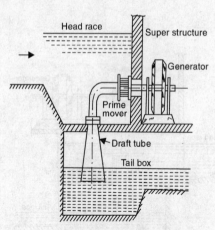

Fig. 11.12

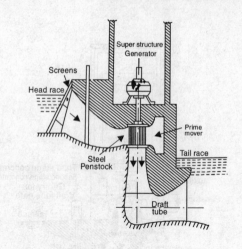

Fig. 11.13

Underground Power House. The conventional hydro-electric power stations are usually located over-ground at the foot of a dam or a hill slope on the banks of a river. The first underground power station *Nerayaz* was built in 1897 in Switzerland. The high capacity underground power plants were built only after second world war. The idea of locating powerhouse underground was suggested not only with the intention of protecting them against air raids but also technical and economical considerations were mainly considered. After second world war,

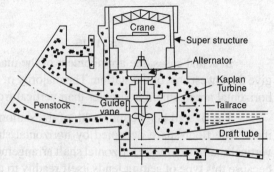

Fig. 11.14

the immunity against air attacks was unquestionably regarded as an important. advantage-of underground power station. A large number of underground power stations have been installed in U.K., U.S.A., Russia, Canada, Japan after second world war and recently in India also.

In all, there are about 300 such stations in service with a total installed capacity of 31 million kW Fig. 11.15. Horizontal setting of Pelton wheel with penstock upto the end of 1963.

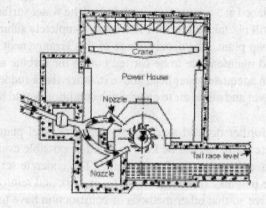

Fig. 11.15

The considerations supporting the construction of underground power stations are stated below :
1. Non-availability of a suitable site for a conventional surface station and good slope for penstock.
2. Danger of falling rocks and snow avalanches particularly in narrow valleys.
3. Availability of underground sound rock and avoidance of a long pressure tunnel and facility for a convenient tail-race outlet.
4. Possibility of elimination of surge tank required for surface station due to long pressure tunnel
5. The rugged topographical features and the difficulties in finding a suitable short and steep slope for pipe lines make it more economical to install the water conduit, the machine, transformer hall and tailrace system underground.
6. Foundation costs for overground power house become excessive in case of poor quality surface layers. The construction of draft tube, spiral case and separating floors in loose weathered rock is again more expensive than the excavation of corresponding parts underground. The costs of underground machine hall are lower than those of the superstructure of a surface powerhouse of similar dimensions.

11.9.1 ADVANTAGES AND DISADVANTAGES OF UNDERGROUND POWER-HOUSE

Advantages
1. Under suitable geological conditions, the underground conduit may prove the shortest and sometimes even straight. The power conduit may be much shorter than the length of power canal used for underground power house as the power canal usually built to follow the contours of the terrain. By locating the power house underground, the number of restrictions as safe topographical and geological conditions along the penstock and sufficient space at the foot of the hill for constructing the power house are completely eliminated.
2. The construction of underground conduit instead of penstock results in considerable saving in steel, the internal pressure is carried partly by the rock if it is of good quality. In sound high quality rock, the penstock is replaced by an inclined or vertical pressure shaft excavated in rock and provided with a steel lining of greatly reduced thickness in comparison with exposed penstock 'roe purpose' of lining in such cases is protection against the seepage losses.
3. The reduced length of the pressure conduit reduced the pressures developed due to waterhammer. Therefore, smaller surge tank is also sufficient.
4. For the economical arrangement, the ratio of the pressure conduit to the tail-race tunnel is also significant. The overall cost of the system is lower if the tail-race tunnel length is relatively large.
5. The construction work at underground power station can continue uninterrupted even under severest winter conditions. The overall construction cost and period of construction is reduced due to continuity of work.
6. Much care is devoted today in many countries to preserve landscape features such as picturesque rock walls, canyons, valleys and river banks in their original beauty against spoiling by exposed penstocks, canal basins and machine halls. There is less danger of disturbance to amenities with an underground power house and pipelines. The other advantages gained by constructing underground power house are listed below. The six advantages mentioned above reduce the constructional difficulties and overall cost of the plant and preserve the original

beauty of landscape. The overall cost is further reduced by the modern techniques in tunnel work and better excavation process.

7. The shorter power conduit of underground power house reduces the head losses.
8. The regular maintenance and repair costs are lower for underground stations as the maintenance required for rock tunnels is less.
9. The power plant is free from landslides, avalanches, heavy snow and rainfall.
10. The useful life of the structures excavated in rock is considerably longer than that of concrete and reinforced concrete structures.
11. It is possible to improve the governing of the turbines with the construction of underground power house.
12. The construction period is reduced mainly due to the possibility of full-scale construction work in winter.
13. Underground power station is bomb-proof and may be preferred for military reasons: They are perfectly protected against air-raids. The military considerations became more predominant with the increased shadow of the war and the building of underground power stations underwent a rapid evolution after second world war.

Disadvantages

1. The construction cost of the underground power house is more compared with the over ground power house :

   (a) The excavation of the caverns required for housing the turbine generator units and auxiliary equipments (machine hall of Koyna project is $00' × 120' × 60' in dimensions) is very expensive.

   (b) The costs of access tunnels are considerable.

   (c) The separate gallery excavated for the inlet valves adds the extra cost.

   (d) The construction of air ducts and bus galleries also adds in total construction costs.

   (e) Special ventilation and air-conditioning equipment required for underground adds in the constructional costs.

   (f) In some cases, the tailrace tunnel of an underground power house requires a more elaborate solution than a tailrace tunnel designed for the surface arrangement. The advantage gained by reducing the pressure conduit would be lost by extending the tailrace tunnel.

   (g) The first cost is also increased by locating the transformer and high-voltage switchgear underground. The above-mentioned constructions increase the capital cost of the plant.

2. The operational cost of the power plant increases due to following reasons :

   (a) The lighting cost.

   (b) The running cost of air-conditioned plant.

   (c) The removal of water seeping may be more costly than for the surface arrangement.

Adequate lighting, proper ventilation, maintenance of uniform climatic conditions within the power houses, provision of the necessary safety equipments against flooding, maintenance of proper acoustical conditions, augmenting the feeling of safety by providing a sufficient number of well placed exit; and finally artistic shaping and outfitting of machine hall increases the overall cost of the underground power house compared with ground surface power house.

The choice of the site for the power house either over ground or underground requires a considerable economical analysis according to the available topography and no thumb rule can be applied for its selection.

Types of underground power stations. There are mainly five different types of underground power stations as per hydraulic characteristics.

1. Free level tailrace tunnel without a downstream surge tank. In this arrangement, the long and steep tailrace tunnel is built to cope with the discharge without putting the tunnel under pressure, both under steady and unsteady flow conditions. This type is more suitable with Pelton-wheel because it does not interfere with the flow in tailrace tunnel. The arrangement is shown in Fig. 11.16.

Fig. 11.16

The underground Innertkirchen power house in Switzerland is the example of such Construction. The head is 672.3 meters, length of pressure tunnel is 10 kilometers and tailrace tunnel is 1294 metre with a slope of 4 : 1.

2. Downstream station arrangement or Alpine type. The arrangement is shown in Fig. 11.17. In this arrangement, the water is carried through a long horizontal pressure tunnel to the point of emergence to the surface, from where a step pressure shaft continues dawn to the power house as shown in figure. A surge tank is provided at the junction of pressure tunnel and pressure shaft as in the case of exposed penstock and surface power station. The valve chamber after the surge tank is also provided underground and the valves are also provided before the prime-mover.

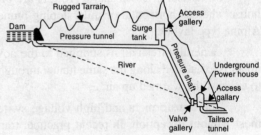

Fig. 11.17

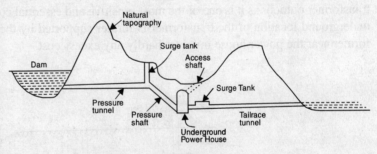

Fig. 11.18

These valves may be located either in the main cavern or a separate valves gallery is excavated for this purpose. Access to both surge tank and power house is provided through horizontal tunnels as shown in Fig. 11.17. The tailrace tunnel is considerably short in length. The arrangement is generally preferred in mountain regions. The downstream surge tank is also used if a tailrace tunnel is long and considerable surges are likely to occur in the tailwater.

3. Intermediate station arrangement. Fig. 11.18 shows the economic site of the power house at an intermediate section of the entire power conduit. The specific characteristic of this arrangement is long headrace tunnel and a long tailrace tunnel. Upstream and downstream surge tanks are necessary as shown in figure to deal with the pressure oscillations in both headrace and tailrace. If the prime mover is impulse type, there is no interference between tailrace and headrace level and, therefore, the dimensions of both surge tanks can be calculated independently according to the usual surge theory. If the prime mover is reaction type, the oscillations in tailrace and headrace interfere with each other and, therefore, the larger areas of surge tank is required than the volume required for surge tank used with impulsive type prime mover.

The Santa Giustina power house in Italy is an example of this type of arrangement. Koyna power project also comes under this class except there is no surge tank to the downstream side as the tailrace tunnel is not very long.

4. Station arrangement without surge tank. If the pressure tunnel and tailrace tunnel are short in length, the upstream and downstream surge tanks can be eliminated from the system. The main points to be considered in this design are the maximum water hammer effect and danger of cavitations in the turbine. The latter can be eliminated by selecting the proper turbine axis level above the draft tube axis level.

A vertical or steep shafts (10 : 1) are provided to access the upstream or Swedish type power house whereas horizontal or mild sloping tunnels prove more favourable. for downstream stations or Alpine type layout.

Generally fresh air as supplied to the power house through the main access tunnel and warmed up air is exhausted along the same tunnel through a separate duct. Sometimes separate air shafts are used to exhaust the warmed up air.

The transformers and high voltage switching equipment were previously located outdoors almost without exception. In recent practice, transformers are also arranged underground in the transformer room excavated in the vicinity of power house. If the decision of the management is in the favour of underground power house against air-raids, no significant gain could be achieved by locating the transformer outdoor as it is one of the most sensitive and essential components of the power project. An underground location of the transformer is further supported by the fact that the location of the transformer near the power house involves hardly any excess cost.

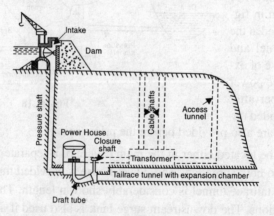

Fig. 11.19

11.10 PRIME-MOVERS

The prime-mover in the hydraulic power plant converts the energy of water into mechanical energy and further into electrical energy. These machines are classified on the basis of the action of water an moving blades.

As per the action of water on the prime-mover, they are classified as impulse turbine and reaction turbine. In impulse type turbine, the pressure energy of the water is converted into kinetic energy when passed through the nozzle and forms the high velocity jet of water. The formed water jet is used for driving the wheel.

In case of reaction turbine, the water pressure combined with the velocity works on the runner. The power in this turbine is developed from the combined action of pressure and velocity of water that completely fills the runner and water passage.

The casing of the impulse turbine operates at atmospheric: pressure whereas the casing of the reaction turbine operates under high pressure. The pressure acts on the rotor and vacuum underneath it. This is why the easing of reaction turbine is made completely leak proof.

The details of few turbines which are commonly used in hydro-electric power plants are given below.

Pelton Turbine. Figure 11.20 shows the layout of the Pelton turbine. This was discovered by Pelton in 1880. This is a special type of axial flow impulse turbine generally mounted on horizontal shaft, as mentioned earlier A number of buckets are mounted round the periphery of the wheel as shown in Fig. 11.20. The water is directed towards the wheel through a nozzle or nozzles. The flow of water through the nozzle is generally controlled by special regulating system. The water jet after impinging on the buckets is deflected through an angle of 160° and flows axially in both directions thus avoiding the axial thrust on the wheel. The hydraulic efficiency of Peltan wheel lies between 85 to 95%. Now-a-days, Pelton wheels are used for very high heads upto 2000 meters.

Arrangement of jets. In most of the Pelton wheel plants, single jet with horizontal shaft is used. The number of the jets adopted depends upon the specific speed required.

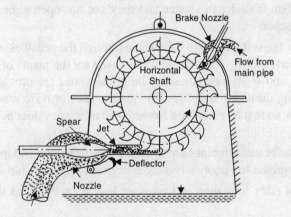

Fig. 11.20

Any impulse turbine achieves its maximum efficiency when the velocity of the bucket at the center line of the jet is slightly under half the jet velocity. Hence, for maximum speed of rotation, the mean diameter of the runner should be as small as possible. There is a limit to the size of the jet which can be applied to any impulse turbine runner without seriously reducing the efficiency. In early twenties, a normal ratio of D/d was about 10 : I. In a modern Turgo impulse turbine, it is reduced upto 4.5 to I. The basic advantage of Turgo impulse turbine is that a much larger jet could be applied to a runner of a given mean diameter. The jet of pelton turbine strikes the splitter edge of the bucket, bifurcates and is discharged at either side.

With the turgo impulse turbine, the jet is set at an angle to face the runner, strikes the buckets at the front arid discharges at opposite side. The basic difference between the two is shown in Fig. 11.21.

The Turgo impulse turbine bridges the gap of specific speed between the Pelton wheel and Francis turbine. Two turgo impulse turbines are used in a power house at Poonch which is 320 km from Jalnmu.

The reaction turbines are further divided into two general types as Francis and Propeller Type. The propeller turbines are further subdivided into fixed blade propeller type and the adjustable blade type as Kaplan Turbine.

Francis Turbine. In Francis turbine, the water enters into a casing with a relatively low velocity, passes through guide vanes located around the circumference and flows through the runner and finally discharges into a draft tube sealed below the tailwater level. The water passage from the headrace to tail race is completely filled with water which acts upon the whole circumference of the runner.

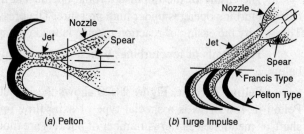

Fig. 11.21

A large part of the power is obtained from the difference in pressure acting on the front and back of the runner buckets, and only a part of total power is derived from the dynamic action of the water.

There are mainly two types of Francis turbines known as open flume type and closed type.

In open flume type, the turbine is immersed under water of the headrace in a concrete chamber and discharges into the tailrace through the draft tube. The main disadvantage of this type is that runner and guide-vane mechanism is under the water and they are not open either for inspection or repair without draining the chamber.

In the closed type, the water is led to the turbine through the penstock whose end is connected to the spiral casing of the turbine. The open flume type is used for the plants of 10 meters head whereas, closed type is preferred above 30 meters head. The guide vanes are provided around the runner to regulate the water flowing through the turbine The guide vanes provide gradually decreasing area of flow for all gate openings, so that no eddies are formed, and efficiency does not suffer much even at part load conditions.

The majority of the Francis turbines are inward radial flow type and most preferred for medium heads. The inward flow turbine has many advantages over outward flow turbine as listed below :

1. The chances of eddy formation and pressure loss are reduced as the area of flow becomes gradually convergent.

2. The runaway speed of the turbine is automatically checked as the centrifugal force acts outwards while the flow is inward.

3. The guide vanes can be located on the outer periphery of the runner, therefore, better regulation is possible.

4. The frictional losses are less as the water velocity over the vanes is reduced.

5. The inward flow turbine can be used for fairly high heads without increasing the speed of the turbine as centrifugal head supports considerable part of supply head.

A comparison of various types of reaction runners of the same power, but of different specific speed. The first three show the sections of Francis runners and the fourth one is a section of propeller runner. It is obvious from the figures that the flow through the runner changes from radially inward to nearly axial as the specific speed of the runner is increased. It is also obvious from the figure that the size of the runner decreases with an increase in specific speed for the same power.

Recent Development in Francis Turbines. The last decade has seen considerable developments in the design of Francis turbines, and the modem trend, is to go in for large sizes of machines with high speeds so as to economise in the cost of plant and civil work and at the same time improve the working characteristics efficiency of the Francis runner.

The largest Francis runner in operation until 1955 was of 147 mW capacity in Sweden. The recent move towards the higher capacities has resulted in sets of 580 mW (680,(I00 B .I3F.) capacity unit at Krasnoyarsk Power Station in Russia. This station has 10 such sets in operation under the head of 103 meters. The Canada Electricity Board has planned to manufacture 11 units of 485 mW capacity to be used at Churchill Fails plant. Me 660 mW capacity unit has been designed in U.S.A. for the Grand Coulee power station and these are the largest Francis turbines in the world so far developed. The water turbines of 650 mW capacity are reported to be under design in Russia, for the Sayano-Shush enkaya station on the river Yenisei in Siberia. It is also said that 800 to 1000 mW hydro sets are also being planned for huge hydro-power station coming up in Siberia.

The largest Francis turbine of 172 mW capacity in India at present is under manufacture for the Dehar project by Heavy Electrical Ltd., Bhopal. The manufacture of 200l2S0 mW capacity units which will be used in hydro projects planned in the Himalayas is also undertaken by the same company. Manufacture of high capacity units in India is largely limited by the lack of transport facilities, the small power grids and long transmission lines.

Propeller Turbine. The propeller runner may be considered as a development of a Francis type in which the number of blades is greatly reduced and the lower band omitted. It is axial flow turbine having a small number of blades from three to six as shown in Fig. 11.22. The propeller turbine may be fixed blade type or movable blades type known as Kaplan Turbine.

The fixed blade propeller type turbine has high efficiency (88°l0) ; at full load but its efficiency rapidly drops with decrease in load.

The efficiency of the unit is hardly 50% at 40% of full load at part load operation. The use of propeller turbine is limited to the installations where the units run at full load conditions at all times. The use of propeller turbine is further limited to low head installations of 5 to 10 meters.

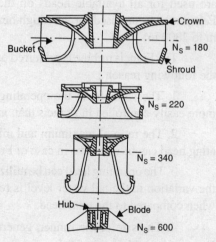

Fig. 11.22

Kaplan Turbine. Great strides are made in last few decades to improve the performance of propeller turbine at part load conditions. The Kaplan turbine is a propeller type having a movable blade instead of fixed one. This turbine was introduced by Dr. Vitkor Kaplan. This turbine has attained popularity and rapid progress has been made in recent years in the design and construction of this turbine:

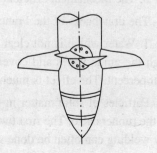

The rotor of the Kaplan turbine is shown in Fig. 11.23. The blades are rotated to the most efficient angle by a hydraulic servo-motor. A cam on the governor is used to change the blade angle with the gate position so that high efficiency is always obtained at almost any percentage of full load.

These turbines are constructed to run at speeds varying from 60 to 220 r.p.m. and to work under varying head from 2 to 60 meters. These are particularly suitable for variable heads and for variable flows and where the ample quantity of water is available.

The specific speed of Kaplan lies in the range of 400 to 1500 so that the speed of the rotor is much higher than that of Francis Turbine for the same output and head or Kaplan turbine having the same size as Francis develops more power under the same head and flow quantity.

The velocity of water flowing through Kaplan turbine is high as the flow is large and, therefore, the cavitations is more serious problem in Kaplan than Francis Turbine. The propeller type turbines have an outstanding advantage of higher speed which results in lower cost of runner, generator and smaller power house substructure and superstructure. The capital and maintenance cost of Kaplan turbine is much higher than fixed blade propeller type units operated at a point of maximum efficiency.

For a low head development with fairly constant head and requiring a number of units, it is always advisable to install fixed blade propeller type runners for most of them and Kaplan type for only one or two units. With this combination, the fixed blade units could be operated at point of maximum efficiency and Kaplan units could take the required variations in load. Such combination is particularly suitable to a large power system containing a multiplicity of the units.

Francis Versus Pelton. The Francis turbines are used for all available heads on the other hand. Pelton wheels are used for very high heads only (200 m to 2000 m).

The Francis turbine is preferred over Pelton for the following reason :

1. The variation in the operating head can be more easily controlled in Francis than in Pelton.

2. The ratio of maximum and minimum operating head can be even two in case of Francis turbine.

3. The operating head can be utilized even when the variation in the tail water level is relatively large when compared to the total head.

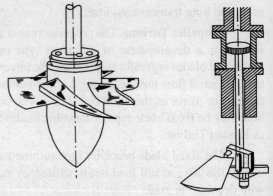

Fig. 11.24. Kaplan Turbine.

4. The size of the runner, generator and power house required is small and economical if the Francis is used instead of Pelton for the same power generation.

5. The mechanical efficiency of Pelton decreases faster with wear than Francis.

The drawbacks of the Francis compared with Pelton are listed below :

1. Water which is not clean can cause very rapid wear in high head Francis turbine. In passing through the guide vanes and cover facings, it can quickly reduce overall efficiency of the turbine by several percent. The effect is much more serious in turbines of small diameter than in large ones.

Particles of solid matter in the water will wear the lip of the spear, the nozzle and after several years the runners also. The first two are easily removable, renewable and repairable. The runner repairing by welding can often be done without removing the runner from the shaft or casing.

2. The inspection and overhaul of a Francis is much more difficult job than that of the equivalent Pelton turbine. The badly worn-out parts will have to be replaced by new ones and it will take a considerable time.

3. Cavitations is an ever-present danger in Francis as well as in all reaction turbines. The raising of power house floor level to reduce the danger of flooding may be followed by endless cavitations troubles.

4. Usually below 60% load, the Pelton is much better as it gives more efficiency than Francis of low specific speed. If there is possibility of running the prime-mover below 50% load for a long period, the Francis will not only lose its efficiency but the cavitation danger will become more serious.

5. The water hammer effect with the Francis is more troublesome than the Pelton turbine. Kaplan versus Francis Turbine.

The advantages of Kaplan over Francis are listed below :

1. It is more compact in construction and smaller in size for the same power developed.

2. Its part-load efficiency is considerably high. The efficiency curve remains more or less flat over the whole load range.

3. The frictional losses passing through the blades are considerably lower due to small number of blades used.

11.11. SPECIFIC SPEED OF TURBINE

The specific speed of a turbine is defined as the speed at which the turbine runs developing one B.H.P. under a head of one metre.

The equation for the specific speed of a turbine can be obtained by using the principle of similarity.

(1) $$V = \frac{\pi DN}{60} \propto \sqrt{H}$$

$$\therefore \quad D \propto \frac{\sqrt{H}}{N},$$ where D and N are diameter and speed of a turbine and H is the head acting on the turbine.

(2) $$Q = \pi DB \cdot V_f \propto D^2 \sqrt{H}$$

where B is the height of the blade and V_f is the velocity of flow.

Substituting the value of D in the above equation,

$$\therefore \quad Q \propto \frac{(H)^{3/2}}{N^2}$$

(3) $$P = \frac{\rho QH}{75}$$

where P is the power developed.

Substituting the value of Q in the above equation, we get

$$P \propto \frac{\rho}{75}\left[\frac{(H)^{3/2}}{N^2}\right] H \propto \frac{(H)^{5/2}}{N^2}$$

$$\therefore \quad N^2 \propto \frac{(H)^{5/2}}{P}$$

$$\therefore \quad N \propto \frac{(H)^{5/4}}{\sqrt{P}} = C\frac{(H)^{5/4}}{\sqrt{P}}$$

where C is knour as constant depending upon the type of the turbine.

If the turbine develops 1 B.H.P. under one metre head then

$$C = N = N_s$$

where N_s is the specific speed as per the definition.

Substituting the value of C in the above equation, we get

$$N_s = \frac{N\sqrt{P}}{(H)^{5/4}} \text{ when } P \text{ is in H.P.}$$

$$= \frac{1.165\, N\, \sqrt{KW}}{(H)^{5/4}} \text{ when the power is in kW.} \qquad \ldots(1)$$

By definition, the specific speed is number of revolutions per minute at which a given runner would revolve if it were so reduced in proportions that it would develop one H.P. under one metre-head.

Sometimes the power developed is given in kilowatts instead of metric H.P., the head being in metre as before.

The specific speed of a single jet Pelton wheel in terms of diameter of runner and diameter of jet in metric units is given by

$$N_s \text{ (single jet petrol)} = 244.75\, \frac{d}{D} \qquad \ldots(2)$$

In a multi-jet pelton wheel, the H.P. is directly proportional to the number of jet if the head remains constant. The specific speed of multi jet Pelton wheel is given by

$$N_s \propto \sqrt{n} \text{ as } N_s \propto \sqrt{P} \text{ and } P \propto n.$$

Therefore, the specific speed of multi jet unit can be calculated by multiplying the specific speed of single jet unit with a factor $\sqrt{n}$ where n is number of jets used.

It is necessary to know a characteristic of an imaginary machine identical in shape for comparing the characteristics of machines of different types. The imaginary turbine is called a specific turbine. The specific speed provides a means of comparing the speed of all types of hydraulic turbines on the same basis of head and horse power capacity.

The overall cost of installation (runner + generator + power house and auxiliary equipments) is lower if a runner of high specific speed is used for a given head and H.P. output. The selection of too high specific speed reaction runner would reduce the size of the runner to such an extent that the discharge velocity of water into the throat of draft tube would be excessive. This is objectionable because a vacuum may be created in the extreme case.

The runner of too high specific speed with high available head increases the cost of turbine on account of high mechanical strength required. The runner of two low specific speed with low available head increases the cost of generator due to the low turbine speed.

An increase in specific speed of turbine is accompanied by lower maximum efficiency and greater depth of excavation of the draft tube. In choosing a high specific speed turbine, an increase in cost of excavation of foundation and draft tube should be considered in addition to the efficiency. The weighted efficiency over the operating range of the turbine is more important in the selection of a turbine instead of maximum efficiency.

Experience has determined that there is a range of heads and specific speeds for each type of turbine. Special conditions may sometimes dictate departure from common practice.

The ranges of heads and specific speeds for different types of turbines are tabulated in Tables 11.3 and 14.4.

Table 11.3

Type of Turbine	Range of Head	Specific speed in metric units
Pelton (1 nozzle)	200 metres	10—20
Pelton (2 nozzles)	to	20—40
Pelton (4 nozzles)	2000 metres	40—50
Turgo impulse turbine	as above	50—100
Francis (low speed)	15 metres	80—120
Francis (medium speed)	to	120—220
Fracis (high speed)	300 metres	220—350
Francis (express)		350—420
Propeller	5 metres to 30 metres	310—1000

Table 11.4

Type of Turbine		N_s in MKS	N_s in SI	N_s in FPS
Axial flow (Kaplan)	Slow	300 – 450	14.8 – 22.2	67.5 – 101.2
	Normal	450 – 700	22.2 – 34.6	101.2 – 157.4
	Fast	700 – 1200	34.6 – 59.3	157.4 – 270.0
Radial and Mixed flow (Francis and Deriaz)	Slow	60 – 150	2.9 – 7.4	5.5 – 33.7
	Normal	150 – 250	7.3 – 12.4	33.7 – 56.2
	Fast	250 – 400	12.4 – 19.8	56.2 – 90.0
Impulse (Pelton)	Slow	4 – 10	0.2 – 0.5	0.9 – 2.3
	Normal	10 – 25	0.5 – 1.2	2.3 – 5.6
	Fast	25 – 60	1.2 – 3.0	5.6 – 13.5

11.12 DRAFT TUBES

Reaction turbines must be completely enclosed because a pressure difference exists between the working fluid (water) in the turbine and atmosphere. Therefore, it is necessary to connect the turbine outlet by means of a pipe known as draft tube upto tailrace level.

1. Output of reaction turbine when the tailrace level is above the turbine (submerged turbine.) The position of the turbine is shown in Fig. 11.25 and energies at all points are measured taking x-y as reference line, considering the energies of unit mass of water at all points, we can write

$$E_a = E_b = H_c + \frac{p_a}{\rho}$$

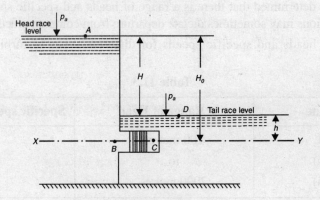

Fig. 11.25

W_1 (Work done per kg of water passing through the turbine)

$$= E_b - E_c = \left(H_o + \frac{p_a}{\rho}\right) - \left(\frac{p_c}{\rho} + \frac{V_c^2}{2g}\right) = \left(H_o + \frac{p_a}{\rho}\right) - \left(\frac{p_a}{\rho} + h + \frac{V_c^2}{2g}\right)$$

as $\quad \dfrac{p_c}{\rho} = \dfrac{p_a}{\rho} + h \quad$ for pressure equilibrium

$$\therefore \quad W_1 = H_o - h - \frac{V_c^2}{2g} = H - \frac{V_c^2}{2g} \quad \ldots(a)$$

where H is the net head between headrace and tailrace level and V_c is the velocity of water leaving the turbine.

2. Output of reaction turbine with draft tube. The arrangement of the turbine with draft tube is shown in fig. 11.26 and energies at all points are measured taking x-y as reference line.

HYDRO-ELECTRIC POWER PLANTS

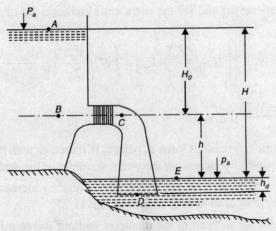

Fig. 11.26

$$E_a = E_b = H + h_d + \frac{p_a}{\rho}$$

$$E_c = h + h_d + \frac{V_c^2}{2g} + \frac{p_c}{\rho}$$

$$E_d = \frac{V_d^2}{2g} + \frac{p_d}{\rho}.$$

W_Z (work done per kg of water passing through the turbine) $= E_b - E_c = E_b - (E_d + h_f)$

where $E_c = E_d + h_f$ where h is the head lost by water passing through the draft tube (friction and other losses).

$$= \left(H + h_d + \frac{p_a}{\rho}\right) - \left(\frac{V_d^2}{2g} + \frac{p_d}{\rho} + h_f\right)$$

$$= \left(H - \frac{V_d^2}{2g}\right) + \left(h_d + \frac{p_a}{\rho} - \frac{p_d}{\rho}\right) - h_f$$

The pressure at the point D and E must be same.

$$\therefore \quad \frac{p_d}{\rho} = \frac{p_a}{\rho} + h_d$$

Substituting this value in the above equation, we get

$$W_2 = \left(H - \frac{V_d^2}{2g}\right) - h_f$$

$$= H - \frac{V_d^2}{2g} \quad \text{in } h_f \text{ is taken as zero} \qquad \ldots(b)$$

Comparing the equations (*a*) and (*b*) the extra work done per kg of water due to draft tube is given by

$$\Delta W = W_2 - W_1 = \left[\left(H - \frac{V_d^2}{2g}\right) - h_f\right] - \left(H - \frac{V_d^2}{2g}\right) = \frac{V_c^2 - V_d^2}{2g} - h_f$$

$$= \frac{V_c^2 - V_d^2}{g} \text{ if } h_f = 0. \qquad \ldots(c)$$

The head on the turbine (*H*) remains same as before, *W* increases with the decrease in velocity V_d. The velocity V_a can be decreased by increasing the outlet diameter of the draft tube.

The outlet diameter of the draft tube can be increased either by increasing the height of the draft tube or by increasing the angle of draft tube as shown in Fig. 11.27.

The increase in height for increasing the diameter without increase in angle is limited by the pressure at the outlet of the runner (at point *C*). This will be discussed later in detail.

An increase in draft tube angle (2*a*) for increasing the diameter without increase in height is limited by the losses in the draft tube.

The flow in the draft tube is from low pressure region to high pressure region. In such flow, there is a danger of water particles separating out from main stream and trying to flow back resulting in formation of eddies which are carried away in main stream causing losses. The maximum value of a is limited to 4. The gain in *work* by increasing an angle a above 4 will be lost in eddy losses and separated flow.

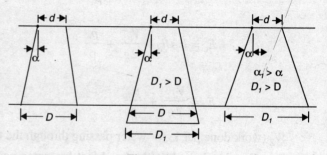

Fig. 11.27

Sometimes in order to decrease the length of draft tube, the diverging angle has to be made more than 4° and under such cases to reduce the losses due to separation, the air is sucked from the inside surface of the draft tube.

Prof. Ackeret has shown that the efficiency of draft tube was raised from 50 to 80% by air sucking process. However, water equal to 5°Ia of the total quantity is also withdrawn with the air. The work done by the draft tube is further increased by decreasing h_f. This is generally done by proper lining the draft tube and by proper designing the shape and size of the draft tube.

The efficiency of the draft tube is given by

$$\eta = \frac{\Delta W}{V_c^2/2g} = \frac{V_c^2 - V_d^2}{V_c^2} = \left[1 - \left(\frac{V_d}{V_c}\right)^2\right]$$

The chief advantages of using draft tube are listed below :

(1) It allows the turbine to be set above the tailrace water level where it is more accessible and yet does not cause any sacrifice in the head of turbine. It also prevents the flooding of generator and other equipment during flood period when the tailrace, water height goes up.

(2) It converts part of the velocity energy of the water leaving the turbine into the pressure energy and increases the overall efficiency of the plant.

HYDRO-ELECTRIC POWER PLANTS

Cavitation and Limitation of Turbine Height above Tailrace Level. The formation of water vapour and air bubbles on the water surface due to the reduction of pressure is known as "Cavitation". When the pressure on the water reduces below the saturation pressure corresponding to the temperature of the water, the rapid formation of water vapour and air bubbles starts. The bubbles suddenly collapse with the violent action and collapsing pressure will be very high. The rapid formation and collapsing of the bubbles causes the pitting of the metallic surface. It also reduces the efficiency of the hydraulic prime mover causing honeycombing of runner and blade contours which reduces the power output.

Referring to Fig. 11.27, we can write

$$E_c = E_d$$

$$\left[\frac{V_c^2}{2g} + h + h_d + \frac{p_c}{\rho}\right] - h_f = \left[\frac{V_d^2}{2g} + \frac{p_d}{\rho}\right]$$

$$\therefore \quad \frac{V_c^2}{2g} + h + h_d + \frac{p_c}{\rho} - h_f = \frac{V_a^2}{2g} + \frac{p_a}{\rho} + h_d$$

as $\dfrac{p_d}{\rho} = \dfrac{p_a}{\rho} + h_d$ for pressure equilibrium

$$\therefore \quad \frac{V_c^2}{2g} + h + \frac{p_c}{\rho} - h_f = \frac{V_d^2}{2g} + \frac{p_a}{\rho}$$

$$\therefore \quad \frac{p_c}{\rho} = \frac{p_a}{\rho} - \left(h + \left(\frac{V_c^2 - V_d^2}{2g}\right) - h_f\right)$$

$$h = \left(\frac{p_a - p_c}{\rho}\right) - \left(\frac{V_c^2 - V_d^2}{2g}\right) + h_f$$

$$= \left(\frac{p_a - p_c}{\rho}\right) - \frac{V_c^2}{2g} + \left(\frac{V_d^2}{2g} + h_f\right)$$

The equation shows that the pressure at point c (at exit of the turbine) is below atmospheric pressure. The pressure p; should not be below the cavitation pressure which is the saturation pressure of water at the water temperature to avoid the cavitation in turbine.

An increase in height of the draft tube also increases the height of the turbine (h) above tailrace level and reduces the pressure p, and increases the danger of cavitation. The height of the turbine above tailrace level to avoid the flooding of superstructure is also controlled by the occurrence of cavitation danger.

Cavitation Factor. Prof. D. Thowa (Germany) suggested a cavitation factor to determine the zone where the turbine can work without any danger of cavitation.

The critical value of cavitation factor is given by

$$\sigma_c = \frac{[(H_a - H_v) - h]}{H}$$

where
H_a = Atmospheric pressure head in meter of water
H_v = Vapour pressure in metre of water corresponding to the water temperature
H = Working head of turbine (difference between headrace and tailrace level in meters)
h = Height of turbine outlet above tailrace level in meters.

The values of H_a with respect to the altitudes above sea level and the values of H_v with respect to the water temperatures are tabulated in the tables (11.5) and (11.6). The values of critical factor depend upon the specific speed of the turbine. The critical values of cavitation factors with respect to specific speed are tabulated in tables.

Table 11.5. Altitude V_s Atmospheric Pressure

Altitude above sea level in metres		0	1000	2000	3000	4000
Baromatric Pressure	mm of Hg	760	676	595	528	463
	Metres of water	10.35	9.2	8.1	7.2	6.3

Table 11.6. Saturation Pressure as Function of Temperature

Temp °C	Pressure in kg/cm²	Pressure in mm of Hg	Pressure in metres of water
0	0.0012	4.60	0.062
2	0.00716	5.30	0.072
4	0.00829	6.30	0.083
6	0.00953	7.04	0.095
8	0.01093	8.05	0.1095
10	0.01251	9.23	0.1251
12	0.01429	10.50	0.1430
14	0.01629	12.05	0.1630
16	0.01853	13.70	0.1855
18	0.02103	15.50	0.2105
20	0.12383	17.60	0.2385
22	0.02700	19.95	0.2700
24	0.00040	22.40	0.3040
26	0.03430	25.30	0.3430
28	0.03850	28.40	0.3850
30	0.04330	32.00	0.4330
32	0.04850	35.80	0.4850
34	0.05420	40.00	0.5420
36	0.06050	44.70	0.6060
38	0.06760	50.70	0.6760
40	0.07520	55.50	0.7520
42	0.08360	61.70	0.8360
44	0.09280	68.50	0.9280
46	0.10280	76.00	1.0280
48	0.11380	84.00	1.1310
50	0.12580	93.00	1.2580

Table 11.7. ($N_s V_s \sigma_s$)

Francis		Kaplan	
N_s	s_c	N_s	σ_c
50	0.04	300 to 450	0.35 to 0.40
100	0.05	450 to 550	0.40 to 0.45
150	0.07	550 to 600	0.46 to 0.60
200	0.1	650 to 700	0.85
250	0.14	700 to 800	1.05
300	0.2	—	—
350	0.27	—	—

11.12.1 METHODS TO AVOID CAVITATION

1. Installation of Turbine below Tailrace Level. The danger of cavitation increases in case of low head and high speed propeller runner as the value of $(V_0^2 - V_d^2)/g$ is considerably large as mentioned earlier. In order to keep the value of p_c within the cavitation limit, the value of h is made negative keeping the runner below tailrace level. For such installations, the turbines remain always under water. It is **riot** advisable as the inspection and repair of the turbine is difficult. The other method to avoid cavitation zone without keeping the runner under water is to use the runner of low specific speed as mentioned earlier.

2. Cavitation Free Runner. The cavitation free runner can be designed to fulfill the given conditions with extensive research. The shape of the blade, the angle of the blade, the thickness of the blade can be changed and experiments can be conducted to find out the best dimensions of the blade (shape, size, angle. etc.

3. Use of Material. The cavitation effect can be reduced by selecting materials which can resist better the cavitation effect. The cast steel is better than cast iron and stainless steel or alloy steel is still better than cast steel. The pitting effect of cavitation on cast steel can be repaired more economically by ordinary welding. It has been observed that the welded parts are more resistant to cavitation than ordinary ones.

4. Polishing. The cavitation effect is less on polished surfaces than ordinary one. Mat, is why the cast steel runners and blades are coated with stainless steel.

5. Selection of Specific Speed. By selecting a runner of proper specific speed for the *given head* from equation (c) and from Tables (11.5) and (11.6), it is possible to avoid the cavitation.

11.12.2 Types of Draft Tubes

(1) Conical Draft Tube. This is known as tapered draft tube and used in all reaction turbines where conditions permit. It is preferred for low specific speed and vertical shaft Francis turbine. The maximum cone angle of this draft tube is limited to 8° ($a = 4°$) for the cause mentioned earlier. The hydraulic efficiency of such type of draft tube is 90%.

In any event, the draft tube should be made as to secure a gradual reduction of velocity (uniform decease towards the exit of draft tube) from the runner to the mouth. A form that is theoretically good is "Trumpet Shaped".

(2) Elbow Type Draft Tube. The elbow type draft tube is often preferred in most of the power plants, where the setting of vertical draft tube does not permit enough room without excessive cost of excavation. This offers an advantage in the cost of excavation ; specially in the rock.

If the tube is large in diameter ; it may be necessary to make the horizontal portion of some other section than circular in order that the vertical dimension may not be too great. A common form of section used is oval or rectangle.

If some other section instead of circular is used, then the tube is so made that the area increases at a similar rate to what it would if it were circular. The other forms are generally used when the draft tubes are moulded in concrete.

If the cross-section is gradually changed from circular at wheel discharge to rectangular or elliptical at the mouth of the tube, there is considerable saving in the excavation cost. Such types of draft tubes give better hydraulic efficiency than conical elbow type. The efficiency of this tube lies between 60 to 70%.

The horizontal portion of the elbow type draft tube is generally inclined upward to lead the water gradually to the tailrace level and prevent the entry of air from the exit end.

(3) Hydracone or Moody Draft Tube. This type of draft tube is also known as bell-mouth draft tube. The whirl component of the water is large when the turbine works at part load conditions. Turbine runs at high speed under low head conditions.

The high speed runner under low head has high whirling component. As water enters the draft tube from the high speed runner, it whirls in the direction in which the runner rotates. The axis of rotation is the vertical axis of the draft tube. The vortex tends to remain in the same plane because of its gyroscopic properties and therefore, will not follow the center line of the tube. This causes eddies and whirls in the curved portion of the draft tube and only a portion of the discharged area at the exit may be effective. Actually water *flows* back into the part of the draft tube and causes serious eddies and losses.

The central cone arrangement of hydracone draft tube reduces the whirl action of the discharged water and increases the efficiency of this draft tube to 85%.

Two different hydracone type draft tubes are shown in Fig. 11.28(*e*) and Fig.11.28(*f*). The *flow* coming out of the high speed runner is whirling flow and can be considered as free vortex flow. The pressure is minimum at the center of the free vortex, therefore, the cavitation may start by the liberation of air and vaporization of water at the center of the draft tube immediately underneath the runner if the velocity of whirl is large. To prevent the cavitation of runner under above-mentioned condition, the high speed runners are used in low head plants, and the central cone of draft tube has been extended right upto the tube as far as possible as shown in Fig. 11.28 (*g*). The tube shown in Fig. 11.28 (*e*) is used for low speed turbines where the whirl velocity is not high.

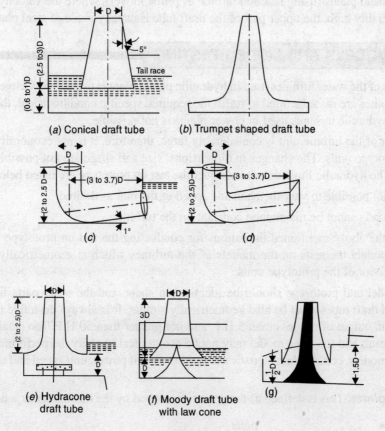

Fig. 11.28

11.12.3 DIFFERENT TYPES OF DRAFT TUBES

In another type of draft tube, a circular plate concentric with draft tube is put into the elbow section as shown in Fig. 11.28 (e) to improve the flow and efficiency of draft tube and for structural reasons in large units. The draft tube plays more efficient role with the use of high specific speed runners like Express Francis and Kaplan. Because the value of velocity energy at the runner exit increases with an increase in specific speed and it will be nearly 45% in case of Kaplan turbine. Greater care is taken in the design and construction of the draft tube when used with a runner of high specific speed.

The different types of draft tubes described are shown in Fig. 11.28 (a) to (g).

Draft Tube Liners. The pitting of the turbine runner and in the upper part of draft tube has occurred in almost all types of runners and kinds of metals used. It is even reported that the runner blades of 5 cm thick have been eaten in less than two years. In avoiding the pitting, the draft head is of greatest importance. A poorly designed runner may be set with a low height above tailrace level and show no pitting, while a properly designed runner may be set high enough above tailrace level but pitting is unavoidable.

It is always desirable that at least the upper portion of the draft tube should be lined with either a cast or plate steel liner to avoid the pitting on the surface of the tube. It has been reported that the unlined concrete draft tube has been eaten by pitting to a depth of 30 cm to 50 cm just below the runner.

In high head plants (using reaction turbine as prime mover), where the velocity in the draft tube may be considerably high, the upper part of the draft tube is usually made of steel plate.

11.13 MODELS AND MODEL TESTING

The size of the water turbines used in hydraulic power plant is usually very large (150,000 H.P. or more). The turbines are manufactured to fulfill the required specific conditions and, therefore, the mass production of hydraulic turbines used in power plants is not possible.

The cost of the turbine unit is considerably large, therefore, it is not economical to conduct the test on the prototype units. The changes in dimensions, size and shape are not possible once the prototype is ready. The hydraulic limitations to conduct the test on prototype are listed below:

1. It is not possible to vary the head and speed of the unit as desired
2. The load cannot be maintained constant on the turbine.

Due to the above-mentioned limitations for conducting the test on prototype unit, it is always necessary to conduct the tests on the models of the turbines which is geometrically similar and can predict the behavior of the prototype units.

The model and prototype should be identical in shape and the other parts like casing, guide mechanism and draft tubes must be also geometrically similar. It is always desirable to make a turbine model having an output of not less than 5 H.P. and not greater than 50 H.P. Too small model may not give accurate result and too large model may not be economical to study the performance of the prototype using the models, certain characteristics of turbine as unit power, unit speed and unit quantity must be known.

1. Unit power. This is defined as the power developed by the turbine under a head of one meter.

$$P = \frac{\rho Q H}{75}$$

But $\quad Q = AV$

where A is area through which water flows and V is the velocity of water.

$$\therefore \quad P = \frac{\rho A V H}{75} = \frac{\rho A H}{75}\sqrt{2gH}$$

as $\quad V = \sqrt{2gH}$

$$\therefore \quad P \propto H^{3/2}$$

or $\quad P = K_1 H^{3/2}$

where K_1 is the coefficient which varies with the speed and gate opening

when $\quad H = 1, P = K_1 = P_u \quad$ (unit power by its definition

$$\therefore \quad P = P_u H^{3/2}$$

$$\therefore \quad P_u = P/H^{3/2}$$

2. Unit speed. This is defined as the speed of the turbine under a head of 1 metre

$$V = \frac{\pi D N}{60}$$

$$N = \frac{60}{\pi D} \cdot V = \frac{60}{\pi D}\sqrt{2gH}$$

$\therefore \qquad N \propto \sqrt{H}$

$\therefore \qquad N = K_2\sqrt{H}$

where K_2 is the coefficient which varies with the conditions of running.

If $H = 1$, then $N = K_2 = N_u$ (unit speed by its definition)

$\therefore \qquad N = N_u\sqrt{H}$

$\therefore \qquad N_u = \dfrac{N}{\sqrt{H}}$

3. Unit quantity. This is defined as the volume of water passing through the turbine under a head of 1 metre.

$$Q = AV$$

$$Q \propto \sqrt{H} \quad \text{as } V = \sqrt{2gH}$$

and A is constant for given turbine

$\therefore \qquad Q = K_3\sqrt{H}$

If $H = 1$, then $Q = K_3 = Q_u$ (unit quantity by its definition)

$\therefore \qquad Q = Q_u\sqrt{H}$

$\therefore \qquad Q_u = \dfrac{Q}{\sqrt{H}}$

If the question of reducing the performance of a turbine under head H to its performance under any other head H, is required, then we can use the following equations.

$$\frac{P_0}{P} = \left(\frac{H_0}{H}\right)^{3/2}$$

$$\frac{N_0}{N} = \sqrt{\frac{H_0}{H}}$$

and
$$\frac{Q_0}{Q} = \sqrt{\frac{H_0}{H}}$$

The principle of similarity is applied to the turbines in order to predict the performance of actual prime movers from the tests on the model.

The vane angle at inlet and outlet will be same for model and prototype. The velocity triangles will also be identical for model and prototype when they are running under certain conditions.

The velocities are proportional to H for all similar turbines and hence :

(a) *Speed* $\qquad v = \dfrac{\pi d n}{60} \propto \sqrt{h}$ for model.

and
$$V = \frac{\pi DN}{60} \propto \sqrt{H} \text{ for prototype.}$$

∴ From the above two equations, we can write

$$\frac{DN}{dn} = \sqrt{\frac{H}{h}}$$

or
$$\frac{N}{n} = \frac{d}{D}\sqrt{\frac{H}{h}}$$

(b) *Quantity* $\quad q = \pi db\, v_f \propto d^2 \sqrt{h}$ as $b \propto d$ and $v_f \propto \sqrt{h}$

and $\quad Q = \pi DBV_f \propto D^2 \sqrt{H}$

∴
$$\frac{Q}{q} = \left(\frac{D}{d}\right)^2 \sqrt{\frac{H}{h}} = \left(\frac{D}{d}\right)^2 \cdot \frac{DN}{dn} = \left(\frac{D}{d}\right)^3 \cdot \frac{N}{n}$$

(c) *Power* $\quad p = \dfrac{\rho_m qh}{75} \cdot \eta_m \propto \rho_m \cdot d^2 \sqrt{h} \cdot h \cdot \eta_m$

and $\quad p = \dfrac{\rho_p QH}{75} \cdot \eta_p \propto \rho_p \cdot D^2 \sqrt{H} \cdot H \cdot \eta_p$

∴
$$\frac{p}{p} = \frac{\rho_p}{\rho_m} \cdot \left(\frac{D}{d}\right)^2 \cdot \left(\frac{H}{h}\right)^{3/2} \cdot \frac{\eta_p}{\eta_m}$$

$$= \frac{\rho_p \eta_p}{\rho_m \eta_m} \cdot \left(\frac{D}{d}\right)^2 \cdot \left(\frac{H}{h}\right)^{3/2} = \frac{\rho_p \eta_p}{\rho_m \eta_m} \cdot \left(\frac{D}{d}\right)^2 \cdot \left(\frac{DN}{dn}\right)^3$$

$$= \frac{\rho_p \eta_p}{\rho_m \eta_m} \cdot \left(\frac{D}{d}\right)^5 \left(\frac{N}{n}\right)^3$$

If $\quad \rho_p = \rho_m$

∴
$$\frac{P}{p} = \frac{\eta_p}{\eta_m} \cdot \left(\frac{D}{d}\right)^5 \left(\frac{N}{n}\right)^3$$

If $\quad \eta_p = \eta_m$ which is not the general case

∴
$$\frac{P}{p} = \left(\frac{D}{d}\right)^2 \cdot \left(\frac{H}{h}\right)^{3/2} = \left(\frac{D}{d}\right)^5 \left(\frac{N}{n}\right)^3$$

(d) The specific speed for model and prototype should also be same

∴ $\quad N_s = n_s$

$$\therefore \quad \frac{N\sqrt{P}}{(H)^{5/4}} = \frac{n\sqrt{P}}{(h)^{5/4}}$$

The capital notations are used for prototype turbine whereas the small notations are used for model. The above five equations are generally used for deciding the quantities required for model or the quantities for prototype if the test data of the model is available.

11.14. SELECTION OF TURBINE

The major problem confronting the engineering is to select the type of turbine which will give maximum economy. The hydraulic prime-mover is always selected to match the specific conditions under which it has to operate and attain maximum possible efficiency.

The choice of a suitable hydraulic prime-mover depends upon various considerations for the given head and discharge at a particular site of the power plant. The type of the turbine can be determined if the head available, power to be developed and speed at which it has to run are known to the engineer beforehand.

The following factors have the bearing on the selection of the right type of hydraulic turbine which will be discussed separately.

(1) Rotational Speed.
(2) Specific Speed.
(3) Maximum Efficiency.
(4) Part Load Efficiency.
(5) Head.
(6) Type of Water.
(7) Runaway Speed.
(8) Cavitation.
(9) Number of Units.
(10) Overall Cost.

1. Rotational speed. In all modern hydraulic power plants, the turbines are directly coupled to the generator to reduce the transmission losses. This arrangement of coupling narrows down the range of the speed to be used for the prime-mover. The generator generates the power at constant voltage and frequency and, therefore, the generator has to operate at its synchronous speed. The synchronous speed of a generator is given by

$$N_{sysn} = \frac{(60 \times f)}{p}$$

where f = Frequency and p = Number of pairs of poles used. For the direct coupled turbines, the turbine has to run at synchronous speed only. There is less flexibility in the value of N_{sysn} as f is more or less fixed (50 or 60 cycles/sec). It is always preferable to use high synchronous speed for generator because the number of the poles required would be reduced with an increase in N_{sysn} and the generator size gets reduced. Therefore, the value of the specific speed adopted for the turbine should be such that it will give synchronous speed of the generator.

The problems associated with the high speed turbines are the danger of cavitation and centrifugal forces acting on the turbine parts which require robust construction. No doubt, the overall cost of the plant will be reduced adopting higher rotational speed as smaller turbine and smaller generator are required to generate the same power. The constructional cost of the power house is also reduced.

2. Specific speed. The equation indicates that a low specific speed machine such as impulse turbine is required when the available head is high for the given speed and power output. On the other hand, propeller turbines with high specific speed are required for low-heads.

The specific speed can be calculated using the equations and if the available head is known. The specific speed versus head are shown in Fig. 11.29 for different turbines.

It is obvious from Fig. 11.29 that there is a considerable latitude in the specific speed of runners which can be used for given conditions of head and power provided that the height of the runner above tailrace level is such as to avoid the danger of cavitation as discussed earlier.

In all modern power plants, it is common practice to select a high specific speed runner because it is more economical as the size of the turbo-generator as well as that of power house will be smaller.

High specific speed is essential when the available head is low and power output is high because otherwise the rotational speed will be very low and it will increase the cost of turbo-generator and the power house as the sizes of turbine, generator and power house required at low speed will be large. On the other hand, there is no need of choosing high specific speed runner when the available head is sufficiently large because even with low specific speed, high rotational speeds can be attained.

Now it has been shown with the above discussion that if the speed and power under a given head are fixed (N_s is fixed), the type of the runner required is also fixed.

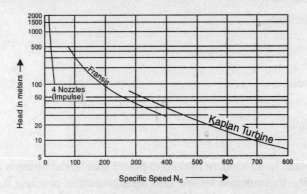

Fig. 11.29

In practice it may be possible to vary the specific speed through a considerable range of values. The speed and power required may be varied for a single runner and the choice is made wider.

Suppose turbine of a given power runs at 120 r.p.m. or at 900 r.p.m. and say available head is 200 meters, if the power is developed in a single unit at 120 r.p.m. is 18000 H.P. the required specific speed of the runner is given by

$$N_s = \frac{120\sqrt{18000}}{(200)^{5/4}} = \frac{120 \times 189.6}{750} = 30.4$$

Now if the same power is developed at 900 r.p.m. in two runners, the required specific speed of the runner is given by

$$N_s = \frac{900\sqrt{18000}}{(200)^{5/4}} = \frac{900 \times 134}{750} = 161$$

The above calculations show that the required power can be developed either with one impulse turbine (Pelton) or two reaction turbines (Francis).

It is customary to choose a speed between certain limits, as neither a very low nor a very high r.p.m. is desirable. The number of units into which a given power is divided is also limited. Nevertheless considerable latitude is left concerning the choice of the prime-mover and number of units used. Ultimately the choice of prime-mover is a matter of extensive experience instead of paper calculation.

3. Maximum Efficiency. The maximum efficiency, the turbine can develop, depends upon the type of the runner used.

In case of impulse turbine, low specific speed is not conducive to efficiency, since the diameter of the wheel becomes relatively large in proportion to the power developed so that the bearing friction and windage losses tend to become too large in percentage value. The value of N_S for highest efficiency is nearly 20.

The low specific speed of reaction turbine is also not conducive to efficiency. The large dimensions of the wheel at low specific speed contribute disc friction losses. In addition to this, the leakage loss is more as the leakage area through the clearance spaces becomes greater and the hydraulic friction through small bracket passages is larger. These factors tend to reduce the efficiency as small values of specific speed are approached.

The high specific speed reaction turbines are associated with large discharge losses ($V_c^2/2g$) as mentioned earlier. The friction and leakage losses are reduced with an increase in specific speed but the discharged losses increase rapidly and the net effect of increase in specific speed is to decrease the efficiency total loss (friction, leakage and discharge) is minimum at medium specific speed. Therefore, it is always preferable to select the reaction turbines of medium specific speed if they operate at constant load conditions. Me effect of specific speed on the maximum efficiency is shown in Fig. 11.30.

Higher efficiencies have been attained with reaction turbines than with Pelton wheels. The maximum *recorded* efficiency till now for reaction turbine is 93.7% but quite a large units have shown efficiencies over 901-W the highest recorded value of efficiency for impulse Turbine is 89% but usual maximum is 82%.

The efficiency of the Pelton wheel is not dependent on its size like reaction turbine. Hence the Pelton wheel may have higher maximum efficiency than the reaction turbine for smaller powers.

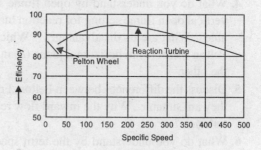

Fig. 11.30

4. Part Load Efficiency. Full load is defined as the load under which a turbine develops its maximum efficiency anything above that is known as overload and anything below that is known as part load.

The part load efficiency differs greatly for different specific speed and types of turbines. Fig. 11.31 shows the variations in part load efficiencies with different types of wheels.

In case of Pelton wheel, only the jet diameter through which the water flows is reduced by the governing *mechanism* when the load on the turbine is reduced below full load. The velocity diagrams at inlet and outlet remain practically unaltered in shape at all loads except for very low and very high loads. Thus the absolute velocity at inlet does not change and discharge loss remains same. Therefore, the part load efficiency curve is more flat in case of Pelton turbine.

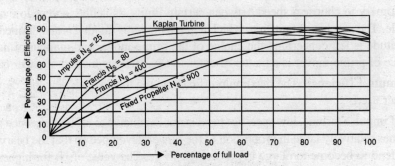

Fig. 11.31

1. Draw a neat sketch of a power house and describe the main features of sub-structure and superstructure.

2. Describe the advantages and disadvantages of underground power stations compared with overground power stations.

3. What topographical and other conditions decide the setting of turbine either vertical or horizontal (*a*) in case of reaction turbine (*b*) in case of Pelton wheel ?

4. What do you understand by open flume setting ? Draw the neat sketches of horizontal and vertical open flume setting for reaction turbines. When the open flume setting is more preferable ? What are its disadvantages ? Which factors are considered in deciding the setting of Pelton wheel (*a*) in horizontal plane (*b*) in vertical plane. Discuss the advantages of one over the other.

5. Discuss the differences between Pelton, Francis and Kaplan turbines and type of power plants they are suitable . Why the inward flow reaction turbines have superseeded the outward flow turbines ?

6. What do you understand by the term specific speed of a water turbine ? What information does it give and how it is made use in practice ? Indicate how the form of a reaction turbine depends upon "specific speed".

7. Find out the expression for the specific speed of a water turbine in terms of H.P. developed, the speed and the head available.

   Further show that the specific speed of a Pelton wheel is 2450 *d/D*, where *d* and *D* are the diameter of jet and diameter of mean circle bucket of Pelton wheel in meters respectively.

   Assume that mean bucket speed = 0.46.

   Maximum efficiency = 0.88% and coefficient of velocity of jet = 1.

HYDRO-ELECTRIC POWER PLANTS

8. Explain why the discharge conditions for a high specific speed runner are less favourable than those for a low specific speed runner both being assumed to be running at their points of maximum efficiency 4.18. Explain why the discharge conditions at part load are less favourable for the high specific speed runner than for the low specific speed runner.

9. For a given head and stream flow available at a certain power plant, what quantities may be changed so as to permit the use of various types of turbines ? Which type of turbine will give the smallest number of units in the plant ? Which type will run at the lowest speed ?

10. How does the maximum efficiency of a reaction turbine vary with the type of the turbine ? For what type it is the highest ? Why ? For what type it is lowest ? Why ?

11. What are the disadvantages of a very low specific speed reaction turbine ? What are its advantages ? How does the efficiency of the Pelton wheel vary with its speed ? Why ?

12. What effects the efficiency of a reaction turbine on part-load ? Is the part load efficiency is a function of specific speed ?

13. What are the advantages and disadvantages of very high specific speed turbine runners ?

14. What are the advantages of a high speed runner under very low heads. What are the advantages of medium speed runner under the same conditions ?

15. What are the advantages of Pelton for very high heads ? What are the disadvantages of low speed reaction turbine for the same conditions ?

16. For the same power under the head compare impulse wheels and reaction turbines with respect to efficiency, speed, space occupied, freedom from breakdown, ease of repairs and durability with silt laden water. Describe the characteristics of various types of turbines used in hydro-electric power stations with reference to (*a*) head (*b*) part load efficiency and maximum efficiency and (*c*) specific speed, and, state how these factors help in the choice of the turbine.

17. What factors are considered in selecting a prime-mover for a hydro-electric power plant ?

18. What factors are mainly considered in selecting a prime-mover for (*a*) run-off river plant (*b*) storage plant (*c*) pump-storage plant.

When the supply is drawn direct from the river instead of storage reservoir, would you go in for Pelton turbine or a reaction turbine ? State the reason for your choice. The head and flow are same in both cases. How do the rotational speed and cavitation effect the selection of a water turbine for hydro-electric power plants ? What are the advantages of reaction turbine over the Pelton wheel in respect of efficiency, size, cost and maintenance ?

Chapter 12

Electrical System

12.1 INTRODUCTION

Without having knowledge about electrical equipment. Power generation from the power plant is difficult to understand. Hence it is necessary to have an idea about role of electrical equipment. The purpose of this chapter is to introduce the students to the electrical equipments used in power plant. The main electrical equipments are as follows;

(1) Generator and generator cooling

(2) Transformers and their cooling

(3) Bus bars

(4) Excitors

(5) Reactors

(6) Circuit breakers

(7) Switch board

(8) Control room equipment

12.2 GENERATORS AND MOTORS

In a generator, an e.m.f. is produced by the movement of a coil in a magnetic field. The current produced by the e.m.f. Interacts with the field to produce a mechanical force opposing the movement, and against which the essential movement has to be maintained. The electrical power ei is produced therefore from the mechanical power supplied.

In a motor, we may suppose a conductor or coil to lie in a magnetic field. If current is supplied to the coil; a mechanical force is mani-fested and due to this force the coil will move. Immediately that relative movement takes place between coil and field, however, an e.m.f. is induced, in opposition to the current.

To maintain the current and the associated motor action, it is therefore necessary to apply to the coil, from an external source, a voltage sufficient to overcome the induced e.m.f. Thus the motor requires electrical power to produce a corresponding amount of mechanical power.

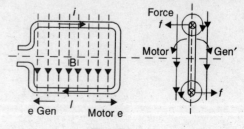

Fig. 12.1

ELECTRICAL SYSTEM

The directions of flux, current and movement in generator and motor action are given in Fig. 12.1. The coil is free to move about the axis O. The component fields are shown, the direction of the mech-anical force, and the directions of rotation for motor and genera-tor action. The direc-tion of the e.m.f. is such as to maintain the current in a generator and to oppose it in a motor. The action is revers-ible: i.e. the same arrangement may act either as generator or motor.

The two-pole and four-pole machines differ considerably in con-struction. At 50 c/s. the former run at 3000 r.p.m. and the latter at 1500. The useful range of two-pole machines has been extended to 300 MVA, and in consequence the four-pole construction is obsolete.

12.2.1 ROTORS

Rotors are most generally made from solid forgings of alloy steel. The forgings must be homoge-neous and flawless. Test pieces are cut from the circumference and the ends to provide information about the mechanical qualities and the microstructure of the material. A chemical analysis of the test pieces is subsequently made. One of the most important examinations is the ultrasonic test, which will discover internal faults such as cracks and fissures. This will usually render the older practice of trepanning along the axis unnecessary.

The rotor forging is planed and milled to form the teeth. About two-thirds of the rotor pole-pitch is slotted, leaving one-third unslotted (or slotted to a lesser depth) for the pole centre.

(*a*) **Windings.** The normal rotor winding is of silver-bearing copper. The heat developed in the conductor's causes them to expand, while the centrifugal force presses them heavily against the slot wedges, imposing a strong frictional resistance to expansion. Ordinary copper softens when hot, and may be subject to plastic deformation. As a result, when the machine is stopped and the copper coals, it contracts to a shorter length than originally. The phenomenon of copper shortening can be overcome by preheating the rotor before starting up. With new machines the use of silver bearing copper, having a much higher yield paint, mitigates the trouble.

Concentric multi-turn coils, accommodated in a slot number that is a multiple of four (e.g. 20, 24, 28 or 32), are used, the slot - pitch being chosen to avoid undesirable harmonics in the waveform of the gap density. The slots are radial and the coils formed of flat strip with separators between turns. The coils may be preformed. The insulation is usually micanite, but bonded asbestos and glass fabric have both been used.

As much copper (or, in some cases, aluminium alloy) as possible is accommodated in the rotor slots, the depth and width of the slots being limited by the stresses at the roots of the teeth, and by the hoop stresses in the end retaining rings. The allowable current depends on cling and expansion. Com-paratively high tempera-ture-rises are allowed: the hot-spot temperature may reach 1400C.

(*b*) **Cooling.** Passage for as much cooling gas as possible is provided, in small machines the main cooling takes place at the outside cylindrical surfaces of the rotor body and retaining rings. It is usual to have a large gap (*e.g.* 45 mm.) allowing for the flow of large quantities of coolant over the rotor surface. For larger machines provision must be made for cooling the bottom of the slot, the con-ventional method being that shown in Fig. 12.2. It is practicable to pass an appreciable volume of gas at high velocity close to the windings, but the temperature-gradient over the slot-insulation is still a dominant factor. For the largest ratings elaborate ventilating arrangements are necessary, and for machines of 100 MVA.

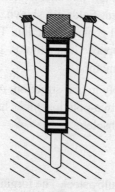

Fig. 12.2. Conventionally Cooled Rotor Slot and Ducts.

Fig. 12.3. Arrangement of Conductors Fop. Direct Cooling (General Electric Co.).

methods are being developed for direct contact between the rotor coils and the gas coolant. One such method is shown in Fig. 12.3. The essential feature is the use of straight rectangular tribes for the slot conductors, ventilated on a cooling circuit separate from that of the stator, the hydrogen gas coolant being circulated through the tubes by a centrifugal impeller mounted on the outboard end of the rotor. The slot tubes are connected at both ends by suitably shaped copper bars forming inlet or outlet ports. The conductors are hard-drawn electrolytic silver-bearing copper, with synthetic-resin bonded glass cloth laminate insulation.

(c) **Overhang.** The appearance and arrangement of the overhang can be seen in Fig. 12.4, which also shows the spigot on which the retaining ring is centred. The conductors are sloped conically inward to provide sufficient space for the thickness of the ring, and are strongly braced to fit closely within the ring without movement other than that due to unavoidable expansion. Non-magnetic, magnesium and magnesium-nickel steels are used for the retaining rings, the non-magnetic property being useful for avoiding excessive magnetic leakage and stray load loss.

(d) **Fans.** The rotor may carry centrifugal or axial (propeller) fans, the former being more common. However, restrictions on rotor diameter and the great length between bearings may make separate fans essential for large machines. They may be coupled to a shaft extension or be separately driven.

Fig. 12.4. Overhang arrangement or 2-Pole Rotor (English Electric).

(e) **Slip Rings.** Slip rings are required for conveying the exciting current to and from the rotor winding. Rings of steel, shrunk over micanite, may be placed one at each end of the rotor, or both at one end, inside or outside the main bearing. Fig. 12.5 shows a typical completed two-pole rotor.

Fig. 12.5. Completed Rotor (Parsons).

ELECTRICAL SYSTEM

12.2.2 STATORS

Large stator housings comprise a series of annular rings flame-cut from steel plate, joined by tubes and longitudinal bars, and carrying ribs to take the stator core laminations. Fig. 12.6 shows a simple stator housing requiring two end plates and four intermediate plates, held apart by tie bars. The core stampings are built up in the frame, the end plates places in position, and the whole stator clamped together by bolts. The frame is then covered with sheet steel.

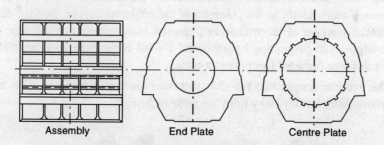

Fig. 12.6. Method of Fabrication Turbo-Generator Stator Housing.

(*a*) **Core.** The active part of the stator consists of segmental lamina-tions of low-loss alloy steel. The slots, ventilation holes and dovetails or dovetail keyholes, are punched out in one operation. The stamp-ings are rather complicated on account of the number of holes and slots that have to be produced.

The use of cold-rolled grain-oriented steel sheet has possibili-ties in machines as well as in transformers, most particularly in two - pole machines where the major loss occurs in the annular part of the core external to the slotting. Here the flux direction is mainly circum-ferential, and by cutting the core- plate sectors in such a way that the pre-ferred flux direction is at right-angles to their central radial axis, Fig. 12.7, a sub-stantial reduction in core-loss can be secured.

It is of great importance that the assembled stator laminations are uniformly compressed during and after building, and that the slots are accurately located. The core plates are assembled between end plates with fingers projecting between the slots to support the flanks of the teeth. The end plates are almost invariably of non-magnetic material, for this greatly reduces stray load loss. The end packets of core plates may be stepped to a larger bore for the same reason.

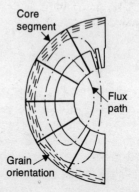

Fig. 12.7. Use of Grain-Oriented Steel.

(*b*) **Windings.** The windings of two-pole machines are comparatively straightforward. The number of slots must be a multiple of 3 (or 6 if two parallel circuits are required). Single-layer concentric or two-layer short-pitched windings may be used.

The single-layer concentric winding is readily clamped in the overhang, but causes a higher load loss because the end-connections run parallel to the stator end-plates. Chording is not possible so that flux harmonics have full effect.

The two-layer winding is more common, chorded to about a 5/6 pitch which practically elimi-nates 5th and 7th harmonics from the open-circuit e.m.f. wave. The end windings are packed, and clamped or tied with glass cord.

It is the invariable practice with two-layer windings to make the coils as half turns and to joint the ends. The conductors must always be transposed to reduce eddy-current losses. The conductors are insulated in many cases with bitumen-bonded micanite, wrapped on as tape, vacuum dried, then impregnated with bitumen under pressure and compressed to size. The process is illustrated in Fig. 12.8. Each copper bar A forming part of a conductor is insulated with mica tape, B and C. A set of bars forming one conductor is assembled and pressed, D. The conductor is insulated with layers of mica, tape, E; then the conductors are assembled to form a. slot bar, F, and pressed to the required dimensions. Fig. 12.9 shows typical conductors. Synthetic resins have now replaced bitumen.

Within the slots, the outer surface of the conductor insulation is at earth potential: in the overhang it will approach more nearly to the potential of the enclosed copper. Surface discharge will take place if the potential gradient at the transition from slot to overhang is excessive, and it is usually necessary to introduce voltage grading by means of a semi-conducting (e.g. graphitic) surface layer extending a short distance outward from the slot ends.

Setting the winding deeply into the slots increases the slot inductance. This has the incidental advantage or the overhang farther away from the rotor end-rings.

Fig. 12.8. Stages in the manufacture of a laminated Conductor
(British Thomson-Houston).

Fig. 12.9. Section or Stator Conductors
(Metropolitan-Vickers).

12.2.3 VENTILATION

Forced ventilation and total enclosure are necessary to deal with the large-scale losses and high rating per unit volume. The primary cooling medium is air or hydrogen, which is in turn passed through a water-cooled heat exchanger.

(a) **Air-Cooling.** The arrangement is that of Fig. 12.10. The water coolers are normally in two sections, so that one can be cleaned while the machine is operating. Fans on the rotor, or separate fans, may be employed, the latter in large machines where bearing-spacing or limitation of the diameter makes integral fans inadequate.

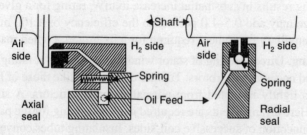

Fig. 12.10. Hydrogen Shaft-seals.

With integral fans mounted on the rotor, the air is fed to the space surrounding the stator overhang, and pipes and channels convey a proportion towards the centre of the stator core. There- from it flows readily inward to the air gap, then axially to the end outlet compartments. With separate fans, however, air can be fed directly to the middle as well as to the ends, as shown in Fig. 12.10.

(*b*) **Hydrogen Cooling.** Compared with air, hydrogen has 1/14 of the density, reducing windage loss and noise; 14 times the specific heat; 1.5 times the heat-transfer, so more readily taking up and giving up heat; 7 times the thermal conductivity, reducing temperature gradients; reduces insulation corona; and will not support combustion so long as the hydrogen/air mixture exceeds 3/1. As a result, hydrogen cooling at 1, 2 and 3 atmospheres absolute can raise the rating of a machine by 15, 30 and 40 per cent respectively.

The stator frame must be gas-tight and explosion-proof. Oil- film gas-seals at the rotor shaft ends are necessary. Two forms are shown in Fig. 12.10, each must accommodate axial expansion of the rotor shaft and stator frame. Oil is fed to the shaft and the flow is split, part towards the interior (gas) side and part to the airside. The latter mingles with the bearing oil, while the former is collected and degassed.

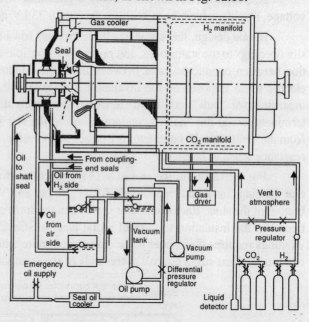

Fig. 12.11. Arrangement of Hydrogen-Cooling System.

Fans mounted on the rotor circulate hydrogen through the ventilating ducts and internally mounted gas-coolers. The gas pressure is maintained above atmospheric by an automatic regulating and reducing valve controlling the supply from normal gas cylinders. When filling or emptying the casing of the machine, an explosive hydrogen-air mixture must be avoided, so that air is first displaced by carbon dioxide gas before hydrogen is admitted: the process is reversed for emptying. It is usual to provide a drier to take up water vapour entering through seals. The hydrogen purity is monitored by measurement of its thermal conductivity.

Turbo-alternators operating at hydrogen pressures just above atmospheric (so that leaks will be outwards) require about 0.03 m^3 per mW of rating per day. This rises to about 0.1 m3 for hydrogen pressure of 2 atm. abs. The gas consumption of synchronous capacitors, which do not need shaft seals, is very much less.

Hydrogen cooling results in substantial increase in mW, rating for a given temperature-rise, and the reduction in windage may add 0.5–1.0 percent to the efficiency of a 100 mW, machine. Fig. 12.11 gives a diagram of the auxiliary equipment required for a hydrogen-cooled machine.

(c) **Direct Cooling.** Direct cooling of stator windings is applied at ratings rather higher than that which makes the method necessary for rotors. Tubular conductors like those of Fig. 12.3 can be used, or thin-walled metal ducts lightly insulated from normal stator conductors. A similar design serves for water-cooling a stator. Here arrangements are required in the overhang for the parallel flow of coolant as well as for the series connection of successive coil-sides. Insulating tubes convey the liquid to and from the water "headers," and the water itself must have adequate resistivity to limit conduction loss. Water-cooling has obvious disadvantages for rotors.

12.2.4 HIGH-VOLTAGE GENERATORS

Although it is usual to combine a generator with a transformer to develop an output at high voltage, machines have been built for feeding a 33 kV network direct.

In one design, the stator has three circular rows of staggered round slots, having narrow radial slot openings to the stator bore, and provided with triple-concentric circular slot conductors. The insulation between conductors, and between the outer conductors and slot walls, is flexible micanite. The electric stress imposed on the insulation is no greater than in a machine built for normal voltage. The innermost (or "bull") conductor in every slot forms that third of the winding connected to the line terminals, while the "outer" conductors are at the star-point end. The conditions for heat dissipation from the central "bull" conductor are rather unfavorable, and a low current density is necessary. The slot reactance tends to be high.

Fig. 12.12 shows the slot of a 33 kV machine of more orthodox design. Three completely separate windings insulated respectively for 11, 22 and 33 kV, (line) are used, the 11 kV, section being at the bottom of the slots. Each conductor is made in the form of a capaci-tor bushing, with conducting shields buried in the insulation to control radial electric stress. The thicker insulation of the higher-voltage windings requires the copper to be deeper and more extensively laminated. The grading shields facilitate longitudinal stress control at the ends of the conductor outside the core.

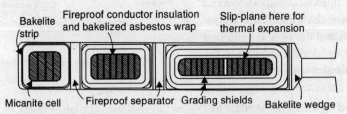

Fig. 12.12. Slots for 33 kV Generator.

The machines discussed are generators designed for water wheel and internal-combustion engine or gas-turbine drives, and salient-pole synchronous motors. Synchronous capacitors may resembl either turbo or salient-pole machines, usually the former.

12.3 TRANSFORMERS

Forces are developed in the transformer but are not allowed to produce movement. Consequentl there is no concern with mechanical power, and only transformer e.m.f. are generated.

ELECTRICAL SYSTEM

12.3.1. CONSTRUCTIONAL PARTS

The transformer is a comparatively simple structure, since there are no rotating parts, or bearings. The chief elements of the construction are :

(1) Magnetic Circuits, comprising limbs, yokes, and clamping structures.

(2) Electric Circuits, the primary, secondary and (if any) tertiary windings, formers, insulation and bracing devices.

(3) Terminals, tappings and tapping switches, terminal insulators and leads.

(4) Tank; oil, cooling devices, conservators, dryers and ancillary apparatus.

Improvements are continually being made in construction, and the practice of different manufacturers depends considerably on the size of unit made, the organization of the factory, and the indivi-duality of the designers. The substance of this chapter is to be taken only as an indication of the construction of modern trans-formers.

The practice in Great Britain and Europe is to concentrate mainly on the single and three-phase core types. Some shell types are built for single-phase use, and somewhat rarely for three phases. A few special constructions are sometimes employed. Attention here will be directed chiefly to the single-phase core and shell and the three-phase core types of construction.

12.3.2. CORE CONSTRUCTIONS

Special alloy steel of high resistance and low Hysteresis loss is used almost exclusively in transformer cores, and all electrical-steel manufacturers have suitable grades. Induction densities up to 1.35-1.55 Wb/m^2 are possible, the limit far 50 c/s being the loss and the magnetizing current.

As the flux in the cores is a pulsating one, the magnetic circuit must be laminated and the separate laminations insulated in order to retain the advantages of subdivision. Paper, Japan, varnish china clay or phosphate may be used. The last-named is able to withstand the annealing temperatures of cold-rolled steels, so that it can be applied to the whole sheet before cutting and annealing.

Burring of the edges of the plates may cause a considerable increase in core loss by providing paths for eddy currents should the sharp edges cut through the insulation and establish contact between adjacent plates. Burrs are removed before core assembly. Silicon alloy steels are hard, and cause wears of the punching tools, so that the removal of burrs needs special attention.

It is found that the magnetic properties of transformer sheet steels vary in accordance with the direction of the grain produced by rolling. Sheets are therefore cut as far as possible along the grain, which is the direction in which the material has a higher permeability. It must not be forgotten that lamination and insulation of core plates reduce the effective or net core area, for which due allowance must be made. In building the core, considerable pressures are used to minimize air gaps between the plates, which would constitute avoidable losses of area and might contribute to noisy operation.

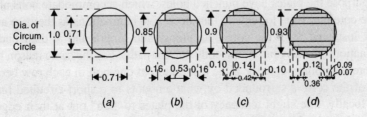

Fig. 12.13. Sections of Core-type Transformer Limbs.

Area % of circum circle	(a) Square	(b) Cruciform	(c) Three-stepped	(d) Four-stepped
Gross	64	79	84	87
Net, Ai	58	71	75	78
No. of packets	1	3	5	7
$A_i = kd^2$ $k =$	0.45	0.56	0.60	0.62

The reduction of core sectional area due to the presence of paper, surface oxide, etc., is of the order of 10 per cent.

1. Core Sections. As has been seen, iron losses make it imperative to laminate transformer cores. The problem of building up these thin sheets into a mechanically strong core is largely a matter for the draughtsman.

With core-type transformers in small sizes, the simple rectangular limb can be used with either circular or rectangular coils. As the size of the transformer increases, it becomes wasteful to employ rectangular coils, and circular coils are usually preferred. For this purpose the limbs can be square, as in Fig. 12.13(a); where the circle represents the inner circumference of the tubular former carrying the coils. Clearly a considerable amount of useful space is wasted, the length of the circumference of the circumscribing circle being large in comparison with the cross-section of the limb. A very common improvement is to employ cruciform limb sections, as in Fig. 12.13(b). This demands at least two sizes of plate. With large transformers further core stepping may be introduced to reduce the length of mean turn and the consequent I2R Loss. Any saving due to core-stepping must, of course, be balanced against the cost of extra labour in shearing several new plate sizes and the reduction in cooling apace between core and coil. A typical core section with three steps is shown in Fig. 12.13(c). The figures indicate the best proportions for the core packets, and the gross area expressed as a fraction of the area of the circumscribing circle. The three-stepped limb is commonly employed and even more steps may be used for very large transformers. In this case, the section is generally too massive to be built solidly: cooling ducts are left between the packets.

Cores for shell-type transformers are usually of simple rectangular cross-section, the Coils being also rectangular.

2. Assemply. A number of methods are available for clamping stacks of stampings together to form cores. In small sizes (*e.g.* below 50 kVA) string or cotton webbing may be employed to bind the plates. Very large cores may also be clamped by strong binding. Cores may also be clamped between iron frames (after the fashion of miniature transformers), but the most usual way is to bolt the plates together. For this purpose the plates have punched holes, which accommodate bolts after assembly, and are used during the core build-ing to register the successive laminations as they are added to the stack. The bolts must be insulated from the core both along their length and at their ends to avoid short-circuiting the laminations, thereby providing eddy current paths. For the same reason it is usual to avoid a double row of core-clamping bolts, for if the insu-lation of one bolt in each row becomes impaired, a considerable area of the core is surrounded by what amounts to a short-circuited turn, and excessive losses Way occur locally. The slight tendency of the plates to "fan" out at their edges; due to central clamping, increases the space between adjacent laminations and provides a safeguard against electrical contact at the sharp edges.

When core-clamping bolts are employed, stiffening or flitch plates are used to give the built-up core more rigidity and to prevent bulging between bolts. Stiffener plates are insulated from the cores and are discontinuous at joints to obviate any tendency for the flux to use them as a conducting path in parallel with the laminations.

For small and medium sized oil-immersed transformers the dissi-pation of the core losses (as heat) is simple, as the surface of the laminations is large compared with their volume and losses.

Very large cores, on the other hand, have a relatively small surface/volume ratio, so that additional ducts must in some way augment the cooling surface. There are two ways of arranging ducts either parallel or perpendicular to the direction of the laminations.

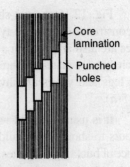

Fig. 12.14. Cooling Ducts.

The first is easy, the second requires special punching. Unfortunately the first method does not present to the oil any additional plate edges. Heat flows twenty times more readily along the laminations to the edges than from plate to plate across the intervening insula-tion, which has naturally low heat conductivity. One method of contriving the exposure of a greater surface of plate-edge is shown in Fig. 12.14, where the upward orientation of the duct facilitates the passage of hot oil, but avoids splitting the core in two.

In any but the smaller sizes it is impracticable to cut complete plates, *i.e.* complete magnetic circuits. If this were attempted, the wastage of sheet would he great, and the difficulty of inserting the coil which must interlink the core-almost insuperable, since each turn would have to be separately threaded through. It therefore becomes necessary to make the coils separately, and to place hem on the cores or to build up the plates through them. This necessitates one or wore joints in the magnetic circuit. Although joints introduce gaps in the continuity of the circuit, the Mates call be suitably arranged to reduce the effect of the gaps on the magnetic conductivity of the joints. As the core is built up, the joints in one layer are arranged at places other than those at which the joints in the previous layer occurred, so that the iron of the preceding and succeeding plates covers the joint in one layer. For example single-phase shell-type transformer of small output may have a core composed of T- and U-shaped stampings. Successive layers are reversed and so arranged that no two joints fall together.

In assembling such a transformer, the coils are made and finished, and the core plates inserted on each side until the complete core is built up, after which it is clamped.

For large shell-type transformers, where the cores are riveted or bolted along their length, the plates are assembled on pins that ensure correct registering. After assembly, the pins are withdrawn, leaving the holes ready for receiving the insulated core-clamping bolts.

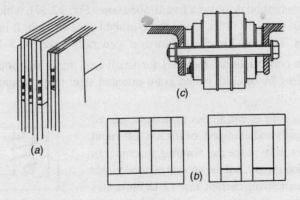

Fig. 12.15. Core CONa Truct 7ON of Core-Type Transforator.

Fig. 12.16 show a shell-type transformer under construction. For core-types the clamps and flitch plates are also arranged in a jig with upright pins, on which the core plates are threaded. The joints between core legs and yokes are invariably interleaved, Fig. 12.15(*a*).

It is usual to build with laminations in threes or fours to shorten the building time and reduce the chance of buckling the plates. Successive sets of plates are arranged as in (*b*) for inter-leaving. Sheets of pressboard may be inserted at intervals in a thick core, (*c*). Small earth clips make electrical connection between packets. For grain-oriented steel intricate mitred joints have been developed.

With the largest cores, the limbs may be built with a central axial slot, and with spacers between the laminations at intervals in the stack, to facilitate cooling.

After the core has been built, the second set of clamps and flitch plates is added and the core tightened, The yoke is removed to admit the coils after the whole core has been stood upright it is then replaced.

The yoke of a core-type transformer is subdivided in a similar way to the limb, and the relative areas of the several packets bear the same relation, to avoid the flux changing from one portion to another

Fig. 12.16. Single Phase Shell Type transformer in Course of Assembly.

at its passage between limb and yoke: such interchange would be productive of eddy-current losses. Sometimes (in core-type transformers) the yokes are made, as a whole, of about 20 per cent greater area than the limbs, thereby reducing the iron loss in parts, which do not involve an increase in the length of copper.

With very large three-phase core types, a limiting factor in the design is the loading gauge of the road or railway route along which the transformer must be hauled. It becomes necessary to reduce the overall height. A common method is to use a five-limbed core (Fig. 12.17), which needs a cross-section in the yokes leas than that required in the usual three-limbed construction; it may be about 30 percent less than that of the limbs. The core losses are, however, generally larger by 5-10 per cent.

A development in constructional methods for small (*e.g.* rural. distribution) transformers employs cores made from lengths of cold. rolled grain-oriented steel strip, wound to shape and impregnated.

Then either (*i*) the core is cut across to make a pair of "C" cores, threaded into coils and clamped; or (*ii*) it is left uncut and the coils are wound on it by a special winding machine. In the former case construction is easy, but the second method yields minimum core loss and magnetizing current. Fig. 12.18 illustrates the method. The three-phase type shown on the right can be produced by method (*i*) only.

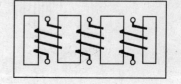

Fig. 12.17. Five-Limbed Core.

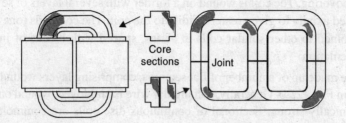

Fig. 12.18. Small Cores for Grain-Oriented Steel.

3. Constructional Framework. Considerable use is made of channel and angle section rolled steel in the framework of core- type transformers. A typical construction is to clamp the top and bottom yokes between channel sections, held firmly by tie-bolts. The bottom pair of channels has cross channels as feet. The upper pair carries clamps for the high and low-voltage connections.

12.3.3 WINDINGS

In addition to the classification as circular or rect. angular, transformer coils can be either concentric or sandwiched. The terms are almost self-explanatory. In Fig. 12.19(a) a single-phase core-type transformer with cylindrical coils is shown (a very common arrangement), and in Fig. 12.19 (b) a single-phase shell-type with sandwich coils. The latter are used almost invariably with shell-type transformers. In Fig. 12.19 the letters L and H refer to the low and high-voltage windings respectively.

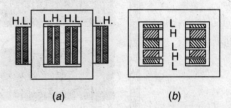

Fig. 12.19. Concentric and Sandwich Winding.

On account of the easier insula-tion facilities, the low-voltage winding is placed nearer to the core in the case of core-typo and on the outside positions in the case of shell-type transformers. The insulation spaces between low- and high-voltage coils also serve to facilitate cooling.

1. Cylindrical concentric helix. Cylindrical concentric helix windings, commonly employed for core-type transformers, can often be built up (generally with axial spacing strips to improve oil circulation between the coil and the tube) on bakelite tubes, which facilitate erection, and form a strong foundation for winding the coils. Wherever possible, simple helical coils are used, preferably in a single layer. Usually the voltage of the low-voltage side is sufficiently small to permit of this, and frequently a helical winding in one or two layers can be used for the high-voltage winding. Where this is not suitable, the coil must be sectionalized in order to reduce the voltage between layers. In this way it becomes unnecessary to put insulation between successive layers over and above that on the wires themselves. With a sec-tionalized winding the voltage per section is of the order 1000 V or leas, but it is possible to reach 5000 to 6000 V. per coil, unsection-alized. The chief difficulty in the making of large concentric coils is the handling of several hundred pounds of copper in a single coil. Care has to be taken to wind the coils tightly and to keep them perfectly circular. For insulation between high- and low-voltage windings bakelite or elephantide tubes may be used. They can be stressed up to about 20 kV per cm. radially, the oil in the duct being regarded as an additional margin.

2. Cross-over coils. Cross-over coils are suitable for currents not exceeding about 20A. They are used for h.v. windings in comparatively small transformers, and comprise wires of small circular section with double cotton covering. The coil is wound on a former with several layers of several turns per layer, tape being interleaved axially to give greater rigidity to the coil. The coil ends (one from inside and one from outside) are joined to other similar coils in aeries, spaced with blocks of insulating material to allow of free oil circulation.

Disc coils are made up of a number of flat sections, comprising layers wound spirally from inside outwards as shown in Fig. 12.20. Generally, rectangular wire is employed, wound on the flat side, so that each disc is mechanically strong. Sectional or continuous disc coils are commonly used. Every turn being in contact with the oil, the cooling is good.

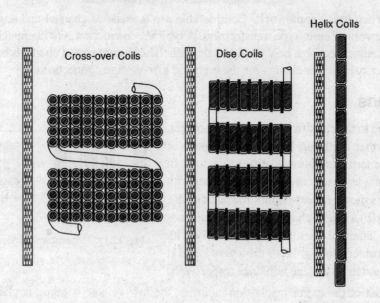

Fig. 12.20. Transformer Coils.

3. Sandwich windings. Sandwich windings commonly employed for shell type transformers, allow of easy control over the reactance. The nearer two coils are together on the same magnetic axis, the greater is the proportion of mutual flux and the less is the leakage flux. If it were possible to accommodate the two coils in the same space, the whole flux would link both windings, and there would be no leakage flux. Subdividing the low- and high-voltage coils can reduce leakage. Each high-voltage section lies between two low-voltage sections. The end low-voltage sections contain half the turns of the normal tow-voltage sections. In order to balance the magneto- motive forces of adjacent sections, each normal section, whether high or low voltage, carries the same number of ampere-turns. The higher the degree of subdivision, the smaller is the reactance.

4. Cross-over of stranded conductors. Conductors of large cross- section are not employed, as being too stiff to handle, and leading to excessive I2R loss. The leakage flux pulsates over the cross-section of the windings, and may induce eddy e.m.f.'s which produce circulating currents and additional losses (often referred to as stray losses). The conductor must consequently be subdivided for the same reason as cores are laminated. A 7.5 mm. square conductor might be approaching the upper limit of size for a 50 c/s transformer. If a larger section is needed, insulated strands in parallel must be used, and balance between all strands attained by transposing their relative positions within the coil.

5. Insulation. The insulation between the h.v. and l.v. windings, and between l.v. winding and core, comprises bakelite-paper cylinders or elephantide wrap. The insulation of the conductors may be of paper, cotton or glass cape, glass tape being used for air-insulated transformers. The paper is wrapped round the conductor in a suitable machine, preferably without overlap of adjacent turns. Paper is not flexible, and a "half-lap" wrapping would cause it to buckle. The wrapped conductor is lashed with cotton strands wound openly, to give some mechanical protection.

Fig. 12.21. Core and Windings of three-phase 200 HVA, 50 cycle 6000/440 V Core-type Transformer (Bruce Peebles).

Paper insulation usually necessitates the use of round coils, while the crossover of the several strands in a conductor must be properly shaped, and not merely twisted.

The high-voltage winding is separated from the low-voltage winding by a series of, ducts and Bakelite cylinders or barrels. Details of the high-voltage winding, which is sectionalized, are shown. It will be seen that the end turns, i.e. those turns in coils 3, 4, and 6, are more heavily insulated than Nos. 6 and 7 to 23. The reason for this is connected with phenomena occurring during switching operations, or with line disturbances. Owing to strain on the insulation between turns at the line end of the high-voltage winding, about 5 percent of the turns are reinforced with extra insulating material.

For large h.v. Transformers the end-turn reinforcement is a matter of careful design. Merely increasing the insulation thickness may result in markedly raising the impact thereon of impulse voltages. The winding insulation has to be coordinated with the means adopted for controlling the distribution of impulse electric stresses, Fig. 12.22(a).

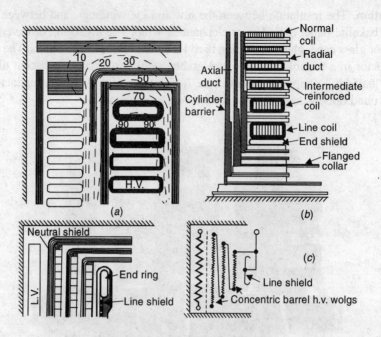

Fig. 12.22. Transformer Insulation Details.

Unimpregnated paper insulation readily absorbs oil when the transformer is inserted in its tank. Small thin-wire coils may be varnished for adhesion. A difficulty, however, is introduced by the shrinking of insulation after a few months of service, resulting in a loosening of the windings. This predisposes to movement of the coils and breakdown on sudden short-circuit. The coils can be pre-shrunk under pressure before assembly to simulate service conditions. They may then be assembled and tightened up without danger of undue further shrinkage.

The permittivity of transformer oil is about 2-2; of bakelite cylinders about 4-4. The electric stress is therefore twice as great on the oil in the annular ducts as on the bakelite cylinders.

In transformers for very high line voltages *e.g.* 150-220 kV. The radial width of the oil ducts is determined by the electric stress and has to be made much wider than would be necessary from considerations of cooling. On such transformers the ducts may be entirely filled with insulating paper, bent over at the coil ends to form an earth barrier. The paper, oil-saturated, has a high dielectric strength, so that the h.v.—l.v. spacing can be reduced. The greater freedom of choice of winding separation permits of better control over the leakage flux. See Fig. 12.22(*b*).

The multi-layer h.v. barrel winding in Fig. 12.22(*c*) is a special helix arrangement for transformers of exceptionally high line voltage. The innermost layer is wound over a neutral shield. Succeeding layers are shortened, giving the additional clearance to the yokes appropriate to the voltage of the layer to earth. The outermost layer, enclosed by a static shield, is connected to the line terminal, and the successive layers joined in series to give the electric stress distribution effect of a capacitor bushing to surge voltages applied to the line terminal. The line and neutral shields may comprise close-wound helices of conducting strip material applied over the appropriate insulated paper layers: the strips are connected electrically, but arranged so that they do not form a short-circuited turn. Tappings are usually required on modern transformers.

6. Leads and Terminals. The connections to the windings are copper rods or bars, insulated wholly or in part, and taken to the bus bars directly in the case of air-cooled transformers, or to the insulator bushings on the tank top in the case of oil-cooled trans-formers. The shape and size of the conductors are of importance in very high voltage systems, not on account of the current-carrying capacity, but because of dielectric stresses, corona, etc., at sharp bends and corners with such voltages.

7. Bushings. Up to voltages of about 33 kV, ordinary porcelain insulators can be used, which do not require special comment. Above this voltage the use of condenser and of oil-filled terminal bushings, or, for certain eases, a combination of the two, has to be considered. Of course, any conductor can be effectively insulated by au provided that it is at a sufficient distance from other con-ducting bodies and sufficiently proportioned to prevent corona phenomena. Such conditions are naturally unobtainable with transformers where the conductor has to be taken through the cover of the containing tank, although common enough with over-head transmission lines.

The oil filled bushing consists of a hollow porcelain cylinder of special shape with a conductor (usually a hollow tube) through its centre. The space between the conductor and the porcelain is filled with oil, the dielectric strength of which is greater than that of air. The dielectric field strength is greatest at the surface of the conductor, and this breaks down at a much lower voltage in air than in oil. Oil is fed into the bushing at the top, where there is a glass cylinder to indicate the oil-level and to act as an expansion chamber for the oil when the bushing temperature rises.

Under the influence of the electric field, foreign substances in the form of dust, moisture or metallic particles, have a tendency to arrange themselves in radial lines, giving rise to paths of low dielectric strength, with consequent danger of breakdown. To prevent such action by unavoidable impurities in the oil, Bakelite tubes are used to surround the conductor concentrically. The effect is to break up radial chains of semi-conducting particles.

Provision must be made in oil-filled bushings for the differing coefficients of linear expansion of porcelain and the metal conductor. The capacitor-type hushing is constructed of thick layers of bakelized paper alternating with thin graded layers of tin-foil.

The result is a series of capacitors formed by the conductor and the first tin-foil layer, the first and second tip-foil layers and so on. Their length and the radial separation of their tin-foil plates control the capacitance of the capacitors. Fig. 12.23 illustrates this. If the thickness of bakelized paper separating successive tin-foil layers is kept constant, and the capacitances of the capacitors are kept constant by successively reducing the length of the tin-fail layers proceeding outward, then the voltage across each capacitor will be the same, giving a practically uniform dielectric stress throughout the radial depth of the insulator. By arranging, the dielectric stress to come within the limits of the material used, a hushing can be built to withstand any desired voltage. In Fig. 12.23 the short stepped end is oil-immersed beneath the tank cover, the smooth long end projecting outwards. For use in out door substations, porcelain rain-shed, the annular space between the rain-shed and the bushing proper being filled up with bitumen cover the bushing. The rain-sheds are corrugated circumf6rentially to accord with the estimated electric field distribution and to provide a long leakage path.

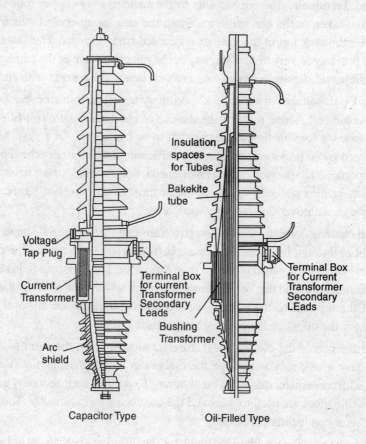

Fig. 12.23

The oil-immersed ends of h.v. bushings may be of re-entrant form, reducing the immersed length and permitting a more uniform distribution of the axial and radial electric stress components.

12.4 COOLING OF TRANSFORMERS

Consider a transformer with k times the linear dimensions of another smaller but otherwise similar unit. Its core and conductor areas are k^2 times greater and its rating (with the same flux and current densities) increases k^4 times, The lasses increase by the factor k^3 but the surface area is multiplied only by the factor V. Thus the loss per unit area to be dissipated is increased k times. Large transformers are therefore more difficult to coal than small ones, and require more elaborate methods.

The cooling of transformers differs from that of rotary machinery in that there is no inherent relative rotation to assist in the circulation of ventilating air. Luckily the losses are comparatively small, and the problem of cooling (which is essentially a problem of preserving the insulation-solid and liquid-from deterioration) can in most cases be solved by reliance on natural self-ventilation. The various methods are.

12.4.1 SIMPLE COOLING

AN: Natural cooling by atmospheric circulation, without any special devices. The transformer core and coils are open all round to the air. This method is confined to very small units at a few kV at low voltages.

AB: In this case the cooling is improved by an air blast, directed by suitable trunking and produced by a fan.

ON: The great majority of transformers are oil-immersed with natural cooling, i.e. the heat developed in the cores and coils is passed to the oil and thence to the tank walls, from which it is dissipated. The advantages over air-cooling include freedom from the possibility of dust clogging the cooling ducts, or of moisture affecting the insulation, and the design for higher voltages is greatly improved.

OB: In this method the cooling of an ON-type transformer is improved by air blast over the outside of the tank.

OFN: The oil is circulated by pump to natural air coolers.

OFB: For large transformers artificial cooling may be used. The OFB method comprises a forced circulation of the oil to a refrigerator, where it is cooled by air-blast.

OW: An oil-immersed transformer of this type is cooled by the circulation of water in cooling tubes situated at the top of the tank but below oil-level.

OFW: Similar to OFB, except that the refrigerator employs water instead of air blast for cooling the oil, which is circulated by pump from the transformer to the cooler.

12.4.2 MIXED COOLING

ON/OB: As ON, but with alternative additional air-blast cooling. ON/OFN, ON/OFB, ON/OFW, ON/OB/OFB, ON/OW/OFW: Alternative cooling conditions in accordance with the methods indicated.

A transformer may have two or three ratings when more than one method of cooling is provided. For an ON/OB arrangement these ratings are approximately in the ratio 1/1.5; for ON/OB/OFB in the ratio 1/1.5/2.

12.4.3 NATURAL OIL COOLING

The diagram in Fig. 12.24 is drawn to indicate on the left the thermal flow of oil in a transformer tank. The oil in the ducts, and at the surface; of the coils and cores, takes tip heat by conduction, and rises cool oil from the bottom of the tank rising to take its place. A continuous circulation of oil is completed by the heated oil flowing to the tank sides (where cooling to the ambient air occurs) and falling again to the bottom of the tank. Oil has a large coefficient of volume expansion with increase of temperature, and a substantial circulation is readily obtained so long as the cooling ducts in the cores and coils are not unduly restricted.

Fig. 12.24 also shows on the right a curve typical of approximate temperature distribution, the figures quoted being rises in degrees centigrade. On full load with continuous operation, the greatest temperature-rise wills probably he in the coils. The maximum oil temperature may be about 10° less than the coil figure, and the mean oil temperature another 15° less.

The best dissipater of external heat is a plain blacked tank. But to dissipate the loss in a large transformer a plain tank would have an excessively large surface area and cubic capacity, and would require a great quantity of oil. Both space and oil are expensive.

Artificial means for increasing the surface area without increasing the cubical contents have, therefore, been developed. These comprise special tank constructions such as

    (a) Fins, welded vertically to the tank sides;

    (b) Corrugations;

    (c) Round- or elliptical-section tubes;

    (d) Auxiliary radiator tanks.

These are illustrated in Fig. 12.25. Little need be said of (a) or (b), the former being not very effective and the latter rather difficult in construction, although formerly used in Europe. Method (c) is extremely common for a wide range of sizes, while (d) is used when there is insufficient room to accommodate all the tubes required by a large transformer.

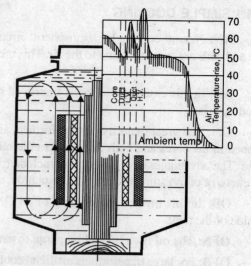

Fig. 12.24. Oil-Circulation and Temperature Distribution.

The tabbed tank provides considerable cooling surface, and the tubes being connected with the tank at the top and bottom only provide a head sufficient to generate a syphoning action; which improves the oil circulation quite apart from enhanced cooling. The tubes may sometimes be "gilled," i.e. wound with a strip-on-edge metal helix, to increase cooling surface and the eddying airflow that more effectively removes heat. The baffling action, however, also tends to restrict the total sir flow so that the net gain is not commensurate with the added cooling surface.

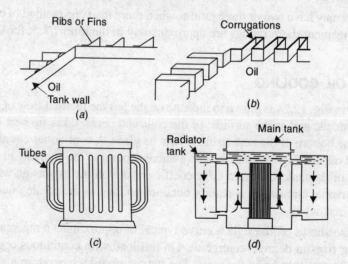

ELECTRICAL SYSTEM

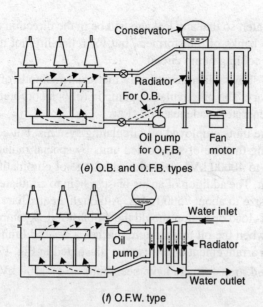

(e) O.B. and O.F.B. types

(f) O.F.W. type

Fig. 12.25. Cooling Methods.

For the largest sizes the radiator type of cooling is used, where separate radiator tanks with fins or corrugations, connected at top and bottom to the main tank, dissipate the heat by oil circulation. One such arrangement is shown diagrammatically in Fig. 12.25(d).

The limit of output with oil-insulated, self-cooled transformers is reached when the tank becomes too large and costly. Another limit, which obtains in some cases, is the railway-loading gauge, which pre-cludes the transport of transformer F and tanks exceeding a certain size. In carriage by road the available routes with their grades, bridges, etc., may decide the type of transformer and tank that can be used. For larger units, transport in parts must be resorted to, with erection on site. A tank dissipating about 50 kW is regarded by some manufacturers as the limit for self-cooling.

12.4.4 FORCED OIL COOLING

When forced cooling becomes necessary in large high-voltage oil-immersed transformers. The choice of the method of cooling will depend largely upon the conditions obtaining at the site. Air-blast cooling can be used, a hollow-wailed tank being provided for the transformer and oil, the cooling air being blown through the hollow space. The heat removed from the inner walls of the tank can be raised to five or six times that dissipated naturally, so that very large transformers can be cooled in this way.

A cheap method of forced cooling where a natural head of water is obtainable is the use of a cooling coil, consisting of tubes through which cold water is circulated, inserted in the top of the tank. This method has, however, the disadvantage that it introduces into the tank a, system containing water under a head greater than that of the oil. Any leakage will, therefore, be from the water to the oil, so that there is a risk of contaminating the oil and reducing its insulating value. Fins are placed on the copper cooling tubes to assist in the conduction of heat from the oil, since heat passes three times as rapidly from the copper to the cooling water we from the oil to the copper tubing. The inlet and outlet pipes are lagged to avoid water from the ambient air condensing on them and getting into the oil.

For large installations the best cooling system appears to be that in which oil is circulated by pump from the top of the transformer tank to a cooling plant, returning when cold to the bottom of the tank. When the cooling medium is water, this has the advantage that the oil can be arranged to work at a higher static head than the water, so that any leakage will be in the direction of oil to water. The system is suitable for application to banks of transformers, but for reliability not more than, say, three tanks should be con-nected in one cooling pump circuit.

Fig. 12.25(*e*) and (*f*) shows diagrammatically the usual methods of cooling employed where separate radiators are necessary. The oil circulation pump in (*e*) is incorporated only if the natural thermal head is insufficient to generate an adequate oil flow.

Until recently all large units employed oil-circulating systems, but considerable advances have been made towards increasing the size of self-cooled units by special radiators. It is possible to build entirely self-cooled units up to 40000 kVA, with the advantage of eliminating breakdown risks due to auxiliary pumping equipment. The addition of an air-blast system to circulate cooling air over the radiators permits the increase of size to about 75000 kVA. Although an auxiliary fan is involved, the transformer is still capable of half- load operation should the air blast fail. A temperature device can be used to bring the fan into action when the oil temperature reaches a desired limit; this improves the overall efficiency at small loads. An arrangement of this type is illustrated in Fig. 12.26.

Tank-less, air-insulated. transformers have been built up to 1500 kVA, but larger sizes require forced air circulation.

Core and windginds of ON/OFB 22.5 MVA, 132/33 kV Transformer.

Fig. 12.26. Complete unit OF ON/OFB 22.5 MVA, 132/33 kV Transformer.

12.4.5 INTERNAL COOLING

The heating of the coils depends on their thermal conductivity, which is itself a function of (*a*) th thickness of the winding, and (*b*) the external insulation.

A coil design, which allows the copper heat to flow radially out-wards with little cross insulation in the path of the flow, leads to economical rating in that a high current density can be employed for a given temperature rise without sacrifice of efficiency. The strip-on-edge winding, consisting of a single layer of copper of rect-angular section wound on edge on a bakelite cylinder with one edge bare and in contact with oil, dissipates heat most effectively. In some designs the flow of heat can be so much improved that the transformer output entails a larger size of tank.

With cores, ageing is not to be feared when modern steels are used and correctly handled, but heating and cooling, with the accom-panying expansion and contraction, lead sometimes to a loosening of the core construction. Owing to the laminated nature of the cores, and the presence (on the surfaces of the plates) of oxide films and paper, varnish, etc., the flow of heat in cores is almost wholly towards the edges. On account of the rather greater exposure of iron to the oil in shell-type transformers, these are better than the core-type as regards the cooling of the iron. On the other hand, the exposed coils of the core type will cool more readily than those of the shell type.

1. Tanks. Small tanks are constructed from welded sheet steel, and larger ones from plain boiler-plate. The lids may be of cast iron, a waterproof gasket being used at the joint. The fittings include thermometer pockets; drain cock, rollers or wheels for moving the transformer into position, eyebolts for lifting, conservators and breathers. Cooling tubes are welded in, but separate radiators are individually welded and afterwards bolted on.

Conservators are require to take up, the expansion and contraction of the oil with changes of temperature in service without allowing the oil to come in contact with the air, from which it is liable to take up moisture. The conservator may consist of an airtight cylindrical metal drum supported on the transformer lid or on a neighbour-ing wall, or of a flexible flat corrugated disc drum. The tank is filled when cold, and the expansion is taken up in the conservator. The displacement of air due to change of oil volume takes place through a breather containing cal-cium chloride or silica gel, which extracts the moisture from the air. Some tank details are illustrated in Fig. 12.27.

Fig. 12.27. 2000 kVA, 20/6.6 kV, 50 cycle Transformer.

2. Transformer Oil. Oil in transformer construction serves the double purpose of cooling and insulating. For use in transformer tanks, oil has to fulfill certain specifications and must be carefully selected. All oils are good insulators, but animal oils are either too viscous or tend to form fatty acids, which attack fibrous materials (e.g. cotton) and so are unsuitable for transformers. Vegetable oils (chiefly resinous) are apt to be inconsistent in quality and, like animal oils, tend to form destructive, fatty acids. Of the mineral oils, which alone are suitable for electrical purposes, some have a bituminous and others a paraffin base. The crude oil, as tapped, is distilled, producing a range of volatile spirits and oils rang-ing from the very light to the heavy, and ending with semi-solids like petro-leum jelly, paraffin wax, or bitumen.

In the choice of oil for transformer use the following char-acteristics have to be considered.

Viscosity. This determines the rate of cooling, and varies with the temperature. A high viscosity is an obvious disadvantage because of the sluggish flow through small apertures which it entails.

Insulating Property. It is usually unnecessary to trouble about the insulating properties of oil, since it is always sufficiently good. A more important matter, however, is the reduction of the dielectric strength due to the presence of moisture, which must be rigorously avoided. A very small quantity of water in oil greatly lowers its value as an insulator, while the presence of dust and small fibers tends to paths of low resistivity.

Flash Point. The temperature at which the vapour above an oil surface ignites spontaneously is termed the flash point. A flash point of not less than about 160°C is usually demanded for reasons of safety.

Fire Point. The temperature at which an oil will ignite and continue burning is about 25 percent above the flash point, or about 200°C.

Purity. The oil must not contain impurities such as sulphur and its compounds. Sulphur when present causes corrosion of metal parts, and accelerates the production of sludge.

Sludging. This is the most important characteristic. Sludging means the slow formation of semi-solid hydrocarbons, sometimes of an acidic nature, which are deposited on windings and tank walls. The formation of sludge is due to heat and oxidation. In its turn it makes the whole transformer hotter, thus aggravating the trouble, which may proceed until the cooling ducts are blocked and the transformer becomes unusable owing to overheating. Experience shows that sludge is formed more quickly in the presence of bright copper surfaces. The chief remedy available is to use oil, which remains without sludge formation after long periods of heating in the presence of oxygen, and to employ expansion chambers to restrict the contact of hot oil with the surrounding air.

Acidity. Among the products of oxidation of transformer oil are CO, volatile water-soluble organic acids, and water. These in com-bination can attack and corrode iron and other metals. The provision of breathers not only prevents the ingress of damp air, but also helps on out-breathing to absorb any moisture produced by oxida-tion of the oil. Oil conservators are desirable to avoid the con-densation of water-soluble acids on the under surface of the tank lid from which acidic droplets may fall back into the oil.

3. Inhibited Oil. The deterioration of oil during its working life can be retarded by the use of anti-oxidants, particularly oxidation "inhibitors." The latter, which are usually of the phenolic or amino type, convert chain-forming molecules in the oil into inactive molecules, being gradually consumed in the process. Inhibitors greatly prolong the phase in the service life of the oil, which precedes the onset of deterioration, and during which the acid and sludge formations are substantially zero.

4. Synthetic Transformer Oil. This has been developed to avoid the risk of fire and explosion, present always with normal mineral oils. Chlorinated diphenyl, a synthetic oil suitable for transformers, is chemically stable, non-oxidizing, rather volatile, and heavier than water. Its dielectric strength is higher than that of mineral oil, and moisture has a smaller tendency to migrate through it. The permittivity is 4.5, compared with about 2.5. This high figure is roughly the same as the permittivity of the solid insulating material used in a transformer, so that the distribution of electric stress will differ markedly from that when mineral oil is used, the stress in the oil being relieved at the expense of the solid insulation. The oil is a powerful solvent of most varnishes, gums, binders and paints, which must consequently be barred from transformers designed for synthetic oil cooling. When decomposed by electric arc, hydrogen chloride gas is the chief product: this may combine with water to form hydrochloric acid.

5. Temperature Rise. The temperature rises permitted in the British Standard Specification for power and lighting transformers.

12.5 BUS-BAR

When a number of generators or feeders operating at the same voltage have to be directly connected electrically, bus-bars are used as the common electrical component. Bus-bars are copper rods or thin walled tubes and operate at constant voltage. We shall discuss some important bus-bars arrangements used for power stations and sub-stations. All the diagrams refer to 3-phase arrangement but are shown in single-phase for simplicity.

12.5.1 SINGLE BUS-BAR SYSTEM

The single bus-bar system has the simplest design and is used for power stations. It is also used in small outdoor stations having relatively few outgoing or incom-ing feeders and lines.

Fig. 12.28 shows the single bus-bar system for a typical power station. The generators, outgoing lines and transformers are connected to the bus-bar. Each generator and feeder is controlled by a circuit breaker. The isolators permit to isolate generators, feeders and circuit breakers from the bus-bar for maintenance. The chief advantages of this type of arrangement are low initial cost, less maintenance and simple operation.

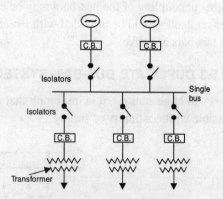

Fig. 12.28. A Single Bus-bar System.

Disadvantages

Single bus-bar system has the following three principal disadvantages

(1) The bus-bar cannot be cleaned, repaired or tested without de-energizing the whole system.

(2) If a fault occurs on the bus-bar itself, there is complete interruption of supply.

(3) Any fault on the system is fed by all the generating capacity, resulting in very large fault currents.

12.5.2 SINGLE BUS-BAR SYSTEM WITH SECTIONALISATION

In large generating stations where several units are installed, it is a common practice to sectionalise the bus so that fault on any section of the bus-bar will not cause complete shut down.

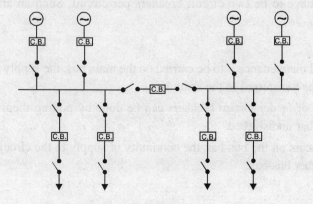

Fig. 12.29

This is illustrated in Fig. 12.29 which shows the bus bar divided into two sections connected by a circuit breaker and isolators. Three principal advantages are claimed for this arrangement. Firstly, if a fault occurs on any section of the bus-bar, that section can be isolated without affecting the supply to other sections. Secondly, if a fault occurs on any feeder, the fault current is much lower than with unsectionalized bus-bar. This permits the use of circuit breakers of lower capacity in the feeders. Thirdly, repairs and maintenance of any section of the bus-bar can be carried out by de-energizing that section only, eliminating the possibility of complete shutdown.

It is worthwhile to keep in mind that a circuit breaker should be used as the sectionalizing switch so that uncoupling of the bus-bars may be carried out safely during load transfer. Moreover, the circuit breaker itself should be provided with isolators on both sides so that its maintenance can be done while the bus-bars are alive.

12.5.3 DUPLICATE BUS-BAR SYSTEM

In large stations, it is important that breakdowns and mainte-nance e should interfere as little as possible with continuity of supply.

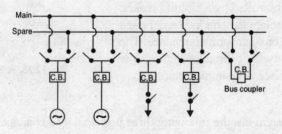

Fig. 12.30

In order to achieve this objec-tive duplicate bus-bar system is used in important stations. Such a system consists of two bus-bars, a "main bus-bar and a "spare" bus-bar (see Fig. 12.30). Each generator and feeder may be connected to either bus-bar with the help of bus coupler which consists of a circuit breaker and isolators.

In the scheme shown in Fig. 12.30, service is interrupted during switch over from one bus to another. However, if it were desired to switch a circuit from one to another without interruption of service, there would have to be two circuit breakers per circuit. Such an arrangement will be too expensive.

Advantages

(1) If repair and maintenance it to be carried on the main bus, the supply need not be interrupted as the entire load can be transferred to the spare bus.

(2) The testing of feeder circuit breakers can be done by putting them on spare bus-bar, thus keeping the main bus-bar undisturbed.

(3) If a fault occurs on the bus-bar, the continuity of supply to the circuit can be maintained by transferring it to the other bus-bar.

ELECTRICAL SYSTEM

12.6 BUSBAR PROTECTION

Busbars and lines are important elements of electric power system and require the immediate atten-tion of protection engineers for safeguards against the possible faults occurring on them. The meth-ods used for the protection of generators and transformers can also be employed, with slight modifi-cations, for the busbars and lines. The modifications are necessary to cope with the protection problems arising out of greater length of lines and a large number of circuits connected to a busbar. Although differential protection can be used, it becomes too expensive for longer lines due to the greater length of pilot wires required. Fortunately, less expensive methods are available which are reasonably effective in providing protection for the busbars and lines. In this chapter, we shall focus our attention on the various methods of protection of busbars and lines.

Busbars in the generating stations and sub-stations form important link between the incoming and outgoing circuits. If a fault occurs on a busbar, considerable damage and disruption of supply will occur unless some form of quick-acting automatic protection is provided to isolate the faulty busbar. The busbar zone, for the purpose of protection, includes not only the busbars themselves but also the isolating switches, circuit breakers and the associated connections. In the event of fault on any section of the busbar, all the circuit equipment's connected to that section must be tripped out to give complete isolation.

The standard of construction for busbars has been very high, with the result that bus faults are extremely rare. However, the possibility of damage and service interruption from even a rare bus fault is so great that more attention is now given to this form of protection. Improved relaying methods have been developed, reducing the possibility of incorrect operation. The two most commonly used schemes for busbar protection are:

(1) Differential protection

(2) Fault bus protection.

12.6.1 DIFFERENTIAL PROTECTION

The basic method for busbar protection is the differential scheme in which currents entering and leaving the bus are totalised. During normal load condition, the sum of these currents is equal to zero. When a fault occurs, the fault current upsets the balance and produces a differential current to operate a relay.

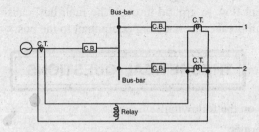

Fig. 12.31

Fig. 12.31 shows the single line diagram of current differential scheme for a station busbar. A generator and supplies load to two lines feeds the busbar. The secondaries of current transformers in the generator lead, in line 1 and in line 2 are all connected in parallel. The protective relay is connected across this parallel connection. All CTs must be of the same ratio in the scheme regardless of the capacities of the various circuits. Under normal load conditions or external fault conditions, the sum of the currents entering the bus is equal to those leaving it and no current flows through the relay. If a fault occurs within the protected zone, the currents entering the bus will no longer be equal to those leaving it. The difference of these currents will flow through the relay and cause the opening of the generator, circuit breaker and each of the line circuit breakers.

12.6.2 FAULT BUS PROTECTION

It is possible to design a station so that the faults that develop are mostly earth-faults. This can be achieved by providing earthed metal barrier (known as fault bus, surrounding each conductor throughout its entire length in the bus structure. With this arrangement, every fault that might occur must involve a connection between a conductor and an earthed metal part. By directing the flow of earth-fault current, it is possible to detect the faults and determine their location. This type of protection is known as fault bus protection.

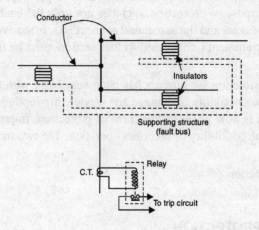

Fig. 12.32

Fig. 12.32 show the schematic arrangement of fault bus protection. The metal supporting structure or fault bus is earthed through a current transformer. A relay is connected across the secondary of this CT. Under normal operating conditions, there is no current flow from fault bus to ground and the relay remains inoperative. A fault involving a connection between a conductor and earthed supporting structure will result in current flow to ground through the fault bus, causing the relay to operate. The operation of relay will trip all breakers connecting equipment to the bus.

THEORETICAL QUESTIONS

1. Write short notes on the following:
   (*a*) Power transformer
   (*b*) Voltage regulation
   (*c*) Transmission of electrical power.
2. Discuss the factors to be considered while deciding the suitability of a transformer.
3. What are the properties of materials used for conductor? Name the materials used for conductors

4. What is the function of a reactor? Describe the various arrangements used for location of reactors.

5. What is a circuit breaker? What are the different types of circuit breakers that are employed in typical power stations?

6. Describe the various generator-cooling methods.

7. Give a list of various electrical protective equipment used in a power station.

8. Describe the various methods of controlling the voltage at the consumer terminal used in power supply system.

9. Write short notes on the following:

   (a) Over-head and underground power transmission system.
   (b) System of electrical energy transmission
   (c) Practical working voltage
   (d) Disposition of conductors
   (e) Characteristics of relays
   (f) Control room
   (g) Types of insulators.

Chapter 13

Pollution and its Control

13.1 INTRODUCTION

The atmosphere consists of a mixture of gases that completely surround the earth. It extends to an altitude of 800 to 1000 kms above the earth's surface, but is deeper at the equator and shallow at the poles. About 99.9% of the mass occurs below 50 km and 0.0997% between 50 and 100 km altitude. Major polluting gases/particles are confined to the lowermost layer of atmosphere known as Troposphere. That extends between 8 and 16 kms above the earth surface.

The main sources of atmospheric pollution may be summarized as follows:

(*a*) The combustion of fuels to produce energy for heating and power generation both in the domestic sector as well as in the industrial sector.

(*b*) The exhaust emissions from the transport vehicles that use petrol, or diesel oil etc.

(*c*) Waste gases, dust and heat from many industrial sites including chemical manufacturers, electrical power generating stations etc.

13.2 ENVIRONMENT POLLLUTION DUE TO ENERGY USE

A considerable amount of air pollution results from burning of fossil fuels. Fuels are primarily derived from fossilized plant material and consist mainly of carbon and/or its compounds. The household sector is the largest consumer of energy in India, accounting for 40-50% of the total energy consumption. As per a report of Planning Commission the share of the household sector in the final use of energy declined although retaining its dominant share at 58.9% in 1987. The most abundantly used fossil fuel for cooking is the wood, which is almost 61% of the total fuel demand for cooking. Burning of traditional fuels introduces large quantities of CO_2 when the combustion is complete, but if there is incomplete combustion and oxidation then Carbon monoxide (CO) is produced, in addition to hydrocarbons. Incomplete combustion of coal produces smoke consisting of particles of soot or carbon, tarry droplets of unburnt hydrocarbons and CO. Fossil fuels also contain 0.5-4.0% of sulphur which is oxidized to SO_2 during combustion.

The environmental effects of various fuels namely coal; oil, nuclear, etc. are of growing concern owing to increasing consumption levels. The combustion of these fuels in industries and vehicles has been a major source of pollution. Coal production through opencast mining; its supply to and consumption in power stations; and industrial boilers leads to particulate and gaseous pollution, which can cause pneumoconiosis, bronchitis, and respiratory diseases. Another major impact of coal mining is land degradation, especially of forest areas.

The consumption of petroleum products in vehicles, industries and domestic cooking activities results in the emission of pollutants in large quantities. Radioactive emissions from nuclear power plants are of grave concern as they can cause serious impact both in terms of spatial and inter-generational concerns. In addition, two key problems are long-term waste disposal and the eventual decommissioning of plants. Due to limited reserves of petroleum, main emphasis needs to be given to non-conventional energy sources such as wind energy, solar energy and ocean energy.

13.3 ENVIRONMENT POLLUTION DUE TO INDUSTRIAL EMISSIONS

Air borne emissions emitted from various industries are a cause of major concern. These emissions are of two forms, viz. solid particles (SPM) and gaseous emissions (SO_2, NO_x, CO, etc.). Liquid effluents, generated from certain industries, containing organic and toxic pollutants are also a cause of concern. Heavily polluting industries were identified which are included under the 17 categories of highly polluting industries for the purpose of monitoring and regulating pollution from them. The Ministry of Environment and Forests has, over the last two decades, developed standards for regulating emissions from various industries and emission standards for all the polluting industries including thermal power stations, iron and steel plants, cement plants, fertilizer plants, oil refineries, pulp and paper, petrochemicals, sugar, distilleries and tanneries have been prescribed. The industrial units in India are largely located in the States of Gujarat, Maharastra, Uttar Pradesh, Bihar, West Bengal and Madhya Pradesh. The highest concentration of sulpher dioxide and oxides of nitrogen is therefore often found in cities located in these states. Some other industrial estates in Delhi, Punjab, Rajasthan and Andhra Pradesh are also becoming critical.

13.4 ENVIRONMENT POLLLUTION DUE TO ROAD TRANSPORT

Road vehicles are the second major source of pollution. They emit CO, HCs, NO_x, SO_2, and other toxic substances such as TSP and lead. Diesel engines are much less polluting than petrol engines. Both types of engines are not very efficient converters of fuel energy. However, diesel types, with a conversion efficiency of around 30%, must be more efficient and use less fuel than petrol types with 15-20% conversion efficiency. Both types of engines have incomplete combustion of fuel so the major pollutant is CO, amounting to 91% by weight of all vehicle emissions.

The primary pollutants produced in vehicle emissions undergo a series of complex interrelated chemical reactions in the troposphere and lower stratosphere to form secondary products.

Four factors make pollution from the vehicles more serious in developing countries.

(1) Poor quality of vehicles creating more particulates and burning fuels inefficiently.

(2) Lower quality of fuel being used leads to far greater quantities of pollutants.

(3) Concentration of motor vehicles in a few large cities

(4) Exposure of a larger percentage of population, that lives and moves in the open.

13.5 HARMFUL EFFECTS OF EMISSIONS

The high concentration of particulates in the atmosphere over large urban and industrial areas can produce a number of general effects. Smoke and fumes can increase the atmospheric turbidity and reduce the amount of solar radiation reaching the ground. The overall effect of air pollution upon the biosphere and the built environment can be broadly considered under 3 headings: The effect upon:

(1) Buildings and materials
(2) Soil, vegetation, crops and animal life
(3) Human beings

13.5.1 BUILDING AND MATERIALS

The fabric of buildings, that are surrounded by heavily polluted air for years undergo chemical changes. Gradual erosion takes place and this is only too evident when grimy upper surface is removed. A good example is that of the famous historical monument 'Taj Mahal' at Agra, which, on account of reaction of Sulphur-di-oxide, emitted from neighbouring industries, with the limestone has slowly, started turning yellow. As a result, on Court's directives, a number of measures have been taken to protect our national heritage monument *e.g.* closure of neighbouring heavy polluting industries, operation of only non-polluting vehicles like battery busses, tonga in the vicinity of Taj Mahal etc.

13.5.2 SOIL, VEGETABLE AND ANIMAL LIFE

The presence of gaseous pollutants in the air and deposition of particulates on to the soil can affect plants. It can effect the cattle and animals too as they have been found to develop breathing difficulties and suffer from low yield of milk, lameness and joint stiffness in a polluted environment.

13.5.3 HUMAN BEINGS

Smoke and SO_2 cause the general and most widespread effects of air pollution on people. Atmospheric smoke contains potentially carcinogenic organic compounds similar to those that occur in cigarette tobacco smoke. The CO affects the cardiovascular system, NO_xs affect the respiratory system, Ozone causes increased sensitivity to infections, lung diseases, irritation in eyes, nose and throat etc.

13.6 STEPS TAKEN SO FAR AND THEIR IMPACT

With the alarming increase in the s, especially in the big cities, Government has taken some important initiatives in the recent years. To start with the emphasis and implementation has been primarily in the big cities but gradually to spread throughout the country. These relate to the progressive tightening of the auto-emission norms (1991,1996,1998 & 2000) and fuel quality specifications (1996) as recommended by the Central Pollution Control Board (CPCB).

Till early 1994 ambient air quality standards in India were based on 8 hourly average times only. In April 1994, these standards were revised and 24 hourly standards were also prescribed. National ambient air quality standards are prescribed for three distinct areas viz.

(*a*) Industrial,
(*b*) Residential, rural and other areas and
(*c*) Sensitive areas.

Following steps have been taken so far:

(1) Unleaded Petrol. With the gradual reduction of lead content in petrol and finally supply of unleaded petrol for all vehicles from Sept. 1998 in the capital city of Delhi, a lethal pollutant from vehicular exhaust has been removed. The lead content in the atmosphere near traffic intersections of Delhi has reduced by more than 60% with this measure.

(2) Sulphur in diesel. The sulphur content in the diesel supplied has been reduced from 0.5% in 1996 to 0.25% in 1997 so as to meet the EURO-II norms.

(3) Tightening of the Vehicular Emission Norms. From 1995 new passenger cars were allowed to register only if they were fitted with catalytic converters. Emission norms for such cars were tightened by 50% as compared to 1996 norms. With the recent directions of the Hon'ble Supreme Court, passenger cars (both petrol and diesel) are required to meet atleast EURO-I norms in June 1999 and from Apr. 2000 only such vehicles meeting EURO-III norms will be permitted to register in the NCR of Delhi. CNG operated vehicles are also permitted by the Supreme Court directions.

(4) 2-T Oil for Two Stroke Engines. From 1.04.99, on the recommendations of CPCB, the low smoke 2T oil became effective. To prevent the use of 2T oil in excess of the required quantity premixed 2T oil dispensers have been installed in all the petrol filling stations. Sale of loose 2T oil has also been banned from Dec. 1998.

(5) Phasing out of Grossly Polluting Vehicles. On CPCB's recommendations initially 20 yr. old vehicles were prohibited from plying from Dec.1998, followed by phasing out of 17 yr. old vehicles from Nov. 1998 and 15 yr. old from Dec. 1998.

13.7 NOISE POLLUTION AND ITS CONTROL

Of late, noise has been recognized as a pollutant which until recently was considered only as a nuisance. The Central Pollution Control Board (CPCB) has notified the ambient noise standards in 1987 under section 20 of the Air (Prevention and Control of Pollution) Act, 1981. The noise standards specify limits as 55dBA and 45dBA as limits for day and night time respectively for residential areas, 75 dBA and 70dBA in the day and night time for industrial areas, and 50 dBA and 40 dBA in the day and night for silence zones. Special campaign for reduction in use of fire crackers in Delhi have resulted in reduced pollution levels during Diwali in 1999.

The creation of noise creates the pollution by increasing the sound level of the atmosphere. It is estimated that the environmental sound level is doubling in loudness after every 10 years. The increase in sound level of the atmosphere is not the major problem presently but desirable sound level is essential in power plants and every step should be taken to reduce the sound level of the power plants to a tolerable level.

Heavy noise environment has extremely unpleasant effects on people exposed to them. Continuous exposure to noise level above 100 dBA has adverse effect on hearing ability within a short time. Therefore, in world energy conference of 1971, a study of noise suppression in thermal power plants occupied a major percentage of the, seminars conducted.

The main sources of noise in a power plant are turbo-alternators, fans and power transformers. The simple sound proofing system consists of adequate insulation of the turbine, body and piping, with a dashpot on the air-inlet to the fans. A moderate treatment includes a complete casing for turbo-alternator. An intensive treatment includes building around the turbine group and fans and complete sound-proofing of the transformers and switch gears.

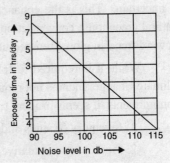

Fig. 13.1. Toleraable Hours in a Day for Different Noise Level.

The tolerable sound level and time of exposure are equally important in the design of allowable noise level in power plant. The working hours of the worker must be reduced in a highly noised area.

90 decibels of a sound level is considered maximum tolerable level for 8 hours exposure. The noise level in more than 50% power plants in U.S.A. is above 90 decibels. The length of exposure versus tolerable noise level for the workers is shown in Fig. 13.1.

Presently enough attention has been given in many developed countries to reduce the noise level in the power plants to the tolerable level.

The curves derived from hearing tests as shown in Fig. 13.2 made on workers show that within 10 years, over 10% of employees subject to 8 hours a day of 95 dBA noise level will suffer from hearing impairment and that of 40 years, 20°/v will be affected to the same extent by the noise level of 92 dBA.

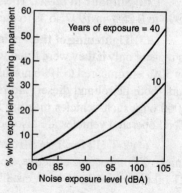

Fig. 13.2

13.8 GREEN HOUSE GASES AND THEIR EFFECTS

The greenhouse effect plays a crucial role in regulating the heat balance of the earth. It allows the incoming short-wave solar radiation to pass through the atmosphere relatively unimpeded; but the long-wave terrestrial radiation emitted by the earth's surface is partially absorbed and then re-emitted by a number of trace gases in the atmosphere. These gases known as GHGs (greenhouse gases) are: water vapor, carbon dioxide, methane, nitrous oxide and ozone in the troposphere and in the stratosphere. This natural greenhouse effect warms the lower atmosphere.

If the atmosphere were transparent to the outgoing long wave radiation emanating from the earth's surface, the equilibrium mean temperature of the earth's surface would be considerably lower and probably below the freezing point of water. Mere incidence of GHG's in the atmosphere, by itself, is no concern. What is more important is that their concentration should stay within reasonable limits so that global ecosystem is not unduly affected. However, by increasing the concentrations of natural GHG's and by adding new GHG's like chloroflouro carbons the global average and the annual mean surface-air temperature (referred to as the global temperature) can be raised, although the rate at which it will occur is uncertain. This is the enhanced greenhouse effect, which is over and above that occurring due to natural greenhouse concentration. Such a rise in the atmospheric concentration of GHG's has led to an upward trend in global temperature.

While it is required to follow the general commitments under the Framework Convention on Climate Change, India is not required to adopt any GHG reduction targets. Irrespective of international commitments, it seems prudent to ready with

1. Inventory of sinks and sources of GHG emission.
2. Predict the cumulative impact of national and international GHG emissions to plan for temperature and sea level rise.
3. Devise landuse plans for the coastal areas likely to be affected.
4. Devise water and land management strategies especially agricultural sector.

13.9 FOSSIL FUEL POLLUTION

The exhaust gases and particulate matter emitted from combustion systems affect the environment in several ways. The major classifications are:

13.9.1 URBAN AIR POLLUTION

It including:

1. Photochemical smog

The reactants are nitric oxide (NO), unburned hydrocarbons (UHCs), and sunlight (*i.e.*, photons). The products (after a few hours of time) are oxidants such as ozone (O_3) and peroxyacetyl nitrate (PAN = $CH_3CO_3NO_2$),

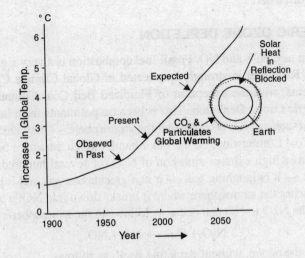

Fig. 13.3 Increasing global temperature due to green house effect.

aldehydes (RCHO, where R = a hydrocarbon radical, *e.g.*, the methyl radical, CH_3), and aerosol haze. An intermediate product is nitrogen dioxide (NO_2), which gives a brownish color to the atmosphere and reaches a peak concentration about half way through the reaction process, at the time at which the original NO is significantly converted to NO_2. The smog products are eye irritants and they diminish lung capability. The different hydrocarbon gases have significantly different smog forming potential. Methane, for example, is very unreactive, whereas ethylene (C_2H_4) and propylene (C_3H_6) are quite reactive. Thus, the smog impact of the hydrocarbon emitted is determined not only by its concentration, but also by its photochemical reactivity.

2. Carbon monoxide (CO)

3. Sulfur dioxide (SO_2)

4. Nitrogen dioxide (NO_2)

5. Toxic gases, vapors, and heavy metals

6. Particulate matter

13.9.2 ACID RAIN

The precursors of which are nitric oxide (NO) and sulfur dioxide (SO_2). Over hours and days the NO and SO_2 oxidize to NO_2 and SO_3, respectively, which subsequently form acids of nitrogen and sulfur. Because of the relatively long time for the chemical transformations to occur, the impact of the acid is generally felt several hundred kilometers downwind of the sources. The sulfur also forms sulfate aerosol. Sulfate aerosol reflects sunlight and is thought to be keeping some industrialized parts of the world cooler than they would otherwise be. That is, these industrialized parts of the world are not receiving proper warming by the greenhouse effect.

13.9.3 GLOBAL CLIMATE CHANGE

It also knows as the Greenhouse Effect. The major greenhouse gases are CO_2, CH_4, nitrous oxide (N_2O), and chloro-fluoro-carbon species (*i.e.*, CFCs), though ozone and soot also play a role. The Greenhouse Effect is already discussed.

13.9.4 STRATOSPHERIC OZONE DEPLETION

Due to gases such as CFC's and NO. Fossil fuel combustion is a very significant cause of Urban Air Pollution and Acid Rain, and is strongly implicated in Global Climate Change. However, land-based combustion systems, with the exception of Fluidized Bed Coal Combustors, are not strongly implicated in Stratospheric Ozone Depletion. This is because pollutants from land-based sources capable of destroying ozone (such as NO) do not reach the stratosphere — they are destroyed in the troposphere. Fluidized Bed Coal Combustion, on the other hand, though attractive because of its low emissions of SO_2 and NO, has a high exhaust emission of N_2O (of several hundred parts per million). Although N_2O is not toxic — it is laughing gas — it is a greenhouse gas and it has few enemies in the troposphere. Thus, it reaches the stratosphere where it breaks down into NO, which is an ozone depleting gas. Conversion of the N2O to NO can occur as follows in the stratosphere:

$$N_2O + O \longrightarrow NO + NO$$

The NO depletes the ozone, without depleting itself, as follows:

$$O_3 + NO \longrightarrow NO_2 + O_2$$
$$NO_2 + O \longrightarrow NO + O_2$$

Overall, an O_3 is lost, and an O-atom, which could have formed O_3 through the reaction below, is lost.

$$O + O_2 \longrightarrow O_3$$

13.9.4 ACID FOG

A recently noted major acid pollutant is acid fog. Its origin is the same as acid rain or snow, *i.e.*, sulfuric and nitric oxides from power plants and, to a lesser extent, motor vehicles. It forms by the mixing of these pollutants with water vapor near the ground. The acid vapors then begin to condense around very tiny particles of fog or smog, pick up more water vapor from the humid air, and turn into acid fog. When the water in the fog burns off (evaporates) due to the sun or other causes, drops of nearly pure sulfuric acid are left behind. It is these drops that make acid fog so acidic. In Los Angeles and Bakersfield in southern California, the mists have a pH of 3.0 compared with 4 or 4.5 for acid rain. Acid fog 100 times as acidic as acid rain has been detected. Cases have been reported where people had trouble breathing when it was foggy. The problems of fog are now believed by some to be more serious than those of smog in these areas.

Many researchers consider the effects of acid precipitation, especially the changing of soil chemistry, to be irreversible and fear its long-range effects. Monitoring programs of air, soil, and water are being instituted to ascertain these long-range effects. How-ever, the uncertainty about the real extent of the problem adds to the prevailing disquiet regarding it.

13.10 POLUTION DUE TO COMBUSTION OF FUEL

13.10.1 Gas Fuel

When a gaseous fuel, such as natural gas, burns it undergoes a series of processes. These are as follows:

- Either in a premixer or in the combustion chamber, the fuel must mix with air.
- The fuel (and the air) must mix with hot, burning and burnt gases present in the chamber. The burning and burnt gases provide heat and active species (O, H, and OH), which upon mixing with the fresh fuel and air cause the fresh fuel to ignite and react with the air.
- The fuel then undergoes a series of chemical reactions.

13.10.2 METHANE

If the fuel is methane (the primary constituent of natural gas), the following reaction steps occur.

Due to the heat and attack by the active species, the methane reacts to a methyl radical (CH_3), which reacts to formaldehyde (HCHO). The formaldehyde reacts to a formal radical (HCO), which then forms carbon monoxide (CO). Through these steps, the active species are used up and H_2 and H_2O are formed in addition to the CO. Overall, the process is as follows:

$$CH_4 + O_2 \longrightarrow CO + H_2 + H_2O$$

Thus, the original fuel is converted into two new fuels (CO and H_2O) and into one product H_2O). The process occurs very quickly, within a fraction of a millisecond to a few milliseconds, depending on the flame temperature, pressure, and fuel-air ratio. (Actually, methane is one of the slower burning hydrocarbon gases.) The process is called **Oxidative Pyrolysis.**

Following oxidative pyrolysis, the H_2 oxidizes, forming H_2O, replenishing the active species, and releasing heat. This occurs very quickly, usually in less than a millisecond.

$$H_2 + (1/2)\ O_2 \longrightarrow H_2O$$

Finally, the CO oxidizes, forming CO_2 and releasing more heat. This process is generally slower than the other chemical steps, and typically requires a few to several milliseconds to occur.

$$CO + (1/2)\ O_2 \longrightarrow CO_2$$

The combustor may be thought of as having **flame zones**, in which the free radical activity is high, the fuel undergoes oxidative pyrolysis, the hydrogen oxidizes, and the CO begins to oxidize, and **post-flame zone**, in which the CO continues to oxidize and reaches its final exhaust concentration.

13.10.3 ALKANES

When Alkanes (C_nH_{2n+2}) of n = 2 and above are burned, the reaction chemistry is different from that of the methane. First, the Alkanes undergo conversion to one or more alkenes (C_nH_{2n}), especially to ethylene (C_2H_4) and pyropylene (C_3H_6). The alkenes then undergo oxidative pyrolysis reaction to CO, H_2, and H_2O, but though a different chemical mechanism than given above for methane. The overall reactions are:

$$C_nH_{2n+2} \longrightarrow C_nH_{2n} + H_2$$
$$C_nH_{2n} + O_2 \longrightarrow CO + H_2 + H_2O \text{ (not balanced)}$$
$$H_2 + (1/2) O_2 \longrightarrow H_2O \text{ (+ Active Species + Heat)}$$
$$CO + (1/2) O_2 \longrightarrow CO_2 \text{ (+ Heat)}$$

13.11 POLUTION DUE TO GAS COMBUSTION

13.11.1 UNBURNED HYDROCARBONS (UHCS)

Any fuel entering a flame will be reacted. Thus, when unburned fuel is emitted from a combustor, the emission is caused by fuel 'avoiding' the flame zones. For example, in piston engines, fuel-air mixture 'hides' from the flame in the crevices provided by the piston ring grooves. Further, some regions of the combustion chamber may have a very weak flame, that is, they have either very fuel-lean or very fuel-rich conditions and consequently they have a low combustion temperature. These regions will cause intermediate species such as formaldehyde and alkenes to be emitted. Sometimes the term 'products of incomplete combustion,' or PICs is used to describe such species. The term UHC represents the sum of all hydrocarbon species emitted.

13.11.2 CARBON MONOXIDE (CO)

Carbon monoxide is emitted because the temperature is too low to effect complete oxidation of the CO to CO_2, because the time (i.e., the residence time) available in the combustion chamber is too short, or because there is insufficient oxygen present. Usually, it is more difficult to design and operate a combustor for very low CO than for very low unburned hydrocarbons. Exhaust emissions of CO are controlled by providing the combustor with sufficient air to assure oxidation of the CO. However, too much air is 'bad,' since then the post-flame zone will be too cool to oxidize the CO. Catalysts in the exhaust stream are also used to control CO. These provide about a 90% conversion of the CO to CO_2 and typically use platinum (or a mixture of platinum group metals) as the active sites (on a ceramic or metal substrate) to oxidize the CO. Almost all automobiles sold today in the US and in many other countries are equipped with three-way catalysts. By running the engine at stoichiometric fuel-air ratio there is enough O_2 left in the exhaust to effect oxidation of the CO and UHCs in the catalyst, and there is a sufficient quantity of reducing species in the exhaust to effect chemical reduction of the NO_x to N_2. Thus, the three 'ways' are CO, UHCs, and NO_x.

13.11.3 NITRIC OXIDE (NO_X)

Nitric oxide forms by attack of O-atoms on N_2. The predominant mechanism is the extended Zeldovich mechanism:

$$N_2 + O \longrightarrow NO + N$$
$$N + O_2 \longrightarrow NO + O$$
$$N + OH \longrightarrow NO + H$$

When this process occurs in the post flame zone, the resulting NO is called thermal NO, because the amount of NO formed is strongly temperature sensitive. In order to limit the amount of NO formed, it is necessary to reduce the combustion temperature (which is generally very effective because of the strong temperature dependency of the NO formation), reduce the residence time (though this will in general increase the CO emission), or limit the availability of oxygen.

NO also forms in the flame zone. In this case, the O-atom and OH concentrations affecting the formation of NO have much higher concentrations than in the post-flame zone. Other mechanisms also contribute, such as the prompt mechanism:

$$N_2 + CH \longrightarrow HCN + N$$

The hydrogen cyanide (HCN) and N-atom oxidize to NO. Because of the high free radical concentrations in the flame zone, the rate of production of NO is faster in the flame zone than in the post flame zone. However, the time available in the flame zone is generally short compared to the time available in the post-flame zone.

Some of the NO may oxidize to NO_2 in the combustor. Thus, the emission is expressed as $NO_x = NO + NO_2$.

Many methods are used to control the emission of NO_x, and a great deal of research has been done on this. A great deal of money is spent on NO_x control throughout the world. As indicated above, automotive NO_x is controlled though the use of the three-way catalyst. Another name for this is non-selective catalytic reduction (NSCR). Some industrial and utility combustors use a different type of exhaust catalyst. This is a selective catalytic reduction or SCR. In this case, ammonia (NH_3) is injected into the exhaust stream ahead of the catalyst. Across the catalyst, the NO_x and NH_3 react to form N_2. The conversion efficiency is about 80 to 90%.

$$NO_x + NH_3 \longrightarrow N_2 + H_2O + O_2 \text{ (not balanced)}$$

Another method with ammonia injection is selective non-catalytic reduction (SNCR). If ammonia is injected into the exhaust at higher temperatures (*i.e.*, at about 1200K) than used in the SCR process, the ammonia reduces the NO_x without the need for the catalyst. However, if the temperature is too high, the ammonia oxidizes into NO.

Combustion modification is also widely practiced to control NO_x and a whole class of low-NO_x engines and combustors has grown up. These have NO_x emissions anywhere from about 10% to 70% of the 'dirty' pre-NO_x-control combustors. The concepts used to effect the NO_x control are no mystery. Reduction of combustion temperature is very effective, thus, injection of water, or a diluents such as steam or recycled exhaust (or flue) gas, is widely practiced. Another diluents is air, and thus, lean premixed combustion is very effect in controlling NO_x. Another method widely used is staged combustion. That is, the first stage of combustion is conducted fuel rich. This creates a lot of CO and UHCs, but it doesn't create much NO_x because of the lack of O_2. Some heat is transferred from the rich gases (for example, to the working fluid of a steam-electric power-plant burner), and then the remaining air is added into the flame to burn off the CO and UHCs. Now, NO_x formation is limited, because of the reduced combustion temperature.

13.11.4 SOOT

Soot is composed of particles of a few hundred angstroms in size. Generally, soot is a fluffy carbon material, though it is generally not pure carbon. Soot forms under hot, fuel-rich conditions, and, once formed, oxidizes relatively slowly. (Soot requires on order of 100 milliseconds to oxidize at typical combustion temperatures.) Soot is controlled by adding air to the flame zone, thus eliminating the hot, rich pockets of gas, which produce it, or by providing a secondary combustion zone of sufficient O_2, temperature, and time to oxidize the soot.

A low emission combustor would have CO and NO_x exhaust concentrations of less than about 100 ppm (parts per million by volume) each, and an UHC emission of under about 10 ppm. A very low emission system would have CO and NO of about 10 ppm each, and UHC less than a few ppm's. High emission burners have CO and NO_x emissions in the several hundred to a few thousand-ppm ranges.

New combustors and emission control methods are being developed 'every day,' and for some situations the goal of 'zero emissions' is being approached. Of course, the term 'zero emissions' does not mean absolute zero emission. It depends somewhat on the 'reference point.' Generally, it means CO and NO_x emissions in the few ppm range. For example, Honda recently announced a special new exhaust catalyst system, which gives very low emissions.

13.12 POLUTION DUE TO LIQUID FUEL

When a liquid fuel, such as oil, is burned, there are additional aspects of the combustion process and the pollutant formation. Specifically:

13.12.1 ATOMIZATION

It is common practice to inject the fuel into the combustor (or premixer) through a nozzle, which atomizes the fuel. That is, the continuous stream of fuel is broken up into a mist of tiny droplets. There are many types of nozzles, some of which rely on very high feed pressures to atomize the fuel, and some of which rely on assistance from steam and air to effect good atomization. Generally, the finer the spray produced by the nozzle the better the combustion process.

13.12.2 VAPORIZATION

The fuel droplets vaporize as they receive heat by mixing with the hot gases in the combustion chamber. Heat can also be received by radiation from any hot refractory wall of the combustion chamber.

13.12.3 MODES OF COMBUSTION

If the vaporization process is fast compared to the reaction chemistry, the combustion of the liquid fuel occurs mainly as clouds of vapor. Thus, the sequence of processes is atomization, vaporization, mixing, and chemical reaction, and the 4-step chemical mechanism given above under Alkanes is valid. On the other hand, if the vaporization process is slow, droplet burning can occur. That is, individual flames may encircle individual droplets, thereby effecting oxidation of the oil.

Generally, though not necessarily, the UHC, CO, and NO_x emissions increase with oil burning compared to gaseous fuel burning. There are several reasons for this, including:

- The additional time required for vaporization,
- The more complex hydrocarbons involved in the fuel,
- Poorer mixing of the fuel vapor with the air,
- Higher localized temperatures (leading to higher NO_x),
- And the formation of deposits in the combustor which can adversely affect combustion 'goodness' by affecting flame shape and burner aerodynamics.

Also, some liquid fuels contain sulfur, organic nitrogen, and mineral elements, which lead to additional pollution. Generally, distillate fuel oil and diesel fuel are low in these compounds, and gasoline and aviation fuel are even lower yet in these compounds. Other (heavier) fuel oils are relatively high in these compounds. Sulfur in the fuel reacts essentially completely to sulfur dioxide (SO_2) upon combustion. Also, a small fraction of the sulfur oxidizes to sulfur trioxide (SO_3), which is a problem because it will readily form sulfur acid mists if the exhaust temperature is too low. Thus, in order to prevent corrosion of the energy system equipment, the exhaust temperature may be maintained higher than it would be without the sulfur. Thus, thermodynamic efficiency is degraded, and the ratio of CO_2 to unit of electrical energy produced increases. Sulfate particulate emission can also result from the sulfur. (Within the 3-way catalyst of an automotive engine, the small amount of SO_2 present can be reduced to hydrogen sulfide (H_2S). Ever smell this?) Organic nitrogen in the fuel, also called fuel-bound nitrogen, will end up partly as NO_x upon combustion and partly as N_2. Staged combustion generally promotes the N_2 end point over the NO_x end point. Mineral matter in the oil can cause several problems, including deposits and flyash emissions. The deposits can promote reactions, which affect the other pollutants, such as SO_2.

About 65% of the oil used in the US is used by the transportation sector that about 25% is used by industry, and that most of the balance is used for residential and commercial heating. Very little is used in the US for electrical generation. Most of the discussion in the paragraph immediately above pertains to industrial, utility, and R/C burning of oil. However, the biggest sector involved in the control of emissions from oil burning is the transportation sector, because the fuels burned by this sector (*i.e.*, gasoline and diesel fuel) are mainly derived from oil. Thus, advancements in engine combustion technology and catalytic exhaust treatment are very important to the control of emissions from oil burning. A major challenge is the development of automotive exhaust treatments which work well for fuel-lean engine combustion. This is important because engines are thermodynamically more efficient when operated with excess air. However, the present 3-way catalyst has a poor NO_x reduction efficiency when the engine is operated lean. Another major challenge is control of NO_x and soot particulate from diesel engines. The oil problem is being addressed with advanced combustion technology and exhaust particle traps.

13.13 POLUTION DUE TO SOLID FUEL

The solid fuel of primary interest is coal. In power plant generally coal is burned for electrical generation.

Coal is burned in several ways, depending on coal particle size. Lumps of coal, including coal particles larger than about 0.25 inch in diameter, are spread on a grate and burned. Conceptually, this is not unlike the burning of wood logs in a fireplace, though industrial and utility coal grate burners are substantially more sophisticated than a fireplace. There is a limit to the size of a grate, (around the order 10 meters square is maximum). Grate coal burners are called Stokers. Stokers come in several forms, and are used for small and medium size coal combustion systems.

Coal particles of about 1/8th inch diameter are burned in fluidized bed coal combustors (FBC). In this system, the coal particles are injected into a bed of limestone or dolomite particles strongly churned or agitated by blowing air through the bed from below. The bed has the appearance of "boiling." The fluidizing medium is limestone (or dolomite), rather than ordinary sand (which could be used), so that the sulfur from the coal is taken up by the limestone and converted into calcium sulfate particles ($CaSO_4$), which are periodically or continuously removed from the bed. Also, thermal NO formation is very low because of the relatively low combustion temperature of the bed (about 800 to 900°C). In addition, flyash emission is controlled to some extend by retaining the ash in the bed. However, the nitrogen contained in the coal, *i.e.*, the fuel-bound organic nitrogen, converts to nitrous oxide (N_2O). If the bed temperature were higher than 800-900°C, the fuel-bound nitrogen would convert to NO_x. Fluidized bed coal combustion is a relatively new commercial technology, which appears to be favored more in Europe than the US.

Although the Stoker and Fluidized Bed technologies are important, the majority of the coal burned in steam-electric power plants is burned in Pulverized Coal Combustion Furnaces. Pulverized coal is coal, which has been ground (*i.e.*, pulverized) into a fine dust, of about 70 micrometers (*i.e.*, microns) mean diameter. Pulverized coal combustors are suspension burners - that is, the coal dust is carried by the furnace air and gases and burned in suspension.

In all solid fuel burners, the fuel undergoes heating and devolatilization as the first stage of the burning process. Devolatilization is analogous to the vaporization process for the liquid fuel. Devolatilization means that part of the solid fuel decomposes and forms gases and tars upon heating. The fraction of the fuel, which forms volatiles, and the composition of the volatiles depend on the nature of the fuel and the particle heating process. Typically, a combustion coal is about 50% volatile matter by weight. The volatiles released from the coal are made up of the following components: CO, H_2, light hydrocarbon gases (such as methane, ethane, ethylene, and propane), oxygenated hydrocarbons, medium molecular weight hydrocarbons, high molecular weight hydrocarbons called tars (which are vapor at furnace temperatures), and inert gases (such as CO_2 and H_2O). The volatiles burn via a mechanism similar to that described above for the Alkanes. Typically, devolatilization occurs within 100 milliseconds, and the volatiles burning occur within a few milliseconds. The particles remaining after devolatilization are composed of char (*i.e.*, mainly carbon) and ash (*i.e.*, mineral matter).

Following the release of volatiles from the solid fuel particle, it is possible for oxygen to diffuse to the surface and oxidize the char particle. Char particle oxidation requires about 100 to several hundred milliseconds of time. The furnace volume has to be big enough to accommodate this. The following reactions happen at the char particle surface (including surfaces created by fissures in the particle):

$$C + (1/2) O_2 \longrightarrow CO$$
$$C + O_2 \longrightarrow CO_2$$

As the char particle burns away, the mineral matter imbedded in the coal as small inclusions gets very hot, becomes molten, and fuses together to form liquid ash particles, which ultimately solidify. Typically, 3 to 5 'big' ash (*i.e.*, flyash) particles form per original pulverized coal particle. Additionally, the volatile mineral matter vaporizes during the devolatilization and char burning stages, and forms tiny, sub-micron particles upon nucleation and condensation. The sub-micron particles can be more of a problem than the 'big' (1 to 10 micron) particles, because the sub-micron particles tend to carry disproportionate amounts of the toxic heavy metals found in coal. A toxic heavy metal emitted as a gas is mercury.

It should be noted that coal contains about every element found in nature. Although C, H, and O are the major elements found in coal, there can also be significant amounts of S and N. Some coals have as much as 10% S by weight. The sulfur is contained in both the organic and inorganic fractions of the coal. It is possible to remove some of the sulfur containing 'rocks' found with the coal using gravity separation methods. Processes have been studied for removing additional amounts of S from coal; however, these processes are not widely used commercially.

Coal contains about 0.5 to 1.5% by weight N bound into the organic structure of the coal. When the coal is heated, much of the N is released during the devolatilization stage as hydrogen cyanide (HCN), other cyano species, and as ammonia (NH_3). These species are rapidly converted into NO or N_2, and thus are not emitted directly from the combustor.

Some coals (*e.g.*, British coals) contain chlorine, which upon combustion can be emitted as hydrogen chloride (HCl).

Typically, coal has about 10% by weight mineral matter. This mineral matter is high in Si, Al, Fe, and Ca and these are the primary elements found in the 'big' flyash particles. However, many other inorganic elements are associated with the mineral matter of coal, including Mg, Na, and K. Toxic metals are found in trace amounts in coal, including Ni, Pb, Cd, Cr, As, Se, and Hg. Generally, the toxic metals are found associated with the sub-micron flyash, though Hg will be emitted as a gas. Radioactive isotopes are also emitted from coal combustors. It has been estimated that more radioactivity is emitted from a coal combustion power plant than from a nuclear power plant. Emission control is practiced as follows for Pulverized Coal Furnaces:

A significant fraction of the **installed** cost of a coal-fired electrical generating station is devoted to environmental control — about 30 to 40%. Also, a significant fraction of the **operating and maintenance** cost is devoted to environmental control. Sulfur (*i.e.*, SO_2 emission) is controlled by burning a low sulfur coal, by pre-combustion coal cleaning, and by flue gas desulfurization.

NO_x is controlled by the use of low-NO_x burners, by SNCR, and by SCR. Note that NO_x is formed from both the air nitrogen and the fuel-bound nitrogen.

Flyash is controlled by stack gas particulate removal using electrostatic precipitators (ESPs) or baghouses (*i.e.*, the stack gas is filtered).

Coal is also burned by gasifying it and burning the gases in a gas turbine engine, which is part of a combined cycle. There are about five integrated gasification combined cycle (IGCC) power plants operating in the US. The combustible compounds present in the gas (*i.e.*, the synthetic gas) are mainly H_2 and CO, though in some cases CH_4 is also present in significant amounts. Other gases present are CO_2, H_2O, and N_2. If the gasifier is 'air blown' there is a lot of N_2 present (about 30 to 50% by volume), and thus the gas has a low heating value (of about 100 to 200 Btu/scf). If an oxygen-blow gasifier is used, the heating value is in the medium Btu range (about 1/4 to 1/2 that of natural gas), though the power-plant now must also include an air separation plant for making O_2 — actually, oxygen enriched air is produced. The gas is cleaned to remove sulfur (as H_2S) and particulate matter before entering the gas turbine. Some gasifier product streams carry several thousand ppms of NH_2 (formed from the coal-bound nitrogen), which needs be controlled either by scrubbing or by special combustion methods, so that it does not react to NO_x in the combustor. However, other gasifiers emit only about 200 ppm ammonia, and when this gas is burned in a gas turbine, the sum of the fuel-NO_x and thermal NO_x leaving the burner is only about 30 to 40 ppm. For reference, when natural gas is used as the fuel in a gas turbine engine equipped with state-of-the-art, commercially available, lean premixed combustors, the NO_x emission is about 9 to 25 ppm (depending on the make and model of the engine.) The coal IGCC system has an overall efficiency in the low 40% range, whereas the natural gas fired combined cycle is about 50 to 58% efficient (as discussed earlier). The IGCC system has higher CO_2 per kW-hr than the natural gas fired system.

13.14 AIR POLLUTION BY THERMAL POWER PLANTS

The environmental pollution by thermal power plants using fossil fuels poses a serious health hazard to modern civilization. Air pollution by thermal plants is a contributing factor in the cause of various respiratory diseases and lung cancer and causes significant damage to the property in addition to causing annoyance to the public.

The thermal power plants burning conventional fuels (coal, oil or gas) contribute to air pollution in a large measure. The combustible elements of the fuels are converted to gaseous products, and non-combustible elements as ash. The common gaseous products of interest are sulphur dioxide, nitrogen oxide, carbon dioxide and carbon monoxide, and large quantities of particulate materials as fly ash, carbon particles, silica, alumina and iron oxide.

The energy industries are one of the largest sources of environmental pollution. A 350 mW coalfired station emits about 75 tons of SO_2, 16 tons of nitrogen oxide, and 500 tons of ash per day if no safeguards is adopted. All steam-generating plants also discharge nearly 60% of heat produced back to the atmosphere irrespective of the fuel used.

Due to large emissions from the thermal power plants, air pollution has become an international problem. This problem is mainly faced by 11 countries in the world, which share 80% of the world's fossil-fired generating capacity. Emissions from their power plants have grown to point where we and all of them now must think for controlling the pollution contributing to a common atmosphere.

Many countries have unique air pollution problems. These are due to fuel characteristics, unfavorable topographical conditions, concentration of power plants in limited area and high population densities. The production capacities of 11 countries, which share 80% world-electric generation. The major pollutants given off by fossil fuel combustion are particulates, SO_2 and other gases and it will be sufficient to discuss about these pollutants.

13.15 WATER POLLUTION BY THERMAL POWER PLANTS

Another serious problem is the water pollution caused by thermal power plants. The water pollution is caused by discharging hot condenser water and water discharged into the river carrying the ash of the plant. The discharge of polluted water causes hydrological and biological effects on the surrounding ecology. The biological study should determine the types of aquatic organisms in the area and their adaptability to the environmental variations.

Thermal pollution of water is very important for the fish cultivation, as their growth is very susceptible to the temperature changes.

Another important constituent in the discharge of cooling water is residual chlorine as chlorine or sodium hypochlorite is used to prevent fouling of the condensers.

Another serious problem associated with the discharged water is the ash carried by the water. The ash gets spread over the large cultivated area along the path of the river and affects the agricultural growth very much. This is because; the ash has high alkaline characteristics, which are injurious for the growth of many agricultural products. The ash destroys the fertility of the land forever. Such phenomenon was badly experienced when the ash from Koradi thermal power station in Maharashtra was discharged in the river. The wastewater from water demineralization plant contains large quantities of chlorides of Ca, Mg, Na and K. This wastewater is channeled out to some river or to an ash pond along the fly ash. On the way to river or ash pond, these salts percolate in the nearby soil and make the ground water salty. In the ash pond, the situation is worse as there is continuous accumulation of these salts and the pond reaches a saturation level of these salts. The process of salt saturation in the pond is further accelerated by solar evaporation of the water. The wells on the area covering a few kilometers from the ponds become salty and polluted water from these wells becomes harmful for human consumption as well as for irrigation purposes. Discharging these salts with the wastewater aggravates the pollution problem but also loses them, even though; their recovery is simple and economical

The wastewater can be treated first with lime, to precipitate magnesium hydroxide and then with soda ash to get precipitated calcium carbonate and the resulting sodium chloride solution can be reused far regeneration of softeners. The above-mentioned reactions are listed below.

$$MgCl_2 + Ca(OH)_2 = Mg(OH)_2 \downarrow + CaCl_2$$
$$CaCl_2 + Na_2CO_3 = CaCO_3 \downarrow + 2\,NaCl.$$

13.16 ENVIRONMENT CONCERNS AND DIESEL POWER PLANTS

With the emergence of liquid fuel based power stations in India, the question of environment pollution has become a matter of raging debate. The coal based thermal power stations, in its earlier stages of inception, were far more polluting? It was because of the combination of sulphur-based pollutants, nitrogen based gaseous matter and also particulate matter with very high ash content being released in the atmosphere.

Globally, environmental regulatory authorities are increasingly concerned with NO_x and SO_x emissions and are liable to consider introducing stringent regulatory standards in the future. While the levels of SO_x emissions is the function of sulphur content inhered in the fuel being used for combustion? NO_x is created by the chemical activity between atmospheric oxygen and nitrogen during combustion. The level of NO_x depends on the combustion conditions.

Optimal combustion in a diesel engine depends upon the achievement of the right balance of equation between compression/combustion pressure, compression ratio, air-to-fuel ratio and mean effective pressure. The toughest of the emission standards currently being considered by various national and international agencies, calls for limitation of NO_x emissions to 600 ppm(15% O_2) for generator sets operating on ocean bound vessels. The shore-based power stations shall demand for further lower limits due to proximity to the human inhabitation.

Burning heavy fuel in diesel engine is convenient mainly due to economics of residual fuel combustion for power generation. Diesel engine designers' world over will increasingly come under pressure to introduce superior combustion features for producing lower levels of SO_x and NO_x.

The exhaust gas composition of emissions or pollutants given above is for using furnace oil of different grades and varying sulphur contents. The exhaust gas of medium speed engines comprises of a host of constituents. In the case of combusting heavy fuel like furnace oil, these emanate either from combustion air and fuel used, or they are reaction products, which get formed during the combustion process. Only some of these are considered to be pollutants for the atmosphere:

Typical Exhaust Gas Emission Values For Modern 4-stroke Diesel Power Plants Using Heavy Fuel

Fuel	FO-kV(2.1% sulphur)	FO (4.5% sulphur)
Load(%)	100.0	100.0
Mech, output (kW/cyl.)	990.0	990.0
Speed (rpm)	500.0	500.0
O_2 (volume %,dry)	12.7	12.9
NO_x (ppm,dry, 15% O_2)	1045.0	1600.0
CO (ppm,dry, 15% O_2)	60.0	95.0
HC (ppm,dry, 15% O_2)	155.0	155.0
SO_2 (ppm,dry, 15% O_2)	405.0	1200.0
TSP(mg/m^3,15% O2)	65.0	90.0

Carbon dioxide (CO_2): CO_2 actually is not noxious as a product of combustion of all fossil fuels. It is now considered to be one of the main causes of the greenhouse effect. A reduction of CO_2 emission can only be achieved by improving the engine efficiency or by using fuels containing lower concentration of carbon such as natural gas.

Sulphur oxides (SO_x): Sulphur oxides are formed due to the combustion of sulphur contained in the fuel. They are one of the primary causes of acid rain. The sulphur oxide emission is primarily influenced by the amount of sulphur contained in the fuel used. Much less influence can be taken by the fuel consumption of engine. The major part (> 95%) of sulphur oxides contained in the exhaust gas of the diesel engines is SO_2.

Nitrogen oxides NO_x (NO, NO_2, N_2O): Nitrogen oxides which are generally referred to as NO_x in the case of internal combustion engines comprise nitrogen monoxide-NO (colourless, water insoluble gas), nitrogen dioxide-NO_2 (reddish brown gas, highly toxic) and dinitrogen monoxide-N_2O (laughing gas, colourless gas previously used as a narcotic). Nitrogen oxides, together with the sulphur oxides are the main causes of acid rain. They also contribute essentially to ozone formation in the air and ground level.

The high temperatures and pressures produced in the combustion space of an IC engine stimulate the nitrogen content in the air and also in the grades used (such as heavy fuel oil) to react with oxygen in the combustion air. In this reaction mechanism, the formation of nitrogen oxides proportionally increases with the temperature rise. This behavior unfortunately combats the efforts of improving on engine efficiency because conversion of energy at the highest possible temperature level is to be aimed for to reach the optimal efficiencies of combustion processes.

The NO_x formation during combustion in the diesel engine is predominantly NO and which is converted to a minor extent to NO_2 by oxidation either in the combustion space or in the exhaust gas systems downstream (exhaust gas piping, exhaust gas turbo charger etc.). In general, exhaust gas leaving the engine is 95% NO and approximately 5% NO_2. To simulate the process of NO oxidation, to form NO_2 in the atmosphere, practically, all the legislation stipulate that in the calculation of NOx mass flow emitted, the entire NO_x must be taken as NO_2. The N_2O concentration in the exhaust gas of medium speed diesel engines, burning heavy fuel is limited to a few ppm. Therefore, it can be neglected from the viewpoint of environmental protection.

Carbon Monoxide (CO): It is a colourless, highly noxious gas which forms where the combustion of fuels containing carbon proceeds under(possibly local) air starvation. In modern DG sets, optimisation of air/fuel mixture formation and use of constant pressure type turbo charging, successfully reduces the CO content of exhaust gases even with the poorest qualities of fuel grades. This type of engine design meets even the stringent standards set by such environmental agencies like TA-Luft of Germany.

Non combusted Hydro carbons(HC): Hydro carbons contained in the exhaust gas consists of a multitude of various organic compounds. However, the HC contents of exhaust gases for modern 4-stroke diesel engines burning heavy fuel are very low and are not a matter of environmental debate.

Soot, Dust: Solids contain in exhaust gases of diesel engines burning heavy fuel not only consist of soot (carbon) resulting due to incomplete combustion of the fuel but also due to dust and ash particles from the fuel and the lube oil, the quality of the combustion air and from the abrasion products. Even though, these constitutes the major source of visible dark coloration, of exhaust gases, soot particles only account for a relatively low percentage of total dust concentration. Based on the ash content of the fuel and the lube, the soot quantity also varies as shown in the table below.

Fuel	Gas Oil	Heavy Fuel Oil
Ash content — Fuel%	0.01	0.10
Ash content — Lube%	1.50	4.00
Soot (Carbon) mg/m^3	15.00	15.00
Fuel ash mg/m^3	4.00	40.00
Lube oil ash mg/m^3	3.00	8.00

Overall analysis of environmental laws will take us into two pronged environmental considerations.

I. Long term consideration on implementation of actual system emission limits. This should take into account existing technology, cost competitiveness, consideration to burn only low grade, tertiary fuels, demand technology life-cycle, nature of project and plant gestation and country objectives.

II. Short term aspects mainly centred on maintaining desired ambient air qualities. This will bring us to the debate emission levels. The ground level dispersion of emission components are easily met far below existing standards by the modern 4-stroke diesel engines while burning heavy fuel.

In view of the above, adequate chimney/stack heights for guiding the exhaust gases away from the ground level can easily ensure low dispersions at ground level after emission at relevant designed chimney heights based on sulphur contents in the fuel.

13.17 NUCLEAR POWER PLANT AND THE ENVIRONMENT

In the United States, and doubtless in almost all countries constructing nuclear power plants, federal licensing proceedings for each plant require the inclusion of detailed environmental statements to be issued as public documents. In the United States, these should be in accordance with the National Environmental Policy Act of 1969 (NEPA). Such statements must assess not only the impact upon the environment that is associated with the construction and the operation of the power plant, but also the effect of the transportation of radioactive materials to and from that plant.

Besides thermal pollution, which it shares with almost all types of power plants, nuclear power's effects on the environment stem mainly from

(1) the nuclear fuel cycle,

(2) low-level dose radiations from nuclear-power plant effluents, and (3) low and high-level dose radiations from wastes.

13.17.1 THE FUEL CYCLE

Most nuclear power plants in operation or under construction in the world today are using, and will continue to use for the near future, ordinary (light) water cooled and moderated reactors: the Pressurized Water Reactor (PWR) and the Boiling Water Reactor (BWR). A small number use the heavy water cooled and moderated reactor (PHWR). The expectations are that the fast-breeder reactor power plant and perhaps an improved version of the gas-cooled reactor power plant will come on line in increasing numbers in the twenty-first century. Almost all-current water reactors use slightly enriched uranium dioxide, UO_2, fuel. The fuel has to go through a cycle that includes prereactor preparation, called the front end, in-reactor use, and post reactor management, called the back end.

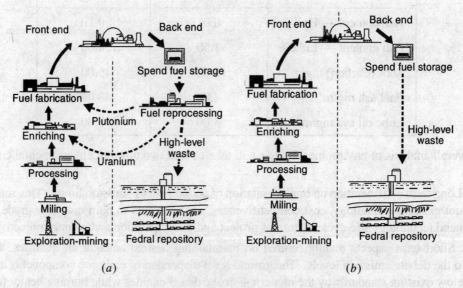

Fig. 13.4. A typical nuclear fuel cycle (a) with reprocessing and (b) without reprocessing.

The different process are briefly explained below:

1. Mining of the uranium ore.
2. Milling and refining of the ore to produce uranium concentrates, U_3O_8.
3. Processing to produce of uranium hexafluoride, UF_6, from the uranium concen-trates. This provides feed for isotopic (U^{235}) enrichment.
4. Isotopic enrichment of uranium hexafluoride to reach reactor enrichment require-ments. This is done invariably now by the gaseous diffusion process.
5. Fabrication of the reactor fuel elements. This includes conversion of uranium hexafluoride to uranium dioxide UO_2, pelletizing, encapsulating in rods, and assembling the fuel rods into subassemblies.

6. Power generation in the reactor, resulting in irradiated or spent fuel.
7. Short-term storage of the spent fuel.
8. Reprocessing of the irradiated fuel and conversion of the residual uranium to uranium hexafluoride, UF_6 (for recycling through the gaseous diffusion plant for reenrichment) and/or extraction of Pu^{239} (converted from U^{238}) for recycling to the fuel-fabrication plant. Reprocessing can reuse up to 96 percent of the original material in the irradiated fuel with 4 percent actually becoming waste.
9. Waste management, which includes long-term storage of high-level wastes.

Step 8, reprocessing, may be bypassed, which results in disposal of both reusable fuel and wastes. This is the current (1982) U.S. Department of Energy process for dealing with irradiated fuel. The fuel assemblies are stored for at least 10 years and then buried. This is the so-called throw-away fuel cycle.

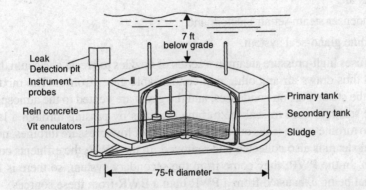

Fig. 13.5. A typical low-level liquid-waste storage tank with double-walled containment.

13.17.2 WASTES

The wastes associated with nuclear power can be summarized as:

1. Gaseous effluents. Under normal operation, these are released slowly from the power plants into the biosphere and become diluted and dispersed harmlessly.

2. Uranium mine and mill tailings. Tailings are residues from uranium mining and milling operations. They contain low concentrations of naturally occurring radio-active materials. They are generated in large volumes and are stored at the mine or mill sites.

3. Low-level wastes (LLW). These are classified as wastes that contain less than 10 nCi (nanocuries) per gram of transuranium contaminants and that have low but potentially hazardous concentrations of radioactive materials. They are generated in almost all activities (power generation, medical, industrial, etc.) that involve radioactive materials, require little or no shielding, and are usually disposed of in liquid form by shallow land burial (Fig. 13.5).

4. High-level wastes (HLW). These are generated in the reprocessing of spent fuel. They contain essentially all the fission products and most of the transuranium elements not separated during reprocessing. Such wastes are to be disposed of carefully.

5. Spent fuel. This is unreprocessed spent fuel that is removed from the reactor core after reaching its end-of-life core service. It is usually removed intact in its fuel element structural form and then stored for 3 to 4 months under water on the plant site to give time for the most intense radioactive isotopes (which are the ones with shortest half-lives) to decay before shipment for reprocessing or disposal. Lack of a reprocessing capacity or a disposal policy has resulted in longer on-site storage, however. If the spent fuel is to be disposed of in a throwaway system (without reprocessing), it is treated as high-level waste.

13.18 RADIATIONS FROM NUCLEAR-POWER PLANT EFFLUENTS

Radiations from nuclear-power plant effluents are low-dose-level types of radiations. The effluents are mainly gases and liquids. Mainly the effects of these radiations on the populations living near the plants prompt environmental concerns about nuclear power plants. Sources of effluents vary with the type of reactor.

In both pressurized-water reactors (PWR) and boiling-water reactors (BWR), two important sources of effluents are

(1) The condenser steam-jet air ejectors and

(2) The turbine gland-seal system.

The ejector uses high-pressure steam in a series of nozzles to create a vacuum, higher than that in the condenser, and thus draws air and other non-condensable gases from it. The mixture of steam and gases is collected, the steam portion condenses, and the gases are vented to the atmosphere. In the gland seal, high-pressure steam is used to seal the turbine bearings by passing through a labyrinth from the outside in so that no turbine steam leaks out and, in the case of low-pressure turbines, no air leaks in. The escaping gland-seal steam is also collected and removed. In the BWR, the effluents come directly from the primary system. In the PWR, they come from the secondary system, so there is less likelihood of radio-active material being exhausted from a PWR than a BWR from these sources.

The primary-coolant radioactivity comes about mainly from fuel fission products that find their way into the coolant through the few small cracks that inevitably develop in the very thin cladding of some fuel elements. Such activity is readily detectable. However, to avoid frequent costly shutdowns and repairs, the system is designed to operate as long as the number of affected fuel elements does not exceed a tolerable limit, usually 0.25 to 1 percent of the total. Also, some particulate matter finds its way into the coolant as a result of corrosion and wear (erosion) of the materials of the primary system components. These become radioactive in the rich neutron envi-ronment of the reactor core. Corrosion occurs because the radiolytic decomposition of the water passing through the core results in free O_2 and free H and OH radicals as well as some H_2O_2. These lower the pH of the coolant and promote corrosion. Finally, radioactivity in the primary coolant may be caused by so-called *tramp uranium*. This is uranium or uranium dioxide dust that clings to the outside of the fuel elements and is insufficiently cleaned off during fabrication. It will, of course, undergo fission, and its fission products readily enter the coolant. Improved processing and quality control are minimizing the problem of tramp uranium.

13.19 IMPACT ON POLLUTION LOAD AND AIR QUALITY IN DELHI

The major impacts have been observed through the implementation of emission norms and fuel quality specifications effective from 1996, as also phasing out of 15-year-old commercial vehicles and leaded petrol in the year 1998 and phasing out of 8-year-old commercial vehicles and 15-year-old two wheelers from 2000 onwards. The ambient air quality as monitored by CPCB during 1999 shows reduction in levels of various pollutants in ambient air as compared to previous year. The reducing trend was observed with respect to Carbon Monoxide, nitrogen dioxide, and lead in residential areas.

13.19.1 ENVIRONMENTAL CONCERNS

In Delhi today, pollution is one of the most critical problems facing the public and concerned authorities. According to the World Health Organization (WHO), Delhi is the fourth most polluted city in the world in terms of suspended particulate matter (SPM). The growing pollution is responsible for increasing health problems. The deteriorating environment is the result of population pressure and haphazard growth. Industrial development has been haphazard and unplanned. Only about 20% of the industrial units are in approved industrial areas; the remainders are spread over the city in residential and commercial areas. Road transport is the sole mode of public transport; there has been a phenomenal increase in the vehicle population, which has increased from 2 lakhs in 1971 to 32 lakh in 1999.

13.19.2 POLLUTION LEVELS

1. Ambient Air Quality. Data from continuous monitoring of air quality reveals that while suspended particulate matter levels still far exceed stipulated standards, there is a significant downward trend.

Noise levels in Delhi exceed permissible levels in all areas except industrial areas, according to a study by the Delhi Pollution Control Committee. Since noise is measured on a logarithmic scale, an increase of every 3-5 dBA has twice the effect on humans. Diesel generating sets and vehicles, particularly auto-rickshaws, have been identified as major sources of noise pollution in Delhi.

2. Air Pollution. The 1991 report by the National Environmental Engineering Research Institute (NEERI), Nagpur documents the amount of pollution that is contributed by different sectors in Delhi:

Pollution by Sector of Origin

	1970-71	1980-81	1990-91	1999-2000 (estimated)	2000-2001 (estimated)
Industrial (including thermal power)	56%	40%	29%	25%	20%
Vehicular	23%	42%	63%	70%	72%
Domestic	21%	18%	8%	5%	8%

In relative terms, the quantum of industrial air pollution has decreased over the years. However, vehicular pollution has increased rapidly. The drop in share of domestic air pollution is due to the increased number of LPG connections in Delhi, which have replaced other forms of fuel.

3. Water pollution. The 48-km stretch of the Yamuna River in Delhi is heavily polluted by domestic and industrial wastewater. The river water upstream of Wazirabad is fit for drinking after it has been treated, but after the confluence of the Najafgarh drain and 18 other major drains, the water quality becomes heavily degraded and is unfit even for animal consumption and irrigation.

4. Domestic Wastewater Pollution. The increase in population has resulted in a corresponding increase in the volume of domestic wastewater that is generated. Sewage treatment capacity is about 344 MGD at present against about 470 MGD wastewater that is generated each day in Delhi. The sewage treatment capacity is not fully utilized due to malfunctioning of the trunk sewer system.

5. Industrial Wastewater. The industrial wastewater generated in Delhi is about 70 MGD. Although some industrial units have provided facilities to treat wastewater, most small-scale industries do not have such facilities.

6. Vehicular Pollution. The steep increase in vehicle population has resulted in a corresponding increase in pollutants emitted by vehicles. Petrol consumption has increased from 133 thousand tons in 1980-81 to 449 thousand tons in 1996-97 and HSD consumption from 377 thousand tons to 1,234 thousand tons. Two-wheelers, which constitute 66% of the vehicles registered in Delhi, are the major source of air pollution.

7. Solid Waste. NEERI estimates indicate that about 8000 M. Tonnes of solid waste is being generated each day in Delhi at present. In addition, industrial hazardous and non-hazardous waste, such as fly ash from power plants, is also generated. MCD and NDMC could mange to clear about 5000-5500 M. Tonnes of garbage each day resulting in accumulation of garbage in the city area.

8. Hospital Waste Pollution. With the increase in the number of hospitals and nursing homes in Delhi, hospital waste has become another area of concern. Private nursing homes and small hospitals do not have arrangements to treat hospital waste. Installing incinerators to burn hospital waste is not an ideal solution since these incinerators add to air pollution.

13.19.3 MEASURES TO COMBAT POLLUTION

1. Vehicular Pollution. Delhi has more vehicles than the three metropolitan cities of Mumbai, Calcutta and Chennai combined. It is the only metropolitan city where commuters are primarily dependent on a single transport system, i.e., road. This has led to an enormous increase in the number of vehicles with the associated problems of traffic-congestion and increase in air and noise pollution. There is an urgent need to strengthen and encourage use of public transport including development of MRTS and better utilization of the existing ring railway.

The Delhi Government has started an incentive scheme to replace old commercial vehicles. The supply of lead-free petrol in Delhi since April 1998 has brought down the lead content in the air. The promotion of CNG as a fuel for buses, cars, taxis and auto-rickshaws is being considered as a method of reducing the level of vehicular pollution. Replacement of old commercial vehicles, no registration of army and government disposed old vehicles, etc. measures also contributed to some extent.

2. Pollution from Thermal Power Plants. Thermal power plants contribute to 13% of air pollution. The main pollutants are stack emissions; fly ash generation and fugitive emission in coal handling. All three thermal power plants need better use of their emission control devices and the fly ash that they generate. There is an immediate need to use beneficiated/washed coal, which has a maximum ash content of 30%, which will reduce fly ash generation by about 25%. It has also been recommended to the Thermal Power Stations to examine the possibility of installing Bag House Filters in order to control emission of particles between the size of PPM 2.5 to PPM-10.

3. Industrial Air Pollution. The air pollution generated from industrial activity in Delhi is about 12% of total air pollution. Although several steps have been taken, industrial pollution needs to be reduced further. More than 1,300 industrial units, that were not allowed to operate under the MPD-2001 norms, have been closed. A scheme has been prepared to relocate industrial units that currently operate in residential areas. About 1,300 acres of land have been acquired and new industrial estates are being developed at Bawana, Holumbi Kalan and Holumbi Khurd. Land available within existing industrial estates is also being used to accommodate such industrial units. Anand Parbat, Shahdara and Samaipur Badli area are being developed as industrial areas. All industries in Delhi using Coal Fired Boilers have been asked to change over to Oil or Gas Fired Boilers in order to reduce air pollution generated from industrial activities. This will also reduce the Fly Ash generated by the approximate 4000-5000 coal fired boilers in the City. All industries are also being advised to control pollution from diesel generating sets. They have been asked to increase the stack height to a level of 2-3 meters above their building height and also take acoustic measures to reduce the noise level from diesel generating sets.

4. Industrial Wastewater Pollution. There are 28 industrial areas in Delhi. Most of the small and tiny industries do not have individual facilities to treat liquid waste. The Hon'ble Supreme Court has ordered that 15 Common Effluent Treatment Plants (CETPs) be constructed. All water-polluting industries in Delhi have been directed to comply with orders of the Hon'ble Supreme Court and ensure that they do not discharge untreated effluent. Action has been taken against 2,300 industrial units in Delhi so far (January, 2000) and is continuing to cover all such water polluting units. Each unit has been asked to install an Effluent Treatment Plant to ensure neutralization of acidity, removal of oil and grease and removal of total suspended solids to the levels specified for each industry by the Central Pollution Control Board or up to sewage standards wherever specific standards have not been laid down.

The breakdown of funding for the CETPs is given below:

(a) 25% by the Delhi Government

(b) 25% by the Government of India

(c) 20% by concerned industries through the CETP society, and

(d) 30% loan financed by IDBI.

The cost of constructing 15 CETPs which was estimated at Rs. 90 crore in 1996-97 is now estimated at about Rs. 190 crore. Progress has been slow due to reluctance on the part of industrial units to contribute their share.

5. Domestic Wastewater Pollution. The present water supply capacity in Delhi is approximately 591 MGD and the sewage treatment capacity is 344 MGD. 16 new sewage treatment plants are at various stages of commissioning and construction. Of the 16 plants, 5 were completed by March 1999, 8 will be completed in 1999-2000 and one in 2000-01. However, since unauthorized colonies and JJ clusters may not be provided with sewerage systems, wastewater from these areas will continue to be discharged through drains. Accordingly, a parallel channel from Wazirabad to Okhla has been proposed. Water and Power Consultancies Services (WAPCOS) are doing the feasibility study for the proposed channel.

6. Industrial Non-hazardous Waste Management. The main industrial non-hazardous waste is fly ash from power plants that emit about 6,000 metric tons of fly ash per day. Until recently, the fly ash was disposed off for earth filling apart from about 100 metric tons per day that was used to manufacture pozzolana cement. A small quantity of fly ash near BTPS is also used to manufacture bricks. Land is now being allotted to three brick manufacturing units near Rajghat and Indraprastha thermal power stations so that additional fly ash from these plants can be utilized. At the same time, the use of beneficiated/washed coal may reduce the amount of fly ash generated by thermal power plants.

7. Hazardous Waste Management. The National Productivity Council, New Delhi has conducted an Environment Impact Assessment study to select a site for the disposal of hazardous waste. A 150 acre site on the Bawana-Narela Road was selected but it has not been made available due to opposition from local residents.

8. Solid Waste Management. The management of solid waste in Delhi is being improved through measures adopted by concerned agencies. The measures include the following:

(1) Construction of dalaos/dustbins;

(2) Purchase of additional front-end loaders, refuse collectors, mechanical sweepers, tipper trucks, dumper placers, etc.;

(3) Use of garbage to make compost with the participation of the private sector;

(4) Development of new sanitary land-fill sites;

(5) Disposal of garbage at the local area level through vermin composting.

(6) Involvement of NGOs and Resident Welfare Association in segregation and collection of garbage from houses.

9. Hospital Waste Disposal. The Delhi Government has constituted a committee to implement the Bio-Medical Waste (management and handling) Rules, 1998. Almost all government hospitals have installed incinerators for the disposal of hospital waste. Sanjay Gandhi Memorial (SGM) Hospital has also installed an autoclave that is used for 97% of its waste disposal. The Centre for Occupational and Environmental Health (COEH) is helping the committee monitor the progress of the programme and ensure that the Bio-Medical Waste Rules 1998 are implemented by all hospitals in Delhi.

10. Other Measures. Several other measures are being taken to control pollution and improve the environment. These include:

(1) Planting of 21 lakh trees/shrubs in 1999-2000;

(2) Public awareness campaigns;

(3) Setting up eco-clubs in schools;

(4) Development and protection of the Ridge area;

(5) Development of a wildlife sanctuary at Bhati-Asola;

(6) Development of old lakes;

(7) 10 City Forest Sites have been identified by the Forest Department. These will be developed by the Forest Department as 'Green Lungs' for various areas.

The Delhi Plastic (Manufacture, Sale and Usage) and Non-Biodegradable Garbage (Control) Bill, 1999 has been moved in the Legislative Assembly for banning the use of plastic bags for food items. This has been referred to a Select Committee of the House in the December 1999 Session of the Legislative Assembly.

Ambient Air Quality in Delhi

Area/Parameters	1995	1998	1999	CPCB Standards
Industrial Area				
Sulphur dioxide (mg/m^3)	24.1	20.2	19.5	80.00
Nitrogen dioxide (mg/m^3)	35.5	34.7	33.5	80.00
Suspended Particulate matter (mg/m^3)	420	367	365	360.00
Lead (mg/m^3)	110	105	58	1.00
Residential Area				
Sulphur dioxide (mg/m^3)	16.5	15.8	16.2	60.00
Nitrogen dioxide (mg/m^3)	32.5	28.6	26.5	60.00
Suspended Particulate matter (mg/m^3)	409	341	351	140.00
Lead (mg/m^3)	155	95	46	0.75

Traffic Intersections

Sulphur dioxide (mg/m^3)	42	25	20	60.00
Nitrogen dioxide (mg/m^3)	66	63	60	60.00
Suspended Particulate matter (mg/m^3)	452	426	418	140.00
Lead (mg/m^3)	335	136	70	
Carbon Monoxide (mg/m^3)	5587	5450	4241	

Source : Department of Environment, Government of NCT of Delhi and Central Pollution Control Board.

Prescribed Ambient Noise Standards

S.No.	Area LeqdB(A)		
		Day Time*	Night Time**
1.	Industrial Area	75	70
2.	Commercial Area	65	55
3.	Residential Area	55	45
4.	Silence Zone***	50	40

Notes:	
*	Day Time -0600 hour to 2100 hour (15 hours)
**	Night Time-2100 hour to 0600 hour (09 hours)
***	Areas upto 100 metres around certain premises like hospitals, education institutions and courts may be declared as silence zones by the competent authority; honking of vehicle horns, use of loudspeaker, bursting of crackers and hawkers' noise should be banned in these zones.

Source: State of the Environment 1995, Ministry of Environment and Forests.

Discharges and Bod Levels in Storm Water Drains

Sl.No.	Description of Drain	Discharge (mld)	BOD (mg/l)
1.	Supplementary Drain	177	22
2.	Najafgarh Drain	1180	125
3.	Magazine Road Drain	4	190
4.	Sweepers Colony Drain	27	88
5.	Kheybar Pass Drain	23	65
6.	Metcalf House Drain	11	85
7.	Qudsia Bagh	24	155
8.	Mori Gate Drain	85	

Sl.No.	Description of Drain	Discharge (mld)	BOD (mg/l)
9.	Moat Drain	2	195
10.	Civil Mill Drain	55	180
11.	Rajghat/Delhi Gate Drain	43	190
12.	Sen-Nursing Home	100	280
13.	Drain No.14	153	320
14.	Bara Pula Drain	255	165
15.	Maharani Bagh Drain	54	370
16.	Kalkaji Drain	27	210
17.	Tehkand Drain	34	310
18.	Tuglakabad Drain	11	150
19.	Trans Yamuna	1471	240
	Total	3651	

Source : DJB Pre-feasibility study report on rehabilitation of Sewer System.

13.20 METHOD FOR POLLUTION CONTROL

The following methods for developing the power generating capacity without pollution to the atmosphere.

1. F.P. Rogers has suggested that it would be safer to set the nuclear power plants underground. This definitely preserves the environment. There would be lot of difficulties in excavation, concreting, roof lining, structural supporting, lowering the reactor equipments and many others. But even then it is suggested that locating the power plant underground would be profitable in the long run.

2. The tidal power must be developed in the coining years that is free from pollution.

3. The thermal discharges to the environment are common from fossil and nuclear-fueled power stations. Significant quantities of particulates and gases from fossil-fueled system, small quantities of radioactive gases from nuclear, have an impact upon an environment. Offshore sitting of power plants mitigates these problems of pollution. Offshore sitting of power stations also isolates the plants from earthquakes and provides the thermal enhancement of the water to increase recreational and commercial values. No doubt, offshore location requires new design consideration and floating platforms in the sea increasing the capital cost of the plant.

4. It was proposed that the thermal pollution of the atmosphere and the generation cost of the plant could be reduced by using the low-grade energy exhausted by the steam. The ideal use for enormous quantity of residual energy from steam power plants requires large demand with unity power factor.

Particularly in U.S.A., many uses of energy are available in winter, but not in summer therefore finding large-scale valuable uses of thermal energy is the key for developing beneficial uses.

It is estimated that the total energy used in U.S.A. for air-conditioning is equivalent to the total energy used for heating the offices and residences. The low-grade energy exhaust by the thermal plants is not readily usable for air-conditioning purposes. It is possible to use this energy by stopping the expansion of steam at a temperature of 95°C to 100°C and use of this energy can be made to drive an absorption refrigeration system such as lithium bromide water system. This will be a definite positive answer to reduce the thermal pollution of environment otherwise caused by burning extra fuel to run the absorption refrigeration system in summer or to run the heating systems in winter.

As for open field irrigation, soil heating with warm water and better cultivation of the fishes in slightly warm water. In short, a combination of uses could consume all heat from a large thermal power station, making conventional cooling unnecessary and reduce the generating cost with minimum thermal pollution of the atmosphere.

5. Use the sun energy for the production of power that is absolutely free from air-pollution.

13.21 CONTROL OF MARINE POLLUTION

Demographic pressure and rapid industrialisation has led to increased generation of wastes, and, these wastes reach the sea either directly or indirectly through rivers. This has caused pollution in the marine environment particularly in the coastal waters. Besides these sources, release of agricultural wastes containing pesticide residues and operational releases of ships and tankers containing oil, also cause pollution in the marine environment.

In order to monitor the levels of marine pollutants in a systematic manner as well as to quantify their transport rates from land-based sources to the sea, a well-knit multi institutional programme on Environmental Monitoring and Modelling has been launched by establishing an Apex Centre at Regional Centre, National Institute of Oceanography, Bombay and ten units at expert institutions such as NIO, Goa and its regional centres at Cochin and Waltair, Central Salt and Marine Chemicals Research Institute (Bhavnagar), Centre for Earth Science Studies (Trivandrum), units of Central Electrochemical Research Institute at Tuticorin and Madras, Regional Research Laboratory (Bhubaneswar), Zonal Office, Central Pollution Control Board (Calcutta) and the DOD Centre in the Andaman and Nicobar Islands. As a supportive measure for the develop-ment of suitable methodology on monitoring and modelling, DOD supported cells at Centre for Mathematical Modelling and Computer Simulation (Bangalore), Regional Research Laboratory (Trivandrum) and National Environmental Engineering Research Institute (Nagpur), have also been established. The programme, after considerable amount of generation of data, is expected to provide knowledge on behaviour of pollutants in the sea and trends on fluctuation of pollutants in the sea, which will be useful in the pollution abatement measures.

THEORETICAL QUESTIONS

1. What do you know about environment pollution due to energy uses?
2. Explain environment pollution due to industrial emissions.
3. Explain environment pollution due to road transport.

4. Define harmful effect of emission.
5. What are the step taken for reduce air pollution ?
6. Write short notes on Noise pollution and its control.
7. What is the green house gases and their effects ? Explain.
8. Briefly explain fossil fuel pollution.
9. What is acid rain, explain ?
10. Write short notes on stratospheric ozone depletion, Acid Fog.
11. Write on pollution due to combustion of fuel.
12. Explain how gas combustion polluted the atmosphere.
13. What do you understand by liquid fuel pollution?
14. What do you understand by solid fuel pollution?
15. What do you understand by thermal pollution, explain the bad effects of thermal pollution?
16. Explain the pollution due to nuclear power plant.
17. Explain the method to reduce the pollution.

Appendix
CONVERSION FACTORS

(1) Length	1 inch	= 2.54 cm
	1 foot	= 30.48 cm
	1 yard	= 91.43 cm
	1 meter	= 100 cm = 3.26 feet = 39.37 inches
	1μ (micron)	= 3.281×10^{-6} ft = 0.0001 cm
(2) Area	1 m^2	= 100×100 cm^2
(3) Weight	1 ton	= 1.016 tonnes
	1 kg	= 2.204 lb
	1 lb	= 453.6 gm
(4) Volume	1 cu.ft	= 0.0283 cu.m
	1 cu.in	= 16.39 cu.cm
(5) Density	1 kg/m^3	= 0.062 lb/ft^3
	1 lb/ft^2	= 16.02 kg/m^2
(6) Work, Energy	1 joule	= 1 Nm
		= 1 watt-sec
		= 2.7778×10^{-7} kWh
		= 0.239 cal
	1 cal	= 4.184 joule
		= 1.1622×10^{-6} kWh
	1 kWh	= 8.6×10^5 cal
		= 3.6×10^6 joule
	1 kgfm	= (1/427) kcal
		= 9.81 joule
(7) Force	1 Newton	= 1 kg-m/sec^2
		= 0.012 kgf
	1 kgf	= 9.81 N
(8) Pressure	1 bar	= 750.06 mm Hg
		= 0.9869 atm
		= 10^5 N/m^2
	1 N/m^2	= 1 Pascal
		= 10^{-5} bar
		= 10^{-2} kg/m-sec^2

1 standard	atmosphere	= 760 mm Hg
		= 1.03 kgf/cm^2
		= 1.01325 bar
		= 1.01325 × 10^5 N/m^2
		= 29.92 inches of mercury
		= 10.332 m H$_2$O
(9) Power	1 watt	= 1 joule/sec
		= 0.86 kcal/sec
	1 h.p.	= 735.3 Watt
	1 kW	= 1000 Watt
		= 860 kcal/h
(10) Temperature	°K	= 273 + °C
	°R	= 460 + °F
		where, °K = Degree Kelvin
		°R = Degree Rankine
		°C = Degree Centigrade
(11) Specific Heat	1 kcal/kg-°k	= 4.18 kJ/kg-k
(12) Thermal conductivity	1 Watt/m-k	= 0.8598 kcal/h-m °C
	1 kcal/h-m-°C	= 1.16123 Watt/m-k
		= 1.16123 joules/s-m-k
(13) Heat Transfer co-efficient	1 watt/m^2-k	= 0.86 kcal/m^2-h-°C
	1 kcal/m^2-h-°C	= 1.163 Watt/m^2-k

Glossary

Anthracite: The highest rank of coal; used primarily for residential and commercial space heating. It is a hard, brittle, and black lustrous coal, often referred to as hard coal, containing a high percentage of fixed carbon and a low percentage of volatile matter. The moisture content of fresh mined anthracite generally is less than 15 percent. The heat content of anthracite ranges from 22 to 28 million Btu per ton on a moist, mineral matter- free basis. The heat content of anthracite coal consumed in the United States averages 25 million Btu per ton, on the as-received basis (*i.e.*, containing both inherent moisture and mineral matter). Note: Since the 1980's anthracite refuse or mine waste has been used for steam electric power generation. This fuel typically has a heat content of 15 million Btu per ton or less.

Asphalt: A dark brown-to-black cement-like material obtained by petroleum processing and containing bitumens as the predominant component; used primarily for road construction. It includes crude asphalt as well as the following finished products: cements, fluxes, the asphalt content of emulsions (exclusive of water), and petroleum distillates blended with asphalt to make cutback asphalts.

ASTM: The American Society for Testing and Materials.

Aviation Gasoline: A complex mixture of relatively volatile hydrocarbons with or without small quantities of additives, blended to form a fuel suitable for use in aviation reciprocating engines. Fuel specifications are provided in ASTM Specification D 910 and Military Specification MIL-G-5572. **Note:** Data on blending components are not counted in data on finished aviation gasoline.

Aviation Gasoline Blending Components: Naphthas that are used for blending or compounding into finished aviation gasoline (e.g., straight-run gasoline, alkylate, and reformate). Excluded are oxygenates (alcohols and ethers), butane, and pentanes plus.

Barrel (petroleum): A unit of volume equal to 42 U.S. gallons.

Barrels per Calendar Day (operable refinery capacity): The amount of input that a distillation facility can process under usual operating conditions during a 24-hour period after making allowances for the following limitations: the capability of downstream facilities to absorb the output of crude oil processing facilities of a given refinery (no reduction is made when a planned distribution of intermediate streams through other than downstream facilities is part of a refinery's normal operation); the types and grades of inputs to be processed; the types and grades of products to be manufactured; the environmental constraints associated with refinery operations; the reduction of capacity for scheduled downtime, such as routine inspection, mechanical problems, maintenance, repairs, and turnaround; and the reduction of capacity for unscheduled downtime, such as mechanical problems, repairs, and slowdowns.

Barrels per Stream Day (operable refinery capacity): The maximum number of barrels of input that a distillation facility can process within a 24-hour period when running at full capacity under optimal crude and product slate conditions with no allowance for downtime.

Bituminous Coal: A dense, black coal, often with well-defined bands of bright and dull material; used primarily as fuel in steam-electric power generation, with substantial quantities also used for heat and power applications in manufacturing and to make coke. Bituminous coal is the most abundant coal in active U.S. mining regions. Its moisture content usually is less than 20 percent. The heat content of bituminous coal ranges from 21 to 30 million Btu per ton on a moist, mineral-matter-free basis. The heat content of bituminous coal consumed in the United States averages 24 million

Btu per ton, on the as-received basis (*i.e.*, containing both inherent moisture and mineral matter). In this report, bituminous coal includes subbituminous coal.

British Thermal Unit (Btu): The quantity of heat needed to raise the temperature of 1 pound of water by 1°F at or near 39.2°F.

Butane: A normally gaseous straight-chain or branched-chain hydrocarbon (C_4H_{10}). It is extracted from natural gas or refinery gas streams. It includes isobutane and normal butane and is designated in ASTM

Specification D1835 and Gas Processors Association Specifications for commercial butane. Isobutane: A normally gaseous branched-chain hydrocarbon. It is a colorless paraffinic gas that boils at a temperature of 10.9°F. It is extracted from natural gas or refinery gas streams.

Normal Butane: A normally gaseous straight-chain hydrocarbon. It is a colorless paraffinic gas that boils at a temperature of 31.1° F. It is extracted from natural gas or refinery gas streams.

Butylene: An olefinic hydrocarbon (C_4H_8) recovered from refinery processes.

Catalytic Cracking: A refining process that consists of using a catalyst and heat to break down the heavier and more complex hydrocarbon molecules into lighter and simpler molecules.

Coal: A readily combustible black or brownish-black rock whose composition, including inherent moisture, consists of more than 50 percent by weight and more than 70 percent by volume of carbonaceous material. It is formed from plant remains that have been compacted, hardened, chemically altered, and metamorphosed by heat and pressure over geologic time. Coals are classified according to their degree of progressive alteration from lignite to anthracite. In the U.S. classification, the ranks of coal include lignite, subbituminous coal, bituminous coal, and anthracite and are based on fixed carbon, volatile matter, heating value, and agglomerating (or caking) properties.

Coal Coke: A solid carbonaceous residue derived from low-ash, low-sulfur bituminous coal from which the volatile constituents are driven off by baking in an oven at temperatures as high as 2,000 degrees Fahrenheit so that the fixed carbon and residual ash are fused together. Coke is used as a fuel and as a reducing agent in smelting iron ore in a blast furnace.

Coke Plants: Plants where coal is carbonized in slot or beehive ovens for the manufacture of coke.

Commercial Sector: An energy-consuming sector that consists of service-providing facilities and equipment of: businesses; Federal, State, and local governments; and other private and public organizations, such as religious, social, or fraternal groups. The commercial sector includes institutional living quarters. Common uses of energy associated with this sector include space heating, water heating, air conditioning, lighting, refrigeration, cooking, and running a wide variety of other equipment.

Conversion Factor: A number that translates units of one system into corresponding values of another system. Conversion factors can be used to translate physical units of measure for various fuels into Btu equivalents.

Cord (wood): A cord of wood measures 4 feet by 4 feet by 8 feet or 128 cubic feet.

Crude Oil (Including Lease Condensate): A mixture of hydrocarbons that exists in liquid phase in underground reservoirs and remains liquid at atmospheric pressure after passing through surface separating facilities. Included are lease condensate and liquid hydrocarbons produced from tar sands, gilsonite, and oil shale. Drip gases are also included, but topped crude oil (residual oil) and other unfinished oils are excluded. Where identifiable, liquids produced at natural gas processing plants and mixed with crude oil are likewise excluded.

GLOSSARY

Crude Oil Used Directly: Crude oil consumed as fuel by crude oil pipelines and on crude oil leases.

Cubic foot (cf), natural gas: The amount of natural gas contained at standard temperature and pressure (60 degrees Fahrenheit and 14.73 pounds standard per square inch) in a cube whose edges are one foot long.

Diesel Fuel: Fuel used for internal combustion in diesel engines; usually that fraction of crude oil that distills after kerosene.

Distillate Fuel Oil: A general classification for one of the petroleum fractions produced in conventional distillation operations. It includes diesel fuels and fuel oils. Products known as No. 1, No. 2, and No. 4 diesel fuel are used in on-highway diesel engines, such as those in trucks and automobiles, as well as off-highway engines, such as those in railroad locomotives and agricultural machinery. Products known as No. 1, No. 2, and No. 4 fuel oils are used primarily for space heating and electric power generation.

Electrical System Energy Losses: The amount of energy lost during generation, transmission, and distribution of electricity, including plant and unaccounted for uses.

Electricity Sales: The amount of kilowatthours sold in a given period of time; usually grouped by classes of service, such as residential, commercial, industrial, and other. 'Other' sales include sales for public street and highway lighting and other sales to public authorities, sales to railroads and railways, and interdepartmental sales.

Electric Utility: A corporation, person, agency, authority, or other legal entity or instrumentality that owns and/or operates facilities for the generation, transmission, distribution, or sale of electric energy for use primarily by the public. Utilities provide electricity within a designated franchised service area and file forms listed in the Code of Federal Regulations, Title 18, Part 141. Note: Facilities that qualify as cogenerators or small power producers under the Public Utility Regulatory Power Act (PURPA) are not considered electric utilities.

Electric Utility Sector: The electric utility sector consists of privately and publicly owned establishments that generate, transmit, distribute, or sell electricity primarily for use by the public and that meet the definition of an electric utility. Nonutility power producers are not included in the electric utility sector.

End-Use Sectors: The residential, commercial, industrial, and transportation sectors of the economy.

Energy: The capacity for doing work as measured by the capability of doing work (potential energy) or the conversion of this capability to motion (kinetic energy). Energy has several forms, some of which are easily convertible and can be changed to another form useful for work. Most of the world's convertible energy comes from fossil fuels that are burned to produce heat that is then used as a transfer medium to mechanical or other means in order to accomplish tasks. Electrical energy is usually measured in kilowatthours, while heat energy is usually measured in British thermal units.

Energy Consumption: The use of energy as a source of heat or power or as an input in the manufacturing process.

Energy Consumption, End-Use: The sum of fossil fuel consumption by the four end-use sectors (residential, commercial, industrial, and transportation) plus electric utility sales to those sectors and generation of hydroelectric power by nonelectric utilities. **Net** end-use energy consumption excludes electrical system energy losses. **Total** end-use energy consumption includes electrical system energy losses.

Energy Consumption, Total: The sum of fossil fuel consumption by the five sectors (residential, commercial, industrial, transportation, and electric utility) plus hydroelectric power, nuclear electric power, net imports of coal coke, and electricity generated for distribution from wood and waste and geothermal, wind, photovoltaic, and solar thermal energy.

Ethane: A normally gaseous straight-chain hydrocarbon (C_2H_6). It is a colorless, paraffinic gas that boils at a temperature of $-127.48°F$. It is extracted from natural gas and refinery gas streams.

Ethanol: An anhydrous, denatured aliphatic alcohol (C_2H_5OH) intended for motor gasoline blending.

Ethylene: An olefinic hydrocarbon (C_2H_4) recovered from refinery processes or petrochemical processes.

Exports: Shipments of goods from within the 50 States and the District of Columbia to U.S. possessions and territories or to any foreign country.

Federal Energy Regulatory Commission (FERC): The Federal agency with jurisdiction over interstate electricity sales, wholesale electric rates, hydroelectric licensing, natural gas pricing, oil pipeline rates, and gas pipeline certification. FERC is an independent regulatory agency within the Department of Energy and is the successor to the Federal Power Commission.

Federal Power Commission (FPC): The predecessor agency of the Federal Energy Regulatory Commission. The Federal Power Commission was created by an Act of Congress under the Federal Water Power Act on June 10, 1920. It was charged originally with regulating the electric power and natural gas industries. It was abolished on September 30, 1977, when the Department of Energy was created. Its functions were divided between the Department of Energy and the Federal Energy Regulatory Commission, an independent regulatory agency.

Fiscal Year: The U.S. Government's fiscal year runs from October 1 through September 30. The fiscal year is designated by the calendar year in which it ends *e.g.,* fiscal year 1992 begins on October 1, 1991, and ends on September 30, 1992.

Fossil Fuel: Any naturally occurring fuel, such as petroleum, coal, and natural gas, formed in the Earth's crust from long-term organic matter.

Fossil-Fueled Steam-Electric Power Plant: An electricity generation plant in which the prime mover is a turbine rotated by high-pressure steam produced in a boiler by heat from burning fossil fuels.

Gasohol: A blend of finished motor gasoline containing alcohol (generally ethanol but sometimes methanol) at a concentration of 10 percent or less by volume. Data on gasohol that has at least 2.7 percent oxygen, by weight, and is intended for sale inside carbon monoxide nonattainment areas are included in data on oxygenated gasoline.

Geothermal Energy: Hot water or steam extracted from geothermal reservoirs in the earth's crust. Water or steam extracted from geothermal reservoirs can be used for geothermal heat pumps, water heating, or electricity generation.

Heat Content of a Quantity of Fuel, Gross: The total amount of heat released when a fuel is burned. Coal, crude oil, and natural gas all include chemical compounds of carbon and hydrogen. When those fuels are burned, the carbon and hydrogen combine with oxygen in the air to produce carbon dioxide and water. Some of the energy released in burning goes into transforming the water into steam and is usually lost. The amount of heat spent in transforming the water into steam is counted as part of gross heat content but is not counted as part of net heat content. Also referred to as the higher heating value. Btu conversion factors typically used in EIA represent gross heat content.

GLOSSARY

Heat Content of a Quantity of Fuel, Net: The amount of usable heat energy released when a fuel is burned under conditions similar to those in which it is normally used. Also referred to as the lower heating value. Btu conversion factors typically used in EIA represent gross heat content.

Heavy Oil: The fuel oils remaining after the lighter oils have been distilled off during the refining process. Except for start-up and flame stabilization, virtually all petroleum used in steam-electric power plants is heavy oil.

Hydroelectric Power: The production of electricity from the kinetic energy of falling water.

Hydroelectric Power Plant: A plant in which the turbine generators are driven by falling water.

Imports: Receipts of goods into the 50 States and the District of Columbia from U.S. possessions and territories or from any foreign country.

Industrial Sector: An energy-consuming sector that consists of all facilities and equipment used for producing, processing, or assembling goods. The industrial sector encompasses the following types of activity: manufacturing; agriculture, forestry, and fisheries; mining; and construction. Overall energy use in this sector is largely for process heat and cooling and powering machinery, with lesser amounts used for facility heating, air conditioning, and lighting. Fossil fuels are also used as raw material inputs to manufactured products. In this report, nonutility power producers are included in the industrial sector.

Internal Combustion Electric Power Plant: A power plant in which the prime mover is an internal combustion engine. Diesel or gas-fired engines are the principal types used in electric power plants. The plant is usually perated during periods of high demand for electricity.

Isopentane: A saturated branched-chain hydrocarbon (C_5H_{12}) obtained by fractionation of natural gasoline or isomerization of normal pentane.

Jet Fuel, Kerosene-Type: A kerosene-based product with a maximum distillation temperature of 400 degrees Fahrenheit at the 10-percent recovery point and a final maximum boiling point of 572 degrees Fahrenheit and meeting ASTM Specification D 1655 and Military Specifications MIL-T-5624P and MIL-T-83133D (Grades JP-5 and JP-8). It is used for commercial and military turbojet and turboprop aircraft engines.

Jet Fuel, Naphtha-Type: A fuel in the heavy naphtha boiling range having an average gravity of 52.8 degrees API, 20 to 90 percent distillation temperatures of 290 degrees to 470 degrees F, and meeting Military Specification MIL-T-5624L (Grade JP-4). It is used primarily for military turbojet and turboprop aircraft engines because it has a lower freeze point than other aviation fuels and meets engine requirements at high altitudes and speeds.

Kilowatthour (kWh): The electrical energy unit of measure equal to one thousand watts of power supplied to, or taken from, an electric circuit steadily for one hour.

Kerosene: A light petroleum distillate that is used in space heaters, cook stoves, and water heaters and is suitable for use as a light source when burned in wick-fed lamps. Kerosene has a maximum distillation temperature of 400 degrees Fahrenheit at the 10-percent recovery point, a final boiling point of 572 degrees Fahrenheit, and a minimum flash point of 100 degrees Fahrenheit. Included are No. 1-K and No. 2-K, the two grades recognized by ASTM Specification D 3699 as well as all other grades of kerosene called range or stove oil, which have properties similar to those of No. 1 fuel oil.

Lease and Plant Fuel: Natural gas used in well, field, and lease operations (such as gas used in drilling operations, heaters, dehydrators, and field compressors), and as fuel in natural gas processing

Lease Condensate: A mixture consisting primarily of pentanes and heavier hydrocarbons which is recovered as a liquid from natural gas in lease separation facilities. This category excludes natural gas plant liquids, such as butane and propane, which are recovered at downstream natural gas processing plants or facilities.

Light Oil: Lighter fuel oils distilled off during the refining process. Virtually all petroleum used in internal combustion and gas-turbine engines is light oil.

Lignite: The lowest rank of coal, often referred to as brown coal, used almost exclusively as fuel for steam-electric power generation. It is brownish-black and has a high inherent moisture content, sometimes as high as 45 percent. The heat content of lignite ranges from 9 to 17 million Btu per ton on a moist, mineral-matter-free basis. The heat content of lignite consumed in the United States averages 13 million Btu per ton, on the as-received basis (*i.e.*, containing both inherent moisture and mineral matter).

Liquefied Petroleum Gases (LPG): A group of hydrocarbon-based gases derived from crude oil refining or natural gas fractionation. They include ethane, ethylene, propane, propylene, normal butane, butylene, isobutane, and isobutylene. For convenience of transportation, these gases are liquefied through pressurization.

Lubricants: Substances used to reduce friction between bearing surfaces, or incorporated into other materials used as processing aids in the manufacture of other products, or used as carriers of other materials. Petroleum lubricants may be produced either from distillates or residues. Lubricants include all grades of lubricating oils, from spindle oil to cylinder oil to those used in greases.

Methanol: A light, volatile alcohol (CH_3OH) eligible for motor gasoline blending.

Miscellaneous Petroleum Products: All finished petroleum products not classified elsewhere-for example, petrolatum, lube refining byproducts (aromatic extracts and tars), absorption oils, ram-jet fuel, petroleum rocket fuels, synthetic natural gas feedstocks, and specialty oils.

Motor Gasoline: A complex mixture of relatively volatile hydrocarbons with or without small quantities of additives, blended to form a fuel suitable for use in spark-ignition engines. Motor gasoline, as defined in ASTM Specification D-4814 or Federal Specification VV-G-1690C, is characterized as having a boiling range of 122 to 158 degrees Fahrenheit at the 10-percent recovery point to 365 to 374 degrees Fahrenheit at the 90-percent recovery point. 'Motor Gasoline' includes conventional gasoline; all types of oxygenated gasoline, including gasohol; and reformulated gasoline, but excludes aviation gasoline. Note: Volumetric data on blending components, such as oxygenates, are not counted in data on finished motor gasoline until the blending components are blended into the gasoline.

Motor Gasoline Blending Components: Naphthas (*e.g.*, straight-run gasoline, alkylate, reformate, benzene, toluene, xylene) used for blending or compounding into finished motor gasoline. These components include reformulated gasoline blendstock for oxygenate blending (RBOB) but exclude oxygenates (alcohols, ethers), butane, and pentanes plus.

Natural Gas (dry natural gas): Natural gas which remains after: (1) the liquefiable hydrocarbon portion has been removed from the gas stream (i.e., gas after lease, field, and/or plant separation); and (2) any volumes of nonhydrocarbon gases have been removed where they occur in sufficient quantity to render the gas unmarketable. Dry natural gas is also known as consumer-grade natural gas. The parameters for measurement are cubic feet at 60 degrees Fahrenheit and 14.73 pounds per square inch absolute.

GLOSSARY

Natural Gasoline: A term used in the gas processing industry to refer to a mixture of liquid hydrocarbons (mostly pentanes and heavier hydrocarbons) extracted from natural gas. It includes isopentane.

Net Interstate Flow of Electricity: The difference between the sum of electricity sales and losses within a State and the total amount of electricity generated within that State. A positive number indicates that more electricity (including associated losses) came into the State than went out of the State during the year; conversely, a negative number indicates that more electricity (including associated losses) went out of the State than came into the state.

Nonutility Power Producer: A corporation, person, agency, authority, or other legal entity or instrumentality that owns or operates facilities for electric generation and is not an electric utility. Nonutility power producers include qualifying cogenerators, qualifying small power producers, and other nonutility generators (including independent power producers). Nonutility power producers are without a designated franchised service area and do not file forms listed in the Code of Federal Regulations.

North American Industrial Classification System (NAICS): A system of numeric codes used to categorize businesses by type of activity in which they are engaged. It replaces the Standard Industrial Classification (SIC). This new structure was developed jointly by the United States, Canada, and Mexico to provide consistent, comparable information on an industry-by-industry basis for all three economies.

Nuclear Electric Power (nuclear power): Electricity generated by an electric power plant whose turbines are driven by steam produced by the heat from the fission of nuclear fuel in a reactor.

Pentanes Plus: A mixture of hydrocarbons, mostly pentanes and heavier, extracted from natural gas. Included are isopentane, natural gasoline, and plant condensate.

Petrochemical Feedstocks: Chemical feedstocks derived from petroleum principally for the manufacture of chemicals, synthetic rubber, and a variety of plastics. The categories reported are Naphthas less than 401°F.

End point and 'Other oils equal to or greater than 401°F end point.'

Petroleum: A broadly defined class of liquid hydrocarbon mixtures. Included are crude oil, lease condensate, unfinished oils, refined products obtained from the processing of crude oil, and natural gas plant liquids. Nonhydrocarbon compounds blended into finished petroleum products, such as additives and detergents, are included after blending has been completed.

Petroleum Coke: A residue high in carbon content and low in hydrogen that is the final product of thermal decomposition in the condensation process in cracking. This product is reported as marketable coke or catalyst coke.

Petroleum Coke, Catalyst: The carbonaceous residue that is deposited on and deactivates the catalyst used in many catalytic operations (e.g., catalytic cracking). Carbon is deposited on the catalyst, thus deactivating the catalyst. The catalyst is reactivated by burning off the carbon, which is used as a fuel in the refining process. That carbon or coke is not recoverable in a concentrated form.

Petroleum Coke, Marketable: Those grades of coke produced in delayed or fluid cokers that may be recovered as relatively pure carbon. Marketable petroleum coke may be sold as is or further purified by calcining.

Petroleum Consumption: The sum of all refined petroleum products supplied. For each refined petroleum product, the amount supplied is calculated by adding production and imports, then subtracting changes in primary stocks (net withdrawals are a plus quantity and net additions are a minus quantity) and exports.

Petroleum Products: Products obtained from the processing of crude oil (including lease condensate), natural gas, and other hydrocarbon compounds. Petroleum products include unfinished oils, liquefied petroleum gases, pentanes plus, aviation gasoline, motor gasoline, naphtha-type jet fuel, kerosene-type jet fuel, kerosene, distillate fuel oil, residual fuel oil, petrochemical feedstocks, special naphthas, lubricants, waxes, petroleum coke, asphalt, road oil, still gas, and miscellaneous products.

Photovoltaic and Solar Thermal Energy: Energy radiated by the sun as electromagnetic waves (electromagnetic radiation) that is converted into electricity by means of solar (photovoltaic) cells or concentrating (focusing) collectors.

Plant Condensate: One of the natural gas liquids, mostly pentanes and heavier hydrocarbons, recovered and separated as liquids at gas inlet separators or scrubbers in processing plants.

Propane: A normally gaseous straight-chain hydrocarbon (C_3H_8). It is a colorless paraffinic gas that boils at a temperature of $-43.67°F$. It is extracted from natural gas or refinery gas streams. It includes all products designated in ASTM Specification D1835 and Gas Processors Association Specifications for commercial propane and HD-5 propane.

Propylene: An olefinic hydrocarbon (C_3H_6) recovered from refinery or petrochemical processes.

Refinery (petroleum): An installation that manufactures finished petroleum products from crude oil, unfinished oils, natural gas liquids, other hydrocarbons, and alcohol.

Renewable energy resources: Energy resources that are naturally replenishing but flow-limited. They are virtually inexhaustible in duration but limited in the amount of energy that is available per unit of time. Renewable energy resources include: alcohol fuels, wood, waste, hydro, geothermal, solar, wind, ocean thermal, wave action, and tidal action.

Residential Sector: An energy-consuming sector that consists of living quarters for private households. Common uses of energy associated with this sector include space heating, water heating, air conditioning, lighting, refrigeration, cooking, and running a variety of other appliances. The residential sector excludes institutional living quarters.

Residual Fuel Oil: The heavier oils, known as No. 5 and No. 6 fuel oils, that remain after the distillate fuel oils and lighter hydrocarbons are distilled away in refinery operations. It conforms to ASTM Specifications D396 and D975 and Federal Specification VV-F-815C. No. 5, a residual fuel oil of medium viscosity, is also known as Navy Special and is defined in Military Specification MIL-F-859E, including Amendment 2 (NATO Symbol F-770). It is used in steam-powered vessels in government service and inshore powerplants. No. 6 fuel oil includes Bunker C fuel oil and is used for the production of electric power, space heating, vessel bunkering, and various industrial purposes.

Road Oil: Any heavy petroleum oil, including residual asphaltic oil, used as a dust palliative and surface treatment on roads and highways. It is generally produced in six grades, from 0, the most liquid, to 5, the most viscous.

Short Ton (coal): A unit of weight equal to 2,000 pounds.

Solar Energy: The radiant energy of the sun that can be converted into other forms of energy, such as heat or electricity.

GLOSSARY

Special Naphthas: All finished products within the naphtha boiling range that are used as paint thinners, cleaners, or solvents. Those products are refined to a specified flash point. Special naphthas include all commercial hexane and cleaning solvents conforming to ASTM Specifications D1836 and D484, respectively. Naphthas to be blended or marketed as motor gasoline or aviation gasoline, or that are to be used as petrochemical and synthetic natural gas (SNG) feedstocks, are excluded.

Standard Industrial Classification (SIC): A set of codes developed by the Office of Management and Budget which categorizes industries into groups with similar economic activities. North American Industry Classification System has replaced it.

Still Gas (refinery gas): Any form or mixture of gas produced in refineries by distillation, cracking, reforming, and other processes. The principal constituents are methane, ethane, ethylene, normal butane, butylene, propane, and propylene. It is used primarily as refinery fuel and petrochemical feedstock.

Subbituminous Coal: A coal whose properties range from those of lignite to those of bituminous coal and used primarily as fuel for steam-electric power generation. It may be dull, dark brown or black, soft and crumbly, at the lower end of the range, to bright, jet black, hard, and relatively strong, at the upper end. Subbituminous coal contains 20 to 30 percent inherent moisture by weight. The heat content of subbituminous coal ranges from 17 to 24 million Btu per ton on a moist, mineral-matter-free basis. The heat content of subbituminous coal consumed in the United States averages 17 to 18 million Btu per ton, on the as-received basis (i.e., containing both inherent moisture and mineral matter). In this report, subbituminous coal is included in bituminous coal.

Supplemental Gaseous Fuels: Any gaseous substance that, introduced into or commingled with natural gas, increases the volume available for disposition. Such substances include, but are not limited to, propane-air, refinery gas, coke oven gas, still gas, manufactured gas, biomass gas, and air or inert gases added for Btu stabilization.

Transportation Sector: An energy-consuming sector that consists of all vehicles whose primary purpose is transporting people and/or goods from one physical location to another. Included are automobiles; trucks; buses; motorcycles; trains, subways, and other rail vehicles; aircraft; and ships, barges, and other waterborne vehicles. Vehicles whose primary purpose is not transportation (e.g., construction cranes and bulldozers, farming vehicles, and warehouse tractors and forklifts) are classified in the sector of their primary use. In this report, natural gas used in the operation of natural gas pipelines is included in the transportation sector.

Unfinished Oils: All oils requiring further processing, except those requiring only mechanical blending. In most cases, these are produced by partial refining or are purchased in an unfinished state for conversion to finished products by further refining.

Unfractionated Streams: Mixtures of unsegregated natural gas liquid components, excluding those in plant condensate. This product is extracted from natural gas.

United States: The 50 States and the District of Columbia.

Value Added by Manufacture: A measure of manufacturing activity that is derived by subtracting the cost of materials (which covers materials, supplies, containers, fuel, purchased electricity, and contract work) from the value of shipments. This difference is then adjusted by the net change in finished goods and work-in-progress between the beginning and end-of-year inventories.

Waste Energy: Garbage, bagasse, sewerage gas, and other industrial, agricultural, and urban refuse used to generate electricity.

Waxes: Solid or semisolid materials derived from petroleum distillates or residues. Waxes are light-colored, more or less translucent crystalline masses, slightly greasy to the touch, consisting of a mixture of solid hydrocarbons in which the paraffin series predominates. Included are all marketable waxes, whether crude scale or fully refined. Waxes are used primarily as industrial coating for surface protection.

Wind Energy: The kinetic energy of wind converted into mechanical energy by wind turbines (*i.e.*, blades rotating from a hub) that drive generators to produce electricity for distribution.

Wood Energy: Wood and wood products used as fuel, including round wood (cord wood), limb wood, wood chips, bark, sawdust, forest residues, charcoal, pulp waste, and spent pulping liquor.

CPCB: Central Pollution Control Board

DBA: Decibel

GHGs: Greenhouse Gases

NEERI: National Environmental Engineering Research Institute

PWR: Pressurized-water reactors

BWR: Boiling-water reactors

Bibliography

1. Giancoli, Douglas, *Physics: Principles with Application.* Prentice Hall: Upper Saddle River, New Jersey. 5th Edition, Volume 1, 1998.
2. A.G. Iyer., *Environment Concerns and Diesel Power Plants.* 24th May, 2000.
3. *Gas Turbines Enhances Power Plant Efficiency* by Steinegger, Sulzer. Technical Review, Vol. 162, Feb. 1980.
4. *Future Fuels for Gas Turbines,* by D.K. Mukherjee, Brown Boveri Review, Dec. 1980.
5. Lecture Notes by Charles D. Sigwart, on Nuclear Power Plant.
6. *Environmental Impact Assessment of Gas Turbines (24n4S)* by Roy & Roy IJPRVD July-Aug. 1988.
7. *Brayten Cycle Challenge Rankine* by J. Makansi.
8. IJPRVD July-Aug. 1988. *Gas Turbine Power Generation Special.*
9. Cogeneration as a Answer to Rising Industrial Energy Costs (Active Conservation Techniques) (2-b) by Stiffer ACT-Nov. 1986.
10. *Hot Prospects for Combined Cycle Power Plants* by Werner Schrider Scimeris Review Jan. 1995.
11. National Seminar on Steam and Gas Based Cogeneration Systems 1993, Hyderabad.
12. *Combined Cycle & Co-generation* by Roy & Agrawal.
13. *Cogeneration-Combined Cycles & Synthetic Fuels* by M. Berman, Power Engineering.
14. *Combined Cycles and Refined Coal* by J. Papanarcos, Power Engineering.
15. *Steam Bottoming Plant for Combined Cycle* by R.W. Foster, Combustion.
16. 'Test Code for Sound Rating Air Moving Devices,' Standard 300, Air Movement and Control Association, Arlington Heights, Ill., 1967.
17. Stern, A.C. (Ed.): 'Air Pollution,' vol. 1, 2d ed., Academic Press, Inc., New York, 1968.
18. Carson, J.E., and H. Moses, *The Validity of Several Plume Rise Formulas,* J. Air Pollut. Central Assoc. vol. 26, no. 11, 1976.
19. Briggs, G.A., 'Plume Rise,' Atomic Energy Commission Critical Review Series, TID-25075, Washington, D.C., 1969.
20. Wark, K., and C.F. Warner, 'Air Pollution, Its Origin and Control,' 2nd ed., Harper & Row Publishers, Incorporated, New York, 1981.
21. Shade, D.H. (Ed.): 'Meteorology and Atomic Energy,' Atomic Energy Commission Publication TID-24190, Washington, D.C., 1968.
22. Lambers, W.S., and W.T. Reid, "A Graphical Form for Applying the Rosin and Hammler Equation for the Size Distribution of Broken Coal," U.S. Bureau of Mines Circular 7346, 1946.
23. Diesel, Rudolf: "*Theory and Construction of a Rotational Heat Motor,*" Bryan Donkin (trans.), London, 1894.
24. Anderson, John, *Pulverized Coal Under Central Station Boilers,* Power, vol. 51, no. 9.

25. Blizard, John, *'Transportation and Combustion of Powdered Coal,'* U.S. Bureau of Mines Bulletin 217, Washington, D.C., 1923.
26. Nusselt, W., *The Combustion Process in Pulverized Coal Furnaces*, VDI Zeitschrift, vol. 68, no. 6.
27. Van Heerden, C., A.P.P. Nobel, and D.W. van Krenelen, *Studies of Fluidization*, The Critical Mass Velocity, Chem. Eng. Sci.
28. White, P.C., R.L. Zahradnik, and R.E. Vener, *'Coal Gasification,'* Office of Fossil Energy, ERDA Quarterly Report, July-Sept 1975.
29. Kydd, P.H., *'Integrated Gasification-Gas Turbine Cycle Performance,'* General Electric Company Technical Information Series Report No. 75CRD021, March 1975.
30. Othmer, D.F., *Energy-Fluid from Solids*, Mech. Eng., November 1977.
31. *'Environmental Development Plan Biomass Energy Systems,'* U.S. Department of Energy Report No. DOE/EDP-0032, September 1979.
32. Church, E.F., *'Stream Turbines,'* 3d ed., McGraw-Hill Book Company, New York, 1950.
33. Baily, F.G., and E.H. Miller, *'Modern Turbine Designs for Water-Cooled Reactors,'* General Electric Company Report No. GER-2452, 34.
34. Cowgill, T., and K. Robbins, *Understanding the Observed Effects of Erosion and Corrosion in Steam Turbines*, Power, September 1976.
35. Hohn, A., *Rotors for Large Steam Turbines*, Brown Boveri Rev., vol. 60, no. 9, , Sept. 1973.
36. *Energy—The Scene in India* (16-17) Career & Competition Times, June 1984.
37. *Energy Transition in Developing Countries* by Yues Rovani, The Indian & Eastern Engineer June 1984.
38. *Parameters & Future Unit Sizes in India,* URJA Feb. 1984.
39. *Trends in Power Station Engineering by* A.K. Sah.
40. *Global Energy Economy* by K.C. Pant.
39. *Some Policy Issues for Rural Energy* by N.S.S. Arokiaswamy.
40. Contours of VIII-Plan (4-11) by Ranjarajan, Yojana.
41. *Problems for Meeting Country's Power Need* by Arokiaswamy UPRVD
42. *Power Development Issues & Options* by B. Chand IJPRVD.
43. *Power Development in VIII-Plan* by Suri and Mukerjea, IJPRVD.
44. *Adhocism in Power Planning & Its Implications* by S.N. Roy.
45. *Next generation of Power Plants Power*, Aug. 1990.
46. *Power Development in VIII-Plan* by N.S. Vasant, URJA.
47. *Problems for Meeting Country's Power Needs* by N.S.S. Arokiaswamy IJPRVD.
48. *Developing Countries Energy as a Source of Life* by A.L. Gorshkov,
49. *Meeting Future Power Needs* by Bill Meade, Independent Energy April 89.
50. *Power Development* by Bahadur Chand.
51. *NTPC Role in Development of Power in India* by P.S. Bami.

BIBLIOGRAPHY

52. Lugand, P. and Arictti, '*Combined Cycle Plants with Frame of Gas Turbine,*' Trans of the AS ME, J. of Engg. for Gas Turbine and Power, Vol. 113, 1991.

53. Scaizo, A.J., Bannister R.L. Decorro M. and Howard, G.S., '*Evolution of Westinghouse Heavy-Duty Power Generation and Industrial Combustion Turbines*,' Trans of the ASME, Vol. 118, 1996.

54. Chiesa P. and Consonni S., '*Natural Gas Fired Combined Cycles with Low CO_2 Emmissions*, Trans of the ASME, J. of Engg. for Gas Turbines and Power.

55. Bannister, R.L., Cheruvu, N.S., Little, D.A. and McQuiggan, G., '*Development Requirements for an Advanced Gas Turbine System*, Trans of the ASME, J. of Engg. for Gas Turbine and Power, Vol. 117, 1995.

56. Rice, L.J., P.E., '*Thermodynamic Evaluation of Gas Tai bine Cogeneration Cycles* : Part I: *Heat Balance Mothod Analysis, Part II: Complex Cycle Analysis*, Trans. of ASME, J. of Engg. for Gas Turbines and Power.

57. El-Masri, M.A., '*GASCAN—An Interactive Code for Thermal Analysis of Gas Turbine Systems*,' Trans. of ASME, J. of Engg. for Gas Turbines and Power, Vol. 110 1988.

58. El-Masri, M.A., *A Modified. High-Efficiency Recuperated Gas Turbine Cycle*, Trans of ASME, J. of Engg. for Gas Turbine and Power, Vol. 110, 1988.

59. Yadav, R. and Singh, L., '*Comparative Performance of Gas/Steam Combined Cycle Power Plants*, ASME, Int. Gas Turbine and Aero-engine Congress and Exposition.

60. Sanjay, Singh and Prasad, B.N., '*Prediction of Performance of Simple Combined Gas/Steam Cycle and Co-generation Plants with Different Means of Cooling.*'

61. Dixon, S.L., *Fluid Mechanics, Thennodynamics of Turbomachinery*, Pergamon Press, 2nd ed, 1975.

62. *Flue Gas Conditioning Upgrades Performance* by L.A. Midkiff.

63. *Reheat Wet Scrubber Stack Gas to Avoid Downstream Difficulties* by Choi and Rosenberg.

64. *Dry SO_Z Scrubbing at Artelope Valley Statioh* by Davis and Others Combustion, Oct. 1979.

65. *Dry Scrubbing Maintains High Efficiency* by R.A. Davis, Power Engineering, Oct. 1979.

66. *Emission Control for Industrial Coal Fired Boilers* by Dimitry and Others, Combustion, Jan. 1980.

67. *Commercial Scale Dry Scrubbing Holds Promise for SO_Z Control Power.*

68. *Spray Tower—The Work House of Flue Gas Desulfuiization* by A. Saleen Power.

69. *Fiber-fabric Selection for SO_2 Dry Removal*, by L. Bergmann, Power Engineering.

70. *Why Scrubbers* by F.F. Ross, Combustion, Oct. 1980.

71. *Understanding Venturi Scrubbers for Air-pollution Control* (53) by Jack D. Brady Plant Engineering, Sept. 30, 1952.

72. *Composite Lining Add Life to Acid Scrubbers* (139) by Thomas Kember Chemical Engineering, Aug. 1992.

73. *Air-Pollution Cleaning Waste-Dry Versus Wet* (131-41) by R.W. Goodwin Journal of Energy Engineering (ASCE) Sept. 83.

74. *SO_2-Control* (51-524) by Jason Makansi Power, Oct. 1982.

75. *Reheat Wet Scrubber Stack Gas to Avoid Down-stream Difficulties* (40-42) by Choi and Rosenber, Power July 79.
76. *Flyash removes SO-, Effectively from Boiler Flue Gas* (61-63) by C. Johnson. 43. Power-Engineering, Vol. 83 June 79.
77. *Instrumentation & Control for Double Loop Lime Stone* (72-75) Power Engineering, Vol. 83, June 79.
78. *Spray Dryer System Scrubs SOI* (29-32) by L.A. Midkiff Power, Jan. 1979.
79. *Improving Scrubber Demister Pcrformance* (33-35) by W. Ellison, Power Jan. 79.
80. Sanjay, Singh Onkar and Prasad, B.N. '*Thermodynamic Performance of Complex Gas Turbine Cycles* 'Proceeding of IJPGC 2002, June 24-26, 2002, Virginia, USA (UPGC—2002-26109).
81. Li, K.W. and Priddy. A.P., '*Power Plant System Design*, John Wiley, 1985.
82. Haywood, R.W., *Analysis of Engineering Cycles*, Pergamon Press, Oxford, 1975.
83. Horlock, J.H., '*Combined Heat and Power*,' Pergamon Press, Oxford, 1984.
84. Horlock, J.H., *Combined Cycle Plants*, Pergamon Press, 1992.
85. Bannister, R.L., 'Newby, R.A., and Diehrl, R.C., *Developing a Direct Coal Fired Combined Cycle*, Mechanical Engineering.
86. El-Wakil, M.M., '*Power Plant Technology*, McGraw-Hill, New York, 1985.
87. Basu, P. and Haldar, P.K., '*Combustion of Single Carbon Particles in a Fast Fluidized Bed of Fine, Solids*, Fuel 68, 1989.
88. Basu, P. and Fraser, S.A., '*Circulating Fluidized Bed Boiler: Design and Operations*,' Butterworth-Heinemann, 1991.
89. Howard, J.R., *Fluidized Bed Technology : Principles and Applications*, Adam Hilger, 1989.
90. Kehlhofer, R., *Combined Cycle Gas and Steam Turbine Power Plants*, Fairmont Press, Lilburn, 1991.
91. Reznikov, I. and Lipov Y.M., '*Steam Boilers of Thermal Power Stations*, Mir Publishers, Moscow, 1983.
92. British Electricity and Generating Board (CEGB), Modern Power Station Practice, Third ed., Vol. 1 to 8, Pergamon Press, 1991.
93. Kotas, T.J., '*The exergy Method of Thermal Plant Analysis*,' Butterworths, 1985.
94. Nag, P.K., *Power Plant Engineering*, 2nd ed., Tata McGraw Hill Co. Ltd, New Delhi, 2002.
95. Arora & Domkundwar, *A Course in Power Plant Engineering*, Dhanpat Rai & Sons. New Delhi.
96. Yadav, R., *Internal Combustion Engincs and Air Pollution*, 1st ed., 2002, Central Publishing House, Allahabad.
97. Yadav R., *Thermodynamics and Heat Engines*, Vol. 1 and 11, Central Publishing House Allahabad.
98. *Energy Analysis of a Combined Cycle Plant* (A 1) by K. Veeranjaneyulu and Others, National Seminar on Engergy Management, Mar. 2-3, 1995 at Motilal Nehru Regional Engg. Coollege, Allahabad.

99. *Heat Pipe as an Energy Saver* (A21) by H.R. Saboo, National Seminar on Engergy Management, Mar. 2-3, 1995 at Motilal Nehru Regional Engg. Coollege, Allahabad.
100. *High Efficiency Gas Turbine Co-generation System* (A 58) by Rajendra Karwas.
101. *Electrical Power Management* (A 154) by A. Naogeswar, National Seminar on Engergy Management, Mar. 2-3, 1995 at Motilal Nehru Regional Engg. Coollege, Allahabad.
102. *Energy Conservation and Non-Conventional Sources* (B 17) by Manoj Karalay, National Seminar on Engergy Management, Mar. 2-3, 1995 at Motilal Nehru Regional Engg. Coollege, Allahabad.
103. *Energy Conservation—You can Do it ?* (C 202) by Kanitakar & Rava, National Seminar on Engergy Management, Mar. 2-3, 1995 at Motilal Nehru Regional Engg. Coollege, Allahabad.
104. *An Insight to Energy Management* (C 314-25) by Jain and Shukla, National Seminar on Energy Management Mar. 2-3, 1995, Regional Engineering College, Allahabad.
105. *Management of Fossil Fuel Resources* (C 252-59) by O.P. Rao, National Seminar on Energy Management Mar. 2-3, 1995, Regional Engineering College, Allahabad.
106. *Energy Conservation in Thermal Power Plants* (47-50) by M. Venkaiah, URJA, April 1992.
107. *Energy Management in Power Sector* by Bahadur Chand, URJA, Mar. 1992.
107. *Energy Conservation* (3-10) by S.H. Miller, Electrical India, 31 Jan. 1992, (27) Indian Power Plants.
108. *Ramagundam Super Thermal Power Station* (59-115) URJA, Vol. XII, Aug. 2, 1982.
109. *Richand Super Thermal Project* (153-54) by S.B. Kapur, URJA, Sept. 1982.
110. *Panipat Thermal Power Project* (30-32) Bhagirath, Jan.-Mar. 1980.
111. *Nuclear Power Plant Cooling System Selection* (55-67) by C. Stephen, Journal of Power Division, (ASCE), July 1975.
112. *Cooling Water of Thermal Stations* by N.G.K. Murthy (Irrigation and Power), Oct. 1978.
113. *Taming Bugs and Sline in Cooling Water System* by J.F. Walko, Plant Engineering, 1979.
114. *Controlling Corrosion in Cooling Water System* by G.A. Cappeline, Plant Engineering,
115. *Air Cooling Sans Water in Arid Area* (94-95) by R.R. Brogden, Power, Mar. 1980.
116. *A Water Conserving Zero Discharge Cooling Technology* (BCT Process) by Sanderson and Others, Combustion, Aug. 1980.
117. *Corrosion Problems in Thermal Power Stations* (31) by Miss P.S. Rajalakshmi, Electrical India, Sept. 15, 1980.
118. *Cooling Water Decarbonization and Treatment* (47-49) by Guiocihio and Others, Sulzer Technical Review, Vol. 63, Feb. 1981.
119. *Treatment of Water for Power Station* by B.V. Savithrr, Science Reporter.
120. *A New Copper alloy for Condenser Tubes* by C .J. Gaffaglio, Power Engineering, Aug. 1982.
121. *Cooling Water Facilities at Nuclear Station* by William and others, Journal of Energy EngineerinQ (ASCE), Mar. 1983.
122. *Water Conservation and Reuse at Coal-Fired Power Plants* (345-59) by F.L. Shomey, Journal of water Resources and Management (ASCE), Oct. 1983.

123. *Development of Cooling Water Intake & Outfall for Atomic & Steam Stations* by Wada, IJPRVI.
124. *Electro-Chemical Instrumentation* (7-13) by K. Ramani, Instrumentation Bulletin Vol. 2, No. 4, 1981.
125. *Safety Valves* (70-74) by G. Dodero, Power, July 79.
126. *Understanding pH Measurement and Control* by J. Hodulik, Plant Engineering, Sept. 15, 1983. (22) Pollution and Control.
127. *Environmental Control and Related Problems* by A. Haldar, Science and Engineering, Aug. 1978.
128. *Cost Estimating for Air Pollution Control* by H.C. Wendes, Heating Piping Air-Conditioning, May 79.
129. *NO, Emission Controlled by Design of Pulverised Coal Burners* by A. Barsin, Power Engineering, Aug. 79.
130. *Effects of SO_2 on Stomala*, Science Reporter, Dec. 79.
131. *Taming Coal Dust Emission* (123-26) by H.E. Soderberg, Plant Engineering, May 15, 1980.
132. *Thermal Pollution and the Environment* by R.V. Wagh, Indian and Eastern Engineer, July 1980.
133. *Understanding the Earth Climate* (35-39) by S.K. Gupta, Science Today, Aug. 1986.
134. *Environmental Pollution* by S.S. Solanki, the Indian and Eastern Engineer, Aug. 1980, 12. Environmental Pollution (329) Invention Intelligence, Aug. 1980.
135. *Air-Pollution* (15-24) by R.N. Trivedi, Industrial India, Sept. 1980.
136. *Air Pollution* by R.N. Trivedi (15) Industrial India, Vol. 31, Sept. 1980.
137. *Sampling Particulate Emission from Exhaust Stacks* by David B. Rimberg, Plant Engineering, Oct. 2, 1980.
138. *Fume Incinerators for Air Pollution Control* by T.F. Meinhold, Plant Engineering, Nov. 13, 1980.
139. *Reducing NO emission* by L. Pruce, Power, Jan. 1981.
140. *Environmental Considerations and Energy Planning in India* by N.L. Ramanathan, URJA, May 1981.
141. *Role of Meteorology in Air Pollution*, Science Reporter, June, 1981.
142. *Emission Factors for Fuel Oil Combustion* by S.C.J. Kale, Journal of Institute of Engineers, Environmental Engineering Division.
143. *Environmental Engineering Monitory Systems* by K.R. Rao.
144. *Some Critical Environmental Issues* by Satish Pha, Energy Management, Oct.-Dec. 1981.
145. *Non-burn Electric Generation* by Makansi and Strauss, Power, July 1992.
146. *Chernobyl Another Meltdown* by Brain Eads, Reader's Digest, July 1995.
147. *Researching Radioactive Waste Disposal* by Feates and Keen, New Scientist, 16th Feb. 1978.
148. *Nuclear Waste Disposal the Geological Aspects* by Chapman and Others, New Scientist.
149. *Nuclear Waste Disposal-Radiological Protection Aspects* by Hill and Grimwood, New Scientist.

BIBLIOGRAPHY

150. *Advanced Light, Water Reactor* by Golay and Todreas Scientific American, April 1990.
151. *Advanced Light Water Reactors* by Michael and Todreas, Scientific American, April 1990.
152. *Worldwide Nuclear Power Status* by Leonard and Robert, IJPRVD Mar.-April 1987.
153. *The Future of Nuclear Power after Chernobyl* by N.L. Char, IJPRVD 1986 Annual Review.
154. *Nuclear Power Generation in India* by M.R. Srinivasan, IJPRVD Oct. 1988.
155. *Grappling with Nuclear Waste* by B.R. Nair, Mechanical Engineering, May 1989.
156. Fast Reactors Revive Nuclear Programme Power, Dec. 1990.
157. *Nuclear Power-Good Safety* by M.R. Srinivasan, IJPRVD Jan.-Feb. 1988.
158. *Nuclear Power in Developing Countries* by M.R. Srinivasan, IJPRVD, June 1988.
159. *Nuclear Fuel—The Thinking Man's Alternative* by N. Chamberlain, Energy World, April 1989.
160. *Nuclear Power Generation in India* by M.R. Srinivasan, IJPRVD, Oct. 1988.
161. *Introduction to Non-Conventional Energy Resources* by A.K. Raja, Manish Dwivedi, Amit Prakash Srivastava.

Index

A

Acid fog 421
Acid vapours 421
Aeroelectric plant 65
Air pollution 415, 429
Alcohol 46
Alkanes 423
Alpha 327
Alpha decay 322
Anaerobic digestion 54
Anderson cycle 100
Anode 74
Anthracites 44
Artificial winds 65
Atmosphere 415, 419, 435, 441
Atmospheric pollution 415
Atom 313
Atomic energy 26
Atomic mass unit 313
Atomic minerals division 339
Atomic number 314
Atomization 425
Auxiliary systems 283
Axial flow compressor 273

B

Batch type biogas plant 57
Bearings 285
Beta decay 322
Beta particles 316, 327
Bhakra Beas Management Board 13
Binary cycle power plant 93
Binary vapour cycle 17
Bio-gas engine 238
Biofuels 53
Biological processes 34
Biological sciences 34
Biomass 34, 54
Biomass power 54
Biomass pyrolysis 76
Biopower 53
Bioproducts 53
Bleeding of steam 16
Block rate tariff 128
Bluff body method 276
Boiler 18
Bolts 390
Booms 351
Brake horse power 241
Breaker 411
Breeder reactor 328
Bus-bars 410, 412
Bushings 402

C

Camshaft 244
CANDU 333
Capacity factor 313
Carbon monoxide 423
Carburetion 235
Carnot cycle 15
Cased turbines 358
Cathode 74, 78
Cavitation Factor 374
Cell potential 74
Cell voltage 76
Central electricity authority 339
Central Electrochemical Research Institute 103
Central flow condenser 20
Central Power Research Institute 13
Centrifugal compressor 273
Centrifugal force 108
Centrifugal type super chargers 249
Centrifuge 337

INDEX

Chain reactions 45
Channel systems 97
Chemical energy 4
Chemical reactions 34
Chemo electricity 73
Chernobyl 339, 340
Chlorinated diphenyl 409
Chronological load curves 133
Civilization 33
Clamped 396
Closed cycle gas turbine plant 270
CNG 43
Coal 44, 428
Coal particles 427
Cogeneration 280, 281
Coil design 408
Cold reserve 122
Combined cycle 297
Combustion 326, 237, 275, 424
Combustion chamber 269
Combustion of fuels 415
Combustion of the fuel 18
Common rail injection 242
Compressor 268
Condenser 19
Conductors 399, 402
Conical draft tube 376
Connecting rod 235
Conservators 408
Control rods 326
Converter reactor 328
Coolant 326, 327, 333
Cooling 388, 285
Cooling of transformers 403
Cooling system 406
Core-type transformers 398
Cores 395
Cost of fuels 125
Cost of nuclear power plant 328
Crankshaft 246
Critical mass 316

Cross-over coils 398
Crude oil 102
Current 387
Cylinder 235, 237
Cylindrical concentric helix windings 398

D

Damodar Valley Corporation 12
Daughter nucleus 322
Dawn flow condenser 20
Decomposition 44
Deep mining 44
Demand 35
Demand factor 122
Demographic pressure 442
Depreciation 124
Deuterium 313, 314
Diesel 253
Diesel cycle 14
Diesel engine 235, 236, 238, 430
Diesel power plant 131
Direct cooling 393
Direct cycle 327
Disc coils 399
Distributors System 243
Dish/engine system 68
Diversity factor 122
Domestic wastewater pollution 436, 438
Double circuit system 327
Downstream surge tanks 362
Draft tube liners 378
Dry steam power plant 91
Dry sump lubrication system 247
Dual cycle 14
Duel fuel engine 238, 239
Dump power 122

E

Economics 297
Economy 132
Effect of variable load on power plant design 135

Efficiency of a diesel engine plant 254
Efficiency of cells 78
Einstein equation 315
Ejector condenser 23
Elbow type draft tube 377
Electric circuits 394
Electric energy 344
Electric vehicles 311
Electrical energy 34
Electrical generator 235
Electricity 4, 5, 311, 335
electrode current density 76
Electrolysis 76
Electromagnetic 34
Electromagnetic 336
Electron 79
Electron volt 317
Emission 79
Empirical formulas 347
Energy 1, 3, 33, 38, 321, 326, 415, 426
Energy conservation process 37
Energy consumed 129
Energy exploited 46
Energy load curve 133
Energy science 33, 34, 36
Energy sources 38
Energy strategies 49
Energy technologies 35, 36
Energy transformations 33
Engines 431
Enrichment 321
Environmental effects 415
Environmental pollution 429
Equinoxes 68
Evaporation condenser 20
Exhaust 237
Exhaust system 250
Exothermic or endothermic 318
Exploitation 28
Exploration 28, 42

F

F.H.P. 241
Fact 26
Fans 389, 392
Farakka 50
Fault bus protection 413
Filter 242, 247
Fire point 409
Firm power 122
First law of thermodynamics 14
Fission 315, 318, 320, 321, 324, 326, 327, 335, 338, 435
Fixed cost 123
Fixed dome type digester 57
Fixed element 130
Flame zones 422
Flash point 409
Flash steam power plant 92
Flat demand rate 128
Fossil fuels-oil 46
Float systems 97
Floating gas holder type 58
Flux 394, 399
Forced circulation cooling system 252
Forced cooling 406
Forced ventilation 391
Fossil fuel 37, 421
Fossil-superheat hybrid systems 94
Four stroke engines 237
Francis turbine 365, 367
Fuel 18, 242
Fuel cell 103, 104
Fuel pumps 238
Fuel system 286
Full pressure system 246
Fusion 318, 319, 320

G

Gamma 316
Gamma particles 316

INDEX

Gamma radiation 316, 322
Gas engines 235
Gas power cycles 15
Gas turbine 268, 274, 275, 276, 283, 286, 290, 293, 297
Gaseous diffusion method 335
Gasoline 46
Gasoline engine 235
Gear 285
Generator 357, 387
Generator cooling 387
Geological resources 45
Geothermal 88
Geothermal-Preheat Hydrid Systems 94
Geysers 89
Glow plug 236
Gravitational potential energy 4
Greenhouse effect 311, 419, 421

H

Half life 317
Hazardous waste management 438
Heat 422
Heat balance 419, 256
Heat balance diagram 254
Heat engine 80
Heat exchanger 274, 278
Heat pumps 311
Heavy hydrogen 314
Heavy water 333
Heliostat field 69
Helium 313
High head plants 354
High lube oil temperature 290
High-level jet condenser 22
High-level wastes 434
Homegrown 86
Hopkinson demand rate 129
Hospital waste disposal 439
Hospital waste pollution 437
Hot dry 88

Hot reserve 123
Hybrid 98
Hybrid geothermal power plant-fossil system 94
Hydel power 7
Hydel-power plants 9
Hydracone or moody draft tube 377
Hydraulic accumulator 284
Hydraulic energy 24
Hydraulic power plant 126
Hydraulic turbines 23
Hydro potentials 50
Hydro-plants 353
Hydrocarbons 42, 43
Hydroelectric generation 308
Hydroelectric power 107
Hydrogen 38, 104
Hydrogen-cooled machine 393
Hydrogen cooling 393
Hydrogen cyanide 424
Hydrogen nucleus 314
Hydrogen pressures 392
Hydrogen storage 105
Hydrograph 348
Hydrological cycle 344
Hydropower 107

I

I.C. engine 14, 235, 248
Ideal cell potentials 75
Ideal fuel cell potentials 75
Igneous rock 87
Ignition 283
Ignition system 284
Impulse turbine 24, 25, 364
Indian atomic energy programme 27
Indicated horse power 240
Indicated mean effective pressure 240
Individual pump injection 243
Industrial air pollution 437
Industrial non-hazardous waste management 438
Industrial production 133

Industrial wastewater 436
Industrial wastewater pollution 438
Initial cost 123
Injection process 236
Injector 236, 287
Innertkirchen power house 362
Innovative heat exchanger 108
Insulating properties of oil 409
Insulation 400
Integrated duration curve 133
Intercooler 274
Internal combustion engine 78
Investor's profit 130
Isolators 411
Isotopes 314

J

Jet condensers 22

K

Kaiga Atomic Power Plant 332
Kakarpar Nuclear Power Plant 332
Kalpakkam Nuclear Power Station 329
Kaplan turbine 366
Kinetic energy 4, 5, 24, 79
Korba 50
Kursh power station 342

L

Labour cost 125
Laminations 395
Laws of chemical combination 313
Light water 332
Light water reactors 332
Lliquid fuel 101, 427
Load curve 122
Load demand 133
Load duration curve 122
Load factor 11, 121
Long wall method 44
Low compression gas engine 238

Low head power plants 354
Low level jet condensers 22
Low lube oil pressure 289
Low wind speed turbines 65
Low-level wastes 434
LPG 46
Lubricant 244
Lubrication system 235, 253, 284, 285
Lubrication system of diesel power plant 243

M

Magnesium 389
Magnetic circuits 394
Magnetic field 336
Main bus-bar 411
Mass curve 348
Mass numbers 318
Maximum demand 129
Maximum efficiency 250, 384
Mechanical protection 400
Medium head plants 355
Meteorology 347
Methane 422
Microturbines 54
Mnemonic 74
Moderated reactor 433
Moderating ratio 325
Moderator 325, 326, 327, 333
Modular systems 54
Modulating control 290
Molecular theory 313
Mother science 34
Motor 387
Multiplication factor 335

N

Naphtha 101
Narora Atomic Power Plant 331
Narora Nuclear Power Station 329
Nathpa Jhakri Hydro-electric Project 14
National Chemical Laboratory 103

INDEX

National economy 33
National Hydroelectric Power Corporation 11
National Thermal Power Corporation 11
Natural circulation cooling system 252
Natural gas engine 238
Natural uranium 323, 332, 45
Neutron 342
NGRI 96
NHPC 96
Nimbkar Agricultural Research Institute 106
Nitric oxide 423
NJPC 13
Noise 418
North-Eastern Electric Power Corporation Ltd. 12
Nozzles 24, 238
NTPC 51
Nuclear 313
Nuclear energy 4, 308, 309, 310
Nuclear duel 9, 323
Nuclear power 45, 308
Nuclear Power Corporation 339
Nuclear Power Corporation of India Limited 28
Nuclear power plant 131, 324, 328, 333, 335, 432
Nuclear power reactors 27
Nuclear power stations 10
Nuclear reactions 318
Nuclear reactor 45, 323, 328, 337

O

Ocean development 29
Ocean regime 28
Ocean thermal energy 97
Oil 40, 42
Oil engines 235
Oil for transformer 408
Oil formation 41
Oil pressure 247
Inductors 390
OPEC 40, 42
Open cycle gas turbine 269
Open-cycle gas turbine power plant 269

Operational costs 123, 125
Organic wastes 55
Otto cycle 14
Output of reaction turbine 371
Output of reaction turbine with draft tube 371
Over potentials 74
Oxidative pyrolysis 422

P

Parabolic-trough systems 68
Part load efficiency 384
Peak load 133
Pelton turbine 364
Percentage method 124
Performance of the diesel engine 240
Petro-geothermal 88
Petrol 46
Petroleum products 47
PHWRs 27
Physics 33
Pilot oil 239
Piston 235
Piston cylinder type 249
Plant capacity factor 121, 123
Plant factor 121
Plant operating factor 121
Plant use factor 123
Polarization 77
Pollution 436
Pollution from thermal power plants 437
Positron 314
Post-flame zone 422
Potential 24
Potential energy 5, 344, 352
Power tower system 68
power 1, 5, 23, 337, 365, 269, 381
Power demand 50
Power engineers training society 13
Power generation 8, 52, 82, 297
Power houses 356

Power industry 49
Power plant 1, 2, 131, 331, 418
Power transformers 418
Precipitation 346
Pressure gauge 246
Pressure ratio 238
Pressure-regulating valve 246
Pressurized-water reactor 433
Primary energies 48
Prime control 287, 288
Prime mover 18, 19, 23, 363
Prime power 122
Process of depletion 46
Propeller turbine 366
Propeller type 61
Protective control 288
Pumped storage plant 108
Pumping plant 355
Pyrolysis 54

R

Radiant energy 4
Radiations 67, 435
Radioactive 315
Radioactive decay 315
Radioactivity 318, 321, 322, 435
Radioisotopes 322
Rain fall 346
Ramagundam 50
Rana Pratap Sagar (Rajasthan) Nuclear Station 329
Rankine cycle 14
Rate of interest 124
Rates 127
Reaction turbine 24, 25
Reactor core 327
Reactor plant 329
Reactor vessel 329
Reciprocating oil engines 235
Reflector 326
Reformation of hydrocarbons 76
Regeneration 16, 277

Regenerator type heat exchanger 274
Regenerators 274
Reheat cycle 16, 279
Reheat-regenerative cycle 17
Renewable 3
Renewable energy 52
Renewable energy technology 311
Reserve capacity 121
Reservoir 92, 108, 349, 350,
Retention 57
Reversible cell 74
Road transport 436
Road vehicles 416
Rockefeller's standard oil trust 40
Roots blowers 249
Rotational speed 382
Rotor 388, 389
Rotor winding 388
Run-of-river plant 355
Run-off 347
Rural Electrification Corporation 12

S

Safeguard 334
Safety system 338
Sandwich coils 398
Sandwich windings 399
Science 25, 33
Scientific research 26
Second law of thermodynamics 14
Sectional or continuous disc coils 399
Semi-pressure system 245
Shell type transformers 396, 399
Short wave radiation 68
Shut down type control 289
Single bus-bar system 410
Single circuit system 327
Single stage continuous type biogas plant 56
Singrauli 50
Sinking fund method 124
Slip rings 389

INDEX

Sludging 409
Small end bearings 246
Smoke 417
Solar collector 71
Solar energy 52
Solar farm power plant 69
Solar photovoltaics 72
Solar radiation 68
Solar thermal applications 72
Solar ventilation system 71
Solid fuel 426
Solid Waste Management 438
Soot 425
Soot, dust 432
Space colling 72
Space heating 71
Spare 411
Spare bus-bar 411
Spark plug 236
Special alloy steel 394
Specific speed 368, 383
Spent fuel 434
Spinning reserve 123
Splash system 245, 246
Standardization 310
Stator 390, 392
Stator windings 393
Steam 18
Steam generator 18
Steam power plant 14, 135
Steam turbines 294
Step meter rate 128
Stirling engine 282
Storage plants 353
Straight line meter rate 128
Straight line method 124
Stresses 388
Structural material 326
Sulphur 418
Sun 38, 68
Sun energy 442

Supercharging 248, 249
Supervision 127
Surface condenser 20, 21
Swirl flow method 276
Synthetic oil 409

T

Tabbed tank 405
Tanks 408
Tapping switches 394
Tarapur Nuclear Power Station 329
Tariff 127
Taxes 127
Technology 25, 35
Temperature gradient 388
Terminals 394
Thermal conductivity 392
Thermal diffusion 336
Thermal efficiency 131, 277, 278
Thermal energy 4, 33
Thermal generation 51
Thermal neutrons 324
Thermal NO 424
Thermal pollution 441
Thermal power 337
Thermal power plants 429
Thermal power stations 430
Thermal reactor 45
Thermionic converter 78
Thermionic generator 80
Thermodynamics 14, 33
Thermonuclear 319
Third law of thermodynamics 14
Three Mile Island 339
Three Mile Island accident 312
Tidal power 441
Total energy 321
Transformer 394, 403
Transformer cores 395
Transformer tanks 408
Transformers 400

Transformers 363, 395, 397, 406
Turbine 23, 268, 342
Turbine bearings 435
Turbine overspeed 289
Turbine overtemperature 289
Turbo type super chargers 249
Turbo alternators 392
Two state continuous type biogas plant 56
Two stroke cycle engines 247
Two stroke engines 237
Types of tariffs 127

U

Underground power house 359
Unit method 125
Unit power 379
Unit quantity 380
Unit speed 379
Uranium 9
Uranium enrichment 335
Uranium fission 316
Utility factor 121

V

Valence electron 314
Valence shell 314
Values 26
Valves 251
Vane blower 249
Vaporization 425

Vaporization process 427
Vapour power cycle 15
Variable load 135
Vegetation 347
Vehicular pollution 437
Viscosity 408
Voltage-current density characteristics 77
Volumetric efficiency 250

W

Waste water 338, 430
Water coolers 391
Water gas reaction 342
Water pollution 436
Water reactor 433
Water turbine 23, 24
Water vapour 421
Water cooling 393
Waveform 388
Wet sump lubrication 246
Wind turbine 59, 61
Windings 390
Windmill 59, 62
Working fluid 98
World energy production 37
World oil reserves 38
Wrapped conductor 400

Z

Zeroth law of thermodynamics 14
Zirconium 337